T0139113

MicroComputed Tomography

MicroComputed Tomography

Methodology and Applications, Second Edition

Authored by

Stuart R. Stock

CRC Press is an imprint of the
Taylor & Francis Group, an **informa** business

Cover:
Three-dimensional (3D) rendering of the scaffold material, bone and blood vessels. The synthetic hydroxyapatite in the scaffold struts is rendered white and is semitransparent; the bone, slightly less attenuating than the synthetic hydroxyapatite, is rendered cream and is also semitransparent. Blood vessels filled with the contrast agent MicroFil are rendered red and can be seen to have grown into all of the macropores of the scaffold. Unpublished data of Hallman, Hsu and Stock (2019). See Fig. 8.6b for more details.

MATLAB® is a trademark of The MathWorks, Inc. and is used with permission. The MathWorks does not warrant the accuracy of the text or exercises in this book. This book's use or discussion of MATLAB® software or related products does not constitute endorsement or sponsorship by The MathWorks of a particular pedagogical approach or particular use of the MATLAB® software.

CRC Press
Taylor & Francis Group
6000 Broken Sound Parkway NW, Suite 300
Boca Raton, FL 33487-2742

First issued in paperback 2022

© 2020 by Taylor & Francis Group, LLC
CRC Press is an imprint of Taylor & Francis Group, an Informa business

No claim to original U.S. Government works

ISBN-13: 978-1-498-77497-0 (hbk)
ISBN-13: 978-1-03-233738-8 (pbk)
DOI: 10.1201/9780429186745

This book contains information obtained from authentic and highly regarded sources. Reasonable efforts have been made to publish reliable data and information, but the author and publisher cannot assume responsibility for the validity of all materials or the consequences of their use. The authors and publishers have attempted to trace the copyright holders of all material reproduced in this publication and apologize to copyright holders if permission to publish in this form has not been obtained. If any copyright material has not been acknowledged please write and let us know so we may rectify in any future reprint.

Except as permitted under U.S. Copyright Law, no part of this book may be reprinted, reproduced, transmitted, or utilized in any form by any electronic, mechanical, or other means, now known or hereafter invented, including photocopying, microfilming, and recording, or in any information storage or retrieval system, without written permission from the publishers.

For permission to photocopy or use material electronically from this work, please access www.copyright.com (http://www.copyright.com/) or contact the Copyright Clearance Center, Inc. (CCC), 222 Rosewood Drive, Danvers, MA 01923, 978-750-8400. CCC is a not-for-profit organization that provides licenses and registration for a variety of users. For organizations that have been granted a photocopy license by the CCC, a separate system of payment has been arranged.

Trademark Notice: Product or corporate names may be trademarks or registered trademarks, and are used only for identification and explanation without intent to infringe.

Publisher's Note

The publisher has gone to great lengths to ensure the quality of this reprint but points out that some imperfections in the original copies may be apparent.

Library of Congress Cataloging-in-Publication Data

Names: Stock, Stuart R., author.
Title: Microcomputed tomography : methodology and applications / by Stuart R. Stock.
Description: Second edition. | Boca Raton, FL : CRC Press/Taylor and Francis, [2020] | Includes bibliographical references and index.
Identifiers: LCCN 2019046116 (print) | LCCN 2019046117 (ebook) | ISBN 9781498774970 (hardback ; acid-free paper) | ISBN 9780429186745 (ebook)
Subjects: LCSH: Tomography. | X-ray microscopy.
Classification: LCC RC78.7.T6 S73 2020 (print) | LCC RC78.7.T6 (ebook) | DDC 616.07/57--dc23
LC record available at https://lccn.loc.gov/2019046116
LC ebook record available at https://lccn.loc.gov/2019046117

Visit the Taylor & Francis Web site at
http://www.taylorandfrancis.com

and the CRC Press Web site at
http://www.crcpress.com

This edition is dedicated to Michala K. Stock, Sebastian R. Stock,

Meredith R. Stock, and especially to M. Christine Stock.

Airy persiflage, indeed.

Contents

List of Figures

List of Tables

Preface

Not long after the first edition went to press, the author, as one expects, started collecting instances where the text was out of date. The field of micro- and nanoComputed Tomography (the nomenclature "microCT" is used for both) has advanced considerably every year since the first edition text was complete. Some years ago – never mind how long precisely – the author decided that he should commence work on a second edition of the book, and the completed revision follows.

In revising a technical book, choices must be made. (1) Should the original structure of the book be retained or changed? (2) How much of the old material should be retained? How much of the old material must be corrected, edited, or removed? (3) How many of the thousands of new papers should be added to the text? What are the criteria for sampling the plethora of papers and selecting representative studies for inclusion? (4) Are the existing figures and tables adequate to "tell the story?" What new figures are desirable or needed, including those replacing outdated figures? The answer to points (1)–(4) follow for the second edition.

1. The author critically examined the structure of the book in the light of changes in the field of microCT over the past ten plus years. The twin goals of the book were and continue to be for beginners, introduction to the imaging modality, and for those already using microCT, coverage of the wide range of approaches used. Clustering applications by type of structure involved and not discipline should help a researcher find new analysis approaches uncommon in his/her subspecialty. The original structure continues to serve the two goals and remains unchanged.
2. Almost all of the original material is retained except for updates to things like pronouncements of future directions or of "state-of-the-art" capabilities. Tables, in particular, are updated when they involve instrumentation.
3. Through the years since the first edition appeared, the author has kept a list of studies to be included in a revision. This list is very idiosyncratic and reflects the author's interests rather than the breadth of activity in microCT. The author also asked various experts for information on "hot topics" in microCT and then folded in appropriate text, references (and in a very few cases new figures). If that were all that was done, the view of microCT activity would be needlessly biased. Broader coverage of the literature was desired, but so much coverage would make a reader inundated. The author decided to do sampling of the recent literature, using literature searches of microCT synonyms in Pubmed and Compendex for the year 2017.
4. The first edition figures were, by and large, judged to current enough for the coverage of the second edition. A few figures were replaced, a few were added, and some tables were edited.

The author hopes the second edition is helpful to readers and wishes them the best of luck in any microCT studies they conduct.

Stuart R. Stock
Wilmette, IL
May 12, 2019

Acknowledgments

In addition to those acknowledged in the first edition, others were very helpful during the preparation of the second edition. Alexander Sasov, recently retired from Bruker/Skyscan, provided many helpful discussions over many years. The author is grateful to Ge Wang (RPI) for helpful discussion on reconstructions and likely future directions including machine learning and to Mike Marsh (Object Research Systems) for discussion of deep learning in image segmentation. Ryan Ross (Rush University) helped the author organize information on Scanco MicroCT-50 performance. Ajay Limaye (Australian National University) and Graham Davis (Queen Mary – University of London) provided very useful illustrations of segmentation and visualization using 2D histograms. Peter Voorhees and my collaborator Claus-Peter Richter (both of Northwestern University) provided information on rapid data acquisition with microCT and on hearing, respectively. Alexander Rack (ESRF), Dula Parkinson (ALS), and Christoph Rau (Diamond Light Source) provided updates on microCT at their respective facilities, and Phil Salmon (Bruker/Skyscan) provided technical information on lab microCT. The discussions of various people located at APS added enormously to this volume: Francesco De Carlo, Vincent De Andrade, Peter Kenesei, Jun Park, Jon Almer, Mark Rivers, and Dean Haeffner. "On the editorial side, the author thanks Nora Konopka and Prachi Mishra for their help and especially their patience."

While writing this book, the author had the nagging feeling that someone whose help was important had not been mentioned here. Apologies are offered if this were the case.

About the Author

Dr. Stuart R. Stock completed his undergraduate degree in 1977 and Masters degree in 1978, in materials science and engineering at Northwestern University and he later completed the post doc in the same field in 1983. His Ph.D. was in metallurgical engineering at the University of Illinois Urbana-Champaign. He was on the materials science faculty at Georgia Tech for over 16 years rising to the rank of Professor.

In 2001, he returned to Northwestern University, as a faculty member to the medical school.

Dr. Stock has used x-ray diffraction for materials characterization for over 40 years and revised Cullity's classic text Elements of x-ray Diffraction (now Cullity and Stock, Elements of x-ray Diffraction, 3rd Ed.). He has employed x-ray imaging for the same length of time. His first synchrotron radiation experiments were over 35 years ago, and he currently collects data at the Advanced Photon Source ten or more times a year. His first paper on microCT appeared in 1986, and microCT studies of inorganic materials and composites and of mineralized tissue have appeared throughout these 33 years.

Stuart R. Stock
Department of Cell and Developmental Biology
Feinberg School of Medicine
Northwestern University
Chicago, IL USA

Abbreviations[1]

η	azimuthal angle along a diffraction ring collected on an area detector
θ	scattering angle, typically Bragg angle, i.e., the angle for x-ray diffraction
λ	x-ray wavelength
μ	linear attenuation coefficient
μ/ρ	mass attenuation coefficient
ρ	density
φ, Φ	x-ray phase angle
0D, …, 4D	zero-dimensional (point), one-dimensional (line), two-dimensional (area), three-dimensional (volume), four-dimensional (volume plus time)
1K, 2K, etc.	number of pixels in a detector array or voxels in a reconstruction, in thousands, i.e., multiples of 210 or 1024
a	lattice parameter
AI	artificial intelligence
ALS	Advanced Light Source
APS	Advanced Photon Source
ART	algebraic reconstruction technique
BMD	bone mineral density
BMP-2	bone morphogenetic protein 2
BSE	back scattered electron (imaging)
BV	bone volume
BV/TV	bone volume fraction
c	lattice parameter
CAF	conductive anodic filaments
CAT	older medical synonym for CT
CCD	charge-coupled device
CMOS	complementary metal oxide semiconductor (device)
Conn.D	connectivity density
CT	computed tomography
CVI	chemical vapor infiltration
d	periodicity of scatterers, typically interplanar spacings in crystals
***d*50**	mean particle diameter
D	collagen D-periodicity
DA	degree of anisotropy
DEJ	dentinoenamel junction
DL	deep learning, a variant of machine learning
DMM	double multilayer monochromator
e−	electron(s)
E	energy of x-ray photon(s)
EBSD	electron backscatter diffraction
EDD	energy dispersive diffraction

[1] Note that abbreviations common to all of the sciences (symbols for chemical elements, units of physical quantities, etc.) are not included in the list.

ESRF	European Synchrotron Radiation Facility
FEM	finite element modeling
FIB	focused ion beam (sectioning)
FTIR	Fourier transform infrared (spectroscopy)
FOV	field of view
GPU	graphic processing unit
hkl	diffraction plane indices, written *hk.l* for hexagonal crystal systems such as calcite and hydroxyapatite
I	intensity
$\mathbf{I_0}$	incident intensity
kVp	electrical potential in kV from filament to target in x-ray tubes
MIL	mean intercept length
ML	machine learning or maximum likelihood
MRI	magnetic resonance imaging
MTF	modulation transfer function
NSLS-II	National Synchrotron Light Source II
OVX	*ovariectomized*
P	projection or profile; lowercase represents a ray of the projection
pixel	picture element
pQCT	peripheral quantitative CT
PSF	point spread function
REV	representative elemental volume(s)
s	beam direction (vector) or ray direction
S	Fourier transform of projection P
SART	simultaneous algebraic reconstruction technique
SANS	small angle neutron scattering
SAXS	small angle x-ray scattering
SEM	scanning electron microscope (or microscopy)
sig	magnitude of signal from an object
SIRT	simultaneous iterative reconstruction technique
SLS	Swiss Light Source
SMI	structure model index
Sp	spacing
SPring-8	Super Photon ring – 8 GeV
SV	surface area per unit volume
TEM	transmission electron microscopy
Th	thickness
TV	total volume
VV	volume fraction
voxel	volume element
WAXS	wide angle x-ray scattering, i.e., x-ray diffraction
Z	atomic number

1

Introduction

X-ray computed tomography (CT) is an imaging method where individual projections (radiographs) recorded from different viewing directions are used to reconstruct the internal structure of the object of interest. An obvious alternative to CT is serial sectioning, which is quite laborious, although automated approaches used in conventional sectioning (Levinson, Rowenhorst et al. 2017) and in focused ion beam sectioning ameliorate this shortcoming. CT, however, is much faster and possesses the advantage of being noninvasive and nondestructive; that is, the same component can be reinstalled after inspection or the same sample can be interrogated multiple times during the course of mechanical or other testing.

X-ray CT is quite familiar in its medical manifestations (CT or CAT scans), but it is less known as an imaging modality for components or materials. CT provides an accurate map of the variation of x-ray absorption within an object, regardless of whether there is a well-defined substructure of different phases or if there are slowly varying density gradients. High-resolution x-ray CT is also termed microComputed Tomography (microCT), or microtomography, and reconstructs samples' interiors with the spatial and contrast resolution required for many problems of interest. The application of microCT to biological and physical science or engineering problems is the subject at hand. The division between conventional CT and microCT is, of course, an artificial distinction, but here microCT is taken to include results obtained with at least 50- to 100-μm spatial resolution. The actual resolution needed for a particular application depends on the microstructural features of interest and their shapes.

No sooner had Röntgen discovered x-radiation than he applied the newly found penetrating radiation to radiological imaging – the first widely disseminated publication of the discovery of x-rays featured a radiological image (Röntgen 1898). Less than 20 years later, this new medical tool was in use across the battlefields of World War I (Hildebrandt 1992). Locating projectiles and shrapnel and checking the reduction of fractures noninvasively were true breakthroughs. One of the advantages of radiological images or radiographs, simplicity, can also be a severe limitation: these images are nothing more than two-dimensional (2D) projections of the variation of x-ray absorptivity within the object under study. While recording stereo pairs allows precise three-dimensional (3D) location of high-contrast objects, this approach is impractical when a large number of similar objects produces a confusing array of overlapping images or when there are no sharp charges of contrast on which one can orient.

A strategy for recovering 3D internal structure evolved prior to digital computers (Webb 1990). It involves translating the patient (or object to be imaged) together with the detection medium (film or other 2D detector) in such a way that only one narrow slice parallel to the translation plane remains in focus. This approach is termed laminography or focal plane tomography.* Features outside of this slice are out of focus and are blurred to the point that

* The word tomography is derived from the Greek word *tomos* for slice, section, or cut, as in common medical usage such as appendectomy, plus -graphy (1987) and appears in print as early as 1935 (Grossman 1935a, 1935b).

they disappear from the image. Sharp images are difficult to obtain, in part, because of the thickness of each slice, and the smearing of images outside the imaging "plane" across the image of the plane of interest seriously degrades contrast within the slice. Laminography continues to be used as an inspection tool for objects whose geometry is impractical for CT; for example, relatively planar objects such as printed wiring boards.

CT, an approach superior to laminography for most applications, became possible with the development of digital computers. Radon established the mathematics underlying CT in 1917 (Radon 1917, 1986), and in 1968, Cormack demonstrated the feasibility of using x-rays and a finite number of radiographic viewing directions to reconstruct the distribution of x-ray absorptivity within a cross-section of an object (Cormack 1963). In the early 1970s, Hounsfield developed a commercial CT system for medical imaging (Hounsfield 1968–1972, 1971–1973, 1973), and the number of medical systems is now virtually uncountable.

In CT of patients, several constraints affected the way in which the apparatus was developed. First, the dose of x-rays received by the patient must be kept to a minimum. Second, the duration of data collection must be limited to several seconds to prevent blurring of the image by involuntary patient movement. These considerations do not apply in general to imaging of inanimate objects, and longer data collection times can be used to improve the signal-to-noise ratio in the data. Early on, engineering components and assemblies were characterized by medical CT scanners, but, because the medical systems were optimized for the range of contrast encountered in the body and not for objects of technological interest, systems for nondestructive evaluation and material characterization were soon marketed.

Industrial CT has gained a measure of acceptance (Bossi and Knutson 1994, Copely, Eberhard et al. 1994), but the high cost of the instrumentation means that it will not replace x-radiography in many nondestructive evaluation applications. Applications where x-ray CT offers significant economic advantages include five areas (Bossi and Knutson 1994): new product development, process control, noninvasive metrology, materials performance prediction, and failure analysis. The information CT provides can drastically shorten the iterative cycles of prototype manufacture, and testing is required to bring new manufacturing processes under control. Evaluation of castings by radiography is very time consuming because of widely varying thicknesses of these components, while CT allows relatively inexpensive inspection; with accurate 3D tomographic measurements, castings with critical flaws can be eliminated before subsequent costly manufacturing steps, and those with anomalies such as voids that can be demonstrated to be noncritical can be retained and not scrapped. Integrating CT data of as-manufactured components into structural analysis programs seems very promising, particularly for anisotropic materials such as metal matrix composites (Bonse and Busch 1996). Final assembly verification, for example, in small jet engines, is a third area where CT appears to be cost-effective.

The extreme sensitivity of CT to density charges can be exploited to follow damage propagation in polymeric matrix composites (Bathias and Cagnasso 1992), even when the microcracks produced cannot be resolved by the most sensitive x-ray imaging techniques. It can be used in studying asphalt recycling (González, Norambuena-Contreras et al. 2018). X-ray CT can be performed with portable units and offers considerable promise for studying the processes active in growing trees and for milling lumber: the environmental effects on a forest of a nearby chemical or power plant can be assessed over a number of years on the same set of trees, daily and seasonal changes in the cross-sectional distribution of water can be obtained, and luck can be removed from the process of obtaining large wooden panels with beautiful ring patterns and without knots or decay (Onoe, Tsao et al. 1984, Habermehl and Ridder 1997). Pyrometric cones used for furnace temperature calibration are produced in the millions annually and in at least 100 compositions, and CT has been applied to understand why certain powder compositions for these dry-pressed, self-supporting cones produce large-density gradients in dies and rejection rates (due to fracture) several times higher than most other compositions (Phillips and Lannuti 1993). Of interest also is comparative work using magnetic resonance imaging and x-ray CT to study ceramics (Ellingson, Engel et al. 1989).

As with any other imaging modality, new applications required resolution of even smaller features, and this became the goal of one branch of workers in CT. If instead of resolving features with dimensions barely smaller than millimeters, as is typical of industrial or medical CT equipment, one were able to image features on the scale of 1–10 μm, then many microstructural features in engineering materials could be studied nondestructively. This size scale is also important in biological structural materials such as calcified tissue. Areas in which microCT has been employed profitably include damage accumulation in composites, fatigue crack closure in metals, and densification of ceramics. The author suspects that there has been and continues to be extensive use of microCT to reverse-engineer components for mechanical systems and for seminconductors and circuitry, but, for obvious reasons, this activity is largely invisible in the open literature.

The advances in microCT imaging since the 1980s have been enabled by consumer and military demand: constantly improving area detectors for cameras and video systems, ever larger and brighter video displays, several generations of more capable data storage and transfer solutions, and exponential increase in computational power without significant increase in cost. Before considering the mathematics and physics of CT and the hardware requirements for microCT, it is constructive to review the early chronology of microCT (or at least the author's perspective of how this developed). The first realization of microCT seems to have been in 1982 (Elliott and Dover 1982); this group used a microfocus x-ray source and a pinhole collimator to collect high-resolution data.† In work published in 1983, Grodzins (1983a, 1983b) suggested how using the tunability of synchrotron x-radiation would allow one to obtain enhanced contrast from a particular element within a sample imaged with CT – that is, by comparing a reconstruction from data collected at a wavelength below that of the absorption edge of the element in question with a reconstruction from a wavelength above the edge. In 1984, Thompson et al. (1984) published low-resolution CT results using synchrotron radiation and the approach advocated by Grodzins. Within a few years, multiple groups had demonstrated microCT using synchrotron radiation (Bonse, Johnson et al. 1986, Flannery, Deckman et al. 1987, Flannery and Roberge 1987, Hirano, Usami et al. 1987, Spanne and Rivers 1987, Ueda, Umetani et al. 1987, Kinney, Johnson et al. 1988, Sakamoto, Suzuki et al. 1988, Suzuki, Usami et al. 1988, Engelke, Lohmann et al. 1989a, 1989b), while others applied x-ray tube–based microCT (Burstein, Bjorkholm et al. 1984, Feldkamp, Davis et al. 1984, Seguin, Burstein et al. 1985, Feldkamp and Jesion 1986, Feldkamp, Kubinski et al. 1988). It is important to emphasize the shift from collecting single slices to collecting volumetric (i.e., simultaneously collected multiple adjoining slices) data.

In the 1990s, there was enough interest in microCT that several companies began to manufacture and sell microCT systems that are well equipped with analysis software. Multiple manufacturers now offer a range affordable, turnkey microCT systems for routine, day-to-day laboratory characterization à la SEM (scanning electron microscopy). Between the release of the first edition and the writing of this edition, many of the original and highly innovative manufacturers have been incorporated into much larger instrumentation

† Sato et al. (1981) presented reconstructions of an optical fiber, claiming 20-μm spatial resolution, but, due to the noise in the image and the unfortunate sample geometry, it appears to the author that the slices are dominated by reconstruction artifacts.

groups; the long-term effect of this consolidation remains to be seen. Recently, commercial nanoCT systems (spatial resolutions substantially below one micrometer for small samples) and in vivo microCT systems (for small animal models of human diseases) can be found in many research laboratories. Dedicated microCT instruments at third-generation synchrotron x-radiation sources (e.g., APS, ESRF, SPring-8) and at other storage rings have multiplied opportunities for 3D imaging at highest spatial resolution and contrast sensitivity, but daily access is not an option.

At the time of this writing, best estimates place the worldwide number of operating microCT systems between 2,500 and 3,000, the majority of these commercial tube-based systems. The exact number is extremely difficult to ascertain because of several factors:

1. Manufacturers keep their sales numbers close to the vest. One can get some idea of the number of systems a particular manufacturer has shipped, but this is only in off-the-record discussions.
2. Many systems installed more than 12–15 years ago are no longer operating, due to mechanical failure or obsolescence leading to replacement with newer systems.
3. The line between microCT and CT is becoming increasingly blurred. Many manufacturers of CT systems for industrial inspection of multi-centimeter-sized components, for peripheral human extremity imaging, or for in vivo imaging of small animals, are now offering "high resolution" options, which extend the scanner capabilities into what the author regards as microCT; namely, tomographic imaging with ~50 μm voxels (volume elements).

The bottom line is that there are many microCT systems operating and producing huge volumes of data. An increase in the annual number of papers published has accompanied the increase of microCT systems operating. Figure 1.1 covers the decade 2004–2013 and shows the annual number of papers published and listed in the abstracting indices PubMed and Web of Science under the search terms "microCT" and "microtomography," respectively (Stock 2014). In 2012 and 2013, these searches found over 1,100 papers annually, a number that had grown substantially since 2004. On the one hand, the numbers in Fig. 1.1 include duplicates between the two databases and thus are an overestimate. On the other hand, the simple search significantly underestimates the papers appearing each year and using microCT: the numbers in Fig. 1.1 do not incorporate papers that would have been found in a given database if the other synonyms were searched; papers in conference

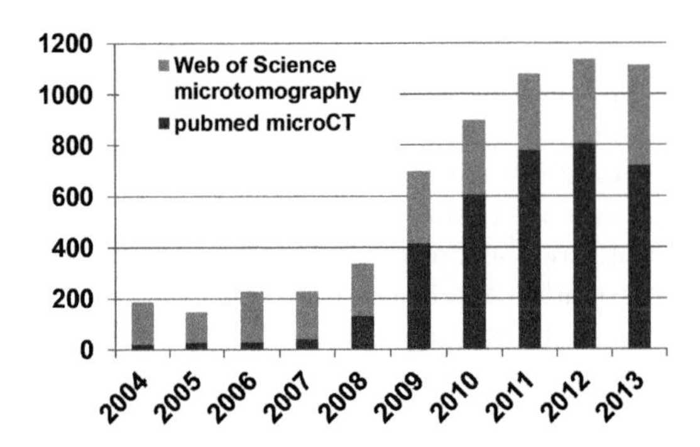

FIGURE 1.1
Annual number of citations for "microCT" in PubMed OR "microtomography" in Web of Science for the years 2004–2013. Adapted from (Stock 2014).

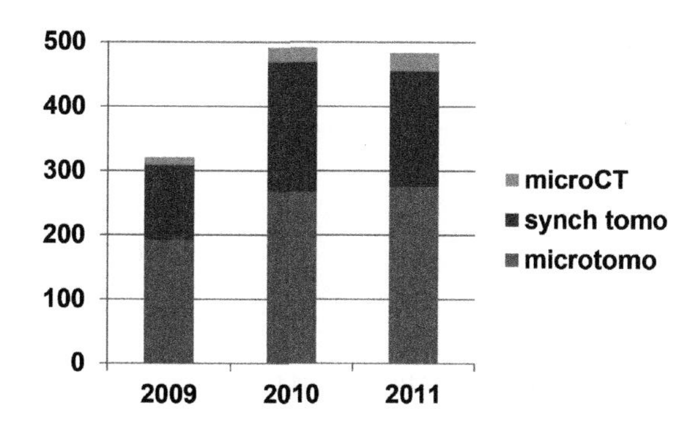

FIGURE 1.2
Web of Science results for 2009, 2010, and 2011 using search terms "x-ray" AND "microtomography," "synchrotron tomography," or "microCT." Adapted from (Stock 2012).

proceedings and chapters in texts are underrepresented (see the search in Compendex discussed below) and, by the end of the decade, as shown in Fig. 1.1, many biomedical papers use microCT as one of many techniques, and this term and its synonyms do not appear in the keywords or abstract. Thus, Fig. 1.1 substantially underestimates rates of publication but certainly not the trends.

In the literature, microCT is synonymous with microtomography and (synchrotron) tomography, with different authors tending to use various specific terms, and Figs. 1.2–1.4 demonstrate that many papers would be missed if only one of the search terms were used. Further, no single search engine reveals all papers in this field, as illustrated by the comparison of results obtained from Web of Science and Compendex, covering engineering and physical sciences, and PubMed, covering biological and medical areas.

Figure 1.2 shows the annual number of papers found by Web of Science for the years 2009–2011. The search was on "x-ray" and one of the three following: microCT, microtomography, or synchrotron tomography. The number of papers published in 2009 were just over 300, whereas in 2010 and 2011 the number approached 500. Some, but far from all, papers appear in more than one search term. Very few "hits" occurred with "microCT." In 2011, Web of Science found 334 unique papers from the total of 483 in the three searches.

Figure 1.3 shows the annual number of papers found by Compendex for the same search terms and years as the Web of Science search shown in Fig. 1.2. Between 200 and 250 total

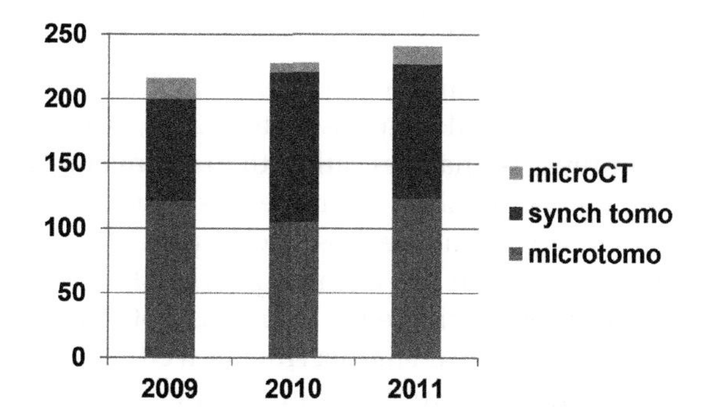

FIGURE 1.3
Compendex results for 2009, 2010, and 2011 using search terms "x-ray" AND "microtomography," "synchrotron tomography," or "microCT." Adapted from (Stock 2012).

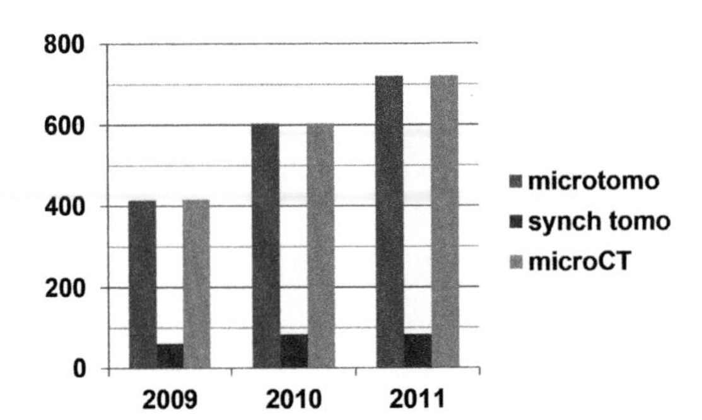

FIGURE 1.4
PubMed results for 2009, 2010, and 2011 using search terms "x-ray" AND "microtomography," "synchrotron tomography," or "microCT." Adapted from (Stock 2012).

papers were found in each year. Of the 241 papers found in 2011, 210 were unique. Similar to the search in Web of Science, only a small fraction of the hits were with microCT.

Figure 1.4 shows the annual number of papers found by PubMed for the same search terms and years as in Figs. 1.2 and 1.3. This plot differs slightly from Figs. 1.1 and 1.2, however, in that the number of results with each search is plotted side by side. In 2011, over 700 papers were identified by "microCT," an equal number by "microtomography," and far fewer by "synchrotron tomography." A comparison of the PubMed microCT and microtomography searches reveals all of the papers appear in both. Once duplicates are removed, PubMed finds just over 800 papers in 2011.

Based on the three individual databases (each of which was edited for duplicates), 2011 yields 1,350 papers published on the x-ray microCT technique at voxel sizes smaller than 50 μm. Removing duplicates between using the Endnote software reduces this number to 1,220. Because the different databases abbreviate journal titles differently, there are further duplicates within the list, and removing these brings the number of papers in 2011 to 1,119.

The number of papers in 2011 seems rather low given that there would have been about 2,000 microCT systems operating worldwide. One would (conservatively) expect each system to have produced at least three publications per year for a total of ~6,000 publications annually, much higher than the number cited in the above paragraph. Two factors certainly contribute to the difference between 1,119 found and the estimate of 6,000 expected. First, some unknown fraction of microCT systems are used for proprietary research that is never published. Second, as mentioned earlier, many biomedical research papers report microCT results but may not explicitly mention these results in a way that database searches will detect.

Searches in the different databases and with different microCT synonyms yield considerably fewer duplicates than one would suppose (Stock 2010, 2012, 2014). This means that those looking to build on previous analyses rather than "reinventing the wheel" must be thorough in how they search. Sadly, based on numerous papers the author has reviewed, many researchers largely ignore all but the most recent papers and fail to reach the standards of papers published one or even two decades previously.

With well over 1,000 papers appearing each year, it is not surprising that microCT is being used increasingly across diverse engineering as well as life and physical science fields. Examples of activity in rather different areas appear in Figs. 1.5–1.7. Figure 1.5 shows the annual number of papers (2005–2013) located via Compendex in three engineering research areas: fracture, batteries, and polymers. There is a gradual increase in polymer and fracture papers, but the number of papers in batteries is more variable, which

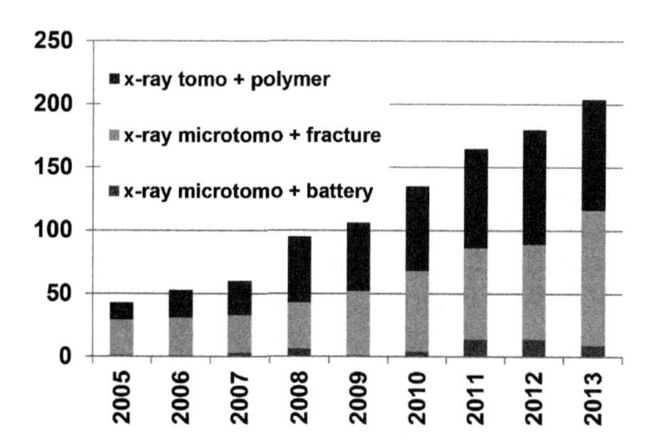

FIGURE 1.5
Number of papers identified in Compendex searches from 2005 to 2013 using the following searches: "x-ray AND tomography AND polymer," "x-ray AND microtomography AND fracture," and "x-ray AND microtomography AND battery." Adapted from (Stock 2014).

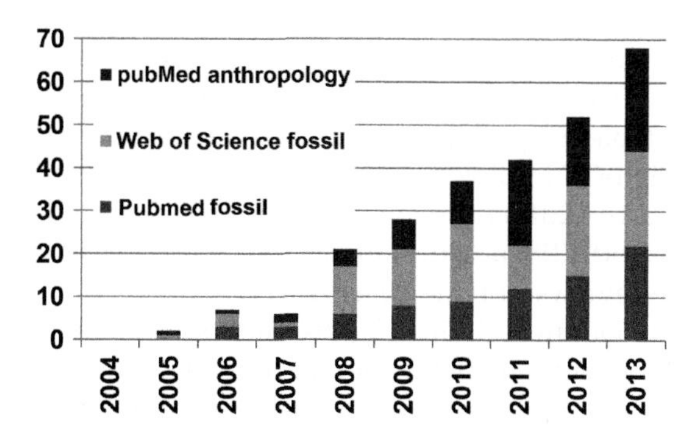

FIGURE 1.6
Annual number of papers from 2004 to 2013 in the fields of anthropology and paleontology in Web of Science (search "microtomography" AND "fossil") and in PubMed (search "microCT" AND "anthropology" and search "microCT" AND "fossil"). Adapted from (Stock 2014).

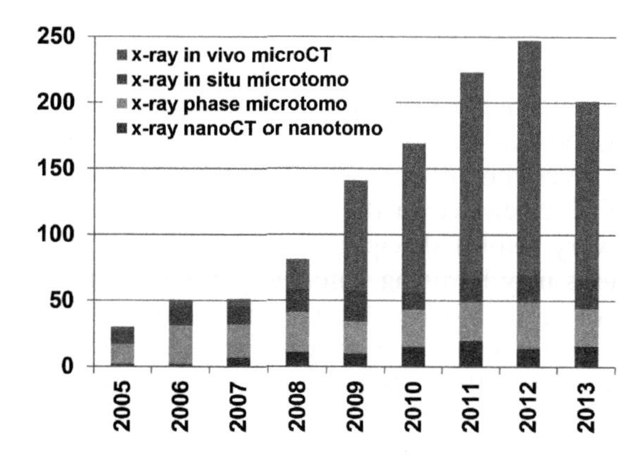

FIGURE 1.7
Annual papers from 2005–2013 found in Compendex searches of "x-ray" AND the following: "microtomography" AND "in situ"; "microtomography" AND "phase"; "microtomography" AND ("nanotomography" or "nanoCT"). A PubMed search on "x-ray" AND "in vivo" AND "microCT" is also plotted. Adapted from (Stock 2014).

is probably a reflection of the narrowness of this topic compared to the other two. Unlike engineering, the fields of anthropology and paleontology appear to have been slower to adopt microCT; searches in PubMed and Web of Science found no papers in 2004, which rose to nearly 70 in 2013 (Fig. 1.6). Figure 1.7 covers 2005 through 2013 and shows some areas (phase microtomography and in situ applications) maintained a steady and constant level, whereas nanoCT/nanotomography and in vivo microCT showed increased publication activity starting in 2007/2008, a reflection of the introduction of dedicated instruments for these two activities. In 2018, over 100 Talbot-Lau phase contrast imaging systems were reported to be operating worldwide at academic centers (Zabler 2018). At the time of writing, at least three manufacturers offered x-ray phase contrast tomography systems. One expects, therefore, that the literature in this area will increase more rapidly than in the past.

There have been quite a few reviews of microCT in the open literature (not to mention many book chapters that are not covered here), and these are mostly focused on a specific aspect of microCT or on microCT in a specific discipline or application. The reviews mentioned here are by no means exhaustive. Comprehensive reviews of microCT of materials appeared after the early era of microCT and about a decade later (Stock 1999, 2008). Others have focused specifically on quantitative microCT, with the latter citation having substantial coverage (Schladitz 2011, Maire and Withers 2014). Pore structure in pharmaceuticals is important for predictable drug delivery and has been examined (Markl, Strobel et al. 2018). Summaries of microCT in biology (Mizutani and Suzuki 2012) and in biomimicry (du Plessis and Broeckhoven 2019) have appeared. MicroCT investigation of paving materials (du Plessis and Boshoff 2019) and of cementitious materials (da Silva 2018) is the focus of other reviews. Three reviews of bone cover different aspects of current research activity: bone and implants (Neldam and Pinholt 2014), nanoCT of bone ultrastructure (Langer and Peyrin 2016), and longitudinal observation of resorption and formation (Christen and Müller 2017). Summaries of digital volume correlation applied to strain measurement in bone (Roberts, Perilli et al. 2014), used with other materials (Fedele, Ciani et al. 2014). and employed for 3D strain mapping (Toda, Maire et al. 2011) have appeared. Results on 3D cavitation in high strength steels (Gupta, Toda et al. 2015) and the extensive literature on synchrotron microCT imaging of failure in structural materials (Wu, Xiao et al. 2017) have been covered.

Given the increase in new microCT users each year, an organized description of the fundamentals and applications of x-ray microCT should be helpful to many. Many aspects are common to both commercial laboratory and synchrotron microCT systems, and comparing and contrasting the two realms is quite valuable. Such a synthesis is particularly valuable (and efficient) given the difficulty of locating discussions of fundamentals in the literature. Most microCT reviews cover the subject as it pertains to a specific discipline, and, while this can be very helpful, discipline-based summaries frequently are quite limiting because other fields may be using valuable approaches that would be missed. In organizing the coverage of prior work by type of analysis approach rather than discipline, the present book on microCT seeks to highlight synergies rather than borders between disciplines. The common thread through the book, therefore, is presentation of microCT as a single modality, each study discussed employing one set of capabilities or experimental designs from a continuum of possibilities.

The following chapters fall into two categories: methodology, which is covered in Chapters 2–6, and applications, which comprise the remainder of the book (Chapters 7–12). Chapter 2 briefly reviews fundamentals of x-radiation and imaging, while Chapter 3 discusses reconstruction from projections. Experimental methods used to perform microCT are the subject of Chapter 4, and Chapter 5 covers microCT in practice. Data analysis and visualization are the subjects of Chapter 6. Chapter 7 introduces simple microstructure quantification and metrology. More complex analyses required for quantifying microstructure in cellular solids and in network specimens are discussed in Chapters 8 and 9, respectively. Microstructural evolution is an important area of microCT research described in Chapter 10. The subject of Chapter 11 is mechanically induced damage and deformation. The topics of Chapter 12 are multimode studies where microCT is integrated with another characterization technique such as x-ray diffraction and studies where reconstruction is based on approaches other than conventional absorption or phase contrast.

References

Bathias, C. and A. Cagnasso (1992). Application of x-ray tomography to the nondestructive testing of high-performance polymer composites. *Damage Detection in Composite Materials*. J. E. Master. West Conshocken, PA, ASTM. ASTM STP **1128**: 35–54.

Bonse, U. and F. Busch (1996). "X-ray computed microtomography (μCT) using synchrotron radiation." *Prog Biophys Molec Biol* **65**: 133–169.

Bonse, U., Q. Johnson, M. Nichols, R. Nusshardt, S. Krasnicki and J. Kinney (1986). "High resolution tomography with chemical specificity." *Nucl Instrum Meth* **A246**: 644–648.

Bossi, R. H. and B. W. Knutson (1994). "The advanced development of x-ray computed tomography applications." United States Air Force Wright Laboratory Publication WL-TR-93-4016.

Burstein, P., P. J. Bjorkholm, R. C. Chase and F. H. Seguin (1984). "The largest and smallest X-ray computed tomography systems." *Nucl Instrum Meth* **221**: 207–212.

Christen, P. and R. Müller (2017). "In vivo visualisation and quantification of bone resorption and bone formation from time-lapse imaging." *Curr Osteoporos Rep* **15**: 311–317.

Copely, D. C., J. W. Eberhard and G. A. Mohr (1994). "Computed tomography part I: Introduction and industrial applications." *J Metals* **46**: 14–26.

Cormack, A. M. (1963). "Representation of a function by its line integrals, with some radiological applications." *J Appl Phys* **34**: 2722–2727.

da Silva, Í. B. (2018). "X-ray computed microtomography technique applied for cementitious materials: A review." *Micron* **107**: 1–8.

du Plessis, A. and C. Broeckhoven (2019). "Looking deep into nature: A review of micro-computed tomography in biomimicry." *Acta Biomater* **85**: 27–40.

du Plessis, A. and W. P. Boshoff (2019). "A review of X-ray computed tomography of concrete and asphalt construction materials." *Constr Building Mater* **199**: 637–651.

Ellingson, W. A., P. E. Engel, T. I. Hertea, K. Goplan, P. S. Wang, S. L. Dieckman and N. Gopalsani (1989). Characterization of ceramics by NMR and x-ray CT. *Industrial Computed Tomography*. Columbus, OH, ASNT: 10–14.

Elliott, J. C. and S. D. Dover (1982). "X-ray microtomography." *J Microsc* **126**: 211–213.

Engelke, K., M. Lohmann, W. R. Dix and W. Graeff (1989a). "Quantitative microtomography." *Rev Sci Instrum* **60**: 2486–2489.

Engelke, K., M. Lohmann, W. R. Dix and W. Graeff (1989b). "A system for dual energy microtomography of bones." *Nucl Instrum Meth* **A274**: 380–389.

Fedele, R., A. Ciani and F. Fiori (2014). "X-ray microtomography under loading and 3D-volume digital image correlation. A review." *Fund Info* **135**: 171–197.

Feldkamp, L. A., L. C. Davis and J. W. Kress (1984). "Practical cone-beam algorithm." *J Opt Soc Am* **A1**: 612–619.

Feldkamp, L. A. and G. Jesion (1986). "3-D x-ray computed tomography." *Rev Prog Quant NDE* **5A**: 555–566.

Feldkamp, L. A., D. J. Kubinski and G. Jesion (1988). "Application of high magnification to 3D x-ray computed tomography." *Rev Prog Quant NDE* **7A**: 381–388.

Flannery, B. P., H. W. Deckman, W. G. Roberge and K. L. D'Amico (1987). "Three-dimensional x-ray microtomography." *Science* **237**: 1439–1444.

Flannery, B. P. and W. G. Roberge (1987). "Observational strategies for three-dimensional synchrotron microtomography." *J Appl Phys* **62**: 4668–4674.

González, A., J. Norambuena-Contreras, L. Storey and E. Schlangen (2018). "Self-healing properties of recycled asphalt mixtures containing metal waste: An approach through microwave radiation heating." *J Environ Manage* **214**: 242–251.

Grodzins, L. (1983a). "Critical absorption tomography of small samples: Proposed applications of synchrotron radiation to computerized tomography II." *Nucl Instrum Meth* **206**: 547–552.

Grodzins, L. (1983b). "Optimum energies for x-ray transmission tomography of small samples: Applications of synchrotron radiation to computerized tomography I." *Nucl Instrum Meth* **206**: 541–545.

Gupta, C., H. Toda, P. Mayr and C. Sommitsch (2015). "3D creep cavitation characteristics and residual life assessment in high temperature steels: a critical review." *Mater Sci Technol* **31**: 603–626.

Habermehl, A. and H. W. Ridder (1997). γ-ray tomography in forest and tree sciences. *Developments in X-ray Tomography*. U. Bonse. Bellingham (WA), SPIE. **SPIE Vol 3149**: 234–244.

Hildebrandt, G. (1992). Paul P. Ewald, the German period. *P.P. Ewald and his Dynamical Theory of X-ray Diffraction*. D. W. T. Cruickshank, H. J. Juretschke and N. Kato. Oxford, International Union of Crystallography: 27–34.

Hirano, T., K. Usami, K. Sakamoto and Y. Suziki (1987). High Resolution Tomography Employing an X-ray Sensing Pickup Tube, Photon Factory, Japanese National Laboratory for High Energy Physics, KEK: 187.

Hounsfield, G. N. (1968–1972). A method of and apparatus for examination of a body by radiation such as X or gamma radiation. UK.

Hounsfield, G. N. (1971–1973). Method and apparatus for measuring x- and γ-radiation absorption or transmission at plural angles and analyzing the data. US.

Hounsfield, G. N. (1973). "Computerized transverse axial scanning (tomography): I. Description of system." *Brit J Radiol* **46**: 1016–1022.

Kinney, J. H., Q. C. Johnson, U. Bonse, M. C. Nichols, R. A. Saroyan, R. Nusshardt, R. Pahl and J. M. Brase (1988). "Three-dimensional x-ray computed tomography in materials science." *MRS Bull* **13**(January): 13–17.

Langer, M. and F. Peyrin (2016). "3D X-ray ultra-microscopy of bone tissue." *Osteopor Int* **27**: 441–455.

Levinson, A. J., D. J. Rowenhorst, K. W. Sharp, S. M. Ryan, K. J. Hemker and R. W. Fonda (2017). "Automated methods for the quantification of 3D woven architectures." *Mater Char* **124**: 241–249.

Maire, E. and P. J. Withers (2014). "Quantitative X-ray tomography." *Inter Mater Rev* **59**: 1–43.

Markl, D., A. Strobel, R. Schlossnikl, J. Bøtker, P. Bawuah, C. Ridgway, J. Rantanen, T. Rades, P. Gane, K. E. Peiponen and J. A. Zeitler (2018). "Characterisation of pore structures of pharmaceutical tablets: A review." *Int J Pharm* **538**: 188–214.

Mizutani, R. and Y. Suzuki (2012). "X-ray microtomography in biology." *Micron* **43**: 104–115.

Neldam, C. A. and E. M. Pinholt (2014). "Synchrotron μCT imaging of bone, titanium implants and bone substitutes - a systematic review of the literature." *J Craniomaxillofac Surg* **42**: 801–805.

Onoe, M., J. W. Tsao, H. Yamada, H. Nakamura, J. Kogure, H. Kawamura and M. Yoshimatsu (1984). "Computed tomography for measuring the annual rings of a live tree." *Nucl Instrum Meth* **221**: 213–230.

Phillips, D. H. and J. J. Lannuti (1993). "X-ray computed tomography for testing and evaluation of ceramic processes." *Am Ceram Soc Bull* **72**: 69–75.

Radon, J. (1917). "Über die Bestimmung von Functionen durch ihre Integralwerte längs gewisser Mannigfaltigkeiten." *Berichte der Sächsischen Akademie der Wissenschaft* **69**: 262–277.

Radon, J. (1986). "On the determination of functions from their integral values along certain manifolds." Translated by P. C. Parks. *IEEE Trans Med Imaging* **MI-5** (#4): 170–175.

Roberts, B. C., E. Perilli and K. J. Reynolds (2014). "Application of the digital volume correlation technique for the measurement of displacement and strain fields in bone: A literature review." *J Biomech* **47**: 923–934.

Röntgen, W. (1898). "Über eine neue Art von Strahlen [Concerning a new type of radiation]." *Ann Phys Chem* **New Series 64**: 1–37.

Sakamoto, K., Y. Suzuki, T. Hirano and K. Usami (1988). "Improvement of spatial resolution of monochromatic x-ray CT using synchrotron radiation." *J Appl Phys* **27**: 127–132.

Schladitz, K. (2011). "Quantitative microCT." *J Microsc* **243**: 111–117.

Seguin, F. H., P. Burstein, P. J. Bjorkholm, F. Homburger and R. A. Adams (1985). "X-ray computed tomography with 50-μm resolution." *Appl Optics* **24**: 4117–4123.

Spanne, P. and M. L. Rivers (1987). "Computerized microtomography using synchrotron radiation from the NSLS." *Nucl Instrum Meth* **B24/25**: 1063–1067.

Stock, S. R. (1999). "Microtomography of materials." *Int Mater Rev* **44**: 141–164.

Stock, S. R. (2008). "Recent advances in x-ray microtomography applied to materials." *Int Mater Rev* **58**: 129–181.

Stock, S. R. (2010). Trends in micro- and nanoComputed tomography 2008-2010. *Developments in X-ray Tomography VII*. S. R. Stock. Bellingham (WA), SPIE. **7804**: 780402.

Stock, S. R. (2012). Trends in micro- and nanoComputed tomography 2010-2012. *Developments in X-ray Tomography VIII*. S. R. Stock. Bellingham (WA), SPIE. **8506**: 850602.

Stock, S. R. (2014). Trends in micro- and nano-computed tomography 2012-2014. *Developments in X-Ray Tomography IX*. S. R. Stock. Bellingham (WA), SPIE. **9212**: 921202.

Suzuki, Y., K. Usami, K. Sakamoto, H. Kozaka, T. Hirano, H. Shiono and H. Kohno (1988). "X-ray computerized tomography using monochromated synchrotron radiation." *Japan J Appl Phys* **27**: L461–L464.

Thompson, A. C., J. Llacer, L. C. Finman, E. B. Hughes, J. N. Otis, S. Wilson and H. D. Zeman (1984). "Computed tomography using synchrotron radiation." *Nucl Instrum Meth* **222**: 319–323.

Toda, H., E. Maire, Y. Aoki and M. Kobyashi (2011). "Three-dimensional strain mapping using in situ X-ray synchrotron microtomography." *J Strain Anal Eng Design* **46**: 549–561.

Ueda, K., K. Umetani, R. Suzuki and H. Yokouchi (1987). A high-speed subtraction angiography system for phantom and small animal studies, Photon Factory, Japanese National Laboratory for High Energy Physics, KEK: 186.

Webb, S. (1990). *From the Watching of Shadows: The Origins of Radiological Tomography*. Bristol, Adam Hilger.

Wu, S. C., T. Q. Xiao and P. J. Withers (2017). "The imaging of failure in structural materials by synchrotron radiation X-ray microtomography." *Eng Fract Mech* **182**: 127–156.

Zabler, S. (2018). "Phase-contrast and dark-field imaging." *J Imaging* **4**: 113.

2

Fundamentals

A certain amount of background is required before one can appreciate the constraints on microCT performance. Limits exist to the rate at which data can be collected and the spatial resolution and contrast sensitivity that can be obtained for a given specimen. Understanding the fundamentals underlying experimental trade-offs is essential to using microCT effectively and efficiently.

2.1 X-radiation

The details of x-ray generation and interaction with matter are covered very briefly in this section under the assumption that most readers will have encountered this subject before. Considerably more detail can be found in texts on x-ray diffraction analysis of materials (Cullity and Stock 2001), or on nondestructive evaluation (Halmshaw 1991).

2.1.1 Generation

X-rays are generated when charged particles are accelerated or when electrons change shells within an atom. Figure 2.1 shows a schematic of an x-ray tube, the source used in lab microCT systems. Electrons flow through a filament (generally W) at a potential kVp relative to the target (generally a metal such as Cu, Mo, Ag, or W). Electrons are emitted from the filament and accelerate toward the target under the effect of the potential. Upon striking the target, the electrons decelerate producing (a) the Bremsstrahlung or continuous spectrum or (b) characteristic radiation (Fig. 2.2). Characteristic radiation arises from electronic transitions generated by the incident electrons, and the high-intensity peaks have a very narrow energy range. As the potential across the tube increases, the intensities of both types of radiation increase and the peak of the continuous spectrum shifts to higher energies. Only a very small fraction of the energy of the electron beam is converted to x-radiation; most of the energy is released as heat. The amount of heat that can be dissipated from the target is what restricts the total electron current, the electron density, and the x-ray intensity. One strategy developed to increase the amount of heat is rotating the target and thereby spreading the heat and increasing the area that can be cooled. Another recent approach is to use a stream of liquid metal as the target (Espes, Andersson et al. 2014). Further limiting the x-ray intensity that is available for imaging is the fact that the x-rays are emitted in all directions (Fig. 2.1) and most never encounter the specimen. A compact laser-driven betatron source has also been used to generate x-rays for laboratory microCT (Cole, Symes et al. 2018).

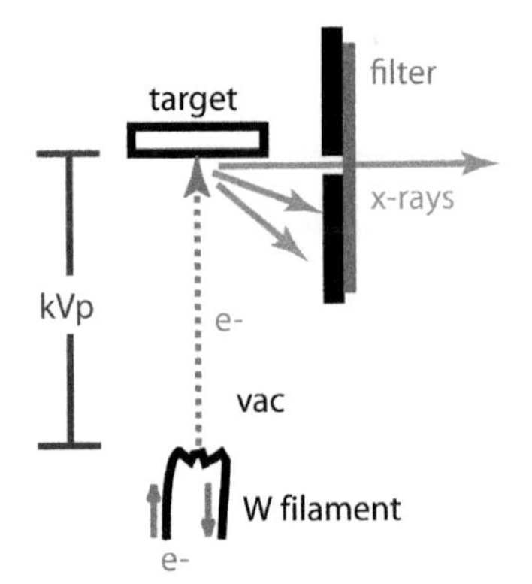

FIGURE 2.1
Schematic of an x-ray tube with electrons (e^-) flowing through a W filament, thermionically emitted from the filament and accelerated within a vacuum (vac) by potential kVp. As drawn, x-rays pass through an aperture and a filter.

Electrons (or positrons) accelerating in storage rings such as synchrotrons are another source of x-radiation (Fig. 2.3). If the electrons travel at relativistic velocities and are deflected by a magnetic field, a continuous spectrum of electromagnetic radiation results spanning from microwaves to very hard x-rays (Fig. 2.4). Several factors give synchrotron radiation an advantage over tube sources for x-ray imaging. First, the intensity of x-rays delivered to a specimen is much greater, and synchrotron radiation can be tuned to a very narrow energy range of wavelength most advantageous for examining a given sample. The relativistic character of the synchrotron radiation emission process confines the resulting radiation to directions very close to the plane of the electron orbit, and the divergence of the beam is very small. Thus, synchrotron radiation not only possesses very high flux but also has much higher brightness and spectral brightness (intensity per unit area of source and intensity per unit area per unit

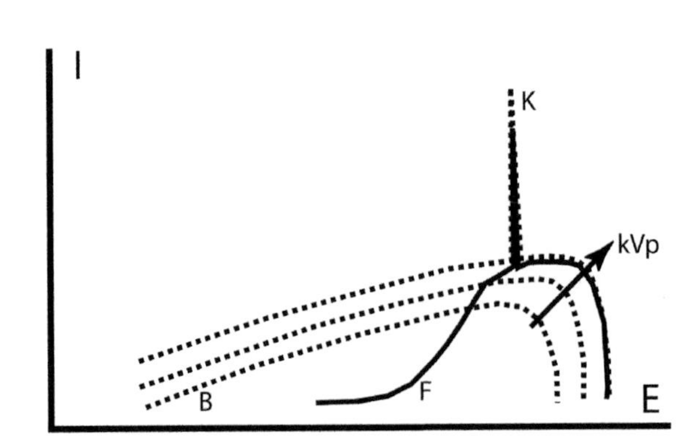

FIGURE 2.2
Schematic of x-ray tube spectra (intensity I as a function of photon energy E) consisting of continuous spectrum B and characteristic lines (K, only one shown). Each curve (dashed line) is the spectrum at a particular potential kVp, and the arrow labeled with kVp shows the shift of the continuous spectrum with increasing kVp. The solid gray curve labeled F shows the effect of the filter (eliminating the softer, lower energy radiation) decreasing the energy range of the x-rays incident on the specimen. The presence of the filter is particularly important in x-ray imaging: the softer radiation would not pass through a specimen but would saturate the detector where the beam passes around the specimen and would increase scatter degrading contrast. Note that the x-ray (photon) energy E is in keV, while the x-ray tube voltage kVp is in kV.

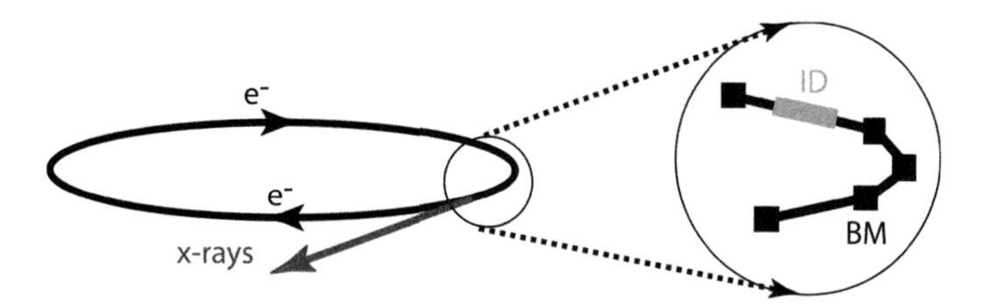

FIGURE 2.3
Schematic of a synchrotron storage ring showing electron bunches e⁻ circulating around the ring. The bunches travel at relativistic velocities (the electron energy is 7 GeV at the APS), and, where the bunches are deflected by bending magnets BM or insertion devices ID, x-rays are emitted. In a sense, the x-ray beam is a highly collimated searchlight blinking on and off.

area solid angle per unit area per unit energy bandwidth, respectively). At some sources of synchrotron x-radiation, the x-ray source size is very small, and the source can have high spatial coherence, something essential for imaging with phase contrast, a subject covered below.

Several types of devices produce the intense magnetic fields required to produce synchrotron radiation (Fig. 2.3). Bending magnets situated periodically around the storage

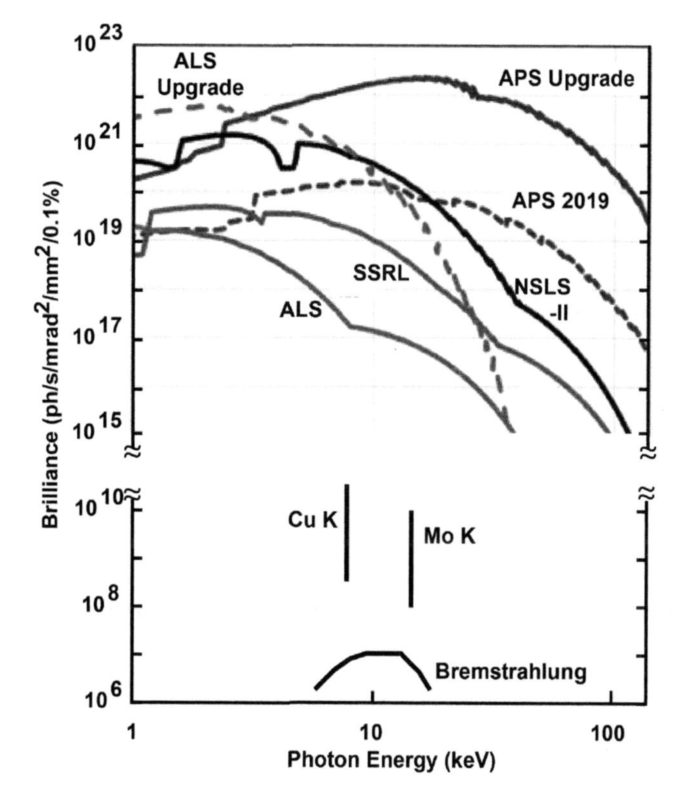

FIGURE 2.4
Relative intensities of x-ray tubes and of synchrotron radiation sources in the US as a function of photon energy. The vertical axis plots brilliance; that is, intensity per unit time per unit area per unit solid angle per unit bandpass. The curves of synchrotron radiation sources are based on calculations by M. Borland, APS.

ring deflect the electrons and force them to circulate within the ring (which is a polygon and not a circle). Insertion devices placed between the bending magnets and consisting of a number of closely spaced magnets are another way synchrotron radiation is delivered to experiments. Each bending magnet or insertion device line at a given synchrotron is optimized for certain operating characteristics, and it is beyond the scope of this section to discuss specifics of x-ray imaging stations at a particular ring. Storage rings for synchrotron radiation are typically very large facilities and are found around the world. Because the characteristics of each differ so markedly and change from time to time, the interested reader is advised to do an internet search for further details. Recent development of tabletop synchrotron radiation sources offers another option for x-ray brightness imaging (Hirai, Yamada et al. 2006).

Figure 2.4 compares synchrotron radiation brilliance to photon energy for different storage rings (insertion devices) and for x-ray tube sources. Two of the plots are projections for performance after upgrades are completed within a few years of writing. The expected brilliance curves of the upgrades of ESRF (European Synchrotron Radiation Facility) and of SPring-8 are slightly lower than that of the Advanced Photon Source (APS). The APS bending magnets (data not shown) produce at least three and one-half orders of magnitude higher brilliance than is obtained from an x-ray tube with a Mo target (lower right side of Fig. 2.4), i.e., at the energy of the Mo Kα characteristic line, 17.44 keV. At higher energies, the difference between x-ray tubes and the APS bending magnets is much greater. The highest brilliance at a storage ring, however, is not produced by bending magnets but rather by insertion devices. In general, synchrotron radiation with energies up to 25–30 keV can be obtained at many stations and many storage rings. Sources for energies above 30 keV are rarer and those above 60 keV are rarer still.

Discussion of how this radiation is conditioned for use in imaging is postponed until later in this chapter. Approaches differ for x-ray tubes and for synchrotron sources, and x-ray interactions with matter must be discussed before the reader can understand these methods.

2.1.2 Interaction with Matter

As discovered by Roentgen (1898), the attenuation of x-rays of wavelength λ is given for a homogeneous object by the familiar equation:

$$I = I_o \exp(-\mu x), \tag{2.1}$$

where I_o is the intensity of the unattenuated x-ray beam, and I is the beam's intensity after it traverses a thickness of material x characterized by a linear attenuation coefficient μ (Cullity and Stock 2001). Typically one finds μ given in cm^{-1}. Rewriting Eq. (2.1) in terms of the mass attenuation coefficient μ/ρ (units cm^2/g) and the density ρ (units g/cm^3) explicitly recognizes that the fundamental basis of the amount of attenuation is the number of atoms encountered by the x-ray beam:

$$I = I_o \exp[(-\mu/\rho)\rho x]. \tag{2.2}$$

Mass attenuation coefficients are a materials property and are a strong function of atomic number of the absorber Z as well as the x-ray wavelength λ (the inverse of energy). Over much of the energy range used most frequently for microCT and except at absorption edges, mass attenuation coefficients can be described by the relationship $\mu/\rho \sim Z^m \lambda^n$, where m equals three or four and n equals (approximately) three. Figure 2.5 is a log-log plot of μ/ρ as a function of x-ray photon energy for several materials of interest (two elemental solids and two multielement solids). In Fig. 2.5, two processes produce the curves shown. The curve's linear sections are from photoelectric absorption, and the flattening of the plots at higher energies results from a change in the dominant absorption mechanism to Compton scattering. More details on these absorption mechanisms appear in texts on nondestructive evaluation (Halmshaw 1991). In Fig. 2.5, the absorption edges for bone (calcium) and titanium are indicated by arrows, and the dashed rectangle roughly indicates on the one hand the energies most frequently used in microCT and on the other the range of mass attenuation coefficients allowing practical imaging.

Two of the substances plotted in Fig. 2.5 (cortical bone and polyethylene) consist of more than one element. One is often not fortunate enough to find values for a specific mixture already tabulated. Values of the linear attenuation coefficient of any mixture or compound can be calculated for a particular energy from first principles using mass attenuation coefficients for the elements and densities for the phases present or the density of the compound in question. Specifically, the linear attenuation coefficient of the mixture or compound equals

$$<\mu> = \sum (\mu/\rho)_i \, \rho_i, \tag{2.3}$$

that is, the weight-fraction-weighted average.

X-rays scatter from the atoms in objects and from very small objects (submicrometer dimensions). Incoherent scattering is generally folded into the linear attenuation coefficients described above. Reinforced scattering can lead to peaks of intensity in certain directions: small-angle scattering from various sources including identically sized, but nonperiodically dispersed submicrometer objects and diffraction from periodic arrays

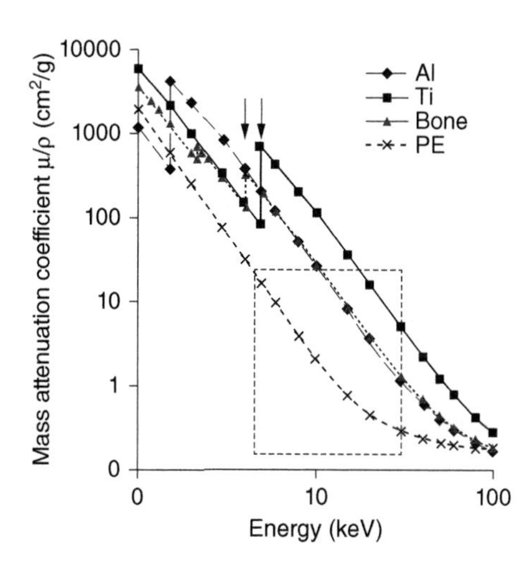

FIGURE 2.5
Mass attenuation coefficient μ/ρ as a function of photon energy for four materials: Al, Ti, cortical bone, and polyethylene (PE). The arrows identify absorption edges for Ti and Ca, the principle absorber in bone. The rectangle bounded by the dashed line very roughly indicates the ranges of the two variables most often encountered in microCT. Plotted from NIST tabulations (Hubbel and Seltzer 2006).

of atoms in crystalline solids. This latter phenomenon is utilized to produce monochromatic radiation for x-ray imaging, improving contrast sensitivity, and is taken up in Section 2.3.

In strict terms, Eqs. (2.1) and (2.2) are incomplete descriptions of x-ray attenuation, but they suffice for most applications. These equations do not consider that x-rays are ever so slightly refracted when passing through solids (indices of refraction differ from one by a few parts per million), enough so that the x-ray wavefronts distort when passing through regions of different electron density [see Fitzgerald (2000) for an introduction]. One situation where a more complete description is needed is imaging with a beam possessing significant spatial coherence. Such coherence can be achieved using (a) an x-ray tube with a very small focal spot or with slits providing a small virtual source size or (b) synchrotron radiation from some (but not all) storage rings. Aside from a brief mention of focusing optics in Section 2.3, further discussion is postponed until Chapter 4.

Soft x-rays have found considerable use in tomographic imaging of subcellular structures. There is a range of soft x-ray energies where water is essentially transparent but carbon (and thus organic molecules) is absorbing; this "water window" extends from the K-edge of carbon at 282 eV to the K-edge of oxygen at 533 eV. Imaging in the water window allows hydrated cells to be imaged, i.e., nearly in their native condition. Soft x-ray microCT and cellular imaging is not the main subject of this volume but is covered in Section 7.2.4.

2.2 Imaging

Various features' visibilities within an object depend on the spatial resolution with which they can be imaged and on the contrast the features have relative to their surroundings. The interplay of contrast sensitivity and spatial resolution defines what can be achieved with CT.

Contrast is a measure of how well a feature can be distinguished from the neighboring background. Frosty the Snowman's eyes of coal show high contrast, while writing with a yellow highlighter marker on a white sheet of paper provides little contrast. Figure 2.6 uses a variable background to illustrate this effect on feature visibility. The more closely spaced pairs of black disks gradually disappear as the background becomes darker.

It is important to be able to quantify the amount of contrast present in an object's image because the smallest change in contrast that can be reliably discerned by the imaging system dictates quantities like detection limits. Contrast is often defined in terms of the ratio of the difference in signal between feature and background to the signal from the background. Thus, the fractional contrast is given by:

$$\text{contrast} = (|\text{sig}_f - \text{sig}_b|)/\text{sig}_b, \tag{2.4}$$

where sig is the signal observed from the object and whose point-by-point variation makes up the image, and the subscripts f and b denote feature and background, respectively.

Spatial resolution describes how well small details can be imaged or small features can be located with respect to some reference point. Figure 2.7 shows two pairs of black disks with different spacing. When imaged with pixel (picture element) size of eight units,

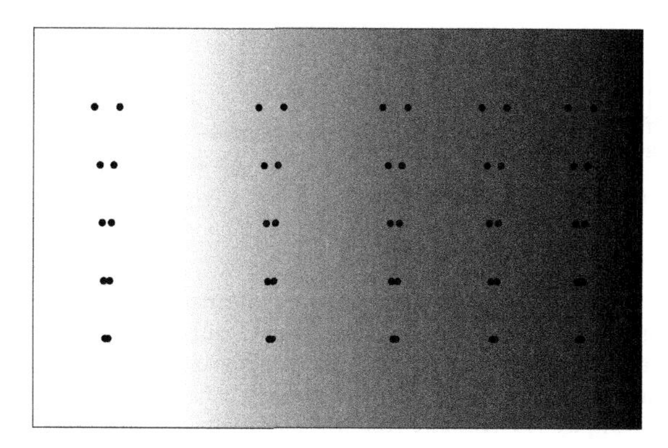

FIGURE 2.6
Influence of contrast on resolution of closely spaced features. The 1D gradient of background gray levels makes it more difficult to resolve the pairs of disks the farther to the right one goes.

there is one open pixel between the pixels containing the disks of the more widely spaced features (Fig. 2.7a), and it is clear that this pair of features can be resolved. The more closely spaced pair of disks, vertically oriented in Fig. 2.7a, occupy adjacent pixels and cannot be resolved. The situation changes, however, when the specimen is imaged with pixel size of four units (Fig. 2.7b): both disk pairs have at least one open pixel of separation and the individual disks are resolved based on this simplistic criterion. As implicitly noted above in Fig. 2.6, a pair of high contrast features can be differentiated at a smaller separation than the same-sized low contrast features. One generally quantifies spatial resolution in terms of the smallest separation at which two points can be perceived as discrete entities.

The presence of noise and inherent imaging imperfection means that quantities such as apparatus performance and image fidelity must be measured in probabilistic terms for a given set of imaging conditions. The point spread function (PSF), for example, describes how the system responds to a point input (i.e., how it images a point), and the modulation transfer function (MTF) represents the interaction of the system (PSF) with multiple features of the object being imaged (i.e., the convolution of all these factors; see the following paragraph). The number of line pairs per millimeter that can be resolved is an often-used simplification. A more accurate approach, reflecting the fact that features must be both detected and resolved, is plotting the contrast required for 50% discrimination of pairs of features as a function of their diameters in pixels; this is termed the contrast-detail-dose curve (1996). The reader is directed to texts on microscopy for further details.

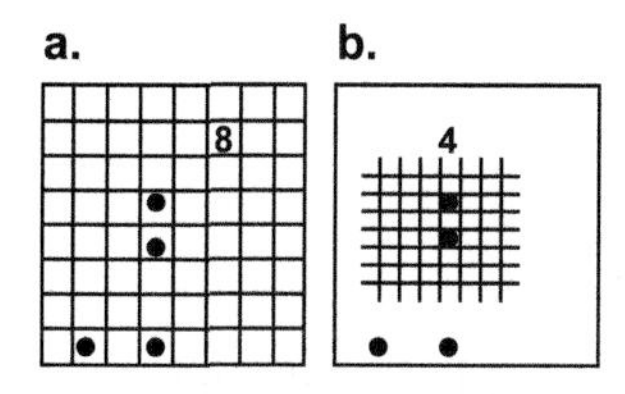

FIGURE 2.7
Two pairs of disks with different spacings. (a) One pair of disks is resolved with pixels of dimensions eight units. (b) Both pairs of disks resolved when the pixel size is four units.

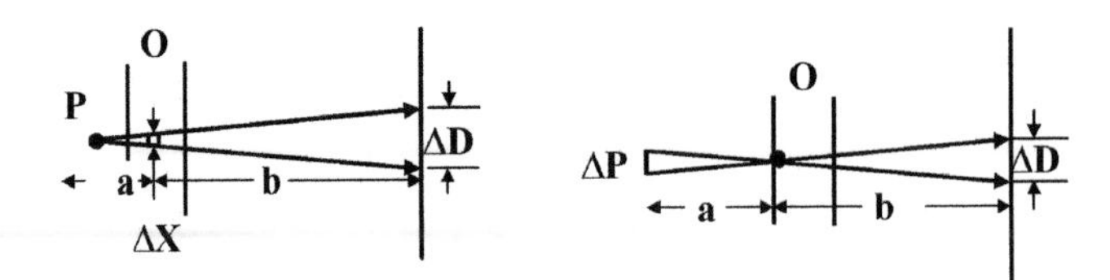

FIGURE 2.8
Illustration of geometrical magnification (left) and penumbral blurring (right). In geometrical magnification, a divergent beam from point source P spreads sample feature at O with width Δx to width ΔD on the detector or film plane. The amount of magnification depends on the ratio of the object to detector separation b to the source to object separation a. Penumbral blurring occurs when there is a finite source size ΔP at a distance a from a point-like feature O in the sample. The crossfire from the source spreads the contrast from O over ΔD on the detector or film b from the object. In both cases, trigonometry is all that is required to calculate the amount of magnification or the level of blurring. Reproduced from (Stock 1999).

Convolution is the mathematical operation of smearing one feature over another feature and is written in one dimension as

$$g(x) = f(x) * h(x) \tag{2.5}$$

$$= \int f(x)\, h(x-u) du \tag{2.6}$$

and in two dimensions as

$$g(x,y) = f(x,y) * h(x,y) \tag{2.7}$$

$$= \iint f(x,y)\, h(x-u, y-v) du dv, \tag{2.8}$$

where the limits of integration are ±∞. As mentioned above, the convolution operation is widely used when considering the PSF and MTF of imaging systems including microCT; it is also the basis of the most widely used reconstruction method, the filtered back projection algorithm, described in Section 3.3.

In closing this section, two geometric effects should be mentioned that pertain to tube-based x-ray imaging. As mentioned in Section 2.1, x-radiation from a tube diverges from the area on the target where the electron beam is incident. If this, the x-ray source size, is small enough, then geometric magnification can be used to see smaller features than otherwise would be the case (Fig. 2.8); the spreading beam can match the feature size to the detector resolution. Source sizes in x-ray tubes used for high-resolution imaging are generally quite small, a few micrometers or larger in diameter, and this, through the effect of penumbral blurring (Fig. 2.8), affects the resolving power of a tube-based microCT system.

2.3 X-ray Contrast and Imaging

Two factors dictate the optimum sample thickness for x-ray imaging (i.e., for greatest contrast). If the specimen is too thick, no x-rays pass through it and no contrast can be seen. If the specimen is very, very thin, no measurable contrast is produced (more precisely, the intensity transmitted through the specimen cannot be distinguished from that passing to either side of the object). The quantity defining x-ray attenuation is the product μx (see Eq. 2.1), and optimum imaging in microCT is found when $\mu x < 2$ for the longest path length through the sample (Grodzins 1983), i.e., greater than 13%–14% transmission through the specimen.

Implicit in the previous paragraph and in the application of Eqs. (2.1) and (2.2) is that the radiation is monochromatic, i.e., the x-ray photons have a single energy. X-rays both from tubes and from synchrotrons are polychromatic before any treatment of the beam is performed. As indicated above in Section 2.1.2, attenuation coefficients are a strong function of x-ray wavelength (energy), and the presence of more than one wavelength complicates analysis of x-ray images. Use of a thin filter between the target and specimen in tube-based imaging (Fig. 2.1) hardens the beam considerably (Fig. 2.2), i.e., preferentially removes the softer radiation and substantially decreases the range of x-ray energies present. Effects such as beam hardening (Section 5.1.7) persist in CT and microCT and can lead the unwary astray. As discussed below, monochromators (crystal or multilayer) can also be used, but this cuts beam intensities too much for practical microCT with x-ray tubes.

Scattering from periodic arrays of atoms or molecules reinforces for certain combinations of x-ray wavelength and angle of incidence; this relationship is described by Bragg's law

$$\lambda = 2d\sin\theta, \tag{2.9}$$

where d is the period of scatterers and θ is the angle of incidence (Fig. 2.9a). The reinforced scattering is termed diffraction and is a basic tool of materials characterization covered in undergraduate texts (Cullity and Stock 2001). X-rays with wavelength λ but incident at an angle θ' not satisfying Bragg's law will not reinforce; x-rays with differing wavelength λ' but incident at the same angle θ will also not reinforce. Atoms within a large single crystal are the basis of a crystal monochromator, and d is the Bragg plane spacing. Multilayer monochromators are another option and consist of alternating low Z and high Z layers whose period d is produced by a technique such as vapor deposition. Because of the high brilliance of synchrotron x-radiation, both types of monochromators are very effective for conditioning beams used in synchrotron microCT. When extremely rapid data acquisition is needed, synchrotron microCT is performed with polychromatic radiation, i.e., with a

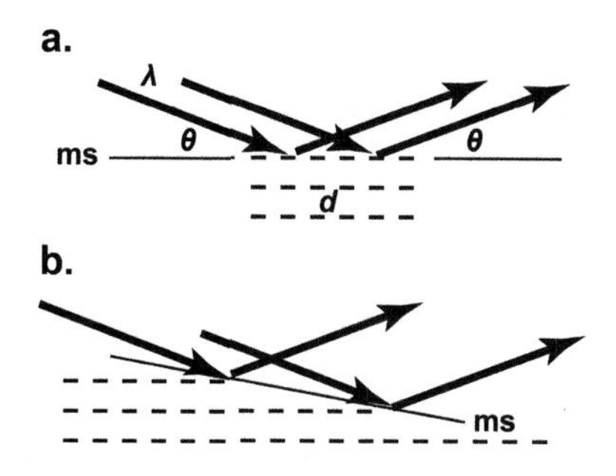

FIGURE 2.9
Monochromators. (a) X-rays with wavelength λ incident at angle θ to the direction of periodicity d produce reinforced scattering (diffraction) along the direction shown. Here the periodicity is normal to the monochromator surface (ms). (b) When the monochromator surface is inclined relative to period of the scatterers, the beam is spread spatially. The orientation and spacing of the scatterers and the width and orientation of the incident beam are the same as in (a), but the orientation of ms has changed. The brightness (intensity per unit area) of the spread beam is decreased as a result.

"pink" beam where filters have removed the softer portion of the spectrum that is not useful for imaging.

Until the last couple of decades, it was a truism that x-ray focusing was largely ineffective; that is, x-ray optics except of the most crude sort were impractical. One exception was use of asymmetrically cut crystal monochromators as beam spreaders (Fig. 2.9b). If such a beam spreader were placed in a synchrotron (parallel) x-ray beam that had passed through a specimen, then the image could be enlarged before the detector was reached. Development of Fresnel optics (zone plates) and of refractive optics has proceeded to the point where they can be used for x-rays hard enough for small specimens of some of the materials discussed in Chapters 7–12. X-ray optics, although of great interest, are only a small aspect of micro- and nanoCT and will not be examined in depth.

The previous paragraph considered ways of extending spatial resolution in x-ray imaging, and some options exist for extending the range of contrast that can be discriminated. As mentioned above, imaging with monochromatic radiation is preferable to imaging with polychromatic radiation if high contrast sensitivity is required: values of the linear attenuation coefficient returned by the reconstruction algorithm are not convoluted with the spread of wavelengths in the monochromatic case. Assuming reasonable mechanical and source stability, increased counting can improve the signal-to-noise ratio in an image, and this is discussed in Chapter 4. Imaging the same specimen at energies above and below the absorption edge of an element within specimen (and numerically comparing the images) is particularly useful for enhancing detection limits for that particular element when it is in low concentration or when other phases are present which have similar linear attenuation coefficients and which lack the element in question. As is seen in Fig. 2.5, the difference in μ/ρ can be greater than five across the absorption edge.

References

ASTM Standards (1996). E 1441-95 Standard guide for computed tomography (CT) imaging. *1996 Annual Book of ASTM Standards*. Philadelphia, ASTM. **03.03**: 704–733.

Cole, J. M., D. R. Symes, N. C. Lopes, J. C. Wood, K. Poder, S. Alatabi, S. W. Botchway, P. S. Foster, S. Gratton, S. Johnson, C. Kamperidis, O. Kononenko, M. De Lazzari, C. A. J. Palmer, D. Rusby, J. Sanderson, M. Sandholzer, G. Sarri, Z. Szoke-Kovacs, L. Teboul, J. M. Thompson, J. R. Warwick, H. Westerberg, M. A. Hill, D. P. Norris, S. P. D. Mangles and Z. Najmudin (2018). "High-resolution μCT of a mouse embryo using a compact laser-driven X-ray betatron source." *Proc Natl Acad Sci U S A* **115**: 6335–6340.

Cullity, B. D. and S. R. Stock (2001). *Elements of X-ray Diffraction*. Upper Saddle River (NJ), Prentice Hall.

Espes, E., T. Andersson, F. Björnsson, C. Gratorp, B. A. M. Hansson, O. Hemberg, G. Johansson, J. Kronstedt, M. Otendal, T. Tuohimaa and P. Takman (2014). Liquid-metal-jet x-ray tube technology and tomography applications. *Developments in X-Ray Tomography IX*. S. R. Stock. Bellingham (WA), SPIE. **9212:** 92120J.

Fitzgerald, R. (2000). "Phase sensitive x-ray imaging." *Phys Today* **53**: 23–26.

Grodzins, L. (1983). "Optimum energies for x-ray transmission tomography of small samples: Applications of synchrotron radiation to computerized tomography I." *Nucl Instrum Meth* **206**: 541–545.

Halmshaw, R. (1991). *Non-destructive Testing*. London, Edward Arnold.

Hirai, T., H. Yamada, M. Sasaki, D. Hasegawa, M. Morita, Y. Oda, J. Takaku, T. Hanashima, N. Nitta, M. Takahashi and K. Murata (2006). "Refraction contrast 11x-magnified x-ray imaging of large objects by MIRRORCLE-type table-top synchrotron." *J Synchrotron Rad* **13**: 397–402.

Hubbel, J. H. and S. M. Seltzer. (2006). "Tables of x-ray mass attenuation coefficients and mass energy-absorption coefficients." NISTIR 5632. Retrieved Jan 22, 2007, from http://physics.nist.gov/PhysRefData?XrayMassCoef/cover.html.

Röntgen, W. (1898). "Über eine neue Art von Strahlen (Concerning a new type of radiation)." *Ann Phys Chem* **New Series 64**: 1–37.

Stock, S. R. (1999). "Microtomography of materials." *Int Mater Rev* **44**: 141–164.

3

Reconstruction from Projections

Understanding the principles of tomographic reconstruction is essential to understanding what CT and microCT can and cannot do and what causes certain artifacts. The treatment here is limited to absorption tomography; reconstruction with x-ray phase contrast is covered separately in Section 4.10.

Reconstruction algorithms have traditionally been classified into iterative or analytic methods. Recently, a third type of approach has gained prominence: data-driven/learning-based artificial intelligence (AI). The characteristic of AI-based reconstruction is the use of training data, and, depending on the methods used, these third approaches can be described as machine learning or deep learning depending on how the reconstruction software is trained.

The basic concepts of reconstruction from x-ray projections are the subject of the first section of this chapter. Section 3.2 covers iterative reconstruction and is illustrated by a simple example of the algebraic reconstruction technique (ART) covered without resorting to the underlying mathematics; for more details see Kak and Slaney (2001) and Natterer (2001). Brief descriptions of other iterative algorithms also appear in Section 3.2. Section 3.3 examines one popular analytic reconstruction approach, the convolution back projection method; this discussion focuses on physical explanations with only a small amount of mathematics. Section 3.4 introduces a second popular analytic method, Fourier-based reconstruction, and requires some application of mathematics; mathematical depth beyond that presented can be found elsewhere (Newton and Potts 1981, Kak and Slaney 2001, Natterer 2001). Section 3.5 introduces AI-based/machine learning approaches to tomographic reconstruction. Section 3.6 discusses performance limits for tomographic reconstruction, and the last section introduces alternative methods for 3D mapping of structure.

3.1 Basic Concepts

Equations (2.1) and (2.2) reveal what is observed after attenuation is complete, and writing the differential form of these equations focuses attention on what occurs within each small thickness element dx:

$$dI/I = -(\mu/\rho)\rho\, dx. \tag{3.1}$$

The size of dx into which the path can be divided varies from instrument to instrument and sample to sample, but, on the scale of the minimum physically realistic thickness element dx, $(\mu/\rho)\rho$ is regarded as a constant and is written simply as μ. Figure 3.1 illustrates how each voxel (volume element) with attenuation coefficient μ_i along path **s** contributes to the

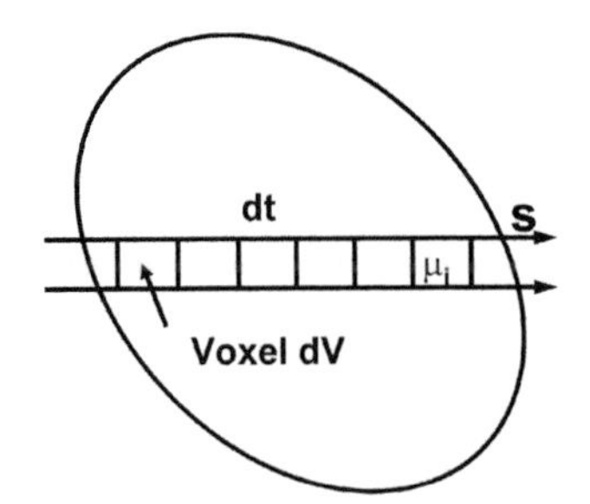

FIGURE 3.1
Contribution of each voxel dV (dimension dt × dt × dt) with linear attenuation coefficient μ_i to the total x-ray absorption along ray **s**.

total absorption. Adding the increments of the attenuation along the direction of x-ray propagation yields the more general form

$$I = I_0 \exp\left[-\int \mu(s)ds\right], \tag{3.2}$$

where μ(s) is the linear absorption coefficient at position s along ray **s**. Assigning the correct value of μ to each position along this ray (and along all the other rays traversing the sample) knowing only the values of the line integral for the various orientations of **s**, that is,

$$\int \mu(s)ds = \ln(I_0/I) = p_S(s), \tag{3.3}$$

is the central problem of computed tomography.

Locating and defining the different contributions to attenuation requires measuring I/I_0 for many different ray directions **s**. Measuring I/I_0 for many different positions for a given **s** is also required: a radiograph measures exactly this quantity, the variation of I/I_0 as a function of position for a given projection or ray direction. Thus, a set of high-resolution radiographs collected at enough well-chosen directions **s** can be used to reconstruct the volume through which the x-rays traverse.

The reality of being able to reconstruct volumes can be illustrated simply by considering how the profile or projection of x-ray attenuation P(**s**) from a simple object changes with viewing direction.* The low absorption rectangle within the slice pictured in Fig. 3.2 casts a spatially narrow but deep "shadow" in the attenuation profile seen along one viewing direction and a spatially wide and shallow "shadow" along the second viewing direction. For the views parallel to the sides of the rectangle, the corresponding changes in the profile are quite sharp but, for views oriented between the two pictured in Fig. 3.2, the changes in the profile would be more complex. Note that paths through the cylinder are relatively short near its edges and are much larger at or near its center; the result is the curved profile outside of the "shadows" which, for simplicity, is not reproduced within the shadowed area. For this simple case, the two correctly chosen views suffice to define the location of the rectangle and the change of μ between the cylinder and rectangle.

* Note that P(**s**) consists of the set of individual projection rays $p_s(s)$, that is, the values of attenuation at each position along the profile.

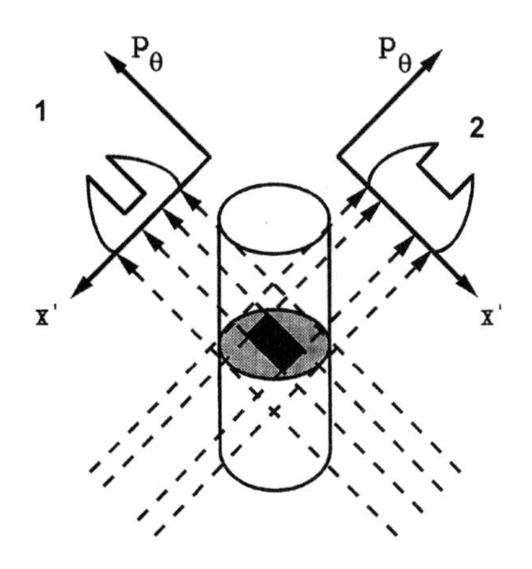

FIGURE 3.2
Illustration of how internal structure of objects can be determined from projections. For simplicity only a single plane and parallel x-radiation are pictured, and the rotation axis for collecting views (absorption profiles P_θ) along different directions θ is vertical and in the center of the cylindrical sample (Stock 1990). © ASME.

In general, the internal structure of specimen is much more complex than that shown in Fig. 3.2, and a more general reconstruction approach will be required. At one extreme, the images of an array of many small objects embedded in a matrix overlap to such an extent that they cannot be distinguished from view to view. At the other extreme, the linear attenuation coefficient can vary continuously across the specimen, with no sharp internal interfaces present. Sections 3.2–3.4 describe approaches for reconstructions with parallel beam. As is described in Chapter 4, most tube-based microCT systems collect data in a fan-beam or cone-beam geometry, and reconstruction with these geometries is briefly covered in Chapter 4.

3.2 Iterative Reconstruction Illustrated by the Algebraic Reconstruction Technique (ART)

ART is an iterative approach to reconstruction, that is, a mathematical trail-and-error approach. Iterative algorithms can be deterministic; examples include ART, simultaneous iterative reconstruction technique (SIRT), and simultaneous algebraic reconstruction technique (SART). Stochastic iterative approaches are a second group, and an example is the maximum likelihood (ML) method. Both deterministic and stochastic iterative algorithms are particularly suited for incorporating prior knowledge about the specimen or for accommodating incomplete data such as significant angular gaps in the series of projections available for reconstruction. It used to be that iterative methods were rarely used because of computational cost and delay but, with the ever-increasing computational power, including graphic processing units (GPU), the iterative methods are gaining in popularity. Techniques beyond ART (SART, ML) are described briefly at the end of this section.

The algebraic method is best illustrated with an example of a 2 × 2 voxel object and projections along six rays and four directions (Fig. 3.3). The numbers within each voxel

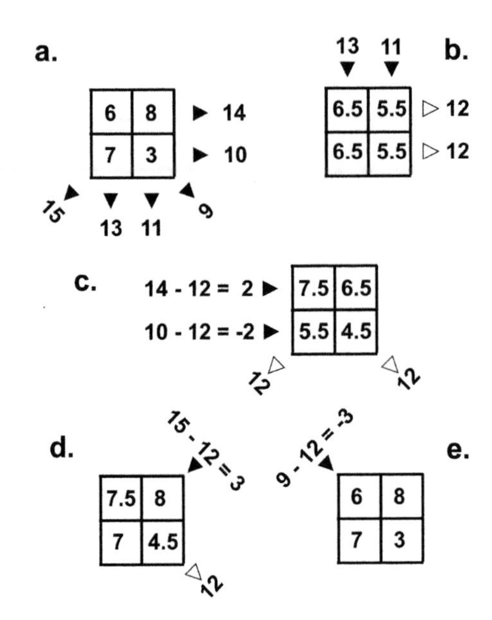

FIGURE 3.3
Algebraic reconstruction. (a) 2 × 2 voxel object. The sums of the voxel values are indicated along the directions shown. These values are the projections from which the unknown voxel values are determined in (b–e). See the text for the details.

in Fig. 3.3a are summed along the three directions; these sums are the projections. For example, in Fig. 3.3a, the top row sums to 14, and the upper left to lower right diagonal sums to 9. For the first guess (not required to be particularly good), choose constant values in each column such that each column yields the correct sum (Fig. 3.3b). The arrows to the right of the object show the resulting (incorrect) sum for each row. The top row of Fig. 3.3b is too low by two units, and the bottom row is too high by two units (left side of Fig. 3.3c). Figure 3.3c shows the correction of the rows: in the top row, 1 is added to the value of each voxel (i.e., one-half of the total difference between measured value and the value calculated for the previous iteration), etc. At this point, the sums of the columns and of the rows are correct, but the values of the diagonals are incorrect (12 instead of 15 from upper right to lower left, 12 instead of 9 from upper left to lower right). Figure 3.3d shows correction along one diagonal: the value of each voxel is increased by one-half of the difference between the measured value and the value in the previous iteration. Figure 3.3e shows the correction of the second diagonal, and the four voxels' values exactly match those in the object.

The example of Fig. 3.3 is quite simplified but should give an idea of the approach. As mentioned above, one situation where an iterative reconstruction technique can be quite useful is when projections are missing (i.e., a certain range of angles is missing, perhaps because they are being blocked by an opaque object like the posts of a load frame which is outside of the field of view [FOV] of interest, or the aspect ratio of a plate-like object prevents some radiographs from being collected). A second instance is real-time microCT of rapid processes where severe angular undersampling is traded for speed and where incorporating *a priori* information can greatly improve the resulting reconstructions; one example of this might occur in the study of rapid dendritic solidification where the known values of the linear attenuation coefficients of the solid and melt could be utilized.

One deterministic and one stochastic iterative method were mentioned at the start of this section: SART and ML. In conventional ART, each ray of a projection is corrected sequentially, but in SART, all of the rays of a given projection are corrected simultaneously, decreasing the noise within the reconstruction (Andersen and Kak 1984). The approach of ML explicitly recognizes the stochastic nature of the x-ray transmission and of x-ray imaging, that is, the realistic assumption that x-ray counts follow a Poisson process; the measured "incomplete" data set is compared with a postulated (and unobserved) "complete" data set in order to maximize the likelihood function of the measured data (Browne and Holmes 1992). An example of the effectiveness of this approach is use of diffracted intensity as the input for reconstructing a bone specimen (Gursoy, Bicer et al. 2015).

Much of the development of iterative methods is driven by clinical imaging, so the medical CT literature is a good source for the latest developments. Two recent examples of the approach for microCT include (Lin, Andrew et al. 2018, Van de Casteele, Perilli et al. 2018).

3.3 Analytic Reconstruction – Back Projection

The principles behind reconstruction via the back-projection method are illustrated in Figs. 3.4 and 3.5. Figure 3.4 shows an array of circular features, one of which is enlarged at the lower left. These features have a lower density at their centers than at their outer borders; the monotonic density variation was chosen for the convenience of producing a uniform, constant absorption projection along any viewing direction. Views along directions *i, ii,* and *iii* produce the profiles shown. In view *i*, for example, the projection consists of peaks, from left to right, of 1, 3, 3, 4, and 2 units of absorption matching the number of circular features aligned along the viewing direction.

The three profiles in Fig. 3.4 can be used to demonstrate how reconstruction via back projection is performed. In Fig. 3.5a, profile *i* is back projected along the viewing direction along which it was obtained. The number of thin lines is used to indicate the amount of

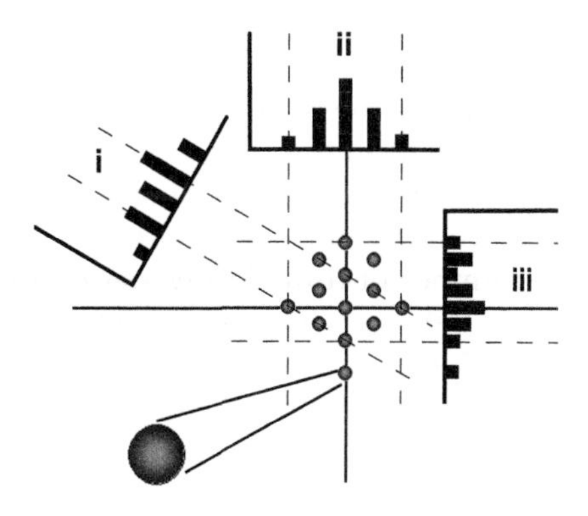

FIGURE 3.4
Absorption profiles of an array of circular objects along three projection directions *i–iii*. These projections are used in Fig. 3.5 to illustrate back projection.

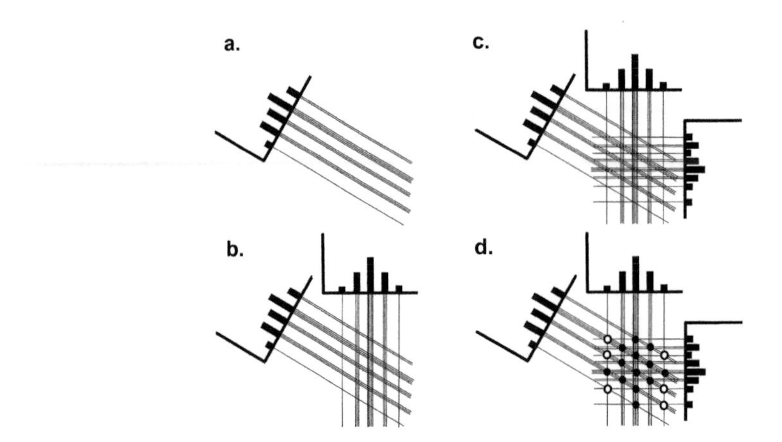

FIGURE 3.5
Back projection illustrated using the "data" of Fig. 3.4. (a) Back projection of radiograph *i*. The number of lines in each ray represents the amount of absorption. (b) Back projection of *ii* added to *i*. (c) Back projection of *i*–*iii*. (d) Identification of possible object positions (circles) based on the three projections. The open circles would be eliminated based on additional views.

absorption along each ray. Figure 3.5b adds projection *ii* to projection *i*, and Fig. 3.5c adds projection *iii* to *ii* and *i*. The correct orientations are maintained in Fig. 3.5b and c. Positions where three rays intersect are possible positions for the circular features, and these positions are labeled in Fig. 3.5d with two kinds of disks: solid disks where the features actually are, and open disks where the presence of features cannot be excluded based on the only three projections available. Incorporation of additional data (projections recorded at different angles) would eliminate the open disks.

In the cartoon representation of Figs. 3.4 and 3.5, the specimen and viewing directions were selected for a simple illustration of back projection and reconstruction. The circular features within the specimen were positioned in well-defined rows and columns, and the views in Fig. 3.4 were chosen along directions where all of the circular features aligned into discrete peaks in the projection profiles. Views at an angle between *i* and *ii*, for example, will not show well-separated peaks. Displacement of a few of the circular features will similarly blend the peaks in the absorption profiles for projections *i*, *ii*, and *iii*. Real specimens will rarely be this convenient.

In the general case, the attenuating mass in each projection is back projected onto a grid, and the contributions added within the space covered by the projections. Different positions along the profile P have different levels of absorption; that amount of attenuation is added to each voxel within the object space that lies along the ray direction in question (represented by the number of thin closely spaced lines in Fig. 3.5). Figure 3.6 shows a reconstruction grid, the cells of which have been filled with the numerical values for two projections from Fig. 3.4. Each value is the sum of the value along the column and the value along the row. Larger values result where both projections have nonzero values.

The careful observer will have noted in both Figs. 3.5 and 3.6 that mass has been distributed across regions where, in fact, no mass is present. Use of *filtered* back projection alleviates this shortcoming and has become a standard algorithm. Figure 3.7 illustrates the filtering process graphically. In Fig. 3.7a, the blurred reconstruction of a point is shown, that is, amplitude A as a function of distance d from the center of the point mass.

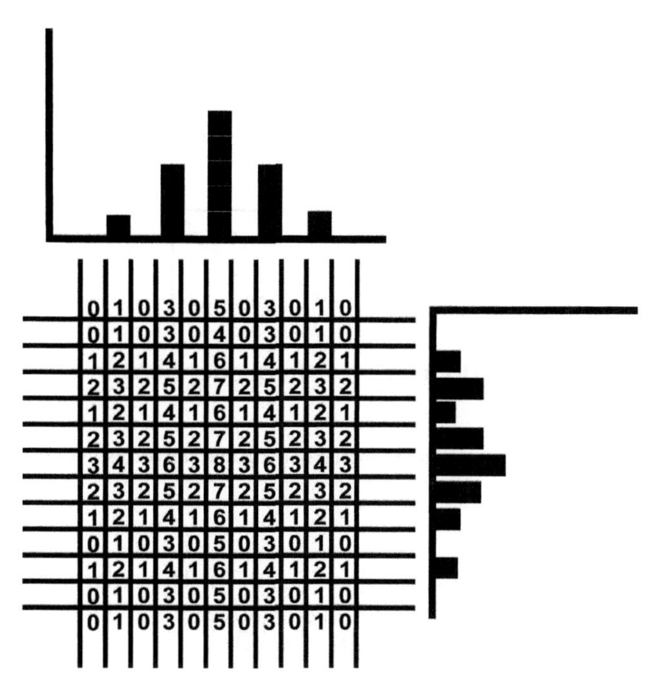

FIGURE 3.6
Back projection onto a grid (see text).

Correction is accomplished using a sharpening filter, and this is done mathematically by convoluting the object's projection with the filtering function. Setting aside the mathematics for a moment, the application of a filter to the blurred profile produces the distance-vs-amplitude plot shown in Fig. 3.7b; when back projected, the negative tails cancel the blur shown in Fig. 3.7a. Consider how this result extends to a profile of several peaks (Fig. 3.7c);

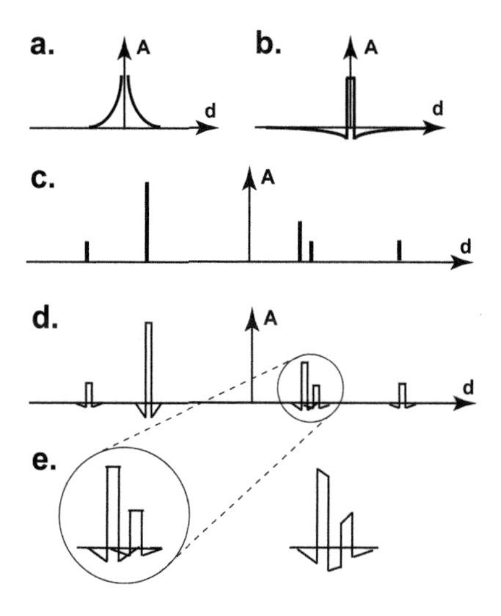

FIGURE 3.7
Illustration of filtering employed in filtered back projection. (a) Blurred reconstruction of a point. (b) Application of a filter to remove the reconstruction-related blur. (c) Profile of several peaks, some of which are closely spaced. (d) Overlapping tails of filtered profiles. (e) Enlargement of overlapping peaks (left) and resultant profile (right).

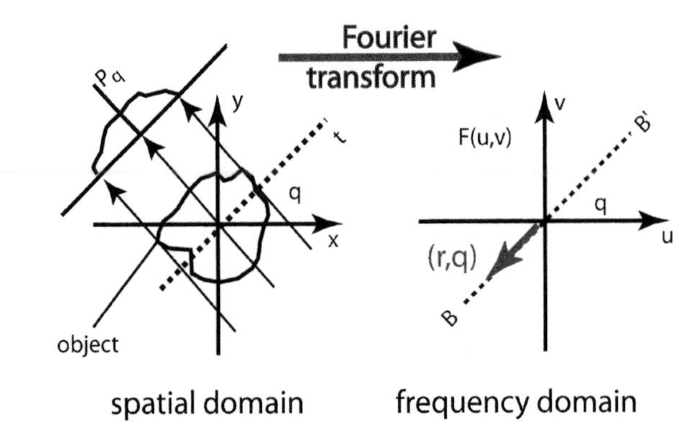

FIGURE 3.8
Relationship between the profile measured in the spatial domain and the corresponding representation in frequency domain (see text).

in some positions the negative tails overlap (Fig. 3.7d). The enlargement of two closely spaced peaks (Fig. 3.7e) shows how overlapping negative tails add to make the total profile. Note that filtering is applied to each absorption profile.

Mathematically, filtered projections are produced by convoluting the filter function with the projection in question. Although the choice of the filter function is extremely important in microCT, it will be ignored here in favor of a more general (brief) description of filtered back projection. Generally, reconstruction is done in polar coordinates, with x replaced by t in a coordinate system **t** rotated by angle q from **x** (left hand side of Fig. 3.8), using Eqs. (2.5) and (2.6) instead of Eqs. (2.7) and (2.8).

Now consider how the convolution operation applies to the reconstruction problem. The object is given by $\mu(x,y)$; see Eq. (3.3) for the relationship between μ and I/I_0 in polar coordinates. The function f(t) in Eq. (2.6) is then the projected ray $p_q(t)$ of Eq. (3.3), and the projection P_q is made up of the individual p_q(t). One obtains the map of μ by performing several steps:

1. Calculating the Fourier transform† S_q of measured projection P_q for each angle q.
2. Multiplying S_q by the value of weighting function (the convolution of two functions is equivalent to simple multiplication of their Fourier transforms, so the weighting function is the transform of the filter) to obtain S_q'.
3. Calculating the inverse Fourier transforms of S_q' and summing over the image plane (direct space or spatial domain) which is the back-projection process.

Note that the summation of the smeared projections is conducted in direct space, as are any interpolations; this differs from the algorithm presented in the following section and relies on the Fourier slice theorem described elsewhere (Kak and Slaney 2001).

The performance of analytic algorithms, like back projection and Fourier-based reconstruction, which is covered in the following section, can suffer in certain instances. Divergent geometries such as encountered in cone beam tomography (Section 4.1) can be problematic. The truncated (i.e., incomplete) data sets encountered in helical

† The Fourier transform of a function and the inverse transform are defined mathematically in the next section.

tomography (Section 4.1) in region-of-interest (a.k.a local tomography or interior tomography, Section 3.8) and in scans with missing or sparse projections can be severe challenges for analytic approaches.

3.4 Analytic Reconstruction – Fourier-Based Reconstruction

Reconstruction can be performed in direct space, that is, in the 3D space in which we move and live; this approach was just described in Section 3.3. Other spaces can be more convenient (more efficient, more robust) for certain operations including reconstruction. In x-ray or electron diffraction, for example, reciprocal space representations are often more instructive than the corresponding direct space data of the single crystal or polycrystalline specimens (Cullity and Stock 2001).

Figure 3.8 indicates schematically how Fourier transforms of absorption profiles P_q (in the spatial domain) can be used to populate the frequency domain. The Fourier components (frequencies and amplitudes along line **t**, projected along **s** in the spatial domain) of the absorption profile provide points along line B-B′ in the frequency domain. Each frequency is plotted at radius *r* along the line shown, and the amplitude of that frequency component provides the numerical value for that point. Data from projections at different angles *q* from 0° to 180° are used to populate frequency space. Because the frequency space representation is as valid as the direct space version of the object (provided, of course, that it is adequately populated with observations), Fourier transformation of the frequency data will produce a valid reconstruction the object in the spatial domain. As the amplitude-frequency representation of an arbitrary profile (or, in fact, any curve) may not be familiar to all readers, a brief digression is appropriate at this point.

Consider for a moment a square wave (Fig. 3.9); this might represent the absorption profile of the rectangle in Fig. 3.2. In terms of Fourier components (i.e., of sine waves of different frequencies and amplitudes), the expression for a square wave is

$$f(x) = 4/\pi\{\sin(\pi[x/L]) + (1/3)\sin(3\pi[x/L]) + (1/5)\sin(5\pi[x/L]) + \cdots\}, \quad (3.4)$$

where *L* is the period and the amplitude is one. Fig. 3.9a plots the amplitudes of the different components (areas of the disks) as a function of frequency. Figure 3.9b pictures the idealized square wave and the first term of this series, which, one can see, is hardly an accurate representation. Figure 3.9c, however, shows the sum of the first seven terms of the series; the square wave is evident, although more terms would be required to damp out the small oscillations (Weisstein 2007).

The two-dimensional Fourier transform $F(u,v)$ of an object function $f(x,y)$ is

$$F(u,v) = \iint f(x,y)\exp(-2\pi i[ux + vy])\,dx\,dy, \quad (3.5)$$

where the limits of integration are ±∞. A projection $P_q(t)$ and its transform $S_q(w)$ are related by a similar equation, and, for parallel projections, some mathematical manipulation yields the relationship

$$F(u,0) = S_{q=0}(u), \quad (3.6)$$

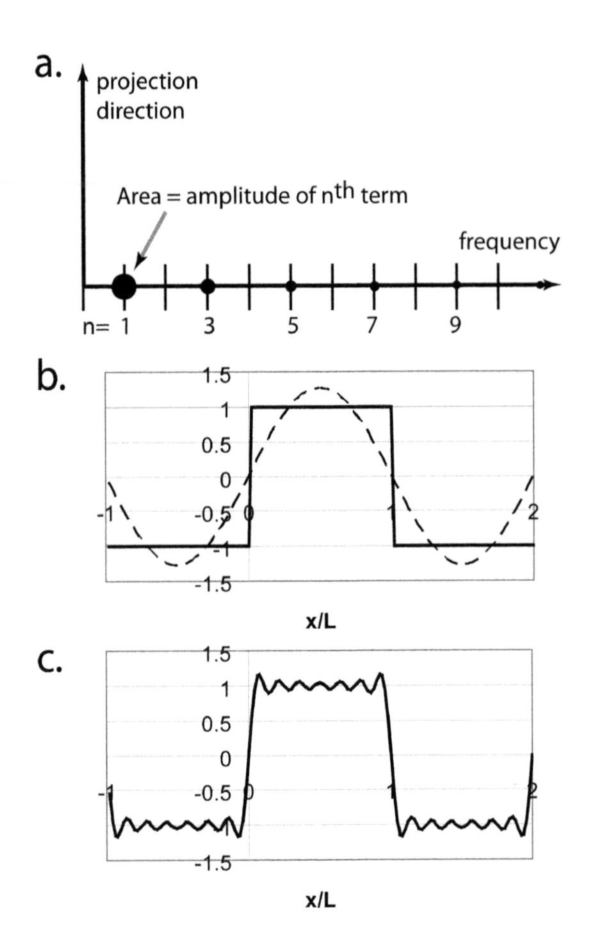

FIGURE 3.9
Illustration of how frequency/amplitude can represent a specific absorption profile, here a square wave. (a) Amplitudes (represented by areas of the disks) as a function of frequency of the terms of the square wave. (b) Idealized square wave, first term of this series. (c) Sum of the first seven terms of the series.

a result that is independent of orientation between the object and coordinate system. This is a form of the Fourier slice theorem which can be stated as:

> The Fourier transform of a parallel projection of an image f(x,y) taken at an angle q gives a slice of a two-dimensional transform F(u,v) subtending an angle q with the u-axis. In other words, the Fourier transform of $P_q(t)$ gives values of F(u,v) along line B-B′ in Fig. 3.8 (Kak and Slaney 2001).

Collecting projections at many angles fills the frequency domain as shown in Fig. 3.10, and the inverse Fourier transform

$$f(x,y) = \iint F(u,v)\exp\left(2\pi i\left[ux + vy\right]\right)du\,dv, \tag{3.7}$$

integration limits again at ±∞, can be used to recover the object function f(x,y). Typically, the fast Fourier transform algorithm is used for these operations, and interpolation between the spokes of data in the frequency domain is required. Inspection

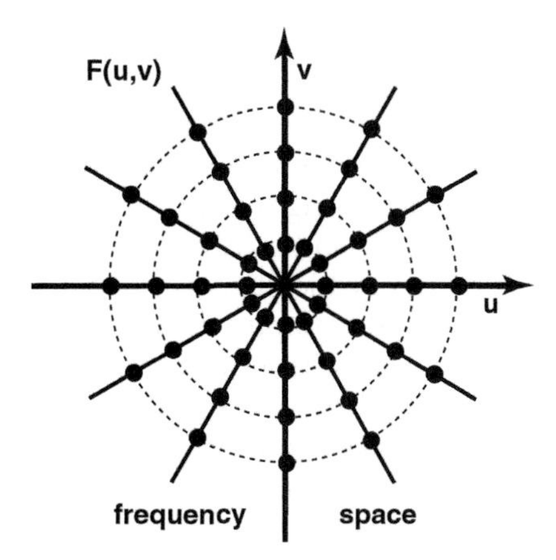

FIGURE 3.10
Frequency space filled with data from many different projections.

of Fig. 3.10 shows sparser coverage farther from the origin of the frequency space; therefore, the high-frequency components are more subject to error than the lower spatial frequencies.

3.5 Reconstruction Employing Machine Learning and Deep Learning

In difficult-to-reconstruct specimens when projections are noisy, truncated, inconsistent, or otherwise compromised, the typically used analytic algorithms often fail to produce acceptable reconstructions. Iterative reconstruction methods can be used to make great improvements, but determining which algorithm is suitable and adjusting its parameters (and validating these results) can be an enormous undertaking. Artificial intelligence/machine learning (ML)/deep learning (DL) techniques[‡] can be applied to enable data-driven tomographic reconstruction, not only for hard-to-reconstruct specimens but also easier specimens whose projections were collected with radically lowered dose (Wang 2016).

One way individuals commonly encounter ML or DL is through text notifications from credit card companies received when suspicious activity is observed on the cardholder's account. The volume of such illegitimate (and the accompanying legitimate) transactions vastly exceeds the ability of humans to monitor, and ML/DL is required to try to stem fraud. Only the briefest of coverage of ML/DL is warranted in this book, and the reader is directed elsewhere for simple descriptions of ML (Theobald 2017), for a roadmap for the development of ML techniques for tomographic reconstruction (Wang 2016), and for the first book of its kind on ML for tomographic imaging (Wang, Zhang et al. 2019).

The non-ML/DL algorithms are "input command" approaches to reconstruction: the input information (projections) is acted upon by a defined command set (for example, the specific implemented filtered back projection software) and the resulting output is the reconstruction. If the reconstruction needs improvement, the command set is altered.

[‡] Deep learning is a subset of machine learning which in turn is a subset of artificial intelligence.

The ML/DL approaches are "input data;" that is, data-driven, where the data (input such as projections and known output; that is, benchmark reconstructions) are fed into the model so that the model is trained to produce output (reconstructions) consistent with the input. The model, once trained, can map new input to a desirable output, which is often superior to counterparts reconstructed using traditional analytic or iterative algorithms.

Often artificial neural networks are used for ML/DL. The building block of an artificial neural network is the artificial neuron, which linearly combines data (inputs) and nonlinearly transforms weighted sums into the output. Interconnections between neurons offer extremely powerful tools for data analysis, including image reconstruction and processing. In a deep neural network (i.e., where DL occurs), there are many hidden levels; that is, the approach is a kind of nonlinear version of multiresolution wavelet analysis. Various ML/DL algorithms are widely used for image processing and computer vision, and DL-based tomographic reconstruction is likely to grow in prominence over time.

3.6 Performance

In understanding the various experimental approaches to microCT (described in the following chapter), it helps first to consider the requirements for reconstructing an $M \times M$ object (i.e., a planar slice through an object consisting of M voxels in one direction and M voxels along a second direction perpendicular to the first). A set of systematically sampled line integrals $\ln (I_0/I)$ must be measured over the entire cross-section of interest such that the geometrical relationship between these measurements is precisely defined. The quality of reconstruction depends on how finely the object is sampled (i.e., the spatial frequencies resolved in the profiles $P(\mathbf{s})$ and the number of viewing directions), on how accurately individual measurements of $\ln (I_0/I)$ are made (i.e., the levels of random and systematic errors), and on how precisely each measurement can be related to a common frame of reference.

The number of samples per projection and the number of views needed depend on the reconstruction method and on the size of features one wishes to resolve in the reconstruction. For an $M \times M$ slice, a minimum of $(\pi/4)\, M^2$ independent measurements are required if the data are noise-free, but faithful reconstruction can still be obtained with sampling approaching this minimum, even in the presence of noise (1996). Features down to one-tenth of the reconstructed voxels can be seen if contrast is high enough (Breunig, Elliott et al. 1992, Breunig, Stock et al. 1993, 1996), and metrology algorithms can measure dimensions to about one-tenth of a pixel with a three-sigma confidence level. The number of samples per view is generally more important than the number of views, errors in I/I_0 of 10^{-3} are significant, and both place important constraints on detectors for CT and microCT. The details of the various reconstruction algorithms lie outside the scope of this review (Newton and Potts 1981, Kak and Slaney 2001).

The precision with which the linear attenuation coefficients can be determined can be expressed in terms its variance

$$\sigma^2{}_0 = \text{const}\, v \left(M_{\text{proj}} \langle N_0 \rangle \right)^{-1}. \tag{3.8}$$

where v is the spatial sampling frequency, M_{proj} is the number of views, and $\langle N_0 \rangle$ is the mean number of photons transmitted through the center of the specimen (Kak and

Slaney 2001). To be strict, Eq. (3.8) applies only to the center voxel of the specimen, but this equation provides important guidance in terms of how changes in several parameters affect reconstructed data. For example, consider the mean value of the linear attenuation coefficient <μ> for a region encompassing a significant number of voxels. Reconstructions produced from projections recorded for time t_0 (with N_0 counts per voxel) would have a standard deviation of the linear attenuation coefficient σ', whereas those recorded for time $4t_0$ ($4N_0$ counts) would be expected to have standard deviation equal to $0.5\sigma'$, if counting statistics were the sole contribution to the variance. Similarly, for the same counting time and x-ray source, if one were to collect data with two sampling dimensions (voxel sizes) v' and $v'/2$, one would expect the standard deviation of the latter measurement to be substantially larger. Other contributions to broadened distributions of linear attenuation coefficients can be substantial and should not be ignored; these include partial volume effects (voxels partly occupied by two very different phases given rise to an intermediate value of μ), which are discussed in Chapter 5.

All of the "exact" reconstruction algorithms require a full 180° set of views, although approximate reconstructions can be obtained where views are missing, for example, where opacity and sample size limits the directions along which useful views may be obtained (Tonner, Sawicka et al. 1989, Haddad and Trebes 1997); the cost is a degraded quality reconstruction. Another approximate data collection approach is spiral or helical tomography (Kalender, Engelke et al. 1997, Wang, Cheng et al. 1997), and it has received considerable attention because it affords increased speed and lower patient x-ray dosage. Only those details important in a particular data collection strategy and those reconstruction artifacts important in the examples will be discussed.

3.7 Sinograms

One of the methods of representing projections for reconstructing a slice is the sinogram. Figure 3.11 schematically illustrates the information contained in a sinogram, a plot of intensity within the projection of a slice as a function of rotation angle. Essentially, the sinogram is the plot of the absorption data for the reconstruction of a single slice, and it gets its name from the fact that projections of objects follow sinusoidal paths as they move around the rotation axis. Note that sinograms appear fairly infrequently in the literature but are quite useful in illustrating certain aspects of the projection data. For example, a method of correcting for mechanical imperfections in a microCT rotation stage relies on properties of the sinogram to refine reconstructions (Section 5.2.1).

3.8 Related Methods

Recording radiographic stereo pairs is often used to precisely triangulate sharply defined features: that is, views of the same sample are recorded along two view directions separated by a precisely known angle, typically between 5° and 10°. This very rapid approach to three-dimensional inspection is of little use and gives way to CT when there are so many similar overlapping objects that individuals cannot be distinguished, when contrast does

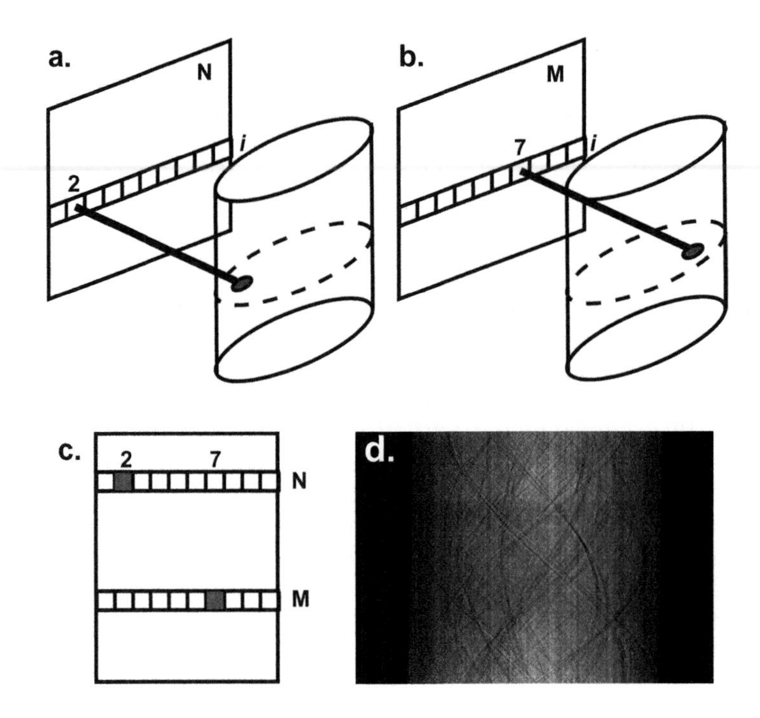

FIGURE 3.11
Illustration of the information contained in a sinogram. The gray object in the cylindrical sample projects onto the second column of the *i*th row of the digital radiograph N in (a). (b) After rotation about the vertical axis (the cylinder's axis, not shown), the gray object now projects onto the seventh column of the *i*th row of the digital radiograph M. (c) The rows of the sinogram for slice *i* (i.e., the slice to be reconstructed from the *i*th row of the radiograph) consist of the successive absorption profiles for the *i*th row derived from radiographs ..., N, ...M, ... The plot is termed a sinogram because the gray object (and all others) trace a sinusoidal path in this representation. (d) Experimental sinogram from an Al corrosion specimen. From Rivers and Wang (2006) but with the original linear gray scale altered for visibility.

not vary sharply within the sample, or when the features to be imaged are so anisotropic that they produce significant contrast only along certain viewing directions which cannot be determined *a priori* (e.g., a crack).

Sometimes the specimen diameter is larger than the detector can cover for the required resolution, and, during rotation, parts of the specimen will pass into and out of the beam (Fig. 3.12a, b). One approach to effectively increasing the detector width is to offset the rotation axis from the center of the beam and rotate through 360° instead of 180°; the range from 180° to 360° covers a different portion of the specimen (Fig. 3.12c, d), and the two sets of data are combined to produce a wider FOV. Another alternative is, at each projection angle, to translate the specimen across the beam and collect two or more offset projections (with some overlap); stitching these projections together covers a wider FOV, and the stitched projections are used as input for reconstruction (see Chapter 5).

Region-of-interest or local tomography is an approach where portions of the specimen pass out of the FOV during rotation (Fig. 3.12). The effect of the missing mass can be corrected by stitching together lower-resolution data for the missing areas of the project or by using known sample composition and geometry and calculating corrected views (Lewitt and Bates 1978, Nalcioglu, Cho et al. 1979). Uncorrected local tomography reconstructions are necessarily approximate, but the extent to which their fidelity is degraded (geometry,

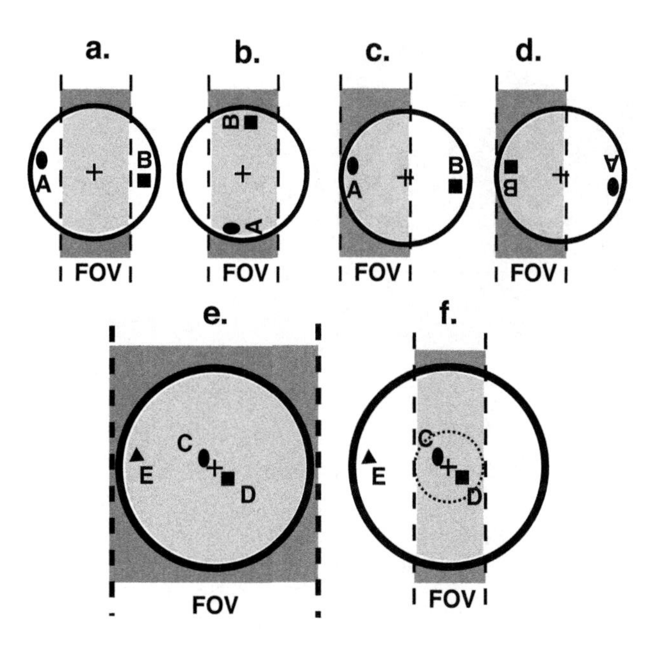

FIGURE 3.12
Field of view (FOV) and specimen diameter. The x-ray beam illuminates the area shaded gray. (a, b) Points A and B rotate into and out of the FOV. (c, d) Illustration of how placing the center of rotation to one side of the FOV and rotating through 360° provides data missing in (a) and (b). (e) The entire specimen diameter is within the FOV, but the smallest voxel size is limited by the number of detector elements. Points C, D, and E remain in the FOV. (f) Local or region-of-interest tomography where the FOV is much smaller, points C and D remain in the FOV while E moves in and out and only the region within the dotted line is reconstructed. Here the voxel size (region diameter divided by the number of detector elements) can be much smaller than in (e) (Stock 2008).

linear attenuation coefficient values) depends on many factors. Errors will become more important as more mass remains longer out of the FOV, and one expects *a priori* that specimens with anisotropic cross-sections will provide the greatest problems. In general terms, the internal geometries in local reconstructions will be reproduced with good fidelity, but if there is significant mass outside of the FOV, dynamic range may be suppressed and/or linear attenuation coefficients affected in directionally anisotropic way (Xiao, Carlo et al. 2007); see Chapter 5 for an example of these artifacts. For specimens with complex, highly anisotropic cross-sections or with high-frequency, anisotropic internal structure, it is essential to check for the presence of artifacts (Kalukin, Keane et al. 1999).

A number of groups/facilities routinely use local tomography. In a custom-built lab microCT system, local tomographic reconstruction compared well with reconstruction with the complete FOV (Jorgensen, Demirkaya et al. 1998). Local tomography is routinely used at ESRF and at APS, so much so that it is sometimes only mentioned in passing (Peyrin, Bonnet et al. 1999). Local tomography is particularly effective in specimens with relatively low absorption such as foams; it has been applied to good effect to study deformation of an Al foam (Ohgaki, Toda et al. 2006). Figure 3.13 shows an extreme example of local tomography with successively smaller fields of view; the highest resolution scan covers only one-sixth of the total diameter. No corrections were applied for missing mass in the reconstruction of Fig. 3.13c, and the enlargement of the area around one tooth shows some fluctuations in background intensity.

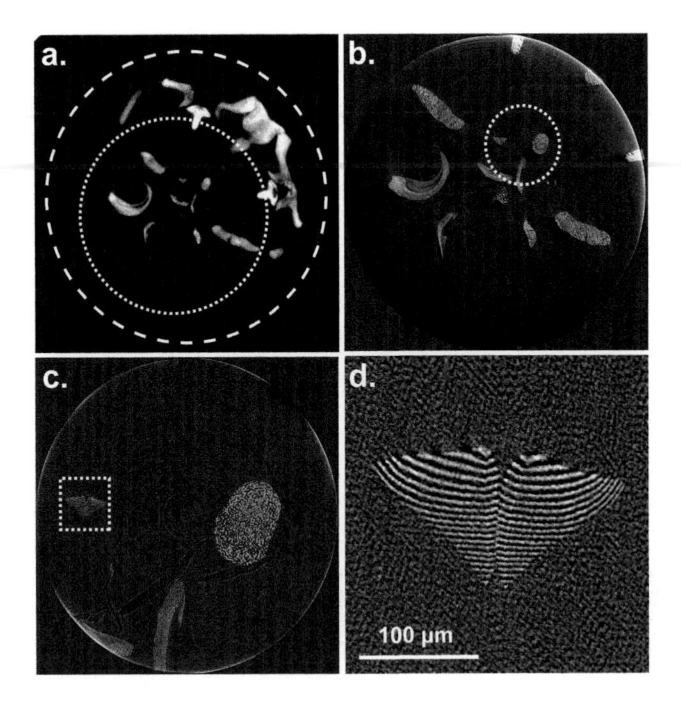

FIGURE 3.13
Local tomography of an intact sea urchin (*Lytechinus variegatus*) oral apparatus or jaw cast in plastic. (a) Lab microCT slice covering the entire specimen. The dashed circle shows the outer diameter of the plastic enclosing the calcified tissue (8.9 mm). The smaller diameter dotted circle (5.9 mm) shows the field of view (FOV) in (b). (b) Synchrotron microCT slice of the sample at the same approximate position as the slice in (a); each isotropic voxel is 2.9 µm on an edge. The dotted circle (1.5 mm diameter) shows the FOV in (c). (c) Synchrotron microCT slice with 1.5 mm FOV and 0.75 µm isotropic voxels. The dotted box shows the area enlarged in (d). (d) Enlargement of one of the five teeth within the sea urchin jaw. S.R. Stock unpublished lab microCT data and synchrotron microCT data collected at 2-BM, APS.

An alternative to local tomography for samples too large to fit in the detector FOV is to record several slightly overlapping projections using translations across the sample diameter. This approach of stitching projections together to build a wider projection spanning the whole sample can be relatively easy to implement if there are discernible features allowing easy registration of the projections. In order to satisfy the angular sampling requirements, however, one would need to collect more projections than in a single FOV-wide data set. The resulting reconstructions can be quite good (Liu, Roddatis et al. 2017).

Partial view reconstruction, where an angular range of projections is unavailable, is related to local tomography in the sense that information is missing. Interpolation of the missing views from the existing data seems to produce tolerable reconstructions (Brunetti, Golosio et al. 2001), but this sort of approximation should be avoided if at all possible. If only one or two adjacent projections are interpolated within an otherwise complete set of views, one is hard pressed to see the effect of the missing data.

Laminography, also termed tomosynthesis, is an alternative approach and is particularly valuable for specimens whose aspect ratios are impractical for conventional microCT (e.g., plate-like specimens). Recent digital methods have been reviewed (Dobbins III and Godfrey 2003), albeit from a clinical and not a microimaging perspective, and Fig. 3.14 illustrates one method of determining three-dimensional positions from a series of views

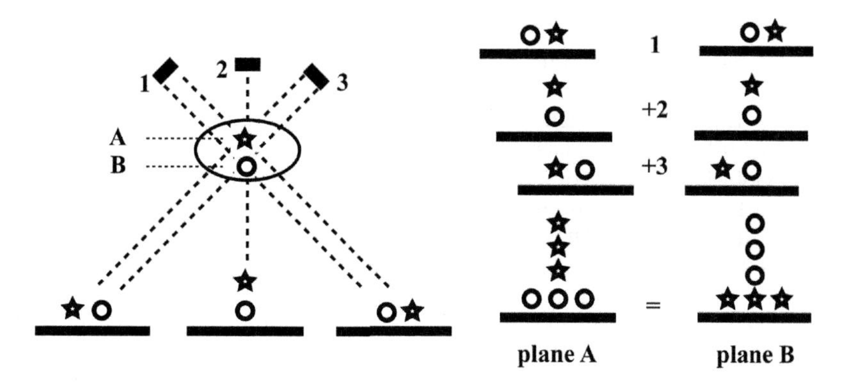

FIGURE 3.14
Illustration of tomosynthesis via the add-and-shift method for parallel rays. (left) Image positions of features (on planes A and B) on the detector plane are shown for source positions 1, 2, and 3 relative to the object. (right) Images shifted to reinforce objects in plane A (star) or to reinforce those in plane B (circle). The features out of the plane of interest are smeared out across the detector, and sharp images occur only for the focal plane. In this illustration, the amount of shift depends on experimental quantities such as the separation between specimen plane and detector and the angle of incidence of the x-ray beam (Stock 2008).

limited to one side of the specimen. There is a cost in terms of degraded contrast by methods such as the shift and add algorithm illustrated in Fig. 3.14. Tomosynthesis has been applied in the microscopic imaging regime in recent studies of perfusion (Nett, Chen et al. 2004), of integrated circuits (Helfen, Baumbach et al. 2006), of NDE (nondestructive evaluation) of long objects (Huang, Li et al. 2004), and of void growth and coalescence in an Mg alloy (Kondori, Morgeneyer et al. 2018). Tomosynthesis has also been applied with phase contrast imaging (Szafraniec, Millard et al. 2014). In situations where displacement of well-defined features can be followed vs rotation, the relative translations of each resolvable point can be converted in depth from one of the specimen surfaces. In this approach. termed stereometry, use of 8–10 views allows a feature's depth to be determined to higher precision than in simple two-view triangulation; the three dimensional fatigue crack surface positions determined with stereometry were in excellent agreement with conventional microCT (Ignatiev 2004, Ignatiev, Lee et al. 2005).

A new variant of coherent diffraction imaging, ptychography, has made great advances. This is covered in the section on phase contrast imaging.

References

ASTM Standards (1996). E 1441-95 Standard guide for computed tomography (CT) imaging. *1996 Annual Book of ASTM Standards*. Philadelphia, ASTM. **03.03**: 704–733.

ASTM Standards (1996). E 1570-95a Standard practice for computed tomographic (CT) examination. *1996 Annual Book of ASTM Standards*. Philadelphia, ASTM. **03.03**: 784–795.

Andersen, A. H. and A. C. Kak (1984). "Simultaneous algebraic reconstruction technique (SART): A superior implementation of the ART algorithm." *Ultrasonic Imaging* **6**(1): 81–94.

Breunig, T. M., J. C. Elliott, S. R. Stock, P. Anderson, G. R. Davis and A. Guvenilir (1992). Quantitative characterization of damage in a composite material using x-ray tomographic microscopy. *X-ray Microscopy III* A. G. Michette, G. R. Morrison and C. J. Buckley. New York, Springer: 465–468.

Breunig, T. M., S. R. Stock, A. Guvenilir, J. C. Elliott, P. Anderson and G. R. Davis (1993). "Damage in aligned fibre SiC/Al quantified using a laboratory x-ray tomographic microscope." *Composites* **24**: 209–213.

Browne, J. A. and T. J. Holmes (1992). "Developments with maximum likelihood X-ray computed tomography." *IEEE Trans Med Imaging* **11**(1): 40–52.

Brunetti, A., B. Golosio, R. Cesareo and C. C. Borlino (2001). Computer tomographic reconstruction from partial-view projections. *Developments in X-ray Tomography III*. U. Bonse. Bellingham (WA), SPIE. **SPIE Proc Vol 4503**: 330–337.

Cullity, B. D. and S. R. Stock (2001). *Elements of X-ray Diffraction*. Upper Saddle River (NJ), Prentice Hall.

Dobbins III, J. T. and D. J. Godfrey (2003). "Digital x-ray tomosynthesis: Current state of the art and clinical potential." *Phys Med Biol* **48**: R65–R106.

Gursoy, D., T. Bicer, J. D. Almer, R. Kettimuthu, F. D. Carlo and S. R. Stock (2015). "Maximum a posteriori estimation of crystallographic phases in X-ray diffraction tomography" *Phil Trans Roy Soc (Lond) A* **373**: 20140392.

Haddad, W. S. and J. E. Trebes (1997). Developments in limited data image reconstruction techniques for ultrahigh-resolution x-ray tomographic imaging of microchips. *Developments in X-ray Tomography*. U. Bonse. Bellingham (WA), SPIE. **3149**: 222–231.

Helfen, L., T. Baumbach, P. Pernot, P. Mikulik, M. Di Michiel and J. Baruchel (2006). High resolution three-dimensional imaging by synchrotron radiation computed laminography. *Developments in X-Ray Tomography V*. U. Bonse. Bellingham (WA), SPIE. **SPIE Proc Vol 6318**: 63180N-63181–63180N-63189.

Huang, A., Z. Li and K. Kang (2004). The application of digital tomosynthesis to the CT nondestructive testing of long large objects. *Developments in X-ray Tomography IV*. U. Bonse. Bellingham (WA), SPIE. **SPIE Proc Vol 5535**: 514–521.

Ignatiev, K. I. (2004). Development of x-ray phase contrast and microtomography methods for the 3D study of fatigue cracks. Ph.D. thesis, Georgia Institute of Technology.

Ignatiev, K. I., W. K. Lee, K. Fezzaa and S. R. Stock (2005). "Phase contrast stereometry: Fatigue crack mapping in 3D." *Phil Mag* **83**: 3273–3300.

Jorgensen, S. M., O. Demirkaya and E. L. Ritman (1998). "Three-dimensional imaging of vasculature and parenchyma in intact rodent organs with x-ray microCT." *Am J Physiol 275 (Heart Circ Physiol 44)* **275**: H1103–H1114.

Kak, A. C. and M. Slaney (2001). *Principles of Computerized Tomographic Imaging*. Philadelphia, SIAM (Soc. Industrial Appl. Math.).

Kalender, W. A., K. Engelke and S. Schaller (1997). Spiral CT: Medical use and potential industrial applications. *Developments in X-ray Tomography*. U. Bonse. Bellingham (WA), SPIE. **3149**: 188–202.

Kalukin, A. R., D. T. Keane and W. G. Roberge (1999). "Region-of-interest microtomography for component inspection." *IEEE Trans Nucl Sci* **46**: 36–41.

Kondori, B., T. F. Morgeneyer, L. Helfen and A. A. Benzerga (2018). "Void growth and coalescence in a magnesium alloy studied by synchrotron radiation laminography." *Acta Mater* **155**: 80–94.

Lewitt, R. M. and R. H. T. Bates (1978). "Image reconstruction from projections: III: Projection completion methods (theory) and IV: Projection completion methods (computational examples)." *Optik* **59**: 189–204 and 269–278.

Lin, Q., M. Andrew, W. Thompson, M. J. Blunt and B. Bijeljic (2018). "Optimization of image quality and acquisition time for lab-based X-ray microtomography using an iterative reconstruction algorithm." *Adv Water Res* **115**: 112–124.

Liu, C., V. Roddatis, P. Kenesei and R. Maaß (2017). "Shear-band thickness and shear-band cavities in a Zr-based metallic glass." *Acta Mater* **140**: 206–216.

Nalcioglu, O., Z. H. Cho and R. Y. Lou (1979). "Limited field of view reconstruction in computerized tomography." *IEEE Trans Nucl Sci* **NS-26**: 546–551.

Natterer, F. (2001). *The Mathematics of Computerized Tomography*. Philadelphia, SIAM (Soc. Industrial Appl. Math.).

Nett, B. E., G. H. Chen, M. S. Van Lysel, T. Betts, M. Speidel, H. A. Rowley, B. A. Kienitz and C. A. Mistretta (2004). Investigation of tomosynthetic perfusion measurements using the scanning-beam digital x-ray (SBDX). *Developments in X-Ray Tomography IV.* U. Bonse. Bellingham (WA), SPIE. **SPIE Proc Vol 5535**: 89–100.

Newton, T. H. and D. G. Potts (1981). *Radiology of the Skull and Brain: Technical Aspects of Computed Tomography.* St Louis, Mosby.

Ohgaki, T., H. Toda, M. Kobayashi, K. Uesugi, M. Niinom, T. Akahori, T. Kobayashi, K. Makii and Y. Aruga (2006). "In situ observations of compressive behaviour of aluminium foams by local tomography using high-resolution tomography." *Phil Mag* **86**: 4417–4438.

Peyrin, F., S. Bonnet, W. Ludwig and J. Baruchel (1999). Local reconstruction in 3D synchrotron radiation microtomography. *Developments in X-ray Tomography II.* U. Bonse. Bellingham (WA), SPIE. **SPIE Proc Vol 3772**: 128–137.

Rivers, M. L. and Y. Wang (2006). Recent developments in microtomography at GeoSoilEnviroCARS. *Developments in X-ray Tomography V.* U. Bonse. Bellingham (WA), SPIE. **SPIE Proc Vol 6318**: 63180J-1–63180J-15.

Stock, S. R. (1990). X-ray methods for mapping deformation and damage. *Micromechanics - Experimental Techniques.* J. W.N. Sharpe, ASME. **AMD 102**: 147–162.

Stock, S. R. (2008). "Recent advances in x-ray microtomography applied to materials." *Int Mater Rev* **58**: 129–181.

Szafraniec, M. B., T. P. Millard, K. Ignatyev, R. D. Speller and A. Olivo (2014). "Proof-of-concept demonstration of edge-illumination x-ray phase contrast imaging combined with tomosynthesis." *Phys Med Biol* **59**: N1–N10.

Theobald, O. (2017). *Machine Learning For Absolute Beginners: A Plain English Introduction* (Second Edition) (Machine Learning For Beginners Book 1). O. Theobald.

Tonner, P. D., B. D. Sawicka, G. Tosello and T. Romaniszyn (1989). Region-of-interest tomography imaging for product and material characterization. *Industrial Computerized Tomography.* Columbus, ASNT: 160–165.

Van de Casteele, E., E. Perilli, W. Van Aarle, K. J. Reynolds and J. Sijbers (2018). "Discrete tomography in an in vivo small animal bone study." *J Bone Miner Metab* **36**: 40–53.

Wang, G. (2016). "A perspective on deep learning." *IEEE Access* **4**: 8914–8924.

Wang, G., P. Cheng and M. W. Vannier (1997). Spiral CT: current status and future directions. *Developments in X-ray Tomography.* U. Bonse. Bellingham (WA), SPIE. **3149**: 203–212.

Wang, G., Y. Zhang, X. Ye and X. Mou (2019). *Machine Learning for Tomographic Imaging.* Bristol (UK), IOP Publishing.

Weisstein, E. W. (2007). Fourier Series – Square Wave, MathWorld – A Wolfram Web Resource.

Xiao, X., F. D. Carlo and S. Stock (2007). "Practical error estimation in zoom-in and truncated tomography reconstructions." *Rev Sci Instrum* **78**: 063705.

4

MicroCT Systems and Their Components

This chapter begins with a brief description of different absorption microCT methods. Section 4.2 describes x-ray source characteristics that impact the performance of microCT systems, and Section 4.3 covers detectors. Discussion of the third important component of systems, sample positioning and rotation subsystems, appears in Section 4.4. Tube-based microCT systems prior to 2008 are covered in Section 4.5, and from 2008 to present are discussed in Section 4.6. Synchrotron microCT systems prior to 2008 are summarized in Section 4.7, and after 2008 are reviewed in Section 4.8. The reason for the partition in both tube-based and synchrotron radiation-based microCT systems is to explicitly highlight the changes that took place in the decade-plus between editions. The author felt that this partition was not necessary in Sections 4.1–4.4 because describing changes in single components is much more straightforward. Full-field nanoCT and lens-based nanoCT are discussed in Section 4.9. MicroCT with x-ray phase contrast instead of absorption is the subject of Section 4.10. MicroCT using x-ray fluorescence from the specimen is covered in Section 4.11. Use of x-ray scattering as the input signal for microCT is the subject of Section 4.12. The final section of this chapter is intended mainly for those new to microCT and discusses how to determine which commercial system or synchrotron microCT facility is best suited for the intended applications. Topics including artifacts found in actual systems, precision and accuracy of reconstructions, and challenges and speculations for the future are covered in Chapter 5.

4.1 Absorption MicroCT Methods

Most microCT systems employ one of four geometries shown in Fig. 4.1. While two arrangements are the same as two of the four generations of scanners into which the CT literature classifies apparatus, the other two are different.

In the first-generation or pencil beam systems (Fig. 4.1a), a pinhole collimator C and a point-like source P produce a narrow, pencil-like beam across which the object O is scanned along $\mathbf{x}_1$; the successive measurements comprise a view (projection) of that physical slice of the specimen; successive views are obtained by rotation about $\mathbf{x}_2$. Only a simple zero-dimensional x-ray detector D is required, perhaps with some scatter shielding S. Energy-sensitive detectors are readily available and, if used instead of gas proportional or scintillation detectors, allow reconstruction with very accurate values of linear attenuation coefficients. Successive views are obtained by rotating the sample and repeating the translation. Obtaining volumetric data (i.e., a set of adjacent slices) borders on infeasible because of the long scan times required, but this is balanced by the inherent simplicity and flexibility of such apparatus and by a relatively greater immunity to degradation of contrast due to scatter. Pencil beam microCT continued to be used with laboratory x-ray sources through the 1990s (Elliott and Dover 1982, 1984, 1985, Borodin, Dementyev et al. 1986, Bowen,

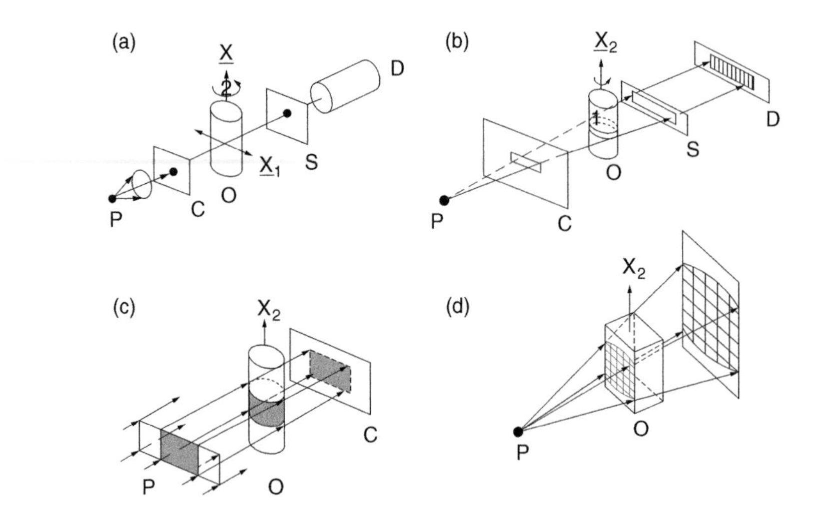

FIGURE 4.1
Four experimental methods for microCT. (a) Pencil, (b) fan, (c) parallel, and (d) cone beam geometries. P is the x-ray source, C is the collimator, O is the object being imaged, x_2 is the specimen rotation axis, x_1 is a translation axis perpendicular to the x-ray beam and the rotation axis, S is the scatter slit, and D is the x-ray detector. Reproduced from Stock (1999).

Elliott et al. 1986, Stock, Guvenilir et al. 1989, Breunig, Elliott et al. 1990, 1992, Breunig, Stock et al. 1993, Elliott, Anderson et al. 1994a, 1994b, Stock, Dollar et al. 1994, Mummery, Derby et al. 1995, Davis and Wong 1996), but, except for x-ray fluorescence tomography and for x-ray scattering tomography, it is now rare to find studies done using first-generation instruments. Very high spatial resolution has been achieved in small samples using synchrotron radiation (Spanne and Rivers 1987, Connor, Webb et al. 1990, Ferrero, Sommer et al. 1993).

Fan beam systems (Fig. 4.1b; that is, third-generation apparatus) use a rotate-only geometry: a flat fan of x-rays defined by collimator C and spanning the sample, originates at the point-like source P, passes through the sample and scatter shield S, and is collected by the one-dimensional x-ray detector. These systems are often used with laboratory microfocus-generated x-radiation. This detector consists of an array of discrete elements that allows the entire view to be collected simultaneously. One to two thousand detectors are typically in the array, making fan beam systems much more rapid than pencil beam systems, but data for only one slice are recorded at a time. Incorporating a linear or area detector makes the system much more susceptible to scatter (than pencil beam systems), that is, the redirection of photons from the detector element on a line-of-sight from the x-ray source into another detector element. In severe cases, this greatly affects the fidelity of a reconstruction. Further, it is necessary to normalize the response of the different detector elements; even with careful correction, ring artifacts from various nonuniformities can still appear in reconstructions.

When examining slices from fan beam systems, it is not only important to note the dimensions of the voxels in the plane of reconstruction but it is also important to ascertain the thickness of the slice: systems collecting data for one slice at a time often are used with a detector width (perpendicular to the reconstruction plane) and slice thickness substantially larger than the voxels' dimensions in the reconstruction plane. This certainly improves signal-to-noise ratio in the reconstruction and is very effective when imaging samples with slowly varying structure along the axis perpendicular to the reconstruction plane (London, Yancey et al. 1990). This approach sacrifices sensitivity to defects much smaller than the slice thickness but also significantly decreases the sample's x-ray dose.

In situations where spatially wide, parallel beams of x-rays are available, the parallel-beam geometry (Fig. 4.1c) allows straightforward and very rapid data collection for multiple slices (i.e., a volume) simultaneously. A parallel beam from a source P (with a certain cross-sectional area) shines through the sample and is collected by a two-dimensional detector array. Since the x-ray beam is parallel, the projection of each slice of the object O on the detector D (i.e., each row of the array) is independent of all other slices. In practice, this must be done at storage rings optimized for the production of the synchrotron x-radiation (Flannery, Deckman et al. 1987, Kinney, Johnson et al. 1988). High-performance area detectors are required, but there is an enormous increase in data collection rates over the geometries described above (see Section 5.4). Because most area detectors consist of a square detector element, slices, are generally, but not always, reconstructed with isotropic voxels (i.e., the voxel dimensions within the reconstruction plane equal the slice thickness).

The cone beam geometry (Fig. 4.1d), the three-dimensional analog of the two-dimensional fan beam arrangement, is a fourth option; it is especially well suited for volumetric CT employing microfocus tube sources (Feldkamp, Davis et al. 1984, Feldkamp and Jesion 1986, Feldkamp, Kubinski et al. 1988). The x-rays diverge from the source, pass through the sample, and are recorded on the area detector. In this geometry, each detector row, except the central row, receives contributions from more than one slice, and the effect becomes greater the farther one goes from the plane perpendicular to the rotation axis. The cone beam reconstruction algorithm is an approximation, however, and some blurring is to be expected in the axial direction for features that do not have significant extent along this direction. Nonetheless, reconstruction of the same 8-mm cube of trabecular bone from data collected with orthogonal rotation axes show only minor differences when the same numerical sections are compared (Feldkamp, Kubinski et al. 1988). With an x-ray source size of 5 μm or smaller, system resolution is limited by that of the x-ray detector array and by penumbral blurring (Fig. 2.8), and can be considerably better than 20 μm. Only the portions of the sample that remain in the beam throughout the entire rotation can be reconstructed exactly. As noted at the end of this section, the greater the cone angle, the larger the reconstruction errors, particularly along the direction parallel to the specimen rotation axis.

Fan beam and cone beam apparatus generally employ point x-ray sources, and this allows geometrical magnification to match desired sample voxel size to detector pixel size (Fig. 2.8). The incorporation of time delay integration (mechanically coupled scanning of sample and detector) into a microCT system (Davis and Elliott 1997) has allowed reconstruction of specimens larger than the detector imaging area or the x-ray beam; directional correlation of noise in large aspect ratio samples remains a problem because this introduces streak artifacts (Davis 1997). Time delay integration has been quite successful in eliminating ring artifacts caused by nonuniform response of the individual detector elements (Davis and Elliott 1997). Multiple frame acquisition with detector translations is a hardware-based approach used to reduce ring artifacts.

Increasing use is being made of spiral or helical scanning in CT or microCT. In this approach, the specimen is translated at a constant rate along the tomographic rotation axis *while* projections are collected. Dedicated helical microCT scanners are commercially available, and this type of data acquisition can be very useful for samples in which a fluid flow front or reaction zone is moving down the length of the specimen. Reconstruction errors need to be carefully considered, however, when setting up scans.

Use of an asymmetrically cut crystal (see Fig. 2.9), positioned between sample and x-ray area detector and set to diffract the monochromatic synchrotron radiation incident on the sample, has been demonstrated to reject scatter and improve sensitivity as well as to provide, through beam spreading, magnification of the x-ray beam prior to its sampling by the x-ray detector (Sakamoto, Suzuki et al. 1988, Suzuki, Usami et al. 1988, Kinney, Bonse et al. 1993). This is an adaptation of a commonly used method in x-ray diffraction topography that allows one to overcome limitations of the detector; that is, to approach resolutions inherent to the x-ray source.

NanoCT, that is, tomography with systems designed to produce voxel sizes substantially below 1 μm, requires much higher precision and accuracy of the various components as well as much longer counting times or much brighter x-ray sources than in microCT (operating in the one-micrometer or greater voxel-size regime). Parallel-beam and cone-beam geometries are used, the latter employing zone plate or other x-ray optics. To a first approximation, the level of mechanical performance required of a nanoCT apparatus scale with the voxel size used, and the details appear in Section 4.9.

Implicit in the above discussion is that the sample moves/rotates and not the detector or x-ray source. This is a practical consideration because x-ray sources and detector can be quite large (and storage rings even larger). There are circumstances where microCT is performed with a gantry design where the source and detector rotate about a stationary specimen. Vertical gantries (horizontal rotation axes) are the rule for in vivo small animal imaging systems where relative motions of the different organs must be prevented, and there is at least one commercial horizontal gantry system available at the time of writing.

Before leaving the discussion of the generalities of microCT systems and considering the characteristics of systems' individual components, a brief digression on reconstructions with fan- and cone-beam is useful. Explaining this earlier in the text, in the chapter on reconstructions, would have been confusing prior to describing the experimental geometries.

Consider first the fan beam geometry and the rays projected from the point source through the specimen and onto the detector (Fig. 4.2a). Because each ray follows a different, albeit predictable, angle relative to the reference direction, the spatial frequencies within each projection cannot be placed along a single spoke in frequency space like they would in a parallel beam projection. Each voxel within the physical slice irradiated by the x-ray beam does, however, contribute to each projection (unlike the case of the cone beam; see below), but, with each ray describing a different projection direction, a coordinate transformation or other procedure is required transform the projections into a form convenient for reconstruction. A simple trigonometric relationship between the two coordinate systems can be used, although the mathematics required to propagate this transformation through the basic reconstruction integrals is somewhat involved. The interested reader is directed elsewhere for these details (Kak and Slaney 2001).

The situation in the cone beam geometry is more complicated than that with fan beam data. Here, one must differentiate the situation on the central plane on the one hand and those of the planes above and below on the other. In the central plane, details of the reconstruction follow those of the fan beam. Off this plane, rays pass from one slice plane to another, and voxels from more than one "slice" contribute to the projection at points like b in Fig. 4.2b. The farther a given plane is from the central plane (i.e., the larger the cone angle κ), the greater the potential errors in the reconstruction. Figure 4.2c shows a simulation of a cone beam reconstruction of several spheres stacked along the rotation axis; here a numerical section perpendicular to the slice planes shows the increasing blurring with distance away from the central plane. Structures with periods along the specimen rotation axis are particularly prone to errors, but other structures reconstruct accurately.

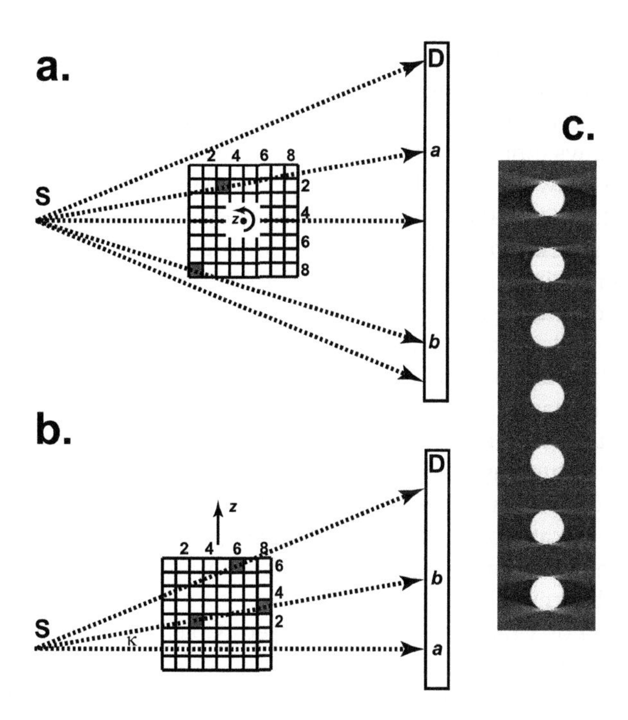

FIGURE 4.2
Illustration of fan beam and cone beam reconstruction. (a) Fan beam geometry. The source S, detector D, and rotation axis *z* are shown, and all of the small squares within the large square object represent voxels within one physical slice of the object. The voxel position is denoted by the numbers outside the specimen (rows, columns). The voxels shown in gray (third column, second row; first column, eighth row) project onto the detector at positions *a* and *b*, and all of the voxels within the physical slice contribute to projected profile of the slice recorded on the detector, that is, all voxels within the physical slices, are sampled in each projection. (b) Cone beam geometry showing a plane perpendicular to the slice plane, that is, parallel to the rotation axis *z*. One-half of the specimen and rays are shown; the other half (below the central ray *a*) is not because it provides no additional information to the figure. Each row of squares in the object represents the voxels within a physical section of the specimen, a section that will be reconstructed into a slice. As pictured, the ray striking the detector at point *b* contains contributions from voxels in rows 2 and 3 (e.g., the gray voxels in column 3, row 2, and column 8, row 3). Note that the ray reaching point *a* is within the central plane, a special plane in cone beam reconstructions: this slice can be reconstructed exactly because it is identical to the fan beam situation pictured in (a). The central task of cone beam reconstruction algorithms is the correct reapportionment of absorption within rays like *b* to the correct slices; the greater the cone angle κ, the greater the potential error. (c) Errors (arrows) at larger cone angles illustrated using a simulated reconstruction of stack of seven solid balls. Courtesy of Tom Case, Xradia Inc.

4.2 X-ray Sources

Most microCT with tube sources of x-radiation has been performed using the entire spectrum, Bremstrahlung, and characteristic radiation, because the cost in data collection times is prohibitive in this photon-starved environment. Exceptions include studies done with pencil beam systems. One group used an energy-sensitive detector to correct for polychromaticity (Elliott, Anderson et al. 1994a) and another used a channel cut monochromator (Kirby, Davis et al. 1997). In a tube-based microCT system, the investigator can vary the tube potential and the tube current to affect imaging conditions. Increasing the tube current produces a linear increase in x-ray intensity but no change in the distribution of x-ray energies. With increased electron flux, the incident on the target, an unintended consequence, may be spreading of the x-ray focal spot, with an adverse effect on resolution (i.e., increased penumbral blurring). Altering the tube voltage (kVp) changes spectrum of x-ray energies emitted, and this can be used to optimize contrast in reconstructions. Increased voltage allows more absorbing specimens to be studied. Lower voltages are used to enhance contrast between low absorption phases such as different soft tissue types. Changes in tube voltage can also alter the focal spot on the target.

In synchrotron-based microCT, the cost of discarding most of the x-ray spectrum during monochromatization is insignificant compared to the times required for sample movement, detector readout, and so on: in most cases, the resulting monochromatic beam is intense enough for a view to be collected in a fraction of a second. Most synchrotron microCT facilities allow the user to select the x-ray energy used to image the specimen (of course there are limits based on the optics available); a little thought and a bit of trial and error will allow the user to obtain the optimum available contrast for a given type of specimen. In some applications (e.g., transient phases), a pink (i.e., polychromatic) beam can be used for very rapid data acquisition. As mentioned earlier (Sections 2.1.2 and 2.3), collecting microCT data above and below the absorption edge of an element of interest and comparing the two reconstructions provides considerably improved sensitivity (Nusshardt, Bonse et al. 1991, Dilmanian, Wu et al. 1997).

Betatron generation of x-radiation has been used (Cole, Symes et al. 2018, Dopp, Hehn et al. 2018). Table-top synchrotron radiation sources offer another option for generating x-rays for microCT (Hirai, Yamada et al. 2006, Toepperwien, Gradl et al. 2018), and commercial systems are available. Essentially these are super x-ray tubes with the accelerated electron beam striking a target, and performance can be quite good. Other types of electron accelerators can also be used with targets to generate x-radiation for imaging, but use of these types of sources, because of their complexity and cost, remains a rarity.

The size of the x-ray source affects the spatial resolution that can be obtained. This is normally not a consideration with synchrotron radiation, given the minuscule intrinsic divergence and the typical source to sample distances employed (>10 m). With x-ray tubes, using a small (5–10 μm or even smaller) diameter x-ray source limits the loading of the target (i.e., the amount of energy which may be deposited) and the resulting x-ray intensity.* This must be balanced against use of a larger spot size where penumbral blurring would prevent small features from being seen. As will be discussed in the context of quantification of crack openings (Chapter 11), there are situations where much can be done even in the presence of significant penumbral blurring.

One solution to heat removal is a rotating anode (target); by constantly (and rapidly) changing the point the electron beam hits, the heat is spatially distributed and more heat can be removed from the target. Rotating anode x-ray sources are quite complicated pieces of equipment and are not typically used for microCT. Another solution to heat dissipation is to make the target a stream of metal (Larsson, Lundström et al. 2013).

In closing this section, the reader should note that electron optics and modern filament materials (e.g., LaB_6) such as are used in scanning electron microscopes (SEMs) can produce beam diameters and x-ray sources sizes substantially smaller than 5 μm. These have found their way into commercial microCT systems. Because very small beams are essential in

* The reader should check with system manufacturers for what they reckon their spot size is on the target, whether it is measured and reported for each tube, and how much it changes with time (due to pitting at the target surface, etc.). Values quoted should be taken with a grain of salt, not only because we all want to put our best foot forward but also because good measurements of x-ray sourced sizes are nontrivial and many factors can influence actual values.

SEMs, it is not surprising, therefore, that SEM-based nanoCT systems have been developed and commercially marketed. For high-resolution tube-based microCT where one expects voxel sizes down to 1 μm on small diameter samples, one expects source diameters on the order of 2 μm or smaller. For nanoCT systems where voxel sizes at 100 nm or smaller are expected, submicron x-ray sources sizes are used.

4.3 Detectors

The characteristics of the x-ray detector array used have important consequences for the performance of a given microCT apparatus. In the early days of volumetric microCT, instruments were sometimes based on linear detectors and sometimes on area detectors, but the field has shifted since the early 2000s to using area detectors. Because volumetric work is largely impractical with pencil beam systems, these are not covered here and details appear elsewhere (Elliott, Anderson et al. 1994b). Even though linear detectors are no longer used much for microCT, it is still worthwhile to include them in the discussion.

Most one- or two-dimensional detector arrays are based on semiconductor devices (e.g., photodiode arrays, charge injection devices, charge-coupled devices [CCDs], and complementary metal oxide semiconductor arrays [CMOSs]*), which work efficiently with optical photons and which are not suitable as direct x-ray detectors.† These detectors are quite transparent at photon energies above 10 keV and quickly suffer radiation damage. Instead of detecting the x-ray photons directly, the array images light given off by an x-ray scintillator chosen for the x-ray energies of interest. These x-ray camera systems couple the scintillator to the visible light detector typically through a lens system or fiberoptic channel plate (Fig. 4.3). The former is generally used with synchrotron radiation, and the latter is often employed with tube-based systems. Electronics are needed to read the linear or area detectors, and these are normally integral to the camera system delivered by the vendor. Mechanical or electronic shutters are needed to separate different frames recorded, but further development of this subject exceeds the intended level of detail.

One-dimensional photodiode arrays (Reticon devices, typically 1,024 elements) have been successfully used in several single slice microCT apparatus (Burstein, Bjorkholm et al. 1984, Seguin, Burstein et al. 1985, Armistead 1988, Suzuki, Usami et al. 1988, Engelke, Lohmann et al. 1989a, 1989b, London, Yancey et al. 1990). In these systems, optical fibers were typically used to couple the detector with the phosphor. Some systems had Gd_2O_2S:Tb-based phosphors coated directly on the end of the fibers while another system coupled to 650-μm thick transparent layer in which 4-μm diameter particles of Gd_2O_2S:Tb were embedded (Engelke, Lohmann et al. 1989a). As is noted elsewhere (Kinney and Nichols 1992), photodiode arrays are quite noisy and suffer from significant nonlinearities, which can lead to very serious ring artifacts in reconstruction.

Some two-dimensional detector systems have been based on vidicons (Feldkamp, Davis et al. 1984, Feldkamp and Jesion 1986, Sasov 1987a, 1987b, 1989, Feldkamp, Kubinski

* CCD and CMOS chips have different architectures, and the latter are much less expensive and can be read much more rapidly. Although purists might shudder, the level of this text is such that the differences will be ignored, and both types of detectors will be lumped together as CCDs.

† A new generation of highly absorbing, and in some cases energy sensitive, area detectors have appeared and are covered at the end of the section.

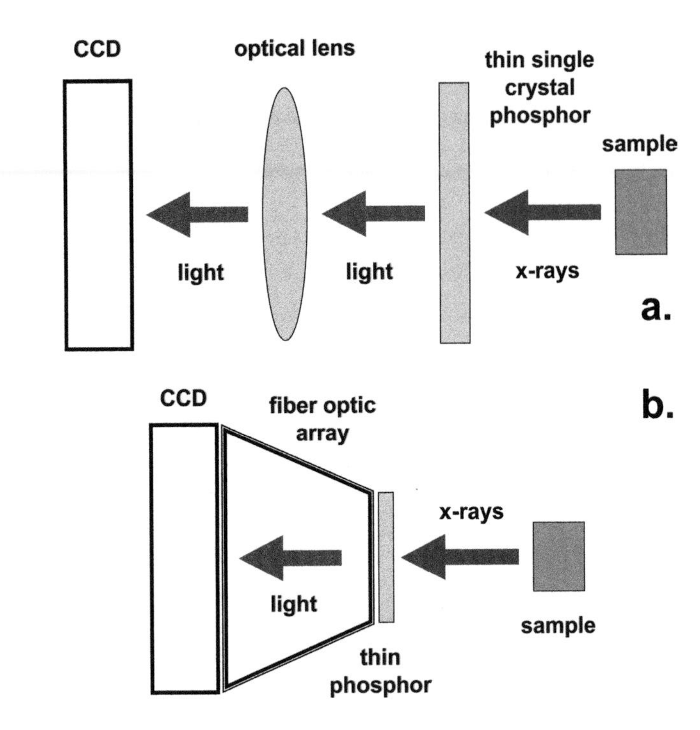

FIGURE 4.3
Scintillator–CCD coupling. (a) Thin crystal scintillator and optical lens. (b) Scintillator screen and fiberoptic taper where the individual light guides direct the light to the CCD detector elements.

et al. 1988, Goulet, Goldstein et al. 1994), but most are based on two-dimensional CCDs. Within each device (metal oxide silicon [MOS]) of the CCD array, an absorbed photon creates a charge pair. The electrons are accumulated in each detector well during exposure, and the individual detector elements are read digitally by transferring charge from pixel to pixel until all columns of pixels have reached the readout register and been stored in the computer system. The larger the detector element area, the greater the number of electrons that can be stored and the greater the dynamic range. Normally one thinks of increased detector element size entailing a sacrifice of spatial resolution, but this need not be so if the number of detector elements sampling the specimen cross-section remains constant.* There is a cost associated with recording greater numbers of electrons: increased CCD readout times (for a given level of readout noise). Multiple frame averages can produce much the same result as using a CCD with larger well depth. At present, most microCT systems employ at least 2K × 2K (2048 × 2048 detector elements) CCDs with 12-bit or greater depths. Table 4.1 lists typical formats, pixel sizes, dynamic ranges, and other characteristics of some of the CCDs used in synchrotron microCT (Bonse, Nusshardt et al. 1991, Kinney and Nichols 1992, Bueno and Barker 1993, Davis and Elliott 1997). Table 4.2 gives area detector specifics for several commercial systems. Some use is being made of 4K or wider detector formats (i.e., in the reconstruction plane). Efforts to develop large detector arrays composed of a mosaic of CCD chips are also of interest (Ito, Gaponov et al. 2007). The most recent detector cameras (below the

* In phosphor-optical lens-detector systems this is achieved by using higher optical magnification.

TABLE 4.1

Detector characteristics for synchrotron microCT.

Manufacturer	Model	Format	Pixel Size (μm)	Depth (bits)	Readout Time (s/frame)	Note
ESRF	FReLoN	2K × 2K	14	14		[a]
Hamamatsu	C4880-10-14A	1K × 1K	12	14	4	[b]
	C4742-95HR	4K × 2.6K	5.9	12	0.6	[b]
Photometrics	Coolsnap K4	2K × 2K	7.4	12	0.33	[c]
Point Grey	Grasshopper 3	1920 × 1200	5.9	12	<0.0125	[d]
pco	edge 5.5	2560 × 2160	6.5	16	<0.001	[d, e, f]
pco	Dimax	2016 × 2016	11	12	<0.001	[e, f]
pco	4000	4008 × 2672	9	12	0.2	[g]
SLS/pco	GigaFRoST	2016 × 2016	11	12	<0.001	[g]
Oryx	MP 5.0	2448 × 2048	3.5	12	<0.006	APS 2-BM

Notes: [a] ESRF ID19 (2006) and (2007). [b] SPring-8 BL20B2 (2007). [c] APS 2-BM (De Carlo, Xiao et al. 2006) and (2008). [d] APS 2-BM and/or elsewhere at APS. [e] ESRF ID19. [f] Diamond I13. [g] TOMCAT, SLS (Mokso, Schlepütz et al. 2017).

This list is intended to be indicative and not exhaustive. The first three detectors are CCD or CMOS camera systems in use in 2007. The cameras below the dividing line are being used or being commissioned for use in 2019. Readout times are rough indicators only: multiple factors can influence the rate at which the camera is read. As wavelength sensitivity is comparable for these Si-based detectors, this information is not included.

dividing line in Table 4.1) allow much more rapid readout than was the case for cameras in use in 2007 (above the dividing line in Table 4.1).

The number of bits in the CCD detector might lead the reader to conclude that there are thousands of levels of meaningful contrast in each reconstruction. This is certainly not the case, at least for routine data collection. In the author's experience, the number of *useful* contrast levels seen in microCT data rarely, if ever, exceeds 8 bits; that is, the range of linear attenuation coefficients can be more than adequately represented on a 0–255 scale. The physics of absorption, counting statistics, and detector noise combine to limit contrast sensitivity, as calculated elsewhere (Rivers 2016). The reader should not, however, conclude that microCT systems would provide adequate contrast if their detectors had 8-bit well depths; 12 or more bits are necessary because, among other things, the direct beam almost always passes to either side of the specimen and one must avoid saturating the detector if proper white field correction is to be applied.

The characteristics of the x-ray-to-light conversion media dictate, to a large extent, what coupling scheme is optimum for a given area detector (Kinney, Johnson et al. 1986, Bonse, Nusshardt et al. 1989). Several scintillator properties are important for efficient microCT and nanoCT. First is the range of wavelengths emitted: this should match that of the peak efficiency range of the detector. Second is absorbing power of the scintillator: high Z,

TABLE 4.2

Detectors in several commercial microCT systems being marketed in 2019.

Manufacturer	Model	Format	Pixel Size (μm)	Depth (bits)
Scanco	MicroCT-50	3.4K × 1.2K	20	14
Bruker	Skyscan 1272	5K × 3K	9	14
Zeiss	Versa 610, 620	2K × 2K	13.5	16

TABLE 4.3
Scintillator characteristics

Material	Name	Form	Z_{eff}	Density (g/cm³)	Emission (max. (nm))	Light Yield (photons/keV)
$Bi_4Ge_3O_{12}$	BGO		75	7.13	480	8
$CdWO_4$	CWO	Crystal	63	7.9	475	15
CsI:Tl		Columnar thin film	54.1	4.51	550	65
Gd_2O_3:Eu			61	7.1	611	19
Gd_2O_2S:Tb	Gadox	Powder	59.5	7.3	545	
$Lu_3Al_5O_{12}$:Ce	LAG	Crystal	61	6.73	535	20
$Lu_3Al_5O_{12}$:Eu	LAG		61	6.73	535	
$Y_3Al_5O_{12}$:Ce	YAG	Crystal	32	4.55	550	40 to 50
$YAlO_3$:Ce	YAP	Crystal				
Y_2O_2S:Eu		Powder				
$Gd_3Ga_5O_{12}$:Eu	GGG		52	7.1		
Lu_2O_3:Eu			68.8	8.4	611	20
Lu_2SiO_5:Ce	LSO		65.2	7.4	420	25
$Lu_3Ga_5O_{12}$:Eu	LGG		58.2	7.4		

(2005, Martin and Koch 2006, 2007).

high-density materials are favored for higher energy photons. Third, the efficiency of emission is an important characteristic: weak light emission exacerbates the photon starvation encountered in tube-based system and also can degrade counting statistics with synchrotron radiation systems. Generally speaking, microCT detector systems operate in an integration mode so that the emission persistence times are not of greatest concern. The scintillator must be defect-free, or at least possess a small number of defects, and emit relatively uniformly over areas of several square millimeters or larger. An ideal scintillator would not be damaged by the large x-ray doses accumulated over time, or at least would degrade slowly, nor would it be affected adversely by the environment (e.g., water vapor, ozone, trace hydrocarbons). Table 4.3 lists some characteristics of scintillators used in x-ray microCT and nanoCT; thin film scintillators are also of interest (Martin and Koch 2006). Cadmium tungstate, for example, has a density of 7.9 g/cm³, an emission peak of 475 nm and FWHM ~ 100 nm, and primary decay time of 14 μs. Cadmium tungstate produces 12–15 light photons/keV (somewhat less than NaI:Tl) (2005) and is a good match for a typical CCD (absolute quantum efficiency >0.4 between 450 and 850 nm for Kodak KAF-4301E, 2K × 2K [2005]). Recently, a dual-layer scintillator crystal detector system was reported for dual energy microCT (Maier, Schock et al. 2017).

Phosphor powders on a screen or embedded in transparent media (Bonse, Johnson et al. 1986, Kinney, Johnson et al. 1988, Engelke, Lohmann et al. 1989b), monolithic or fiberoptic scintillator glasses (Coan, Peterzol et al. 2006), column-oriented polycrystalline thin films (2008), single crystal scintillators (Kinney and Nichols 1992), and lithographically fabricated cellular phosphor array consisting of 2.5-μm spaced close-packed holes filled with plugs of phosphor (Flannery, Deckman et al. 1987) have been used.* All of the phosphor "screens" except the last can be obtained in a straightforward fashion, and the

* Many more citations could be provided here, but those given (somewhat arbitrarily) can, at least, start the interested reader in the direction of additional papers.

reason for going to such extremes in producing a discretized micrometer scale fluorescent screen was to prevent optical crosstalk between adjacent detector elements (Flannery and Roberge 1987). The noise from scatter, however, remains, and little seems to be gained comparatively by the discretization (Kinney and Nichols 1992). On the other hand, a new generation of etched column-filled scintillators was developed (Olsen, Badel et al. 2007). Single-crystal phosphors are probably the most popular for synchrotron microCT, but phosphor development continues including materials formed through thin film processing routes (Koch, Peyrin et al. 1999, Martin and Koch 2006).

As mentioned above, different x-ray-to-light converters provide different strengths. For example, a $CdWO_4$ single crystal plate 0.5 mm thick and an Y_2O_2S:Eu screen about 40 μm thick were compared (Bonse, Nusshardt et al. 1991); the screen produced up to 15 times more light, while the single crystal provided considerable better spatial resolution (80 line pairs/mm at 20% contrast, corresponding to 6-μm resolution). Several commercial inorganic polycrystalline phosphor screens have been compared to fiberoptic glass scintillator arrays, and better performance was observed with the glass (up to 20 line pairs/mm) than the powder scintillators (Bueno and Barker 1993). Corrections for various inhomogeneities are required, both geometrical distortions and variation in light emission, and manufacturers of commercial systems measure these distortions and correct for their effect on the projections.

Microchannel plates have found use coupling powder scintillator screens to CCD cameras (Davis and Elliott 1997). Optical lens systems have been used by many groups to provide an optical link for powder and single-crystal screens and CCD cameras. A particularly effective scheme is to use one or more low-depth-of-focus optical lens(es) combined with a single crystal scintillator (Bonse, Nusshardt et al. 1991, Kinney and Nichols 1992). Low depth of focus restricts contributions to the image (radiograph) to light photons emitted from a narrow range of depths within the scintillator. On the one hand, only a small fraction of the image-forming x-ray photons contribute to the image. On the other, scatter from adjacent layers of the scintillator and contributions from divergently scattered light photons from the in-focus volume are largely eliminated. Therefore, in situations of photon starvation, that is, in tube-based systems, optical coupling of scintillator to area detector is generally not the solution of choice, whereas in photon-rich environments (synchrotron radiation sources), optical coupling is almost always adopted.

Output from an x-ray source is not uniform, nor is detector response, and this must be corrected if there are not to be serious artifacts in the reconstructions. Commonly a flat field correction is applied; that is, each radiograph is corrected on a point-by-point basis with an image recorded under the same conditions as the radiograph but with the specimen removed (i.e., a white-field image). A dark-field correction should be applied as well (an image recorded with no incident x-ray photons). Figure 4.4 shows a raw radiograph and the resulting normalized radiograph using the dark-field and white-field images shown. Despite the high degree of structure seen in the white field (and in the raw radiograph), essentially none of this artificial structure has propagated into the normalized radiograph. Commercial tube-based systems record multiple dark-field and multiple white-field images at the start of the scan; this seems to suffice. One manufacturer has documented a source that is so stable that weekly white-field calibrations suffice to normalize incident beam intensities. In December 2007 at 2-BM, APS, the standard for 2K × 2K reconstructions was to record a white field every 100 projections (by translating the specimen horizontally out of the beam, rotation increment of 0.125°) and to use the average of all white fields and a single dark field recorded at the end of each data set (to prevent thermal drift of the x-ray monochromator) for the correction. Horizontal translation was used because that stage

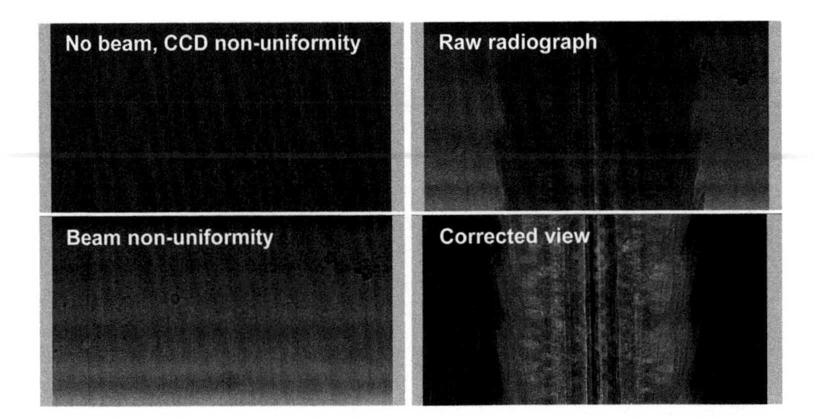

FIGURE 4.4
Correction for beam and detector nonuniformity. The specimen is a sea urchin spine (*Diadema setosum*). The horizontal field of view is 1,024 pixels and the vertical 550 pixels. Data were recorded at 2-BM, APS (18 keV, 300 μm $CdWO_4$ crystal, 4X lens, and ~1.8-μm pixels) by S.R. Stock, K. Ignatiev and F. De Carlo 7/25/2002.

was much faster than the vertical translation stage, but even with the high-quality translation stages, the multiple 3- to 10-mm motions led to lateral displacements in the sample's radiographs and to degraded reconstructions. The current standard at 2-BM, APS, and elsewhere is to collect 10 or more white (and dark) fields before or after the sample projections are recorded (see Fig. 5.6).

There is a limit to the spatial resolution that can be obtained from light-emitting scintillators. This is the wavelength limit described in elementary texts for optical systems. For the materials used in x-ray microCT, this limit is about 0.3 μm, and obtaining better resolving power requires use of x-ray optics before the scintillator (asymmetric crystal beam magnifiers, Fresnel zone plate optics, etc.).

Photon-counting area detectors for x-ray imaging are developing rapidly. These detectors are extremely useful for situations where very low noise is essential or where high-energy sensitivity is required. Photon-counting detectors differ fundamentally from the energy-integrating detectors described above and which record the total energy (large number of photons plus electronic noise) deposited in each pixel during a fixed period of time. The pixels of photon-counting detectors register the interaction of individual photons and the energy that specific photon deposits. A 4 × 16 × 4 pixel photon-counting detector, for example, was recently characterized for use with tomography (Ren, Zheng et al. 2018). A linear photon-counting detector with high energy sensitivity (Rumaiz, Kuczewski et al. 2018) was recently used for energy-dispersive diffraction tomography (Stock, Okasinski et al. 2017). Reviews on photon-counting detectors appear elsewhere (Denesa and Schmittb 2014, Hatsuia and Graafsma 2015).

4.4 Positioning Components

Except for translation of the sample out of the beam to collect white-field images, the specimen rotator and translation along the rotator axis (to image additional FOV) are the mechanical motions during data collection with the typical rotate-only tube-based or synchrotron microCT system. A rotator without wobble (unintended in-plane and out-of-plane translations from perfect circular paths) would be ideal, but measuring the rotator's imperfections and correcting them improves reconstruction quality considerably (De Carlo, Xiao et al. 2006, Rivers and Wang 2006), see Section 5.2.1). Similarly, translation of the specimen into/out of the beam requires accurate repositioning. Stability over time is another requirement.

In general, one requires precision and accuracy substantially smaller than the smallest voxel size that will be reconstructed. As shown by the effect on reconstruction quality by subpixel physical shifts of the rotation axis (Section 5.1.3), imprecision of one-quarter of a voxel can degrade image quality significantly. In other words, if one were assembling a system for microCT with 1-μm voxels, precision and accuracy of one-quarter this value or smaller are certainly needed.

The rotation axis cannot wobble appreciably relative to the detector rows and columns without degrading reconstruction quality. Wobbles of even 0.03° over 2K pixels can shift projected data to an adjacent row of an area detector. For systems designed for reconstruction with 2-μm or larger voxels, high precision mechanical bearings in the rotator are adequate, but when 1-μm voxel size or smaller used, air bearings are required. Translations along the rotation axis also need to be precise and not shift the rotation center.

In summary, high-quality mechanical components are required for microCT systems. Some post-data collection processing can be used to ameliorate the inevitable mechanical imperfections (see Chapter 5), but these do not always work well, and they require a considerable investment of time for each and every specimen. It is better, therefore, to have hardware that does not require the correction steps. Best practices for calibration and certification are described in Section 4.6.

4.5 Tube-Based Systems prior to 2008

Investigators continue to design and build systems for microCT and nanoCT but for those otherwise inclined, the option of purchasing a commercial system has existed since the 1990s. This section focuses on the commercial systems prior to 2008, and specific purpose-built systems are described elsewhere in the book in conjunction with other topics.

Except for microCT systems explicitly employed for one type and size of specimen (e.g., rapid metrology and quality control as part of a manufacturing line for high-precision components), microCT systems must be able to accommodate a range of specimen sizes and x-ray transparencies. Generally, the more flexible a given system and the greater range of specimen types it can accommodate, the greater the requirement for an expert operator. Having presets for resolution and x-ray energy (tube voltage in lab systems) greatly increases the efficiency of data acquisition, even for expert operators, and this is a very important consideration in most labs because of the large throughput of samples required and because a significant number of relatively inexperienced users can be expected.

Consider how specimens with different diameters can be accommodated in a fan beam or cone beam system. For simplicity, consider only the fan plane (or the central plane in a cone beam system). Figure 4.5 shows how placing a small diameter specimen near the x-ray source and farther from the detector allows geometrical magnification to spread the radiograph across the available detector pixels. Placing the same small diameter specimen near the detector would not utilize all of the available detector pixels. With the small

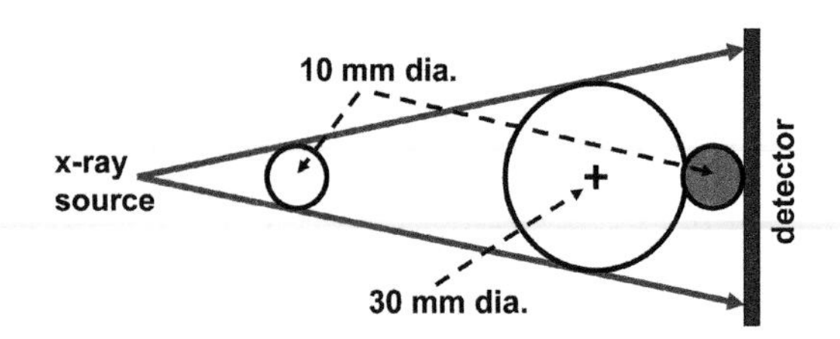

FIGURE 4.5
Sample position relative to detector and x-ray and projection onto an area detector. A 10-mm diameter specimen near the x-ray source (white interior) and far from the detector; that is, in an optimum position for minimizing voxel size in the reconstruction. A second position for the 10-mm diameter specimen (gray interior) near the detector and far from the source. A 30-mm diameter specimen is shown in its optimum position for matching sample diameter and detector width.

sample near the source, the voxel size* *vox* is as small as possible and will be $vox \sim dia/N_{pix}$, where *dia* is the specimen diameter and N_{pix} is the number of detector voxels. With the small specimen away from the source, $vox \sim dia/N_{eff}$, where $N_{eff} \ll N_{pix}$. If a larger diameter sample were to be studied, its optimum position would be closer to the detector (farther from the x-ray source) than the smaller specimen. Here, the cost of imaging a larger diameter is paid in terms of decreased geometrical magnification and greater voxel size (for a given number of detector pixels).

For a 1K detector (number of detector pixels in the plane of reconstruction), one could expect to reconstruct a 10-mm diameter specimen with 10-μm voxels, each containing information unique to that voxel. If effects such as penumbral blurring were large enough, then contrast would be spread over adjacent voxel and resolving power would be worse than expected from the nominal voxel size. With a 2K detector, 5-μm voxels could be expected for the 10-mm specimen. With a 30-mm diameter specimen and a 2K detector, one would obtain 15-μm voxels. For comparison, microCT with synchrotron radiation and a 2K detector might involve a 2-mm diameter specimen reconstructed with 1-μm voxels (see Sections 4.7 and 4.8).

Table 4.4 lists the commercial microCT systems of which the author was aware in 2007, and some of a given manufacturer's models may have been inadvertently omitted. In the time since 2007, some of these companies have been acquired by larger instrumentation firms. Space precludes listing all of the system characteristics that might be of interest, but reconstruction size (in voxels), reconstructed voxel size, and specimen diameter are provided. Generally, these were turnkey systems integrated with radiation shielding, generally a box-like enclosure with interlocks that does not occupy an inordinate amount of floor space.

For purposes of illustration of data acquisition rates and system capabilities, consider the microCT instrument the author has in his laboratory, a circa 2000 Scanco MicroCT-40 system. Use of this example does not imply anything about the relative merit of this apparatus relative to the others listed in Table 4.4, and a version of the MicroCT-40 with upgraded detector is described in the following section. The MicroCT-40 simply is the one commercial system about which the author can speak from considerable firsthand experience.

* The discussion explicitly considers only voxel sizes in reconstructions that contain physical information. One can certainly do the reconstruction with arbitrarily small voxel size, but no information will be contained and there is a significant risk of misrepresentation of what can and cannot be seen in the data if such reconstructions are used.

TABLE 4.4

Commercial laboratory (absorption) microCT systems.

Manufacturer	Model (Application)	Voxel; Reconstruction Sizes	Notes
BIR (BioImaging Research)	MicroCT (specimens)	<50 μm; 1024^2	[a]
Bioscan	NanoSPECT/CT (in vivo animal)	<200 μm	[b]
Biospace	γ IMAGER-S-CT (small animal)	250 μm	[c]
Gamma Medica-Ideas	X-O (small animal)	To 43 μm; 512^3 to 2048^3	[d]
GE [e]	EXplore Vista PET/CT (small animal)	–	–
	EXplore Locus MicroCT (in vivo)	27, 45, or 90 μm isotropic	[f]
	EXplore Locus SP MicroCT (specimen)	To 8 μm isotropic	[g]
	EXplore Locus Ultra CT	–	[h]
	Nittetsu Elex	Ele Scan (specimen)	[i]
	Ele Scan Mini	(specimen)	[j]
Phoenix X-ray	Nanotom	To 0.5 μm	[k]
	v|tome|x| 240	To 4 μm	[l]
	Scanco Medical	XtremeCT (human peripheral in vivo) 41–256 μm; 512^3 to 3072^3	[m]
	vivaCT 40 (in vivo animal)	10–72 μm isotropic; to 2048^2	[n]
	MicroCT 80 (specimens)	10–74 μm isotropic; to 2048^2	[o]
	MicroCT 40 (specimens)	6–72 μm isotropic; to 2048^2	[p]
	MicroCT 20 (specimens)	8–34 μm isotropic; to 1024^2	[q]
Shimadzu	SMX-225CT-SV3 (specimens)	To 4096^2	[r]
Siemens [q]	Inveon multimodality	To 15 μm	[s]
	MicroCAT	To 15 μm; to 4096^2	[t]
Skyscan	1074 (portable)	22 μm; 512^2	[u]
	1076 (in vivo)	<9 μm isotropic, <15 μm	[v]
	1078 (ultrafast in vivo)	47 μm, 94 μm; $(48\ mm)^3$	[w]
	1172 (specimens)	<1 μm, 2 μm, 5 or 8 μm	[x]
	1178 (High-throughput, in vivo)	80, 160 μm; 1024^3	[y]
	2011 (nanotomography)	150, 250, 400 nm	[z]
VAMP	TomoScope 30s (rapid examination)	80 μm	[aa]
Xradia	MicroXCT	1–6 μm; 1024^2	[bb]
	NanoXCT	50–70 nm; 1024^2	[cc]
	NanoXFi	<8 nm	[dd]
XRT	X-AMIN PCX		[ee]
	XuM	<100 nm	[ff]
X-tek	Benchtop CT	5 μm	[gg]
	HMX(ST) CT	Feature detection to 1 μm	[hh]
	Venlo CT		[ii]

Notes

[a] Specimen diameter up to 25 mm, length up to 55 mm.
[b] 1, 2, or 4 SPECT detectors.
[c] Maximum object size 100 mm length, 90 mm diameter.
[d] Maximum object size 97 mm length, 93 mm diameter.

(Continued)

TABLE 4.4 *(Continued)*

Notes

[e] General electric, previously enhanced vision systems.
[f] Specimen diameter up to 85 mm.
[g] Specimen diameter up to 40 mm. Cone beam system.
[h] Diameter up to 140 mm, long axis up to 100 mm/rotation.
[i] Examples of operating parameters given in user reports (Joo, Sone et al. 2003).
[j] Diameter up to 45 mm, length to 50 mm.
[k] Diameter up to 125 mm, length to 150 mm.
[l] Diameter up to 500 mm, length to 600 mm. Other variants of this industrial system are available.
[m] Diameter up to 125 mm, scan length up to 150 mm.
[o] Diameters from 20 to 38 mm, scan length up to 145 mm.
[p] Diameter up to 75.8 mm, scan length up to 120 mm. Cone beam system.
[q] Diameter up to 36.9 mm, scan length up to 80 mm. Stacked (40) fan beam system.
[r] Diameter up to 17.4 mm, scan length up to 50 mm. Fan beam system.
[s] Diameter up to 140 mm.
[t] Previously CTI and Imtek.
[u] PET, SPECT, CT. Diameter to 100 mm.
[v] Cone beam. SPECT option.
[w] Diameter up to 68 mm, scan length up to 200 mm. Cone beam.
[x] Diameter up to 16 mm. Cone beam.
[y] Diameter up to 48 mm, scan length up to 140 mm. Cone beam.
[z] Diameters 20/37 mm or 35/68 depending on version.
[aa] Diameter up to 82 mm, scan length up to 210 mm.
[bb] 0.5–1 mm for maximum resolution, 11 mm maximum diameter (9 μm voxels). Cone beam
[cc] Diameter up to 40 mm, axial length up to 37 mm. Cone beam.
[dd] Diameters 0.5 to 12 mm; 16 slices.
[ee] Phase imaging.
[ff] SEM-based instrument, phase and absorption imaging.
[gg] Field of view (20 μm)^2.
[hh] Diameter up to 50 mm.
[ii] Few details available on line.

(Available 2007 and before) with manufacturer's listed voxel and reconstruction sizes as well as notes on specimen sizes. Adapted from a table compiled and copyrighted by Steven Cool, Radiation Monitoring Devices (used with permission) and supplemented by additional entries.

The circa 2000 Scanco MicroCT-40 system is a turnkey system employing a stacked fan beam geometry (up to 40 slices imaged simultaneously). Presets exist for the principle variables (kVp, tube current, sample diameter, reconstruction resolution), and these are listed in Table 4.5. The investigator has the choice of several different diameter specimen holders that allow reconstruction with different voxel sizes. The smallest reconstructed voxel size (6 μm) is obtained with the 12.3-mm diameter holder and the highest resolution scan parameters (1,000 projections with 2,048 samples). Note that this 6-μm value is *not* the spatial resolution!

In the Scanco MicroCT-40 system, samples are placed into the tubular holder and typically are held in place by foam packing peanuts. The internal temperature of the scanner remains more-or-less constant at about 25°C. Integration times for each projection can be as high as 0.3 s, and the software allows multiple frames to be collected for each projection (i.e., increasing the signal-to-noise ratio in the reconstructions). The time required to collect a high-resolution set of 40 slices with highest definition (0.3 s integration, 1,000 projections of 2,048 samples) was about 22 min. In high-resolution scans of multiple sets of 40 slices, the two DEC Alpha processors (from circa 2000) could

TABLE 4.5

Preset operating parameters for the circa 2001 Scanco MicroCT-40 as a paradigm for turnkey commercial microCT system.

Tube Settings (kVP: μA)	Sample diam (mm)	Resolution
45: 88 or 177	12.3	250 projections, 1024 samples, $(1024)^2$ voxels
50: 72 or 145	16.4	500 projections, 1024 samples, $(1024)^2$ voxels
70: 57 or 114	20.3	1000 projections, 2048 samples, $(2048)^2$ voxels
	30.7	
	36.9	

not reconstruct the last set of 40 slices acquired in the time required to collect the projections for the next 40 slices.

In 2007, in vivo (small animal) microCT systems and nanoCT systems were relatively new commercial products. Systems available in 2007 are identified in Table 4.3, and discussion of nanoCT systems is postponed until Section 4.9. In vivo microCT, like medical CT, is subject to the constraint that the animal or patient must be kept motionless and the tube and detector must be rotated. This means that small animal in vivo microCT systems are generally larger than normal microCT systems and employ a gantry type of design. The relatively large field of view required means that resolution is poorer than in systems designed for smaller specimens. In addition, it is highly desirable to minimize x-ray dose in small animal in vivo microCT, but it is not as essential as in human CT. Nonetheless, in vivo microCT systems are capable of providing more than adequate spatial resolution for murine trabecular bone histomorphometry with tissue doses that are so low that they are reckoned not to interfere with the physiological processes of the imaged tissue (Kohlbrenner, Koller et al. 2001) or to affect tumor growth (Carlson, Classic et al. 2007).

4.6 Tube-Based Systems since 2008

The considerations outlined in the second through fourth paragraphs of Section 4.5 certainly continue to apply. Larger detectors and somewhat brighter x-ray sources are available in current systems compared to those marketed in 2007. Throughput has increased enormously and is discussed in Chapter 5. It is well known that computational power, input/output speeds, and cost of storage media (hard drives) have improved substantially as has reconstruction and analysis software.

The newer generations of tube-based microCT systems can offer voxel sizes below 1 μm, and with this decrease in voxel size comes the increased need for accurate calibration and certification. For systems with x-ray spot sizes several micrometers in size, one manufacturer uses high-precision preloaded ball bearings with 2-μm accuracy and drives the rotation by a stepper motor with backlash-free gearing. For high-resolution systems with submicrometer spot sizes, this manufacturer uses air bearings with direct drives with optical encoders (every rotation stage is individually measured; typical runout is 25–30 nm with maximum acceptable runout and flatness of 50 nm). For translations along the rotation axis (to collect additional FOV), the manufacturer uses multiple calibrations, typically every millimeter of vertical travel, and applies this as a correction.

Since the author installed his Scanco MicroCT-40 system, a new version was released with a much larger detector and cone beam acquisition. Performance for the new-generation Scanco MicroCT-40 and MicroCT-50 is mentioned briefly in Chapter 5 as examples of how performance is changing.

The standard in 2019 for lab microCT is cone beam geometry. Most systems' detectors couple the scintillator to the CCD detector with a fiberoptic, but there is at least one manufacturer that uses optical coupling through a turret with different objectives, each with its own scintillator. The latest generation of systems typically allows stitching to cover specimen widths two or three times larger than the standard horizontal FOV for the resolution selected. Some commercial systems also accommodate local tomography. Most manufacturers have instruments with sample changers, and this allows data acquisition around the clock, a major advantage for labs requiring high-throughput analysis of many specimens per day such as is encountered in phenotyping of different small animal strains. Earlier, multiple specimens could be packed into a sample tube and a batch program run to scan the appropriate FOV for each specimen, but this was much more constraining than what is presently available.

Computational power has changed enormously from 2007 to 2019. In particular, graphics processing units (GPUs), developed for rapid image generation in gaming, are now available with most commercial systems; these are used for reconstructing the data sets. Given the large number of slices collected per unit time, improved input/output rates and data storage solutions were essential, and manufacturers have benefitted by new generations of hard drives (see Section 4.8) and connections through new generations of USB connections and improved Ethernet and wireless accessibility. Apps for commercial microCT systems are also available for smart phones.

4.7 Synchrotron Radiation Systems before 2008

Reviews of microCT at different synchrotron radiation sources appeared periodically before 2008 and were used to highlight new capabilities (see Table 4.6). Because the components required to perform microCT were readily available, many experimental stations occasionally performed microCT in response to their users' requests. Results in the literature were dominated by dedicated imaging/microCT beamlines, however, not just because they awarded many more shifts for microCT but also because the production facilities tailored to a small range of activities were (and are) much more efficient.

Synchrotron microCT (with optical lenses and without x-ray lenses) in the years leading up to 2008 were typically performed with voxel sizes between 1 and 10 μm, although routine operation with voxels sizes below 0.5 μm was possible at certain facilities, and larger voxel sizes were used upon occasion when larger specimen diameters dictated it. Design of microCT systems was (and is) driven by the expected portfolio of specimen types and the features within that need to be resolved. Available resources played (and continue to play) a role in system characteristics, and constant upgrade of capabilities has remained the rule at active synchrotron microCT facilities. The systems of which

TABLE 4.6

Pre-2008 reviews of microCT at different synchrotron radiation sources.

Source	References
APS	(Rivers, Sutton et al. 1999)
	(Wang, De Carlo et al. 1999)
	(De Carlo, Albee et al. 2001)
	(Wang, De Carlo et al. 2001)
	(De Carlo and Tieman 2004)
	(Rivers, Wang et al. 2004)
	(Rivers and Wang 2006)
	(De Carlo, Xiao et al. 2006)
DESY	(Beckmann, Bonse et al. 1999)
	(Beckmann and Bonse 2000)
	(Beckmann, Lippmann et al. 2000)
	(Beckmann 2001)
	(Beckmann, Donath et al. 2004)
	(Beckmann, Vollbrandt et al. 2005)
	(Beckmann, Donath et al. 2006)
ESRF	(Weitkamp, Raven et al. 1999)
	(Schroer, Meyer et al. 2003)
	(Schroer, Cloetens et al. 2004)
	(Di Michiel, Merino et al. 2005)
	(Baruchel, Buffiere et al. 2006)
	(Martin and Koch 2006)
NSLS	(Dowd, Campbell et al. 1999)
SLS	(Stampanoni, Abela et al. 2004)
	(Stampanoni, Borchert et al. 2006)
SPring-8	(Takeuchi, Uesugi et al. 2001)
	(Takeuchi, Uesugi et al. 2002)
	(Uesugi, Tsuchiyama et al. 2003)
Other	(Lopes, Rocha et al. 2003)
	(Thurner, Wyss et al. 2004)

Abbreviations: APS, Advanced Photon Source; DESY, Deutches Elektronen Synchrotron; ESRF, European Synchrotron Radiation Facility; NSLS, National Synchrotron Light Source; SLS, Swiss Light Source; SPring-8, Super Photon ring 8 GeV.

the author was aware were highly modular, and this allowed incremental instrumental improvements.

The typical synchrotron (absorption) microCT system (using the parallel beam directly without focusing optics) consists of specimen rotator, x-ray phosphor (single crystal), optical lens, and CCD detector. The available components have improved in capability and affordability over time.

Consider first the mechanical components and the physical stability required for high-quality reconstructions. Voxel sizes down to 1–2 μm were achieved pre-2008 with (relatively) affordable positioning and optical components. Reconstructions with voxel sizes down to 0.3 μm were not uncommon, but the required stability increased system cost considerably. X-ray detector systems at most synchrotron microCT instruments consisted of commercially available modules: thin single-crystal phosphors, microscope objective lenses, and CCD or other area detectors for optical wavelengths. Cadmium tungstate single crystal phosphors precut and polished to the desired thickness were widely used and were relatively inexpensive. These crystals provide light wavelengths with acceptable efficiencies for CCD detectors. Radiation damage dictates periodic replacement of phosphor crystals (and optical lenses if they are in line with the direct beam or prisms if the optical lenses are placed off the beam axis – damage can be removed by UV treatment).

Most synchrotron microCT instruments can switch quickly between several optical lenses for different FOV and voxel sizes (In 2007, station 2-BM of APS routinely used 1.25X, 2.5X, 4X, and 5X objectives providing FOV of 5.4, 2.7, 1.7, and 1.36 mm, respectively, when used with a 1K detector and twice these FOV when used with a 2K camera [De Carlo and Tieman 2004]); switches involve unscrewing one lens from the CCD camera body and manual replacement with the new lens. The lens must be refocused, and the replacement operation can be completed in a few minutes. A second pre-2008 example is the microCT system at ESRF ID-19. Fields of view between 40 and 0.57 mm are listed (reconstructed voxel sizes of 30 and 0.28 μm, respectively), and a rotation mount for very rapid lens changes was commissioned in 2007 (2007). Most area detectors are $(1K)^2$ or $(2K)^2$ scientific-grade CCDs with depths of 12-bits (De Carlo, Albee et al. 2001, Stampanoni, Borchert et al. 2002a); the specialized FReLoN detector developed at ESRF provided $(2K)^2$ elements with 14-bit depth (see (Weitkamp, Raven et al. 1999)).

Significant advances in beam delivery optics included wide band-pass monochromator systems based on multilayers; these produced (and continue to produce) surprisingly uniform beams and increased throughput dramatically compared to single crystal optics (Chu, Liu et al. 2002, Ferrie, Buffiere et al. 2006, 2007). The system with which the author is familiar is based on a pair of multilayer optics with areas of different layer spacing; tuning to different energies is done by simple translation to the appropriate positions on both optical elements (Chu, Liu et al. 2002). At ID 19 of ESRF, for example, multilayer optics provided $\Delta E/E \sim 10^{-2}$ compared to $\Delta E/E \sim 10^{-4}$ for an Si 111 double crystal monochromator (2007), and the multilayer provided a corresponding increase in intensity.

Decreased voxel sizes in synchrotron microCT are typically achieved by increasing the magnification of the optical lens coupling phosphor to area detector, but there is a limit to what optical magnifications can be used. If the beam passing through the specimen is spread before the phosphor, much smaller voxel sizes can result (at the cost of decreased field of view and increased data collection times for a given brightness incident beam). Placing a perfect crystal in the beam transmitted through the specimen, orienting the crystal to diffract from a Bragg plane inclined with respect to the surface (angle of incidence less than the Bragg angle and exit angle greater than the Bragg angle), and using this diffracted bean for the reconstruction allows smaller voxel sizes for a given lens-area detector combination. This magnification is only along one direction, and use of a second orthogonally oriented crystal is required to magnify along the second direction. Asymmetric Bragg magnifiers have long been used in x-ray diffraction topography (imaging of nearly-perfect crystals using diffraction contrast) (Kohra, Hashizume et al. 1970), and Bragg magnifiers have been used between specimen and phosphor in microCT (Sakamoto, Suzuki et al. 1988, Kinney, Bonse et al. 1993). This approach was used to a limited extent at third-generation synchrotron radiation sources (Spal 2001, Stampanoni, Wyss et al. 2001, Stampanoni, Borchert et al. 2002b, 2003, 2005, 2006), and some of these results are described below in the section on nanoCT. Alignment of these additional optical elements can be time consuming, and at time of writing, the author does not know of any groups still using this approach.

Pre-2008, various synchrotron microCT facilities emphasized (and continue to emphasize) different scientific missions, time domains, or spatial domains. The author's impressions of some of these differences follow (with apologies for its incomplete, subjective nature). The reader should also realize that missions or emphases change with time (as do instrument capabilities). At DESY, the emphasis appeared to be on high-energy microCT and interferometer-based phase imaging (Beckmann, Lippmann et al. 2000). The various facilities at ESRF appeared to emphasize high spatial resolution, high temporal resolution, and phase imaging with the propagation method: for example, polychromatic radiation from a wiggler source can be used from near real-time microCT, down to 10 s per set of projections for one reconstruction (Ludwig, Di Michiel et al. 2005). At SLS, grating-based phase imaging received considerable emphasis. Reports from SPring-8 that have come to the author's attention centered around high spatial resolution and on phase imaging with interferometry. At APS, GSE-CARS focused on geological applications including measurements at high pressure (Wang, Uchida et al. 2005); station 2-BM (De Carlo, Xiao et al. 2006) emphasized rapid throughput (rapid reconstruction via a large dedicated computer cluster, robotic sample changer, facilities for remote access); time-resolved microCT of evolution of fuel spray (5.1 μs temporal and 150 μm spatial resolution) was achieved using the pulsed nature of the storage ring (Liu, Liu et al. 2004).

There has been some interest in laminography or tomosynthesis at synchrotron radiation facilities (see Section 3.7), but, unlike the situation of commercial tube-based systems for applications inspecting circuitry, interest has not grown. By 2008, local (or region of interest) tomography (see Section 3.8) had become well accepted and routinely used for certain classes of samples.

Side-by-side comparison of synchrotron and tube-based reconstructions of the same specimen is a rarity in the literature and might be of interest, particularly to those readers new to microCT. Figure 4.6 shows matching synchrotron and tube-based microCT slices

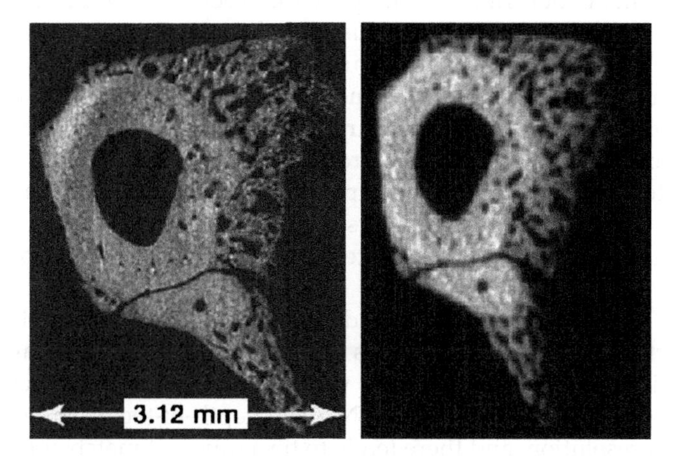

FIGURE 4.6
Synchrotron (left) and tube-based (right) microCT slices of the same sample. The murine tibia (top) and fibula (bottom) in this example of an osteoarthritis model have a considerable growth of bone outside of the cortex. In these and all other slice images (unless otherwise noted), contrast is on a linear gray scale with the lighter the pixel, the higher the voxel's absorption. Synchrotron slice: 21 keV, 0.25° rotation increment, 5-μm isotropic voxels (S.R. Stock, D. Novack, K. Ignatiev, F. De Carlo. April 2004, 2-BM, APS). Lab slice: 45 kVp, 500 projections, 12-μm isotropic voxels (S.R. Stock, D. Novack, Scanco MicroCT-40).

of a mouse tibia. As noted in the caption, neither slice was produced using the highest resolution of which each instrument was capable. The features in the synchrotron slice are sharper than in the lab slice, which is to be expected given the voxel sizes (5 μm vs 12 μm, respectively).

4.8 Synchrotron Radiation Systems since 2008

Between 2007 and 2019, synchrotron microCT changed in many ways, and this has transformed imaging experiments that were challenging in 2007 to routine in 2019 and enabled new experiments that were previously impossible. The rapid evolution has been driven by new generations of area detectors, by increased computational power and improved tools for reconstruction, by accelerated input/output rates (from detectors to hard drive arrays to processor units to portable storage devices), and by new generations of hard drives. The author's observations on the increase in data collection rates from 2007 to 2019 at 2-BM, APS, appear in Chapter 5, and apply in a broad sense to all beamlines which have been investing in their infrastructure. Increasingly, the limits on data collection rates are dictated by the brightness of the source and not by the experimental hardware; with the upcoming upgrades for multiple storage rings, this statement may need to be revised in a few years.

In the previous section, reviews of synchrotron microCT were tabulated. Between 2007 and 2019, additional reviews including synchrotron microCT appeared at a healthy rate. Two important reviews, now a bit dated, are (Stock 2008, Maire and Withers 2014). Instead of compiling another table, the author decided to refer the reader to the websites of the various synchrotron microCT facilities, but with the warning that the listing may be quite out of date or inaccurate.

The detector cameras described in the previous section have been supplanted, and some of the cameras in current use are listed below the horizontal line in Table 4.1. The inexpensive Grasshopper 3 or the equivalent can be used for many applications, and, during "routine" data collection at most beamlines, the time to read each frame (~12.5 ms for 12-bit readout) is small compared to the integration time required to obtain adequate statistics in that frame (on the order of 200–300 ms at 2-BM, APS). The other detectors and interfaces listed are more expensive and are needed for rapid data acquisition. Although Table 4.1 shows detectors somewhat larger than 2K (in the plane of reconstruction), in 2007, the author expected that rather larger detectors would be the rule in 2019.

As lab microCT became widespread and commercial systems' resolution improved, emphasis at synchrotron microCT facilities appears to be shifting toward use of higher magnification objective lenses (5x, 10x, 20x vs 1x, 2x, 2.5x). Voxel sizes at 2-BM, APS are 1.3 μm and 0.65 μm for the first two lenses. Narrow depth-of-focus lenses remain a key to preserving spatial resolution, and there tends to be a mix-and-match approach where different scintillators are used for different spatial or temporal resolutions. At ID19, ESRF, for example, high-resolution imaging (0.15–1.5 μm voxels) uses thin-film scintillators GGG:Eu and LSO:Tb. For larger voxel sizes up to 50 μm, ID19 uses bulk LuAG:Ce, sometimes Gadox, and recently CWO, and for ultra-high speed imaging they use LYSO:Ce. At ALS (Beamline 8.3.2), the microCT instrument uses 20-, 50-, 150-, and 500-μm thick LuAG:Ce.

At 2-BM, APS, LuAG:Ce is used. For synchrotron microCT at 1-ID, APS, and with hard x-rays,* LuAG:Ce has been used and new scintillators such as thin films of LuI_3:Ce have been developed (Marton, Miller et al. 2013).

Imaging with voxel sizes substantially below 2 μm places more stringent requirements on mechanical stability and on tolerances for positioning components (rotational, translational) than was the case with earlier, lower-resolution generations of microCT systems. The trend is to use precision air bearings with magnetic drives instead of ball-based bearings and stepping motor drivers. Systems are often designed to allow continuous rotation in a single direction (i.e., many multiples of 360°) which is very useful for dynamic microCT. Piezoelectric translators above the rotation stage are used for positioning center of sample or center of local volume above the tomography rotation axis.

At 2-BM, APS, the author commonly collects projections after fixed rotation steps of 0.12° (~1,500 projections covering 180°) for 2K × 2K (and more recently 2,560 × 2,560) reconstructions. Many synchrotron microCT facilities collect projections during continuous rotation. The author and others are also making much more use of local tomography; typically, the same data collection parameters are used for local tomography as for scans where the entire sample always remains in the beam. The author also routinely collects a series of 10 or more FOV (with slight overlaps) covering more than a centimeter of specimen length.

The current ideal during "routine" synchrotron microCT is to obtain trial reconstructions within a few seconds of completion of data acquisition. This allows one to check that the data are usable, and the correct FOV has been selected before the next data set is collected. Reconstruction of the full stack of slices for that FOV is completed while acquiring data for the next FOV. Generally, beamlines have enough storage space that writing the reconstructed data to the users' hard drives is not a bottleneck in terms of continuing data acquisition (see Chapter 5 for the author's current data acquisition rates).

Users typically bring hard drives on which to write their raw and reconstructed data. The author sometimes uses a web transfer program to download tens of GB of data from 2-BM, APS, but for TB of data this solution is not ideal. At the time the first edition of this book was being written, the author was storing synchrotron microCT data to externally powered 256 GB USB drives. Currently, he uses 4TB USB-3 drives; these drives do not require a power source beyond that furnished over the USB connection, and six of these drives occupy the same shelf space as one of the 256 GB drives.

Since the first edition was published, many more investigators are doing phase contrast microCT (see Section 4.10 for details), in part because many beamlines feature reconstruction software for single-distance phase retrieval (e.g., TomoPy at 2-BM, APS, (Gürsoy, De Carlo et al. 2014)) using the Paganin algorithm (Paganin, Mayo et al. 2002). Typically, one can quickly alternate between data collection with absorption-dominated contrast and with phase-dominated contrast. The switch at 2-BM, APS, involves increasing/decreasing the specimen to detector separation (moving the detector along a rail nearly parallel to the incident beam direction), resetting the beam-defining slits so that the beam fills the detector, adjusting the rotation axis so that it coincides with the central column of the detector array, and readjusting the detector focus.

* Here, the term hard x-rays is used to refer to experiments performed with photon energies 50 keV and above. Many refer to x-rays with energies of 8 keV and above as "hard x-radiation," but the author prefers to call the range of 8–50 keV (note these limits are arbitrary) simply "x-rays' with the modifiers "soft" and "hard" for the ranges below and above this.

High-energy microCT can be a powerful tool for strongly absorbing specimens such as steel. Insertion devices that are designed to produce high intensities at high energies (E > 50 keV) and which are used primarily for x-ray scattering produce beams that are smaller than those typically used for microCT. At 1-ID, APS, the horizontal beam width is 2 mm, and only small-diameter specimens can be studied unless stitching is employed. Stitching of three (Liu, Roddatis et al. 2017) and four adjacent FOV (Khounsary, Kenesei et al. 2013) has been used at 1-ID. With high-energy photons, x-ray phase contrast dominates images, and this can complicate alignment of adjacent projections unless there are features visible that produce strong contrast.

In addition to a shift toward imaging with smaller voxels, there has also been a shift toward doing more quasi-dynamic and dynamic microCT. Data sets for reconstructions have been acquired in ~1 s, and detectors are deployed that can maintain this rate of data acquisition for many hours (Table 4.1). Often speed is gained by limiting the area of the FOV collected and the number of projections over 180° (decreasing the number of pixels that need to be transferred from the camera). More specifics are included in Chapter 10.

With other microscopy modalities such as scanning electron microscopy, accommodation exists to do the imaging remotely, either by controlling the instrument from elsewhere or by guiding a staff member on site. The author was part of one such remote synchrotron microCT session with Alexander Rack (Stock and Rack 2014). Data were collected in the evening at ID19, ESRF by Rack while, early afternoon, the author was in his office in Chicago. A camera was rigged at ESRF and streamed via Skype to Chicago. Sea urchin teeth were being imaged with local tomography, and it was essential that the remote investigator could identify the subvolume centered on the rotation axis and could direct the on-site investigator to make changes to image portions of the relevant portions of the tooth.

With higher resolution, new and unexpected challenges arise. Lens distortions can affect reconstructions, particularly with small voxel sizes, and these effects are described in Section 5.1.4. When using 10× or 20× objectives, very small sample motions can be detected, and motion artifacts can become important (see Section 5.1.1 for more details). In a long, thin sample (a sea urchin tooth, 1 cm long, 1.5 mm diameter, which is quite rigid) imaged at ID19, ESRF, with a 20× lens and 0.325-μm voxels, sample motion was observed via blurring of the trial reconstructions. The sample was glued to a holder, and the glue was kept out of the beam. The hypothesis was that the blurring was due to air circulation in the hutch. The investigators put a baggie over the sample and the specimen rotator, and the motion disappeared. At 2-BM, APS, the author scanned thin, plate-like samples using a local tomography approach and a 10× lens (0.65-μm voxels). One end of each sample was rigidly clamped in a holder, but sample motion was observed, probably from air flow in the hutch.

4.9 NanoCT (Full-Field, Microscopy-Based)

As of 2007, commercial nanoCT systems are listed in Table 4.3; commercial desktop systems have been described in the literature (Sasov 2004, Tkachuk, Feser et al. 2006). Manufacturers report voxel sizes to 100 nm and perhaps smaller; concomitantly smaller specimen diameters than in microCT are required. Scanning electron microscopes (SEMs) possess many of the attributes required for nanoCT of small specimens, and several groups have modified SEMs for this purpose (Yoshimura, Miyata et al. 2001, Mayo, Davis et al. 2003, Mayo, Miller et al. 2006) or market SEM-based systems. SEMs produce very small diameter electron beams, that is, a very tiny x-ray source, essential for minimizing penumbral blurring and for high spatial coherence for phase imaging.

In 2019, several manufacturers offer tube-based, x-ray microscopy systems for nanoCT. True spatial resolutions of better than 0.5 μm (10% MTF) are reported with pixel sizes on the detector of below 100 nm. Transmission targets in the x-ray tubes are needed to achieve these levels of performance. Mechanical stability is also essential, to the levels cited above, on the order of one-tenth the intended voxel size.

Synchrotron nanoCT reconstructions have been reported using optics to provide submicrometer resolution, and this is a very rapidly developing field. Parabolic x-ray focusing lenses, asymmetric crystal magnifiers, Fresnel zone plates (Fig. 4.7), or Kirkpatrick-Baez optics can be used to achieve the necessary magnification. The reader interested in more details is directed elsewhere (McNulty 2001, Rau, Weitkamp et al. 2001, Schneider, Knöchel et al. 2001, Schroer, Benner et al. 2001, Schroer, Lengeler et al. 2001, Spal 2001, Takeuchi, Uesugi et al. 2001, Uesugi, Suzuki et al. 2001, Stampanoni, Borchert et al. 2002a, 2005, 2006, Schroer, Meyer et al. 2003, Rau, Peterson et al. 2004, Schroer, Cloetens et al. 2004, Rau, Crecea et al. 2006, Schneider, Voide et al. 2006, Withers 2007). The use of 3D nanoCT for studies of bone recently has been reviewed (Langer and Peyrin 2016).

In 2019, transmission x-ray microscopes (TXM) form the basis of synchrotron nanoCT systems. System stability and optics allow projections to be collected with pixel dimensions substantially below 50 nm. If one is imaging with 30-nm voxels, for example, then one would need stability and motion control on the order of 3 nm. In one study, neurites in brain tissue were compared for normal vs schizophrenic individuals using imaging with 26-nm pixels and tomographic reconstruction (Mizutani, Saiga et al. 2019). Ex situ nanoCT data (60-nm resolution, FOV of 75 μm × 75 μm, ~20 min total scan time, 1 s exposure time per projection, and 1,500 projections) complemented nondestructive in operando microCT observation of morphology and water transport processes for platinum group metal-free, polymer-electrolyte fuel cells (PEFCs), and provided data on water distribution in smaller macropores (<1 μm) within the cathode catalyst layer (Normile, Sabarirajan et al. 2018). Websites of some facilities (APS 32-ID and SSRL BL 6-2c, for example) maintain up-to-date bibliographies of studies published from their instruments (Andrade, Wojcik et al. 2019, SSRL 2019).

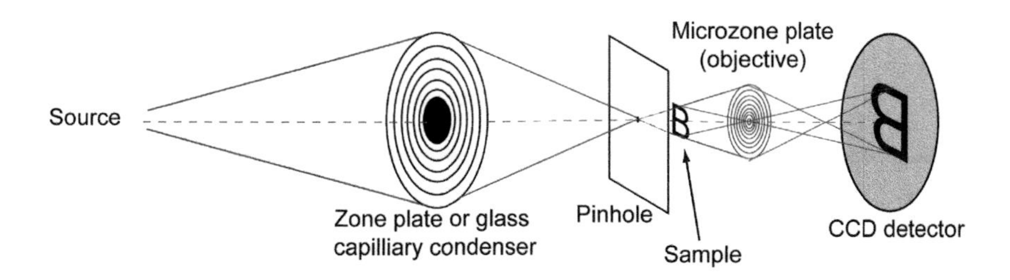

FIGURE 4.7
Lens-based systems for nanoCT employing Fresnel micro-zone plates. The first zone plate acts as a condenser and the second (after the specimen, schematically indicated by the letter "B") serves as the objective lens. A capillary condenser is used in some instruments instead of a zone plate. Adapted with permission from Withers (2007).

4.10 MicroCT with Phase Contrast

MicroCT using phase contrast is rapidly evolving. Covering the different methods of obtaining x-ray phase contrast radiographs and reconstructions in detail could fill an entire volume, and the coverage here is necessarily briefer.

X-rays are ever so slightly refracted when passing through solids (indices of refraction differ from one by a few parts per million), enough so that x-ray wavefronts distort when passing through regions of different electron density (see Fitzgerald [2000] for an introduction). With a suitable x-ray source, that is, one with adequate spatial coherence, it is possible to detect changes in contrast resulting from x-rays traversing volumes with different electron densities. Most frequently, phase imaging is performed at a synchrotron radiation source such as the APS, ESRF, SLS, or SPring-8; imaging can also be performed with x-ray tube sources (Pfeiffer, Weitkamp et al. 2006). Figure 4.8 illustrates four methods where phase effects are used to produce contrast in x-ray images.

In the propagation method (Fig. 4.8a), the detector is placed much farther away from the sample in normal for x-ray imaging (~1 m vs ~1 cm); refracted x-rays "r" diverge and interfere with other x-rays at the detector plane producing detectable fringes in the image at external and internal boundaries between materials with different electron densities. Here contrast is provided by differences in the second derivative of the x-ray phase (Snigerev, Snigereva et al. 1995, Cloetens, Barrett et al. 1996, Cloetens, Ludwig et al. 1999, 2000, Weitkamp, Rau et al. 2001, Ludwig, Buffière et al. 2003, Weon, Je et al. 2006). Images acquired at four or more specimen–detector separations (typically from 5 mm to 1 m or more) are required to extract the phase information (Cloetens, Van Dyck et al. 1999), and this method can be described as an analog of the focus variation method in transmission electron microscopy (Cloetens, Ludwig et al. 1999). The terms "holotomography" and "propagation method" are used in the literature.

In diffraction-enhanced imaging (DEI, Fig. 4.8b), an analyzer crystal is placed in the x-ray beam after the sample; images recorded with different settings of the analyzer isolate changes in the phase angle. This method produces image contrast based on changes in the

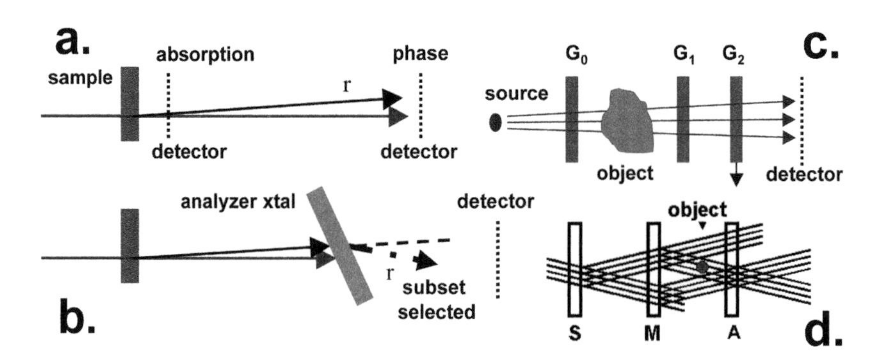

FIGURE 4.8
Methods of x-ray phase imaging. (a) Propagation method. Images with the detector near the sample are dominated by absorption contrast but placing the detector far from the specimens allows refracted x-rays "r" to interfere with transmitted x-rays and to produce edge contrast. (b) Diffraction-enhanced imaging (DEI). (c) Grating-enhanced imaging. Grating G_0 is not required with synchrotron x-radiation but is needed to provide a series of small virtual sources x-ray tube-based imaging. (d) Bonse-Hart interferometer for imaging. "S, M, and A" are crystal beam splitters, mirror, and analyzer, respectively. Reproduced from Stock (2007).

first derivative of the x-ray phase (Davis, Gao et al. 1995, Chapman, Thomlinson et al. 1997). Essentially, the analyzer selects only a small angular fraction of the refracted radiation.

The grating-enhanced imaging method (Fig. 4.8c), also known as Talbot interferometry, is analogous to DEI except that contrast from changes in the first derivative of phase is provided by translation of one analyzer grating relative to a second instead of by rotation of the analyzer crystal and its periodic array (the crystal lattice); see (Momose, Kawamoto et al. 2004, Weitkamp, Diaz et al. 2005, Groso, Stampanoni et al. 2006, Momose, Yashiro et al. 2006, Stampanoni, Groso et al. 2006, Weitkamp, David et al. 2006, Zdora, Vila-Comamala et al. 2017). Results obtained with gratings are used below to illustrate reconstruction of pure phase contrast.

Interferometry for phase imaging is illustrated in Fig. 4.8d. In the Bonse-Hart geometry, a beam splitter "S" produces a reference beam and an imaging beam, the mirror "M" redirects the beams together, the object is placed in one of the beams exiting the mirror, and the analyzer "A" recombines the reference and object-modified beams (Momose, Takeda et al. 1996, 1999, Beckmann, Bonse et al. 1999, Bonse and Beckmann 2001, Momose, Koyama et al. 2003). An alternative is the shearing interferometer (David, Nöhammer et al. 2002, Iwata, Takeda et al. 2004), and the limited FOV of interferometers from monolithic blocks of Si have recently been improved (Takeda, Momose et al. 2000, Yoneyama, Momose et al. 2002, Yoneyama, Takeda et al. 2005). Interferometers allow changes in the x-ray phase to be measured directly, not merely its derivatives.

Some comparisons of the different phase contrast imaging modalities have appeared in the older literature. Kiss et al. (2003) discussed image contrast numerically for absorption vs diffraction enhanced radiography. Pagot et al. (2005) compared radiography for phase propagation and DEI, but it is not clear how their conclusions translate to microCT. Wernick et al. (Wernick, Wirjadi et al. 2003, Wernick, Brankov et al. 2004, Chou, Anastasio et al. 2006) and Paganin et al. (2004) discussed different representations of phase imaging data, but not microCT data, and Mayo et al. (2002) examined these representations in microCT reconstructions.

Grating-based phase imaging provides an illuminating illustration of how phase microCT techniques work. Consider first the situation where no specimen is present, and the spatially coherent x-ray beam passes through phase grating G_1, the lines of which show negligible absorption but substantial phase shift (Fig. 4.8c). Note that grating G_0 is typically present only with imaging with an x-ray tube with large source size. Grating G_1 acts as a beam splitter, producing the two diffracted beams used for image formation. Because the wavelength of the illuminating x-rays (~10^{-10} m) is much smaller than the grating period (~10^{-6} m), the angle between the two beams is so small that the beams overlap almost completely as they propagate away from G_1 and interfere. The interference pattern generated could be imaged directly with an x-ray detector placed an appropriate distance d_g from G_1 (see Weitkamp et al. [2005, 2006] for the relationship of d_g to x-ray wavelength λ, periodicity, and other characteristics of the grating), but lack of spatial resolution of the detector systems has led to an alternative solution, use of an absorption grating G_2 positioned d_g away from G_1. The analyzer grating G_2 acts as a transmission mask for the detector placed immediately behind it and transforms the local interference fringe position into signal intensity variation. Note that the gratings must be parallel.

Placing a specimen upstream of G_1 produces local wavefront distortions $\Phi(x,y)$ and alters the interference pattern. Phase imaging is performed by translating the analyzer grating G_2 by small increments x_g of the fringe periodicity g and recording a radiograph at each position, and Fig. 4.9 shows images of polystyrene spheres for different x_g (Weitkamp, Diaz et al. 2005). The signal intensity $I(x,y)$ at each pixel (x,y) in the detector plane oscillates

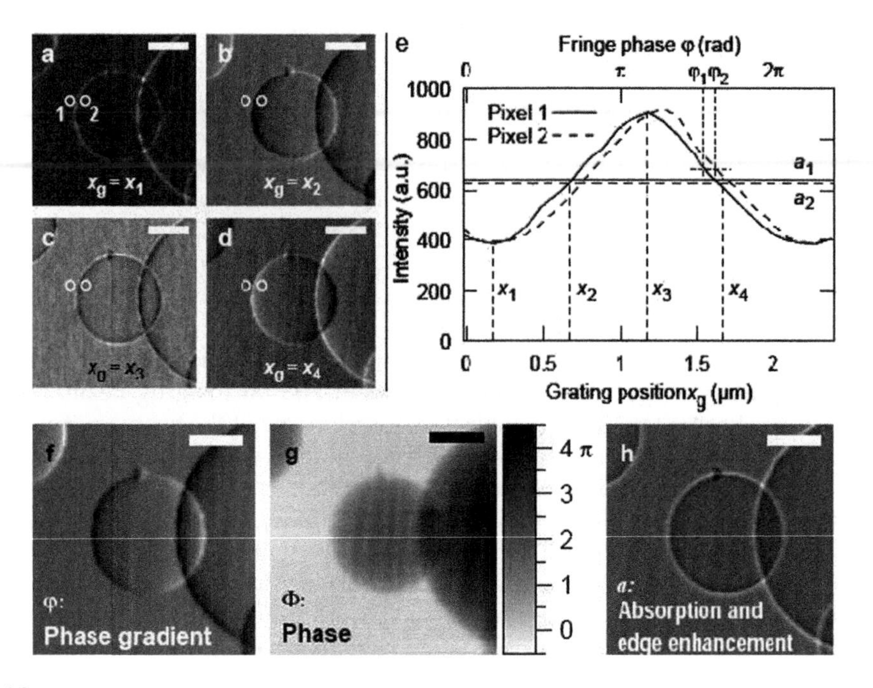

FIGURE 4.9
Principle of phase stepping. (a–d) Interferograms of polystyrene spheres (100 and 200-μm diameter), taken at the different relative positions $x_g = x_1,...,x_4$ of the two interferometer gratings. (e) Intensity oscillation in the two different detector pixels $i = 1, 2$ as a function of x_g. For each pixel, the oscillation phase φ_i and the average intensity a_i over one grating period can be determined. (f) Image of the oscillation phase φ for all pixels. (g) Wavefront phase Φ retrieved from φ by integration. (h) Image of the averaged intensity *a* for all pixels, equivalent to a noninterferometric image. The length of the scale bar is 50 μm. Reproduced with permission from Weitkamp et al. (2005).

as a function of x_g, and the phases $\varphi(x,y)$ of the intensity oscillations in each pixel are related to $\Phi(x,y)$ via:

$$\varphi = \left(\lambda d_g / g_2\right) \partial\Phi / \partial x, \tag{4.1}$$

where g_2 is the period of the absorption grating (Weitkamp, Diaz et al. 2005). The phase profile of the object can be retrieved from $\varphi(x,y)$ by simple one-dimensional integration (Fig. 4.9g). Radiographs at as few as three positions x_g are needed to extract φ if one knows *a priori* that the intensity oscillation is sinusoidal, but the reconstructions in Weitkamp et al. (2005) were obtained using eight phase steps per projection. Once the set of phase radiographs are obtained at the different viewing angles, a pure phase reconstruction can be computed using the normal methods. Recent works show a piece of sandpaper can act as a perfectly acceptable analyzer "grating," simultaneously quantifying phase contrast in two dimensions (Morgan, Paganin et al. 2012).

Synchrotron radiation is not essential for phase microCT (Fitzgerald 2000). The x-ray source size provided by the electron beam in an SEM provides adequate spatial coherence for phase microCT (Mayo, Davis et al. 2003), and modifying an SEM can be an effective way of studying small specimens. The fringe formation underlying the grating method described in the previous paragraph is independent of x-ray wavelength, and, provided

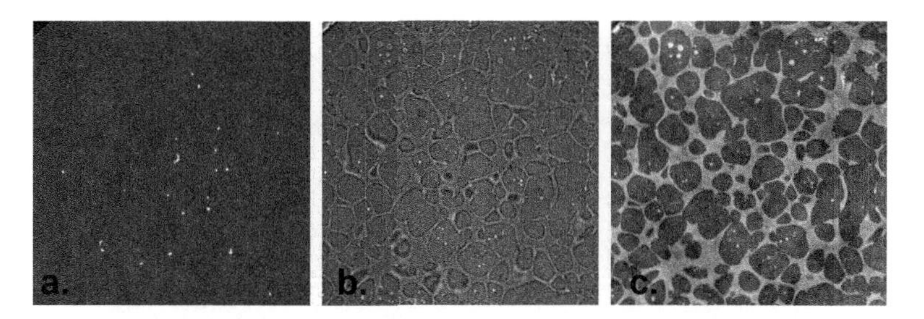

FIGURE 4.10
Matching slices of an Al-Si alloy quenched from a semi-solid state using (a) absorption contrast (sample detector separation DS = 7 mm), (b) phase (propagation) contrast (DS = 0.6 m), and (c) phase contrast (holotomography) with DS = 0.07, 0.2, 0.6, and 0.9 m. Horizontal field of view 0.9 mm, 18 keV, effective voxel size of 1.9 μm. Reproduced from Cloetens et al. (2000) with permission of Lavoisier.

a grating G_0 is used before the specimen (Fig. 4.9c) to provide a small virtual source size (more precisely, a set of independent small sources), a relatively high-power x-ray tube and gratings can be used for phase microCT (Pfeiffer, Weitkamp et al. 2006). Tube-based imaging with the propagation method has also been reported (Bidola, Morgan et al. 2017).

In specimens such as foams where the majority of volume is air, the total phase shift across the specimen varies relatively little, and holotomographic reconstruction can utilize the absolute values of the phase. In solid cylindrical Al-Si specimens such as that used by Cloetens et al. (~1.5 mm dia.), phase shifts will vary over 200 radians at 18 keV (Cloetens, Ludwig et al. 2000), and this dictates that the reconstructions employ the phase variations with respect to the phase introduced by the homogeneous matrix (i.e., the x-ray phase relative to that of the matrix).

Cloetens et al. provide a clear illustration of differences in absorption and phase-enhanced tomography reconstructions produced with the propagation method (Cloetens, Ludwig et al. 2000). Figure 4.10 compares the same slice from an Al-Si specimen (grains of Al embedded in a matrix of very fine Al-Si eutectic) obtained under three different imaging conditions. The radiographs for the first reconstruction were absorption dominated (i.e., they were recorded with a very small specimen detector separation DS); the radiographs for the second with a single, large DS (edge-enhanced interface contrast), and the radiographs at four DS were combined via the holotomography algorithm (see above) for the third. In Fig. 4.10a, absorption contrast does not allow one to distinguish the Al grains and Al-Si eutectic matrix. Edge enhancement allows the two phases to be seen clearly (Fig. 4.10b), but, because the Fresnel fringe intensity varies from position to position, segmentation of the grain and eutectic phases is challenging. The reconstruction with variation in refractive index decrement (Fig. 4.10c) clearly shows the different metallurgical phases whose difference in density is on the order of 0.05 g/cm^3 (Cloetens, Ludwig et al. 2000), and segmentation is quite straightforward. Another comparison was made recently in the context of spine imaging (Hu, Ni et al. 2017).

In interferometer-based phase microCT, the spatial distribution of polystyrene (PS) and poly(methyl methacrylate) (PMMA) in a ~50 vol% mixture were imaged (Momose, Fujii et al. 2005). The polymers are immiscible (although analyses of the values of the refraction indices of both phases suggest immiscibility is not total) and form a phase separated system. The achieved contrast resolution was, in terms of density resolution, <4 mg/cm^3, clearly beyond what is obtainable with absorption-based microCT.

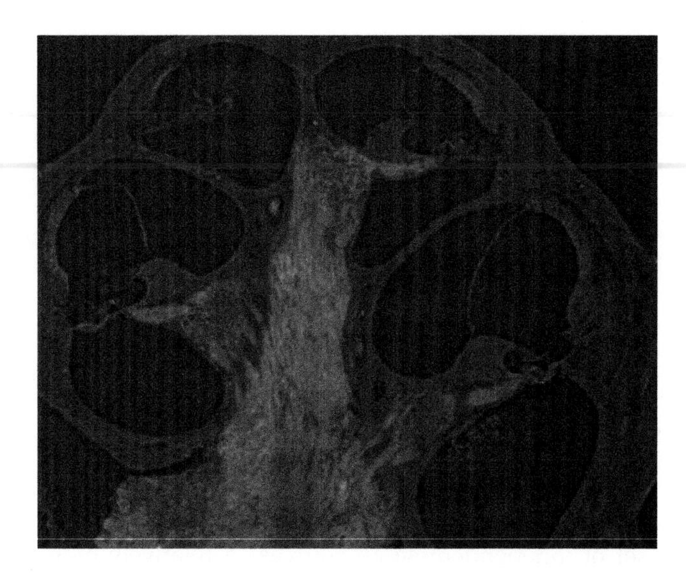

FIGURE 4.11
Phase contrast image of data reconstructed with the Paganin algorithm. The image is a numerical section (normal to the plane of the reconstructed slices) through a demineralized mouse cochlea lightly stained with Os. The turns of the cochlea are visible, and the brightest voxels are the nerve tissue. The horizontal field of view equals 666 µm. (C.P. Richter, S.R. Stock et al. unpublished image from June 2013. Recorded at 2-BM, APS, sample-detector separation of 600 mm, 22.5 keV, 2K × 2K reconstructions with 1.3-µm isotropic voxels, 350-ms integration per frame, 1,500 projections over 180°; see Tan et al. (2018) for more details.)

A relatively straightforward algorithm based on phase retrieval from a single propagation distance is now widely used for synchrotron phase microCT. This algorithm, sometimes termed the Paganin algorithm (Paganin, Mayo et al. 2002), employs the transport of intensity equations and Fourier space filtering based on analysis of these equations to produce reconstructions of projected phase thickness. An example of a Paganin-based reconstruction appears in Fig. 4.11; the image is a numerical section through a demineralized cochlea with nerve tissue lightly stained with Os. The amount of staining would not have produced measurable absorption contrast.

Phase-based microCT has been used in a wide variety of studies, descriptions of which are folded into the various subsections of materials applications. A few examples are mentioned here in closing the subsection, including damage in composites (Cloetens, Pateyron-Salomé et al. 1997, Buffiere, Maire et al. 1999) and structures in biological specimens (Momose, Takeda et al. 1999, Wu, Takeda et al. 2004). Real-time phase radiography of insect respiration produced interesting new insights (David, Nöhammer et al. 2002) and phase microCT, providing the third dimension, will undoubtedly prove very valuable.

Coherent diffractive imaging is another rapidly developing x-ray imaging modality requiring a spatially coherent x-ray source. One form, ptychography, combines diffraction patterns from across the specimen using an iterative algorithm to recover an image of a specimen (Li and Maiden 2018). Although mostly applied in 2D, ptychography can be extended to produce 3D images (Zanette, Enders et al. 2015). Coverage of the detailed mathematics is beyond the scope of this book.

4.11 MicroCT with X-ray Fluorescence

Several interactions can occur between a beam of x-rays and the atoms in the specimen through which the beam passes. If the x-ray photon energy is high enough, atoms can fluoresce, emitting photons with energies characteristic of the electronic shell transition that produced the emitted photons. These characteristic x-rays have well-defined energies, an energy-sensitive detector can measure the intensity of each characteristic peak, and this intensity can be converted to the concentration of these atoms within the irradiated volume. Detection limits (atomic concentration) for x-ray fluorescence are much, much lower than for x-ray absorption, and the elemental specificity is much, much higher for the former; these advantages continue to drive development of fluorescence microCT for applications where small concentrations of elements need to be mapped.

X-ray fluorescence occurs in all directions, and the typical experimental setup positions an energy sensitive x-ray detector to one side of the specimen in order to count the photons emitted from the irradiated volume (in the direction of the detector). Neither an area nor a ribbon-like x-ray beam appears to be practical for use in fluorescence microCT due to the confounding crossfire from different ray paths through the specimen (although receiving slit systems are being developed that may invalidate this conclusion). One consequence of sampling along only a single ray (i.e., of using a pencil beam) is that data collection rates are quite low. Quantification requires correction for absorption of the emitted characteristic x-rays along the path to the detector, and the reader is directed elsewhere for more details (Simionovici, Chukalina et al. 1999, Schroer, Tuemmler et al. 2000).

Focusing optics are typically used to concentrate x-ray flux into a small area; this is a little more complicated than simply focusing the beam into a spot on a surface because the beam divergence as it passes through the specimen must be kept small or accounted for numerically. Zone plate lenses, Kirkpatrick-Baez optics, or capillary condensers can be used. An interesting option is to use a polycapillary focusing optic to localize fluorescence from a single position along the beam path. This appears to be a viable option for 3D mapping, an option that does not require sample rotation to provide a complete map of elemental distribution in a slice (Vincze, Vekemans et al. 2004).

Examples are covered in the applications section of the book, specifically Section 12.2.4.

4.12 MicroCT with Scattered X-rays

X-rays scatter from atoms within solids and from structures with electron densities differing from their surroundings. Scattering from periodic arrays of atoms reinforces intensity along certain directions in the wide-angle x-ray scattering (WAXS) regime, and the distributions of reinforced intensity are referred to as WAXS or diffraction patterns. In the small-angle regime, reinforced scattering can result from arrays of fibrils, particles, etc., that is, peaks in small angle x-ray scattering (SAXS). The larger the size or spacing of the scatterers, the smaller the angle of the scattered intensity. By the author's count, over two dozen papers on scattering tomography have appeared since October 2007 when the author and collaborators published the first of the recent wave of scattering tomography papers (Stock, DeCarlo et al. 2008).

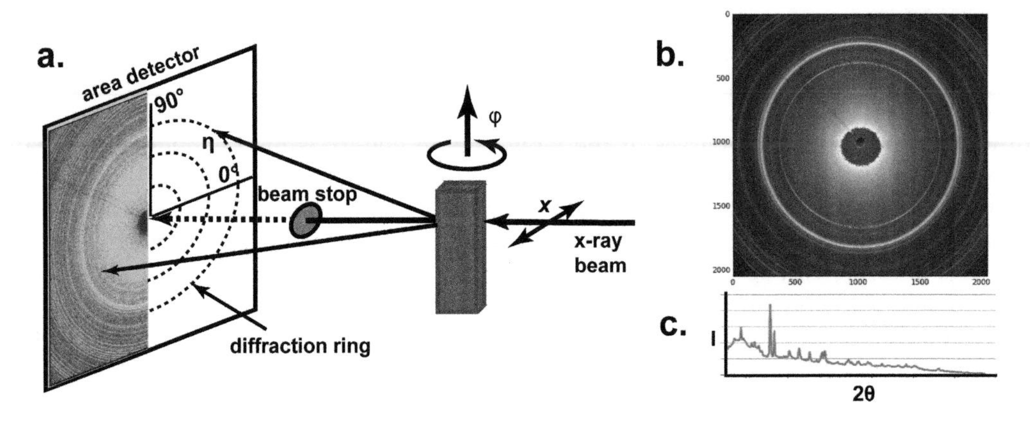

FIGURE 4.12
Illustration of diffraction tomography. (a) Schematic of setup. (b) Typical diffraction pattern recorded on the area detector. (c) Diffraction pattern $I_{2\theta}$ after azimuthal integration. Terms are defined in the text.

Scattering tomography employs a first-generation data collection scheme; that is, a pencil beam is scanned across the specimen (along x) normal to the apparatus rotation axis ω (Fig. 4.12a). In diffraction tomography, the specimen, containing an array of small crystals, interacts with the monochromatic x-ray beam and produces a series of diffraction cones (strong peaks in scattered x-ray intensity) whose angles are related through Bragg's law to the periodicities d_{hkl} of the crystalline phases within the specimen. A diffraction pattern is recorded at each position (the diffraction cones produce rings of high intensity on the area detector, Fig. 4.12b), and translation of the specimen across the x-ray beam builds a projection of the slice. The specimen is then rotated, and the translations are repeated. In the simplest analysis approach, the intensity of different *hkl* peaks integrated (azimuthally along η, around each diffraction ring), and the integrated intensity from a particular *hkl* is the input for reconstruction (Fig. 4.12c). In more comprehensive analysis, the intensity $I_{2\theta}$ at each 2θ of the azimuthally integrated diffraction pattern is used, regardless of the intensity; this results in a separate reconstruction for each 2θ and, for each x,y, the intensity values can be reassembled to give a diffraction pattern $I_{2\theta}$ at each x,y. Integrating only the intensity over a certain azimuthal range (e.g., $\eta = 0° \pm 10°$) and comparing that intensity to that over a similar aximuthal range at, say, 90° from the first sector (e.g., $\eta = 90° \pm 10°$) gives a measure of crystallographic texture (preferred orientation).

As diffraction tomography is a first-generation geometry and because diffracted intensities are much lower than transmitted intensities, data acquisition is extremely slow compared to synchrotron absorption microCT with area beams. The very different information collected in diffraction tomography makes it attractive for certain applications. Diffraction patterns collected with area detectors contain much more information than simply the crystalline phases present and intensities of different peaks of the crystallographic phase(s) (Leemreize, Almer et al. 2013). The other information that can be obtained includes crystallographic texture (Gursoy, Bicer et al. 2015), lattice constants and their variation within the sampled volume (Frølich, Leemreize et al. 2016), macrostrain (e.g., residual stresses), and crystallite size and microstrain (Birkbak, Nielsen et al. 2017). Maps of these quantities can be very informative. In addition, diffraction tomography can also reconstruct the diffraction pattern from each voxel within the slice (Leemreize, Almer et al. 2013). Examples are included in Chapter 12.

SAXS microCT is an ideal approach for studying polymer texture: absorption microCT shows no contrast, but differences in SAXS with position can be pronounced (Schroer, Kuhlmann et al. 2006, Stribeck, Camarilla et al. 2006). The complete SAXS pattern must be recorded for a single ray through the specimen; there would be too much overlap between patterns of adjacent rays if, for example, a ribbon beam were used. In a ~5-mm rod of warm drawn polyethylene, different layers could be resolved with SAXS microCT (see the processing subsection for more details).

The WAXS and SAXS tomography approaches described above employ monochromatic x-radiation. A polychromatic beam and energy dispersive diffraction (EDD) can also be used to perform tomography (Stock, Okasinski et al. 2017). An important advantage is that the specimen does not need to be rotated, making it suitable for mapping large plate-like specimens. The setup for EDD tomography is shown in Fig. 4.13 and requires a linear detector with reasonable spatial resolution and high-energy resolution (Rumaiz, Kuczewski et al. 2018). Reconstructed slices appear in Chapter 12.

4.13 System Specification

There is certainly no one commercial or synchrotron microCT system that is "the best." It all depends on the intended application(s). For someone (new to microCT) considering the purchase of a microCT system, some reasonable steps to go through in the decision-making process and some questions to answer include:

1. Define the specimens of interest (present plus future applications). What is the required field of view and largest specimen dimensions (not just the diameter)? What spatial resolution and contrast sensitivity are needed? What is the needed penetrating power (i.e., the x-ray kVp range) – very different operating parameters would be required for studying Ti samples vs studying soft tissues.
2. Define the software needed. Software for the following are or can be very important for the users: reconstruction of slices, 3D renderings of segmented data and superposition of different thresholds in renderings, numerical analysis of microstructure (including simple 3D manual distance measurements, numerical sectioning, specimen reorientation), movies of spinning renderings or movies paging through stacks of slices, import or export of data sets, image processing beyond simple segmentation (masking, etc.).
3. Identify the computer architecture (PC, Mac, other). How are the data backed up? What is the hard drive capacity and options for downloading data to other applications (ethernet, flash drive, USB-3)?
4. Determine how long it takes to collect data per slice, per data set; reconstruct slices; compute renderings; back up data. Also determine the expected levels of throughput that will be required.
5. Evaluate actual performance on specimens typical of the intended applications. Have the manufacturer run some specimens, consult people who already have the system, "drive" the microCT system for 1–2 days.

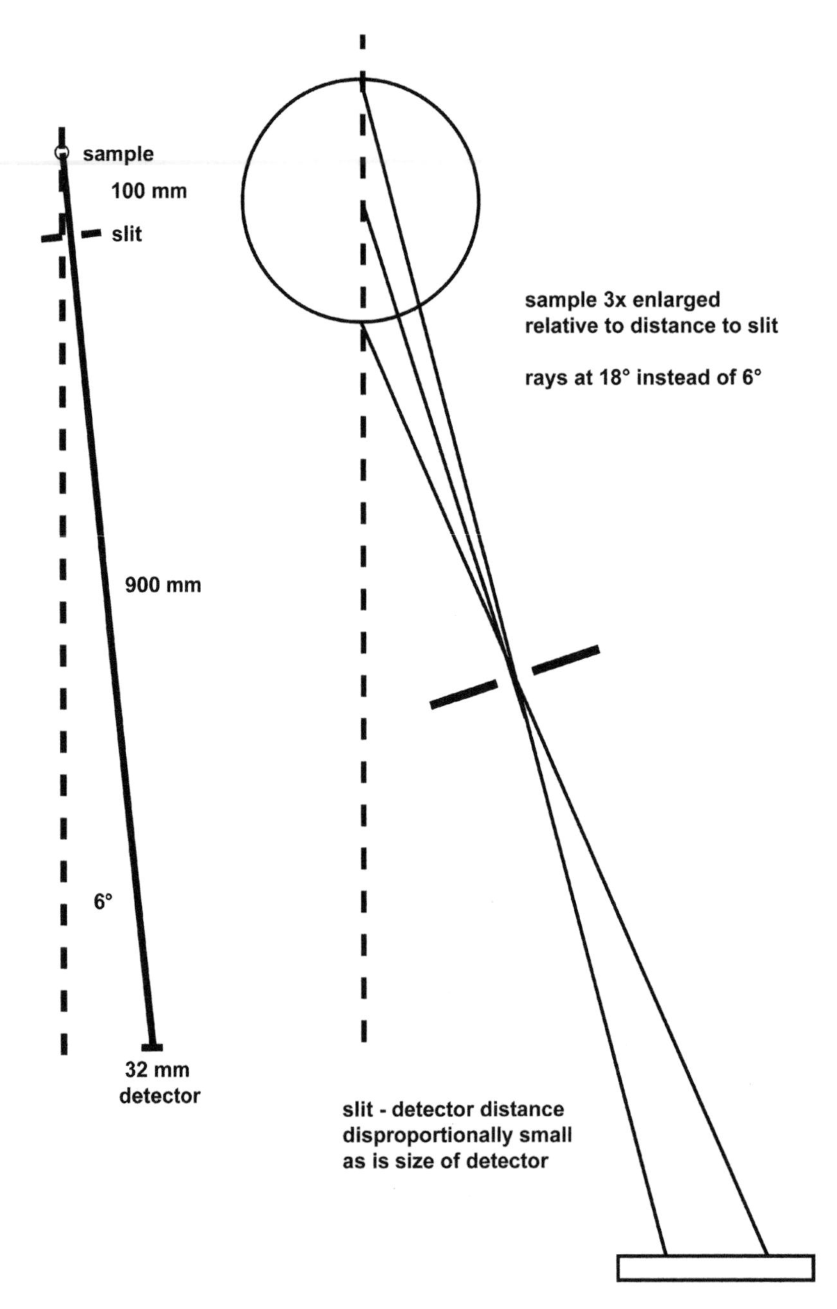

FIGURE 4.13
Geometry for energy dispersive tomography. The schematic at the left shows the relative distances to scale that were used in Stock et al. (2017), and the diagram to the right has scaling changed to illustrate how different depths within the specimen reach different parts of the detector.

6. Determine the required levels of service support. What are previous users' experiences with a given manufacturer?
7. Ascertain how closely the cost of the identified "ideal" system matches the funds available. Can an almost-as-good system give better "bang for the buck?"

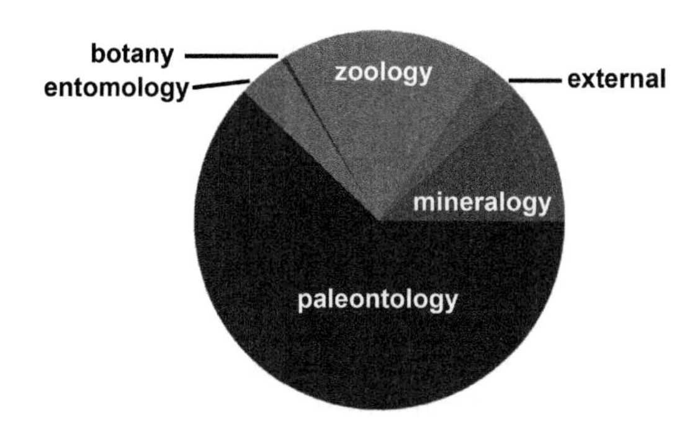

FIGURE 4.14
Scans performed at the Natural History Museum (London) during the first 6 months of 2012, categorized by discipline. Adapted from Stock (2012).

Unless one is replacing an existing lab microCT system or adding an additional instrument, it is very difficult to anticipate the actual range of samples that the scanner will be called upon to image. Data from the Natural History Museum (London) paint an interesting picture for the first 6 months of 2012 (Stock 2012). Figure 4.14 shows the distribution of samples studied, grouped by department furnishing the specimens. "Outside" refers to scans done for commercial firms. The majority of samples fall under the category of paleontology, but many of these were zoological. Figure 4.15 plots voltage (keV) vs volume element (voxel) size. The scan conditions lie mostly in the upper left of the plot; on the average, therefore, the samples require relatively high x-ray energy and have comparatively small relative diameter – this is the only way that small voxel size can be achieved with a given detector. The lack of points in the lower half of the right side of Fig. 4.15 allows one to infer that relatively few very low density, large diameter specimens were scanned in this time period.

The question of whether a tube-based microCT system or synchrotron microCT is required is something each investigator must determine for him/herself. The considerations listed for purchasing a tube-based microCT systems above also apply to the synchrotron vs lab microCT decision. Synchrotron microCT can be done at many storage rings and often at more than one beam line at a single source and deciding which instrument to use can follow the general points listed above. Unlike the author who is less than 1 hour

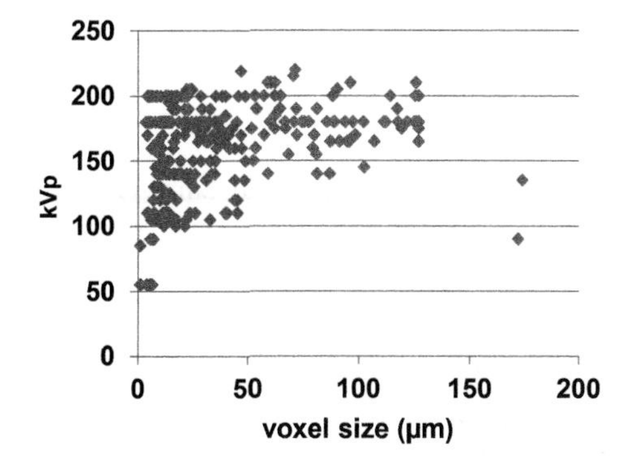

FIGURE 4.15
X-ray tube voltage (kVp) vs voxel size plotted for the data in Fig. 4.14 (Natural History Museum, London, January-June 2012). From Stock (2012).

by car from the APS, most potential users are fairly far from a suitable storage ring, and this affects how often and how long one can get access to the instrument. Travel costs must be factored into any decision to commit to doing synchrotron microCT, but cheap airfares can be obtained, given that schedules are normally formalized several months in advance.

Performing synchrotron microCT involves two steps. First is obtaining beam time, a topic discussed in this paragraph. Second is utilization of that time, something covered in the following paragraph. Obtaining access to a synchrotron microCT system almost always requires writing a fairly short, but focused, scientific proposal. Writing a successful proposal for beam time is not really very difficult, especially if one has experience and if there is a good scientific case for the work. However, first-time users would be well advised to get feedback from disinterested but experienced current users and to consult the texts of others' successful proposals, because what is expected in the proposal may not be evident to a new user. A submitted proposal for beam time is reviewed at several levels, including by users with expertise with the technique of interest, and, if the importance and scientific case warrant, experimental time is awarded. At the APS, for example, there are currently three scheduling cycles over the course of 1 year, and the proposal deadline is currently about 3 months prior to the start of the cycle.

Planning experiments is very different for synchrotron "runs" compared to those done on a lab-based instrument. Access is quite limited (say once per scheduling cycle), and one is typically awarded a number of consecutive 8-h shifts, between 3 and 21 shifts, often 9 shifts for 2-BM, APS. This means that many samples must be prepared or collected in advance and that it is difficult to make good use of the beam time if there are problems with the sample design. One does not have the luxury, when things are not going well, of taking the night off and thinking about what is going wrong (more precisely, one can do this, but the effect on productivity and on obtaining more beam time is quite undesirable). Working away from one's home laboratory also means that one cannot walk down the hall to retrieve key equipment or tools that one may unexpectedly need; many useful items can be borrowed from staff or other users at a storage ring, but locating who has the item in question can be very time consuming or, in the case of late nights or weekends, impossible. Despite the complications mentioned above, the number of samples that can be imaged per unit time (see Section 5.4) and the superior resolution and contrast sensitivity make synchrotron microCT attractive for many, but not all, investigators.

References

(2005). CdWO4 cadmium tungstate scintillation material, Saint-Gobain Ceramics and Plastics.

(2005). KAF-4301E 2084(H) × 2084(V) pixel enhanced response full-frame CCD image sensor performance specification, Eastman Kodak Co.

(2006). FReLoN.

(2007). BL20B2 Image detectors, SPring-8.

(2007). "ID19 High-resolution diffraction topography – microtomography beamline home page." Retrieved 1/24/07, from http://www.esrf.eu/UsersAndScience/Experiments/Imaging/ID19.

(2007). "Scintillation detector applications using Si diodes." Retrieved November 2007, from www.deetee.com.

(2008). Cesium iodide scintillator films, Radiation Monitoring Devices, Inc.

(2008). CoolSNAP K4 monochrome.

Andrade, V. D., M. Wojcik, A. Deriy, S. Bean, D. Shu, K. Peterson, T. Mooney, A. Glowacki, D. Gürsoy, T. Biçer, K. Fezzaa and F. D. Carlo. (2019). "The in-house Transmission X-ray Microscope of APS for in situ nano-tomography." Retrieved April 9, 2019, from https://confluence.aps.anl.gov/display/TXM/TXM+at+32-ID+Home.

Armistead, R. A. (1988). "CT: Quantitative 3-D inspection." *Adv Mater Process Inc Met Prog* (Mar): 41–49.

Baruchel, J., J. Y. Buffiere, P. Cloetens, M. D. Michiel, E. Ferrie, W. Ludwig, E. Maire and L. Salvo (2006). "Advances in synchrotron radiation microtomography." *Scripta Mater* **55**: 41–46.

Beckmann, F. (2001). Microtomography using synchrotron radiation as a user experiment at beamlines BW2 and BW5 of HASYLAB at DESY. *Developments in X-ray Tomography III*. U. Bonse. Bellingham (WA), SPIE. **SPIE Proc Vol 4503**: 34–41.

Beckmann, F. and U. Bonse (2000). Attenuation- and phase-contrast microtomography using synchrotron radiation for the 3-dim. investigation of specimens consisting of elements with low and medium absorption. *Applications of Synchrotron Radiation Techniques to Materials Science V.* S. R. Stock, S. M. Mini and D. L. Perry. Warrendale (PA), MRS. **MRS Proc Vol 590**: 265–271.

Beckmann, F., U. Bonse and T. Biermann (1999). New developments in attenuation and phase-contrast microtomography using synchrotron radiation with low and high photon energies. *Developments in X-ray Tomography II*. U. Bonse. Bellingham (WA), SPIE. **SPIE Proc Vol 3772**: 179–187.

Beckmann, F., T. Donath, T. Dose, T. Lippmann, R. V. Martins, J. Metge and A. Schreyer (2004). Microtomography using synchrotron radiation at DESY: Current status and future developments. *Developments in X-ray Tomography IV.* U. Bonse. Bellingham (WA), SPIE. **SPIE Proc Vol 5535**: 1–10.

Beckmann, F., T. Donath, J. Fischer, T. Dose, T. Lippmann, L. Lottermoser, R. V. Martins and A. Schreyer (2006). New developments for synchrotron-radiation-based microtomography at DESY. *Developments in X-ray Tomography V.* U. Bonse. Bellingham (WA), SPIE. **SPIE Proc Vol 6318**: 631810-1–631810-10.

Beckmann, F., T. Lippmann and U. Bonse (2000). High-energy microtomography using synchrotron radiation. *Penetrating Radiation Systems and Applications II*. F. Doty, H. B. Barber, H. Roehrig and E. J. Morton. Bellingham (WA), SPIE. **SPIE Proc Vol 4142**: 225–230.

Beckmann, F., J. Vollbrandt, T. Donath, H. W. Schmitz and A. Schreyer (2005). "Neutron and synchrotron radiation tomography: New tools for materials science at the GKSS-Research Center." *Nucl Instrum Meth A* **542**: 279–282.

Bidola, P., K. Morgan, M. Willner, A. Fehringer, S. Allner, F. Prade, F. Pfeiffer and K. Achterhold (2017). "Application of sensitive, high-resolution imaging at a commercial lab-based X-ray microCT system using propagation-based phase retrieval." *J Microsc* **266**: 211–220.

Birkbak, M. E., I. G. Nielsen, S. Frolich, S. R. Stock, P. Kenesei, J. D. Almer and H. Birkedal (2017). "Concurrent determination of nanocrystal shape and amorphous phases in complex materials by diffraction scattering computed tomography." *J Appl Crystal* **50**: 192–197.

Bonse, U. and F. Beckmann (2001). "Multiple-beam x-ray interferometry for phase-contrast microtomography." *J Synchrotron Rad* **8**: 1–5.

Bonse, U., Q. Johnson, M. Nichols, R. Nusshardt, S. Krasnicki and J. Kinney (1986). "High resolution tomography with chemical specificity." *Nucl Instrum Meth* **A246**: 644–648.

Bonse, U., R. Nusshardt, F. Busch, R. Pahl, Q. C. Johnson, J. H. Kinney, R. A. Saroyan and M. C. Nichols (1989). "Optimization of CCD-based energy-modulated x-ray microtomography." *Rev Sci Instrum* **60**: 2478–2481.

Bonse, U., R. Nusshardt, F. Busch, R. Pahl, J. H. Kinney, Q. C. Johnson, R. A. Saroyan and M. C. Nichols (1991). "X-ray tomographic microscopy of fibre-reinforced materials." *J Mater Sci* **26**: 4076–4085.

Borodin, Y. I., E. N. Dementyev, G. N. Dragun, G. N. Kulipanov, N. A. Mezentsev, V. F. Pindyurin, M. A. Sheromov, A. N. Skrinsky, A. S. Sokolov and V. A. Ushakov (1986). "Scanning difference microscopy and microtomography using synchrotron radiation at the storage ring VEPP-4." *Nucl Instrum Meth* **A246**: 649–654.

Bowen, D. K., J. C. Elliott, S. R. Stock and S. D. Dover (1986). X-ray microtomography with synchrotron radiation. *X-ray Imaging II*. D. K. Bowen and L. V. Knight. Bellingham (WA), SPIE. **691**: 94–98.

Breunig, T. M., J. C. Elliott, P. Anderson, G. Davis, S. R. Stock, A. Guvenilir and S. D. Dover (1990). Application of x-ray microtomography to the study of SiC/Al metal matrix composite material. *New Materials and Their Applications*. D. Holland. London, Institute of Physics. **111**: 53–60.

Breunig, T. M., J. C. Elliott, S. R. Stock, P. Anderson, G. R. Davis and A. Guvenilir (1992). Quantitative characterization of damage in a composite material using x-ray tomographic microscopy. *X-ray Microscopy III*. A. G. Michette, G. R. Morrison and C. J. Buckley. New York, Springer: 465–468.

Breunig, T. M., S. R. Stock, A. Guvenilir, J. C. Elliott, P. Anderson and G. R. Davis (1993). "Damage in aligned fibre SiC/Al quantified using a laboratory x-ray tomographic microscope." *Composites* **24**: 209–213.

Bueno, C. and M. D. Barker (1993). High resolution digital radiaography and 3D computed tomography. *X-ray Detector Physics and Applications II*. V. J. Orphan. Bellingham (WA), SPIE. **2009**: 179–191.

Buffiere, J. Y., E. Maire, P. Cloetens, G. Lormand and R. Fougeres (1999). "Characterization of internal damage in a MMCp using x-ray synchrotron phase contrast microtomography." *Acta Mater* **47**: 1613–1625.

Burstein, P., P. J. Bjorkholm, R. C. Chase and F. H. Seguin (1984). "The largest and smallest X-ray computed tomography systems." *Nucl Instrum Meth* **221**: 207–212.

Carlson, S. K., K. L. Classic, C. E. Bender and S. J. Russell (2007). "Small animal absorbed radiation dose from serial micro-computed tomography imaging." *Mole Imaging Biol* **9**: 78–82.

Chapman, D., W. Thomlinson, R. E. Johnson, D. Washburn, E. Pisano, N. Gmür, Z. Zhong, R. Menk, F. Arfelli and D. Sayers (1997). "Diffraction enhanced x-ray imaging." *Phys Med Biol* **42**: 2015–2025.

Chou, C. Y., M. A. Anastasio, J. G. Brankov and M. N. Wernick (2006). A comparison of a generalized DEI method with multiple image radiography. *Developments in X-ray Tomography V*. U. Bonse. Bellingham (WA), SPIE. **SPIE Proc Vol 6318**: 631819-1–631819-8.

Chu, Y. S., C. Liu, D. C. Mancini, F. D. Carlo, A. T. Macrander, B. Lai and D. Shu (2002). "Performance of a double-multilayer monochromator at beamline 2-BM at the advanced photon Source." *Rev Sci Instrum* **73**: 1485–1487.

Cloetens, P., R. Barrett, J. Baruchel, J. P. Guigay and M. Schlenker (1996). "Phase objects in synchrotron radiation hard x-ray imaging." *J Phys D* **29**: 133–146.

Cloetens, P., W. Ludwig, J. Baruchel, D. Van Dyck, J. V. Landuyt, J. P. Guigay and M. Schlenker (1999). "Holotomography: Quantitative phase tomography with micrometer resolution using hard synchrotron radiation x-rays." *Appl Phys Lett* **75**: 2912–2914.

Cloetens, P., W. Ludwig, J. P. Guigay, J. Baruchel, M. Schlenker and D. v. Dyck (2000). Phase contrast tomography. *X-ray Tomography in Materials Science*. J. Baruchel, J. Y. Buffière, E. Maire, P. Merle and G. Peix. Paris, Hermes Science: 29–44.

Cloetens, P., M. Pateyron-Salomé, J. Y. Buffière, G. Peix, J. Baruchel, F. Peyrin and M. Schlenker (1997). "Observation of microstructure and damage in materials by phase sensitive radiography and tomography." *J Appl Phys* **81**: 5878–5886.

Cloetens, P., D. Van Dyck, J. P. Guigay, M. Schlenker and J. Baruchel (1999). Quantitative phase tomography by holographic reconstruction. *Developments in X-ray Tomography II*. U. Bonse. Bellingham (WA), SPIE. **SPIE Proc Vol 3772**: 279–290.

Coan, P., A. Peterzol, S. Fiedler, C. Ponchut, J. C. Labiche and A. Bravin (2006). "Evaluation of imaging performance of a taper optics CCD 'FReLoN' camera designed for medical imaging." *J Synchrotron Rad* **13**: 260–270.

Cole, J. M., D. R. Symes, N. C. Lopes, J. C. Wood, K. Poder, S. Alatabi, S. W. Botchway, P. S. Foster, S. Gratton, S. Johnson, C. Kamperidis, O. Kononenko, M. De Lazzari, C. A. J. Palmer, D. Rusby, J. Sanderson, M. Sandholzer, G. Sarri, Z. Szoke-Kovacs, L. Teboul, J. M. Thompson, J. R. Warwick, H. Westerberg, M. A. Hill, D. P. Norris, S. P. D. Mangles and Z. Najmudin (2018). "High-resolution μCT of a mouse embryo using a compact laser-driven X-ray betatron source." *Proc Natl Acad Sci U S A* **115**: 6335–6340.

Connor, W. C., S. W. Webb, P. Spanne and K. W. Jones (1990). "Use of x-ray microscopy and synchrotron microtomography to characterize polyethylene polymerization particles." *Macromol* **23**: 4742–4747.

David, C., B. Nöhammer, H. H. Solak and E. Ziegler (2002). "Differential x-ray phase contrast imaging using a shearing interferometer." *Appl Phys Lett* **81**: 3287–3289.

Davis, G. R. (1997). Image quality in x-ray microtomography. *Developments in X-ray Tomography*. U. Bonse. Bellingham (WA), SPIE. **3149**: 213–221.

Davis, G. R. and J. C. Elliott (1997). "X-ray microtomography scanner using time-delay integration for elimination of ring artifacts in the reconstructed image." *Nucl Instrum Meth* **A394**: 157–162.

Davis, G. R. and F. S. L. Wong (1996). "X-ray microtomography of bones and teeth." *Physiol Meas* **17**: 121–146.

Davis, T. J., D. Gao, T. E. Gureyev, A. W. Stevenson and S. W. Wilkins (1995). "Phase-contrast imaging of weakly absorbing materials using hard x-rays." *Nature* **373**: 595–598.

De Carlo, F., P. Albee, Y. S. Chu, D. C. Mancini, B. Tieman and S. Y. Wang (2001). High-throughput real-time x-ray microtomography at the advanced photon source. *Developments in X-ray Tomography III*. U. Bonse. Bellingham (WA), SPIE. **SPIE Proc Vol 4503**: 1–13.

De Carlo, F. and B. Tieman (2004). High-throughput x-ray microtomography system at the advanced photon source beamline 2-BM. *Developments in X-ray Tomography IV*. U. Bonse. Bellingham (WA), SPIE. **SPIE Proc Vol 5535**: 644–651.

De Carlo, F., X. Xiao and B. Tieman (2006). X-ray tomography system, automation and remote access at beamline 2-BM of the advanced photon source. *Developments in X-ray Tomography V*. U. Bonse. Bellingham (WA), SPIE. **SPIE Proc Vol 6318**: 63180K-1–63180K-13.

Denesa, P. and B. Schmittb (2014). "Pixel detectors for diffraction-limited storage rings." *J Synch Rad* **21**: 1006–1010.

Di Michiel, M., J. M. Merino, D. Fernandez-Carreiras, T. Buslaps, V. Honkimaki, P. Falus, T. Martins and O. Svensson (2005). "Fast microtomography using high energy synchrotron radiation." *Rev Sci Instrum* **76**: 043702-1–043702-7.

Dilmanian, F. A., X. Y. Wu, B. Ren, T. M. Button, L. D. Chapman, J. M. Dobbs, X. Huang, E. L. Nickoloff, E. C. Parsons, M. J. Petersen, W. C. Tomlinson and Z. Zhong (1997). CT with monochromatic synchrotron x-rays and its potential in clinical research. *Developments in X-ray Tomography*. U. Bonse. Bellingham (WA), SPIE. **3149**: 25–32.

Dopp, A., L. Hehn, J. Goetzfried, M. Wenz, H. Gilljohann, H. Ding, S. Schindler, F. Pfeiffer and S. Karsch (2018). "Quick x-ray microtomography using a laser-driven betatron source." *Optica* **5**: 199–203.

Dowd, B. A., G. H. Campbell, R. B. Marr, V. Nagarkar, S. Tipnis, L. Axe and D. P. Siddons (1999). Developments in synchrotron x-ray computed microtomography at the national synchrotron light source. *Developments in X-ray Tomography II*. U. Bonse. Bellingham (WA), SPIE. **SPIE Proc Vol 3772**: 224–236.

Elliott, J. C., P. Anderson, G. R. Davis, F. S. L. Wong and S. D. Dover (1994a). "Computed tomography part II: the practical use of a single source and detector." *J Metals* (Mar): 11–19.

Elliott, J. C., P. Anderson, X. J. Gao, F. S. L. Wong, G. R. Davis and S. E. P. Dowker (1994b). "Application of scanning microradiography and x-ray microtomography to studies of bone and teeth." *J X-ray Sci Technol* **4**: 102–117.

Elliott, J. C. and S. D. Dover (1982). "X-ray microtomography." *J Microsc* **126**: 211–213.

Elliott, J. C. and S. D. Dover (1984). "Three-dimensional distribution of mineral in bone at a resolution of 15 μm determined by x-ray microtomography." *Metab Bone Dis Relat Res* **5**: 219–221.

Elliott, J. C. and S. D. Dover (1985). "X-ray microscopy using computerized axial tomography." *J Microsc* **138**: 329–331.

Engelke, K., M. Lohmann, W. R. Dix and W. Graeff (1989a). "Quantitative microtomography." *Rev Sci Instrum* **60**: 2486–2489.

Engelke, K., M. Lohmann, W. R. Dix and W. Graeff (1989b). "A system for dual energy microtomography of bones." *Nucl Instrum Meth* **A274**: 380–389.

Feldkamp, L. A., L. C. Davis and J. W. Kress (1984). "Practical cone-beam algorithm." *J Opt Soc Am* **A1**: 612–619.

Feldkamp, L. A. and G. Jesion (1986). "3-D x-ray computed tomography." *Rev Prog Quant NDE* **5A**: 555–566.

Feldkamp, L. A., D. J. Kubinski and G. Jesion (1988). "Application of high magnification to 3D x-ray computed tomography." *Rev Prog Quant NDE* **7A**: 381–388.

Ferrero, M. A., R. Sommer, P. Spanne, K. W. Jones and C. Connor (1993). "X-ray microtomography studies of nascent polyolefin particles polymerized over magnesium chloride-supported catalysts." *J Polym Sci A* **31**: 2507–2512.

Ferrie, E., J. Y. Buffiere, W. Ludwig, A. Gravouil and L. Edwards (2006). "Fatigue crack propagation: in situ visualization using x-ray microtomography and 3D simulation using the extended finite element method." *Acta Mater* **54**: 1111–1122.

Fitzgerald, R. (2000). "Phase sensitive x-ray imaging." *Phys Today* **53**: 23–26.

Flannery, B. P., H. W. Deckman, W. G. Roberge and K. L. D'Amico (1987). "Three-dimensional x-ray microtomography." *Science* **237**: 1439–1444.

Flannery, B. P. and W. G. Roberge (1987). "Observational strategies for three-dimensional synchrotron microtomography." *J Appl Phys* **62**: 4668–4674.

Frølich, S., H. Leemreize, A. Jakus, X. Xiao, R. Shah, H. Birkedal, J. D. Almer and S. R. Stock (2016). "Diffraction tomography and rietveld refinement of a hydroxyapatite bone phantom." *J Appl Cryst* **49**: 103–109.

Goulet, R. W., S. A. Goldstein, M. J. Ciarelli, J. L. Kuhn, M. B. Brown and L. A. Feldkamp (1994). "The relationship between the structural and orthogonal compressive properties of trabecular bone." *J Biomech* **27**: 375–389.

Groso, A., M. Stampanoni, R. Abela, P. Schneider, S. Linga and R. Muller (2006). "Phase contrast tomography: An alternative approach." *Appl Phys Lett* **88**: 214104-1–214104-3.

Gursoy, D., T. Bicer, J. D. Almer, R. Kettimuthu, F. D. Carlo and S. R. Stock (2015). "Maximum a posteriori estimation of crystallographic phases in X-ray diffraction tomography." *Phil Trans Roy Soc (Lond) A* **373**: 20140392.

Gürsoy, D., F. De Carlo, X. Xiao and C. Jacobsen (2014). "Tomopy: A framework for the analysis of synchrotron tomographic data." *J Synch Rad* **21**: 1188–1193.

Hatsuia, T. and H. Graafsma (2015). "X-ray imaging detectors for synchrotron and XFEL sources." *IUCrJ* **2**: 371–383.

Hirai, T., H. Yamada, M. Sasaki, D. Hasegawa, M. Morita, Y. Oda, J. Takaku, T. Hanashima, N. Nitta, M. Takahashi and K. Murata (2006). "Refraction contrast 11x-magnified x-ray imaging of large objects by MIRRORCLE-type table-top synchrotron." *J Synchrotron Rad* **13**: 397–402.

Hu, J., S. Ni, Y. Cao, X. Wang, S. Liao and H. Lu (2017). "Comparison of synchrotron radiation-based propagation phase contrast imaging and conventional micro-computed tomography for assessing intervertebral discs and endplates in a murine model." *Spine* **42**: E883–E889.

Ito, K., Y. Gaponov, N. Sakabe and Y. Amemiya (2007). "A 3 × 6 arrayed CCD x-ray detector for continuous rotation method in macromolecular crystallography." *J Synchrotron Rad* **14**: 144–150.

Iwata, K., Y. Takeda and H. Kikuta (2004). X-ray shearing interferometer and tomographic reconstruction of refractive index from its data. *Developments in X-ray Tomography IV*. U. Bonse. Bellingham (WA), SPIE. **SPIE Proc Vol 5535**: 392–399.

Joo, Y. I., T. Sone, M. Fukunaga, S. G. Lim and S. Onodera (2003). "Effects of endurance exercise on three-dimensional trabecular bone microarchitecture in young growing rats." *Bone* **33**: 485–493.

Kak, A. C. and M. Slaney (2001). *Principles of Computerized Tomographic Imaging*. Philadelphia, SIAM (Soc. Industrial Appl. Math.).

Khounsary, A., P. Kenesei, J. Collins, G. Navrotski and J. Nudell (2013). "High energy X-ray microtomography for the characterization of thermally fatigued GlidCop specimen." *J Physics* **Conf. Ser. 425**: 212015.

Kinney, J. H., U. K. Bonse, Q. C. Johnson, M. C. Nichols, R. A. Saroyan, W. N. Massey and R. Nusshardt (1993). X-ray tomographic image magnification process, system and apparatus therefore. U.S. Patent 5245648, USA.

Kinney, J. H., Q. C. Johnson, U. Bonse, M. C. Nichols, R. A. Saroyan, R. Nusshardt, R. Pahl and J. M. Brase (1988). "Three-dimensional x-ray computed tomography in materials science." *MRS Bull* **13**: 13–18.

Kinney, J. H., Q. C. Johnson, U. Bonse, R. Nusshardt and M. C. Nichols (1986). The performance of CCD array detectors for application in high-resolution tomography. *X-ray Imaging II*. D. K. Bowen and L. V. Knight. Bellingham (WA), SPIE. **691**: 43–50.

Kinney, J. H. and M. C. Nichols (1992). "X-ray tomographic microscopy (XTM) using synchrotron radiation." *Annu Rev Mater Sci* **22**: 121–152.

Kirby, B. J., J. R. Davis, J. A. Grant and M. J. Morgan (1997). "Monochromatic microtomographic imaging of osteoporotic bone." *Phys Med Biol* **42**: 1375–1385.

Kiss, M. Z., D. E. Sayers and Z. Zhong (2003). "Measurement of image contrast using diffraction enhanced imaging." *Phys Med Biol* **48**: 325–340.

Koch, A., F. Peyrin, P. Heurtier, B. Ferrand, B. Chambaz, W. Ludwig and M. Couchaud (1999). X-ray camera for computed tomography of biological samples with micrometer resolution using Lu3Al5O12 and Y3Al5O12. *Medical Imaging 1999: Physics of Medical Imaging*. J.M. Boone and J. T. Dobbins III. Bellingham (WA), SPIE. **SPIE Proc Vol 3659**: 170–179.

Kohlbrenner, A., B. Koller, S. Hämmerle and P. Rüegsegger (2001). In vivo microtomography. *Noninvasive Assessment of Trabecular Bone Architecture and the Competence of Bone*. B. K. Bay, S. Majumdar. New York, Kluwer. **Adv Exp Med Biol Vol 496**: 213–224.

Kohra, K., H. Hashizume and J. Yoshimura (1970). "X-ray diffraction topography utilizing double-crystal arrangement of (+, +) or non-parallel (+, -) setting." *Jpn J Appl Phys* **9**: 1029–1038.

Langer, M. and F. Peyrin (2016). "3D X-ray ultra-microscopy of bone tissue." *Osteoporos Int* **27**: 441–455.

Larsson, D. H., U. Lundström, U. K. Westermark, M. Arsenian Henriksson, A. Burvall and H. M. Hertz (2013). "First application of liquid-metal-jet sources for small-animal imaging: High-resolution CT and phase-contrast tumor demarcation." *Med Phys* **40**: 021909.

Leemreize, H., J. D. Almer, S. R. Stock and H. Birkedal (2013). "Three-dimensional distribution of polymorphs and magnesium in a calcified underwater attachment system by diffraction tomography." *J. Roy. Soc. Interface* **10**: 20130319.

Li, P. and A. Maiden (2018). "Multi-slice ptychographic tomography." *Sci Rep* **8**: 2049.

Liu, X., J. Liu, X. Li, S. K. Cheong, D. Shu, J. Wang, M. W. Tate, A. Ercan, D. R. Schuette, M. J. Renzi, A. Woll and S. M. Gruner (2004). Development of ultrafast computed tomography of highly transient fuel sprays. *Developments in X-ray Tomography IV*. U. Bonse. Bellingham (WA), SPIE. **SPIE Proc Vol 5535**: 21–28.

Liu, C., V. Roddatis, P. Kenesei and R. Maaß (2017). "Shear-band thickness and shear-band cavities in a Zr-based metallic glass." *Acta Mater* **140**: 206–216.

London, B., R. N. Yancey and J. A. Smith (1990). "High-resolution x-ray computed tomography of composite materials." *Mater Eval* **48**: 604–608.

Lopes, R. T., H. S. Rocha, E. F. O. D. Jesus, R. C. Barroso, L. F. D. Oliveira, M. J. Anjos, D. Braz and S. Moreira (2003). "X-ray transmission microtomography using synchrotron radiation." *Nucl Instrum Meth A* **505**: 604–607.

Ludwig, W., J. Y. Buffière, S. Savelli and P. Cloetens (2003). "Study of the interaction of a short fatigue crack with grain boundaries in a cast Al alloy using x-ray microtomography." *Acta Mater* **51**: 585–598.

Ludwig, O., M. Di Michiel, L. Salvo, M. Suery and P. Falus (2005). "In-situ three-dimensional microstructural investigation of solidification of an Al-Cu alloy by ultrafast x-ray microtomography." *Metall Mater Trans A* **36**: 1515–1523.

Maier, D. S., J. Schock and F. Pfeiffer (2017). "Dual-energy microCT with a dual-layer, dual-color, single-crystal scintillator." *Opt Express* **25**: 6924–6935.

Maire, E. and P. J. Withers (2014). "Quantitative X-ray tomography." *Inter Mater Rev* **59**: 1–43.

Martin, T. and A. Koch (2006). "Recent developments in x-ray imaging with micrometer spatial resolution." *J Synchrotron Rad* **13**: 180–194.

Marton, Z., S. R. Miller, E. Ovechkina, P. Kenesei, M. D. Moore, R. Woods, J. D. Almer, A. Miceli, B. Singh and V. V. Nagarkar (2013). Ultra-fast LuI3:Ce scintillators for hard X-ray imaging. *J Phys Conf Ser.* IOP Publishing. **425**: 212015.

Mayo, S. C., T. J. Davis, T. E. Gureyev, P. R. Miller, D. Paganin, A. Pogany, A. W. Stevenson and S. W. Wilkins (2003). "X-ray phase-contrast microscopy and microtomography." *Optics Express* **11**: 2289–2302.

Mayo, S. C., P. R. Miller, S. W. Wilkins, T. J. Davis, D. Gao, T. E. Gureyev, D. Paganin, D. J. Parry, A. Pogany and A. W. Stevenson (2002). "Quantitative x-ray projection microscopy: Phase contrast and multi-spectral imaging." *Science* **207**: 79–96.

Mayo, S. C., P. R. Miller, S. W. Wilkins, D. Gao and T. Gureyev (2006). Laboratory-based x-ray microtomography with submicron resolution. *Developments in X-ray Tomography V.* U. Bonse. Bellingham (WA), SPIE. **SPIE Proc Vol 6318**: 63181E-1–63181E-8.

McNulty, I. (2001). Current and ultimate limitations of scanning nanotomography. *X-ray Micro- and Nano-focusing: Applications and Techniques II.* I. McNulty. Bellingham (WA), SPIE. **SPIE Proc Vol 4499**: 23–28.

Mizutani, R., R. Saiga, A. Takeuchi, K. Uesugi, Y. Terada, Y. Suzuki, V. De Andrade, F. De Carlo, S. Takekoshi, C. Inomoto, N. Nakamura, I. Kushima, S. Iritani, N. Ozaki, S. Ide, K. Ikeda, K. Oshima, M. Itokawa and M. Arai (2019). "Three-dimensional alteration of neurites in schizophrenia." *Transl Psychiatry* **9**: 85.

Mokso, R., C. M. Schlepütz, G. Theidel, H. Billich, E. Schmid, T. Celcer, G. Mikuljan, L. Sala, F. Marone, N. Schlumpf and M. Stampanoni (2017). "GigaFRoST: The gigabit fast readout system for tomography." *J Synchrotron Rad* **24**: 1250–1259.

Momose, A., A. Fujii, H. Kadowaki and H. Jinnai (2005). "Three-dimensional observation of polymer blend by x-ray phase tomography." *Macromol* **38**: 7197–7200.

Momose, A., S. Kawamoto, I. Koyama and Y. Suzuki (2004). Phase tomography using an x-ray Talbot interferometer. *Developments in X-ray Tomography IV.* U. Bonse. Bellingham (WA), SPIE. **SPIE Proc Vol 5535**: 352–360.

Momose, A., I. Koyama, Y. Hamaishi, H. Yoshikawa, T. Takeda, J. Wu, Y. Itai, K. Takai, K. Uesugi and Y. Suzuki (2003). "Phase-contrast microtomography using an x-ray interferometer having a 40-μm analyzer." *J Phys IV* **104**: 599–602.

Momose, A., T. Takeda, Y. Itai and K. Hirano (1996). "Phase contrast x-ray computed tomography for observing biological soft tissues." *Nature Med* **2**: 473–475.

Momose, A., T. Takeda, Y. Itai, J. Tu and K. Hirano (1999). Recent observations with phase-contrast computed tomography. *Developments in X-ray Tomography II.* U. Bonse. Bellingham (WA), SPIE. **SPIE Proc Vol 3772**: 188–195.

Momose, A., W. Yashiro, M. Moritake, Y. Takeda, K. Uesugi, A. Takeuchi, Y. Suzuki, M. Tanaka and T. Hattori (2006). Biomedical imaging by Talbot-type x-ray phase tomography. *Developments in X-ray Tomography V.* U. Bonse. Bellingham (WA), SPIE. **SPIE Proc Vol 6318**: 63180T-1–63180T-10.

Morgan, K. S., D. M. Paganin and K. K. W. Siu (2012). "X-ray phase imaging with a paper analyzer." *Appl Phys Lett* **100**: 124102.

Mummery, P. M., B. Derby, P. Anderson, G. R. Davis and J. C. Elliott (1995). "X-ray microtomographic studies of metal matrix composites using laboratory x-ray sources." *J Microsc* **177**: 399–406.

Normile, S. J., D. C. Sabarirajan, O. Calzada, V. De Andrade, X. Xiao, P. Mandal, D. Y. Parkinson, A. Serov, P. Atanassov and I. V. Zenyuk (2018). "Direct observations of liquid water formation at nano-and micro-scale in platinum group metal-free electrodes by operando X-ray computed tomography." *Materials Today Energy* **9**: 187–197.

Nusshardt, R., U. Bonse, F. Busch, J. H. Kinney, R. A. Saroyan and M. C. Nichols (1991). "Microtomography: A tool for nondestructive study of materials." *Synchrotron Rad News* **4**(3): 21–23.

Olsen, U. L., X. Badel, J. Linnros, M. D. Michiel, T. Martin, S. Schmidt and H. F. Poulsen (2007). "Development of a high-efficiency high-resolution imaging detector for 30-80 keV x-rays." *Nucl Instrum Meth* **A576**: 52–55.

Paganin, D., T. E. Gureyev, S. C. Mayo, A. W. Stevenson, Y. I. Nesterets and S. W. Wilkins (2004). "X-ray omni microscopy." *Science* **214**: 315–327.

Paganin, D., S. C. Mayo, T. E. Gureyev, P. R. Miller and S. W. Wilkins (2002). "Simultaneous phase and amplitude extraction from a single defocused image of a homogeneous object." *J Microsc* **206**: 33–40.

Pagot, E., S. Fiedler, P. Cloetens, A. Bravin, P. Coan, K. Fezzaa, J. Baruchel and J. Härtwig (2005). "Quantitative comparison between two phase contrast techniques: Diffraction enhanced imaging and phase propagation imaging." *Phys Med Biol* **50**: 709–724.

Pfeiffer, F., T. Weitkamp, O. Bunk and C. David (2006). "Phase retrieval and differential phase contrast imaging with low brilliance x-ray sources." *Nature Phys* **2**: 258–261.

Rau, C., V. Crecea, C. P. Richter, K. M. Peterson, P. R. Jemian, U. Neuhäusler, G. Schneider, X. Yu, P. V. Braun, T. C. Chiang and I. K. Robinson (2006). A hard x-ray KB-FZP microscope for tomography with sub-100 nm resolution. *Developments in X-ray Tomography V.* U. Bonse. Bellingham (WA), SPIE. **SPIE Proc Vol 6318**: 63181G-1–63181G-6.

Rau, C., K. M. Peterson, P. R. Jemian, T. Terry, M. T. Harris, S. Vogt, C. P. Richter, U. Neuhäusler, G. Schneider and I. K. Robinson (2004). The evolution of hard x-ray tomography from the micrometer to the nanometer length scale. *Developments in X-ray Tomography IV.* U. Bonse. Bellingham (WA), SPIE. **SPIE Proc Vol 5535**: 709–714.

Rau, C., T. Weitkamp, A. A. Snigirev, C. G. Schroer, B. Benner, J. Tümmler, T. F. Günzler, M. Kuhlmann, B. Lengeler, C. E. Krill III, K. M. Döbrich, D. Michels and A. Michels (2001). Tomography with high resolution. *Developments in X-ray Tomography III.* U. Bonse. Bellingham (WA), SPIE. **SPIE Proc Vol 4503**: 14–22.

Ren , L., B. Zheng and H. Liu (2018). "Tutorial on X-ray photon counting detector characterization." *J X-ray Sci Technol* **26**: 1–28.

Rivers, M. L. (2016). High-speed tomography using pink beam at GeoSoilEnviroCARS. *Developments in X-ray Tomography X.* S. R. Stock. Bellingham (WA), SPIE. **9967**: 996733.

Rivers, M. L., S. R. Sutton and P. Eng (1999). Geoscience applications of x-ray computed microtomography. *Developments in X-ray Tomography II.* U. Bonse. Bellingham (WA), SPIE. **SPIE Proc Vol 3772**: 78–86.

Rivers, M. L. and Y. Wang (2006). Recent developments in microtomography at GeoSoilEnviroCARS. *Developments in X-ray Tomography V.* U. Bonse. Bellingham (WA), SPIE. **SPIE Proc Vol 6318**: 63180J-1–63180J-15.

Rivers, M. L., Y. Wang and T. Uchida (2004). Microtomography at GeoSoilEnviroCARS. *Developments in X-ray Tomography IV.* U. Bonse. Bellingham (WA), SPIE. **SPIE Proc Vol 5535**: 783–791.

Rumaiz, A. K., A. J. Kuczewski, J. Mead, E. Vernon, D. Pinelli, E. Dooryhee, S. Ghose, T. Caswell, D. P. Siddons, A. Miceli, J. Baldwin, J. Almer, J. Okasinski, O. Quaranta, R. Woods, T. Krings and S. Stock (2018). "Multi-element germanium detectors for synchrotron applications." *J Instrum* **13**: C04030.

Sakamoto, K., Y. Suzuki, T. Hirano and K. Usami (1988). "Improvement of spatial resolution of monochromatic x-ray CT using synchrotron radiation." *Jpn J Appl Phys* **27**: 127–132.

Sasov, A. Y. (1987a). "Microtomography: I. Methods and equipment." *J. Microsc* **147**: 169–178.

Sasov, A. Y. (1987b). "Microtomography: II. Examples of applications." *J. Microsc* **147**: 179–192.

Sasov, A. Y. (1989). X-ray microtomography. *Radiation methods.* New York, Plenum: 315–321.

Sasov, A. Y. (2004). X-ray nanotomography. *Developments in X-ray Tomography IV.* U. Bonse. Bellingham (WA), SPIE. **SPIE Proc Vol 5535**: 201–211.

Schneider, G., C. Knöchel, S. Vogt, D. Wei and E. H. Anderson (2001). Nanotomography of labeled cryogenic cells. *Developments in X-ray Tomography III.* U. Bonse. Bellingham (WA), SPIE. **SPIE Proc Vol 4503**: 156–165.

Schneider, P., R. Voide, M. Stuaber, M. Stampanoni, L. R. Donahue, P. Wyss, U. Sennhauser and R. Müller (2006). Assessment of murine bone ultrastructure using synchrotron light: Towards nanocomputed tomography. *Developments in X-ray Tomography V.* U. Bonse. Bellingham (WA), SPIE. **SPIE Proc Vol 6318**: 63180C-1–63180C-9.

Schroer, C. G., B. Benner, T. F. Gunzler, M. Kuhlmann, B. Lengeler, C. Rau, T. Weitkamp, A. Snigirev and I. Snigireva (2001). Magnified hard x-ray microtomography: Toward tomography with sub-micron resolution. *Developments in X-ray Tomography III*. U. Bonse. Bellingham (WA), SPIE. **SPIE Proc Vol 4503**: 23–33.

Schroer, C. G., P. Cloetens, M. Rivers, A. Snigirev, A. Takeuchi and W. Yun (2004). "High-resolution 3D imaging microscopy using hard x-rays." *MRS Bull* **29**: 157–165.

Schroer, C. G., M. Kuhlmann, T. F. Gunzler, B. Benner, O. Kurapova, J. Patormmel, B. Lengeler, S. V. Roth, R. Gehrke, A. Snigirev, I. Snigireva, N. Stribeck, A. Almendarez-Camarillo and F. Beckmann (2006). Full-field and scanning microtomography based on parabolic refractive x-ray lenses. *Developments in X-ray Tomography V*. U. Bonse. Bellingham (WA), SPIE. **SPIE Proc Vol 6318**: 63181H-1–63181H-9.

Schroer, C. G., B. Lengeler, B. Benner, T. F. Gunzler, M. Kuhlmann, A. S. Simionovici, S. Bohic, M. Drakopoulos, A. Snigirev, I. Snigireva and W. H. Schroder (2001). Microbeam production using compound refractive lenses: Beam characterization and applications. *X-ray Micro- and Nano-Focusing: Applications and Techniques II*. I. McNulty. Bellingham (WA), SPIE. **SPIE Proc Vol 4499**: 52–63.

Schroer, C. G., J. Meyer, M. Kuhlmann, B. Benner, T. F. Günzler, B. Lengeler, C. Rau, T. Weitkamp, A. Snigirev and I. Snigireva (2003). "Nanotomography based on hard x-ray microscopy with refractive lenses." *J Phys IV* **104**: 271.

Schroer, C. G., J. Tuemmler, T. F. Guenzler, B. Lengeler, W. H. Schroeder, A. J. Kuhn, A. S. Simionovici, A. Snigirev and I. Snigireva (2000). Fluorescence microtomography: External mapping of elements inside biological samples. *Penetrating Radiation Systems and Applications II*. F. P. Doty, H. B. Barber, H. Roehrig and E. J. Morton. Bellingham (WA), SPIE. **SPIE Proc Vol 4142**: 287–296.

Seguin, F. H., P. Burstein, P. J. Bjorkholm, F. Homburger and R. A. Adams (1985). "X-ray computed tomography with 50-µm resolution." *Appl Optics* **24**: 4117–4123.

Simionovici, A., M. Chukalina, M. Drakopoulos, I. Snigireva, A. Snigirev, C. Schroer, B. Lengeler, K. Janssens and F. Adams (1999). X-ray fluorescence microtomography: Experiment and reconstruction. *Developments in X-ray Tomography II*. U. Bonse. Bellingham (WA), SPIE. **SPIE Proc Vol 3772**: 304–310.

Snigerev, A., I. Snigereva, V. Kohn, S. Kuznetsov and I. Schelokov (1995). "On the possibilities of x-ray phase contrast microimaging by coherent high-energy synchrotron radiation." *Rev Sci Instrum* **66**: 5486–5492.

Spal, R. D. (2001). "Submicrometer resolution hard x-ray holography with the asymmetric Bragg diffraction microscope." *Phys Rev Lett* **86**: 3044–3047.

Spanne, P. and M. L. Rivers (1987). "Computerized microtomography using synchrotron radiation from the NSLS." *Nucl Instrum Meth* **B24/25**: 1063–1067.

SSRL, B.-c. (2019). "Publications." Retrieved April 9, 2019, 2019, from https://www-ssrl.slac.stanford.edu/txm/publications.

Stampanoni, M., R. Abela, G. Borchert and B. D. Patterson (2004). New developments in synchrotron-based microtomography. *Developments in X-ray Tomography IV*. U. Bonse. Bellingham (WA), SPIE. **SPIE Proc Vol 5535**: 169–181.

Stampanoni, M., G. Borchert, R. Abela and P. Ruegsegger (2002a). "Bragg magnifier: A detector for submicrometer x-ray computer tomography." *J Appl Phys* **92**: 7630–7635.

Stampanoni, M., G. Borchert and R. Abela (2003). Two-dimensional asymmetrical Bragg diffraction for submicrometer computer tomography. *Crystals, Multilayers, and Other Synchrotron Optics*. T. Ishikawa, A. T. Macrander and J. L. Wood. Bellingham (WA), SPIE. **SPIE Proc Vol 5195**: 54–62.

Stampanoni, M., G. Borchert and R. Abela (2005). "Towards nanotomography with asymmetrically cut crystals." *Nucl Instrum Meth A* **551**: 119–124.

Stampanoni, M., G. Borchert and R. Abela (2006). "Progress in microtomography with the Bragg magnifier at SLS." *Rad Phys Chem* **75**: 1956–1961.

Stampanoni, M., G. Borchert, P. Wyss, R. Abela, B. Patterson, S. Hunt, D. Vermeulen and P. Rüegsegger (2002b). "High resolution x-ray detector for synchrotron-based microtomography." *Nucl Instrum Meth A* **491**: 291–301.

Stampanoni, M., A. Groso, A. Isenegger, G. Mikuljan, Q. Chen, A. Bertrand, S. Henein, R. Betemps, U. Frommherz, P. Bohler, D. Meister, M. Lange and R. Abela (2006). Trends in synchrotron-based tomographic imaging: The SLS experience. *Developments in X-ray Tomography V*. U. Bonse. Bellingham (WA), SPIE. **SPIE Proc Vol 6318**: 63180M-1–63180M-14.

Stampanoni, M., P. Wyss, G. L. Borchert, D. Vermeulen and P. Rüegsegger (2001). X-ray tomographic microscope at the Swiss light source. *Developments in X-ray Tomography III*. U. Bonse. Bellingham (WA), SPIE. **SPIE Proc Vol 4503**: 42–53.

Stock, S. R. (1999). "Microtomography of materials." *Int Mater Rev* **44**: 141–164.

Stock, S. R. (2007). X-ray phase microradiography and x-ray absorption micro-computed tomography, compared in studies of biominerals. *Handbook of Biomineralization*. P. Behrens and E. Bäuerlein. Weinheim, Wiley-VCH. **2**: 389–400.

Stock, S. R. (2008). "Recent advances in x-ray microtomography applied to materials." *Inter Mater Rev* **58**: 129–181.

Stock, S. R. (2012). Trends in micro- and nanoComputed tomography 2010-2012. *Developments in X-ray Tomography VIII*. S. R. Stock. Bellingham (WA), SPIE. **8506**: 850602.

Stock, S. R., F. DeCarlo and J. D. Almer (2008). "High energy x-ray scattering tomography applied to bone." *J Struct Biol* **161**: 144–150.

Stock, S. R., L. L. Dollar, G. B. Freeman, W. J. Ready, L. J. Turbini, J. C. Elliott, P. Anderson and G. R. Davis (1994). Characterization of conductive anodic filament (CAF) by x-ray microtomography and by serial sectioning. *Electronic Packaging Materials Science VII*. P. Børgesen, K. F. Jansen and R. A. Pollak. Pittsburgh, **Mater Res Soc. 323**: 65–69.

Stock, S. R., A. Guvenilir, T. L. Starr, J. C. Elliott, P. Anderson, S. D. Dover and D. K. Bowen (1989). "Microtomography of silicon nitride/silicon carbide composites." *Ceram Trans* **5**: 161–170.

Stock, S. R., J. S. Okasinski, R. Woods, J. Baldwin, T. Madden, O. Quaranta, A. Rumaiz, T. Kuczewski, J. Mead, T. Krings, P. Siddons, A. Miceli and J. D. Almer (2017). Tomography with energy dispersive diffraction. *Developments in X-ray Tomography XI*. B. Mueller. Bellingham (WA), SPIE. **10391**: 103910A.

Stock, S. R. and A. Rack. (2014). Submicrometer structure of sea urchin tooth via remote synchrotron microCT imaging. *Developments in X-ray Tomography IX*. S. R. Stock. Bellingham (WA), SPIE. **9212**: 92120V.

Stribeck, N., A. A. Camarilla, U. Nochel, C. Schroer, M. Kuhlmann, S. V. Roth, R. Gehrke and R. K. Bayer (2006). "Volume-resolved nanostructure survey of a polymer part by means of SAXS microtomography." *Macromol Chem Phys* **207**: 1139–1149.

Suzuki, Y., K. Usami, K. Sakamoto, H. Kozaka, T. Hirano, H. Shiono and H. Kohno (1988). "X-ray computerized tomography using monochromated synchrotron radiation." *Japan J Appl Phys* **27**: L461–L464.

Takeda, T., A. Momose, Q. Yu, J. Wu, K. Hirano and Y. Itai (2000). "Phase contrast x-ray imaging with a large monolithic x-ray interferometer." *J Synchrotron Rad* **7**: 280–282.

Takeuchi, A., K. Uesugi, Y. Suzuki and S. Aoki (2001). "Hard x-ray microtomography using x-ray imaging optics." *Jpn J Appl Phys* **40**: 1499–1503.

Takeuchi, A., K. Uesugi, H. Takano and Y. Suzuki (2002). "Submicrometer-resolution three-dimensional imaging with hard x-ray imaging microtomography." *Rev Sci Instrum* **73**: 4246–4249.

Tan, X., I. Jahan, Y. Xu, S. R. Stock, C. C. Kwan, C. Soriano, X. Xiao, B. Fritzsch, J. García-Añoveros and C.-P. Richter (2018). "Auditory neural activity in congenital deaf mice induced by infrared neural stimulation." *Sci Rep* **8**: 388.

Thurner, P. J., P. Wyss, R. Voide, M. Stauber, B. Müller, M. Stampanoni, J. A. Hubbell, R. Müller and U. Sennhauser (2004). Functional microimaging of soft and hard tissue using synchrotron light. *Developments in X-ray Tomography IV*. U. Bonse. Bellingham (WA), SPIE. **SPIE Proc Vol 5535**: 112–128.

Tkachuk, A., M. Feser, H. Cui, F. Duewer, H. Chang and W. Yun (2006). High resolution x-ray tomography using laboratory x-ray sources. *Developments in X-ray Tomography V.* U. Bonse. Bellingham (WA), SPIE. **SPIE Proc Vol 6318**: 63181D-1–63181D-8.

Toepperwien, M., R. Gradl, D. Keppeler, M. Vassholz, A. Meyer, R. Hessler, K. Achterhold, B. Gleich, M. Dierolf, F. Pfeiffer, T. Moser and T. Salditt (2018). "Propagation-based phase-contrast x-ray tomography of cochlea using a compact synchrotron source." *Sci Rep* **8**: 4922.

Uesugi, K., Y. Suzuki, N. Yagi, A. Tsuchiyama and T. Nakano (2001). Development of submicrometer resolution x-ray CT system at SPring-8. *Developments in X-ray Tomography III.* U. Bonse. Bellingham (WA), SPIE. **SPIE Proc Vol 4503**: 291–298.

Uesugi, K., A. Tsuchiyama, H. Yasuda, M. Nakamura, T. Nakano, Y. Suzuki and N. Yagi (2003). "Micro-tomographic imaging for material sciences at BL47XU in SPring-8." *J Phys IV* **104**: 45–48.

Vincze, L., B. Vekemans, I. Szaloki, F. E. Brenker, G. Falkenberg, K. Rickers, K. Aerts, R. V. Grieken and F. Adams (2004). X-ray fluorescence microtomography and polycapillary based confocal imaging using synchrotron radiation. *Developments in X-ray Tomography IV.* U. Bonse. Bellingham (WA), SPIE. **SPIE Proc Vol 5535**: 220–231.

Wang, Y., F. De Carlo, I. Foster, J. Insley, C. Kesselman, P. Lane, G. von Laszewsk, D. Mancini, I. McNulty, M. H. Su and B. Tieman (1999). Quasi-realtime x-ray microtomography system at the Advanced Photon Source. *Developments in X-ray Tomography II.* U. Bonse. Bellingham (WA), SPIE. **SPIE Proc Vol 3772**: 318–327.

Wang, Y., F. De Carlo, D. C. Mancini, I. McNulty, B. Tieman, J. Bresnahan, I. Foster, J. Insley, P. Lane, G. Von Laszewski, C. Kesselman, M. H. Su and M. Thiebaux (2001). "A high-throughput x-ray microtomography system at the Advanced Photon Source." *Rev Sci Instrum* **72**: 2062–2068.

Wang, Y., T. Uchida, F. Westferro, M. L. Rivers, N. Nishiyama, J. Gebhardt, C. E. Lesher and S. R. Sutton (2005). "High-pressure x-ray tomography microscope." *Rev Sci Instrum* **76**: 073709.

Weitkamp, T., C. David, C. Kottler, O. Bunk and F. Pfeiffer (2006). Tomography with grating interferometers at low-brilliance sources. *Developments in X-ray Tomography V.* U. Bonse. Bellingham (WA), SPIE. **SPIE Proc Vol 6318**: 63180S-1–63180S-10.

Weitkamp, T., A. Diaz, C. David, F. Pfeiffer, M. Stampanoni, P. Cloetens and E. Ziegler (2005). "X-ray phase imaging with a grating interferometer." *Optics Express* **13**: 6296–6304.

Weitkamp, T., C. Rau, A. A. Snigirev, B. Benner, T. F. Gunzler, M. Kuhlmann and C. G. Schroer (2001). In-line phase contrast in synchrotron-radiation microradiography and tomography. *Developments in X-ray Tomography III.* U. Bonse. Bellingham (WA), SPIE. **SPIE Proc Vol 4503**: 92–102.

Weitkamp, T., C. Raven and A. Snigirev (1999). Imaging and microtomography facility at the ESRF beamline ID 22. *Developments in X-ray Tomography II.* U. Bonse. Bellingham (WA), SPIE. **SPIE Proc Vol 3772**: 311–317.

Weon, B. M., J. H. Je, Y. Hwu and G. Margaritondo (2006). "Phase contrast X-ray imaging." *Int J Nanotechnol* **3**: 280–297.

Wernick, M. N., J. G. Brankov, D. Chapman, Y. Yang, C. Muehleman, Z. Zhong and M. A. Anastasio (2004). A preliminary study of multiple image computed tomography. *Developments in X-ray Tomography IV.* U. Bonse. Bellingham (WA), SPIE. **SPIE Proc Vol 5535**: 369–379.

Wernick, M. N., O. Wirjadi, D. Chapman, Z. Zhong, N. P. Galatsanos, Y. Yang, J. G. Brankov, O. Oltulu, M. A. Anastasio and C. Muehleman (2003). "Multiple image radiography." *Phys Med Biol* **48**: 3875–3895.

Withers, P. J. (2007). "X-ray nanotomography." *Mater Today* **10**: 26–34.

Wu, J., T. Takeda, T. T. Lwin, I. Koyama, A. Momose, A. Fujii, Y. Hamaishi, T. Kuroe, T. Yuasa, Y. Suzuki and T. Akatsuka (2004). Microphase contrast x-ray computed tomography for basic biomedical study at SPring-8. *Developments in X-ray Tomography IV.* U. Bonse. Bellingham (WA), SPIE. **SPIE Proc Vol 5535**: 740–747.

Yoneyama, A., A. Momose, I. Koyama, E. Seya, T. Takeda, Y. Itai, K. Hirano and K. Hyodo (2002). "Large area phase contrast x-ray imaging using a two crystal x-ray interferometer." *J Synchrotron Rad* **9**: 277–281.

Yoneyama, A., T. Takeda, Y. Tsuchiya, J. Wu, T. T. Lwin, K. Hyodo and Y. Hirai (2005). "High energy phase contrast x-ray imaging using a two-crystal x-ray interferometer." *J Synchrotron Rad* **12**: 534–536.

Yoshimura, H., C. Miyata, C. Kuzuryu, A. Hori, T. Obi and N. Ohyama (2001). X-ray computed tomography using projection x-ray microscope. *Developments in X-ray Tomography III.* U. Bonse. Bellingham (WA), SPIE. **SPIE Proc Vol 4503**: 166–171.

Zanette, I., B. Enders, M. Dierolf, P. Thibault, R. Gradl, A. Diaz, M. Guizar-Sicairos, A. Menzel, F. Pfeiffer and P. Zaslansky (2015). "Ptychographic X-ray nanotomography quantifies mineral distributions in human dentine." *Sci Rep* **5**: 9210.

Zdora, M. C., J. Vila-Comamala, G. Schulz, A. Khimchenko, A. Hipp, A. C. Cook, D. Dilg, C. David, C. Grünzweig, C. Rau, P. Thibault and I. Zanette (2017). "X-ray phase microtomography with a single grating for high-throughput investigations of biological tissue." *Biomed Opt Express* **8**: 1257–1270.

5

MicroCT in Practice

The preceding chapter discussed microCT systems and the component subsystems from which they are assembled. Systems contain various nonidealities, and their effects on reconstructions and amelioration of these "errors" are the subject of this chapter. Section 5.1 describes reconstruction artifacts, and Section 5.2 discusses the precision and accuracy of reconstructions. The examples are reconstructed using analytic algorithms (Fourier-based reconstructions for the synchrotron microCT examples and filtered back-projection for the lab microCT data), which are by far the most widely used. Sections 5.3–5.5 cover techniques for contrast enhancement, data acquisition challenges, and speculations for future apparatus development, respectively.

5.1 Reconstruction Artifacts

Reconstruction software must cope with various nonidealities intrinsic to the experimental apparatus and to the x-ray sources and return the highest fidelity reconstructions practical. An ideal microCT or nanoCT apparatus would have positioning component errors that are always much smaller than the smallest information-containing voxel size specified in the system design. This ideal system would employ a bright, highly stable x-ray source amenable to flat field correction. It is best to collect the highest quality data possible, but software can ameliorate the effects of instrument nonidealities and from less-than-optimum sampling dictated by experimental requirements.

The performance of reconstruction algorithms has been studied experimentally. For example, Feldkamp and coworkers scanned the same cube of trabecular bone with three orthogonal rotation axes and found the reconstructed structures were in good agreement (Feldkamp, Davis et al. 1984). Sasov used data from narrow cone beam system and compared slices reconstructed with a fan beam algorithm, a cone beam algorithm, and a spiral scan algorithm (Sasov 2001). The quality of slices at the ends of the stack was compared to that at the center, and this study should be considered by those interested in different strategies for rapid data collection.

5.1.1 Motion Artifacts

Small motions of the specimen during collection of the projections for a reconstruction must be avoided. Figure 5.1, of a living sea urchin in seawater, shows an extreme example of the effect of motion. Despite wedging the live sea urchin tightly in place, it moved significantly, smearing the image beyond recognition (compare with Fig. 5.1b, a slice of an intact, fixed sea urchin). Note the image of the specimen holder "SH" and the vial "v" are sharp in Fig. 5.1a. Gluing specimens to a holder and exposing the glue to the x-ray beam should be avoided as it often leads to specimen motion. Motion artifacts can also occur

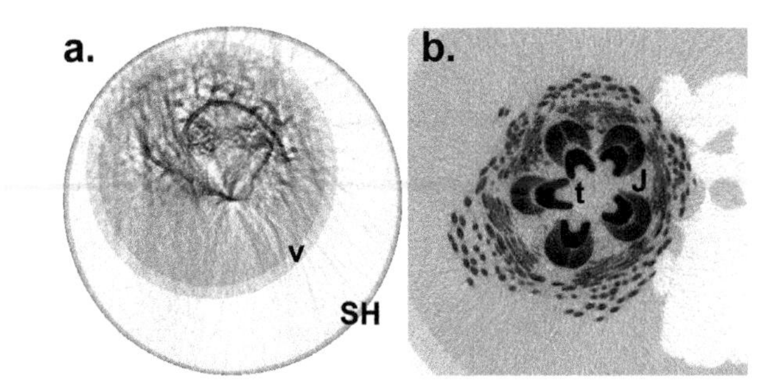

FIGURE 5.1
Motion artifacts illustrated. Here, the darker the pixel, the more absorbing the corresponding voxel. (a) Live sea urchin (*Lytechinus variegatus*) in seawater showing motion of the animal but not of the vial "v" containing the seawater and scanner sample holder "SH." 1,024 × 1,024 voxel field of view, 37 μm isotropic voxels, 250 projections of 1,024 samples, 0.3 s integration per projection, 45 kVp, 88 μA. Scanco MicroCT-40. (b) Fixed sea urchin (*Eucidaris tribuloides*) tooth "t," jaw structure "J" 512 × 512 voxel field of view, 37 μm isotropic voxels, 500 projections of 1,024 samples, 0.3 s integration per projection, 70 kVp, 57 μA. Scanco MicroCT-40.

inside of a specimen, particularly for soft tissues and for flexible foams that can relax (shift positions) under the influence of gravity or slight amounts of drying (or air currents within synchrotron radiation enclosures). Specimens sometimes need to be kept in fluid (e.g., soft tissue in phosphate-buffered saline or bone plus soft tissue in ethanol), and fixing the specimen in place may not be an issue but bubbles can form and move during scans, degrading the reconstructions. Degassing the fluid can help prevent bubble formation (Richter, Young et al. 2018). Removing most of the fluid, leaving only a small amount in the bottom of the tube and out of the beam, may help keep the specimen moist. In extreme cases, x-ray beam damage can distort the specimen during scanning.

In vivo microCT scanning of small animal thoraxes must correct for physiological motion (respiration and rapid heartbeat) because the magnitude of this motion is large enough to blur the surrounding structures. One approach is to use physiological monitoring and to collect multiple frames at each projection angle. After the scan, the projections recorded at specific points during the respiratory or cardiac cycle are gathered and used to reconstruct the structure at that time point. These methods can allow, for example, cardiac function to be computed from the 3D microCT images (Deel, Ridwan et al. 2016).

5.1.2 Ring Artifacts

Figure 5.2 shows ring artifacts in a synchrotron microCT slice, rings that exist even after white-field normalization has been performed. Figure 5.2b shows the result of a robust white-field normalization procedure: most of the rings are eliminated. Unfortunately, ring artifacts do not merely detract from the appearance of an image, but they also can interfere with accurate segmentation and quantification of the amount of phases present and their geometrical properties. Correction with a median filter often does not work particularly well, and use of a purpose-built filter taking advantage of the concentric nature of the rings seems to work well to minimize the rings (Bentz, Quenard et al. 2000). Ring reduction using a 21-point smooth to the average of all rows of the sinogram (2D plot showing transmitted intensity in the radiograph, for one slice, that is, one row of the radiograph, as a function of rotation angle, see Fig. 3.11) offers substantial improvement over uncorrected reconstructions (Rivers and Wang 2006): the actual high-frequency content of the slices does not appear to be affected, only the rings. Sinogram correction algorithms for ring

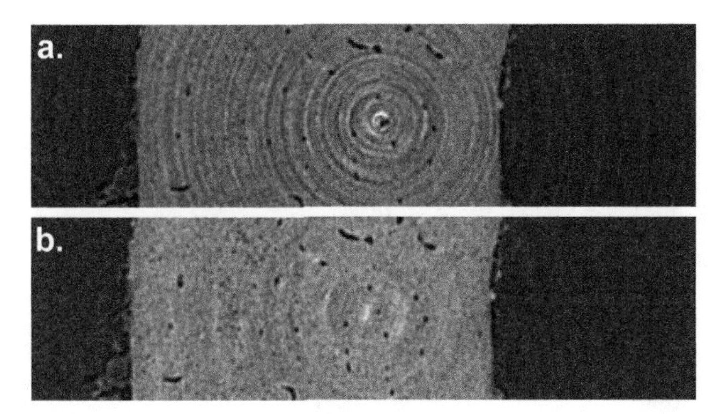

FIGURE 5.2
Section of a $(1K)^2$ reconstructed slice of a small section of bone cut from a rabbit femur showing ring artifacts. (a) Ring artifacts persisting after normal white-field correction. (b) Elimination of most of the ring artifacts after a more robust white-field correction. The lighter the pixel, the greater is the absorption of the corresponding voxel. 21 keV, 0.25° rotation increment, 5 μm isotropic voxels. (S.R. Stock, K. Ignatiev, N.M. Rajamannan, F. De Carlo. March 2003, 2-BM, APS.)

reduction have also been investigated for lab microCT data (Ketcham 2006). These and other artifacts in lab microCT (Davis and Elliott 2006) and in synchrotron microCT (Vidal, Letang et al. 2005) are discussed elsewhere.

5.1.3 Reconstruction Center Errors

Accurate reconstruction requires that the center of rotation be defined very precisely, to within a fraction of a voxel of that in the intended reconstruction. Commercial instruments employ calibration of sample holder centers done when the instrument is installed, and the center is determined for each holder size or type. Typically, manufacturers supply multiple holders of each size/type, and these are fabricated to such tight specifications that all holders of one type/size have identical centers of rotation. In at least one case, the manufacturer builds in a small translation stage on top of the specimen rotator that allows the user to check or repeat the system calibration. For synchrotron microCT instruments, centers of rotation must be determined for each data set for each specimen; this is true even if kinematic mounts are used, and is a consequence of the extreme flexibility required to accommodate many different types of specimens and rapid shifts in the setup for successive users. A recentering algorithm (Brunetti and De Carlo 2004) that seems to work quite well is based on knowledge of the form of artifacts from centering errors: tails of (apparent) mass extending from features like the corners of the specimen or voids or pores within the specimen (Fig. 5.3). These tails increase the number of voxels with nonzero values, and iterating through different trial centers for a representative slice of the volume allows one to select the "best" center for reconstructing the rest of the volume, even in the presence of beam fluctuations, noise, and low contrast. Donath et al. developed metrics for optimization of center-of-rotation corrections (Donath, Beckmann et al. 2006). The center of rotation can also be refined by eye, by using a single trial slice and producing a series of reconstructions with successively greater displacements from the expected rotation center. At 2-BM, APS, in March 2019, automatic center determination failed because local tomography was being performed; trial reconstructions were generated for each of 71 field of view (FOV) using rotation centers ±25 voxels from the nominal center using 0.5 voxel steps and the best center was selected for each FOV in less than 1 min. As Fig. 5.4 shows, even an

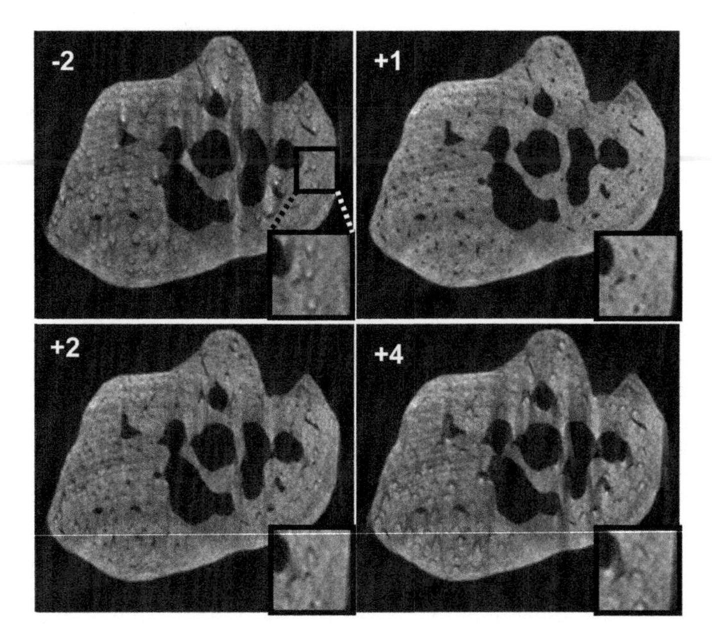

FIGURE 5.3
Illustration of errors introduced by incorrect reconstruction centers. Sections of the same slice reconstructed with translation of the center by the number of voxels indicated in the upper left portion of the image. The correct center (and clearest reconstruction) is with the +1 voxel center. Inset in the lower right of each image is a 2X enlargement of the area shown by the box in the upper left image. Mouse tibia in phosphate-buffered saline and the vertical field of view is 280 voxels. The lighter the pixel, the greater is the absorption of the corresponding voxel. 17 keV, 0.25° rotation increment, 5 μm isotropic voxels. (S.R. Stock, N.M. Rajamannan, X. Xiao, F. De Carlo. December 2005, 2-BM, APS.)

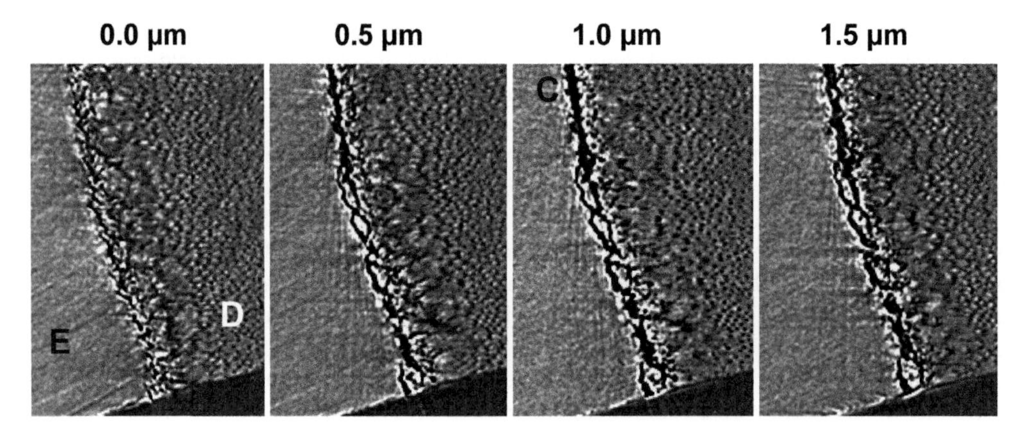

FIGURE 5.4
Effect of subvoxel changes in center of reconstruction illustrated using a section of a slice of the bovine dentinoenamel junction (DEJ). The reconstructed voxel size was slightly smaller than 2 μm, and the physical shifts of the rotation center are given above each image. The lighter the pixel, the more absorbing the voxel is. A crack "C" runs vertically between the enamel "E" and dentin "D." The dentin tubules, open cylinders surrounded by hypermineralized dentin and running nearly perpendicular to the plane of the slice and about the same diameter as the voxel size, are clearest in the reconstruction with the 1.0 μm shift. The horizontal field of view in each image is 220 voxels. 26 keV, 0.125° rotation increment, $(2K)^2$ reconstruction. (K. Ignatiev, S.R. Stock, F. De Carlo. October 2007, 2-BM, APS.)

inaccuracy in the center of rotation less than one voxel can appreciably degrade the quality of a reconstruction (Ignatiev, Stock et al. 2007).

5.1.4 Imperfections in the Optical System and X-ray Source

Nonidealities or imperfections in the components of the microCT optical system can affect reconstructed data. In well-designed systems, these deviations from ideality can be fairly subtle and can go unrecognized. The component coupling the x-ray scintillator to the area detector (lens in synchrotron microCT and fiberoptic plates, or direct application of the scintillator to the CCD, in lab microCT) is covered first followed by scintillators and by x-ray sources.

Distortions in the imaging chain of tube-based microCT systems are typically quantified and corrected (via software) during the manufacturer's initial calibration. As such, this is invisible to users of the instruments. Most synchrotron microCT systems use lenses to couple scintillators to area detectors (CCD or other devices). To this point in the text, nonidealities of these optical lenses have been ignored. Even the high-quality lenses used in microCT systems produce slightly distorted images, and the distortions considered here have radial symmetry. Image magnification can decrease with the distance from the lens axis (barrel distortion) or can increase with radius (pincushion distortion). Lens distortion in projections can be corrected immediately before the sinogram (or equivalent stage) (Nghia, Atwood et al. 2015), but this correction is not routinely performed at all synchrotron microCT imaging facilities because the effect on reconstructions can be small and often irrelevant to the users' applications, and the correction would need to be repeated every time a lens were replaced.

Figure 5.5a shows a cropped slice of a small piece of deer antler; the whiter voxels are areas where hypermineralized cartilage has not yet been replaced with bone, and the small, very dark disks are osteocyte and chondrocyte lacunae contained within the bone-like material. The enlargements of Fig. 5.5b, c show positions where the reconstruction center is correct and where the center is slightly off, respectively. As explained in Fig. 5.5d, e, lens distortion appears to be the origin of the apparent center of rotation difference.

The effect of lens distortion described in the previous paragraph applies to the central diameter of the lens (that is, the diameter perpendicular to the tomographic rotation axis). Features in the specimen that project above or below this plane also suffer radial distortion, but, in this case, the feature's apparent position moves out of its true position in the slice's line of projection, affecting reconstruction fidelity as discussed in Section 5.1.5. In the author's experience, the effect of lens distortion is generally noticeable only in situations where well-defined, small (5–10 voxels wide) structures like osteocyte lacunae are present throughout the reconstructed volume.

In tube-based microCT, the x-ray source position may change; for example, when a sealed x-ray tube is replaced (or the filament or target in a pumped x-ray source), or when thermal drift occurs between the different components of the scanner [particularly when the scanner is first turned on and during the first half hour that the tube is energized, see Sasov et al. (2008)]. For reflective targets, the movement of the emission point can be in the 10–50 μm range during the first quarter hour the tube is energized and is much better thereafter. The resulting reconstructions will be flawed. When tubes are changed (or filaments replaced), one should recalibrate the reconstruction center. For thermal movement, proper warmup time has been established by the manufacturer.

In synchrotron microCT, multiple white fields are typically collected to minimize the effect of random, one-time, high-intensity detector events, termed "zingers." Such zingers can produce strong ring artifacts. The panels of Fig. 5.6 show typical single white field

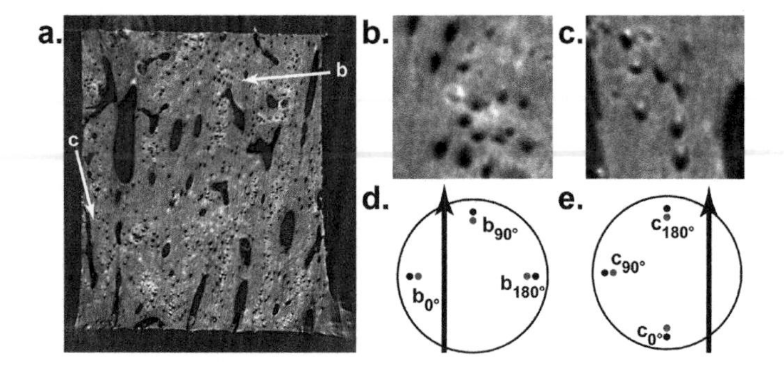

FIGURE 5.5
Images of a sample of deer antler illustrating effect of lens distortion on reconstruction quality. The selected reconstruction center was chosen to produce sharp images over the majority of the slice. (a) Cropped image of slice of the entire antler specimen. The ring artifacts (upper left corner) show the center of rotation. The arrows show the locations of the enlargements of panels (b) and (c). The field of view is 900 × 1,000 voxels (1.170 × 1.300 mm). (b) Enlargement of an area where the reconstruction center yields sharp images of osteocyte lacunae (round black features). (c) Enlargement of an area where lens distortion produces osteocyte lacunae images with characteristic double white tails extending from the sides of the (dark) lacunae, that is, the effect of an incorrect reconstruction center over 180° rotation range (if rotation were over 360°, the artifact would change to a circular white halo around the dark lacuna). The fields of view in (b) and (c) are 100 × 100 voxels (130 × 130 μm). (d) Schematic showing the apparent (dark disk) and true (gray disk) positions of feature "b" at three rotation angles ($b_{0°}$ shows positions at 0° angle, etc.). The arrowed line shows the direction of the x-ray beam. The centers of rotation of the gray and black disks projections onto the same position on the detector. (e) Schematic showing a feature "c" at four rotation angles. Here, due to the different starting position of "c" (relative to "b"), the centers of rotation of the gray and back disk project onto different detector positions, requiring a different rotation center for sharp images compared to that of "b." Imaging conditions were 22.7 keV, 0.12° rotation increment, $(2K)^2$ reconstruction, and 1.30 μm isotropic voxels. (S.R. Stock, H. Kierdorf, S. Gomez, U. Kierdorf, X. Xiao, March 2018, 2-BM, APS.)

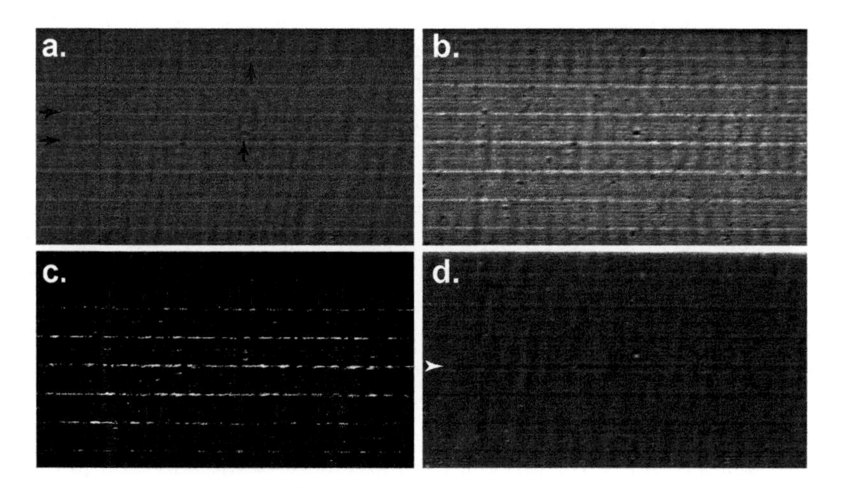

FIGURE 5.6
Synchrotron microCT white fields with lighter pixels indicating higher values. (a) One of the ten white fields collected. The horizontal arrows identify two of the bands of higher intensity from the double multilayer monochromator. The vertical arrows label imperfections in the scintillator. (b) Maximum intensities for all ten white fields projected into a single image and thresholded to show only the most intense pixels. (c) Average white field. (d) Standard deviation of the ten white fields normalized by the average white field. The arrowhead shows the center of the high intensity bands varies very little in intensity, but its borders vary much more. Imaging with 24.9 keV and a 5× lens objective lens. (S.R. Stock, C. Soriano, August, 2017, 2-BM, APS.)

collected for one specimen, the average of the ten white fields collected, the maximum intensity pixels observed in the set of ten white fields, and the ratio of the standard deviation of each pixel's values to the mean pixel value. The horizontal bands of high and low intensity are characteristic of the optics, and there are imperfections in/on the scintillator. As one would expect, the greatest variability occurs at the borders of the high-intensity bands and at the borders of the scintillator imperfections.

5.1.5 Mechanical Imperfections Including Rotation Stage Wobble

In current generations of microCT systems (employing ribbon or area beams), the only required specimen motion is rotation (translation along the rotation axis in order to enlarge the scanned volume can affect the quality of reconstructions, and this is discussed at the end of this section). Specimen axis wobble and rotation axis misalignment can be significant sources of error in reconstructions. Reconstruction software typically uses the pixels of a row of the detector as the input for a single slice's reconstruction. Tilt of the rotation axis from perpendicular to detector rows brings material from adjacent slices into and out of the beam for specific ranges of angles; this degrades the fidelity of the reconstruction. Such tilts are best avoided by very careful alignment (note tilts of even 0.03° over 2K pixels can shift projected data to an adjacent row), but this can be corrected by post-collection rotation of the projection (and resampling of the pixels) to align the rows precisely perpendicular to the actual rotation axis. Automatic routines for this geometric correction are described elsewhere (Weitkamp and Bleuet 2004). Radial lens distortion can also shift images of features from their correct slice.

Samples 80 mm in length or longer sometimes need to be imaged in their entirety, and commercial lab systems typically allow this to be done. The entire length cannot fit into a single FOV, and one needs to translate along rotation axis and collect a new set of projections to image the next FOV and repeat until the entire specimen is covered. Somewhat smaller vertical ranges are typically covered in synchrotron microCT systems. In both cases, keeping rotation axis precisely aligned over centimeters of vertical translation is mechanically challenging. In synchrotron microCT instruments that are constructed for flexibility, one must check the center of rotation for each successive FOV. In commercial lab microCT systems engineered for less flexibility, this does not appear to be an issue once proper calibration has been performed (see Chapter 4). Helical scanning is an alternative to collecting projections after vertical translations and is gaining popularity for situations where data are collected continuously and the area of interest is moving along the axis of the specimen; understanding the artifacts from combined changing structure and helical scanning is complex and, to the best of the author's knowledge, is only beginning to be investigated.

5.1.6 Undersampling

Figure 5.7 shows how angular undersampling can deleteriously affect reconstruction quality. The specimen extends from the lower right corner of the FOV, and air is seen to the left and above the specimen. For this $(1K)^2$ reconstruction, angular sampling is adequate with 0.25° between projections (Fig. 5.7a), and image quality decreases significantly with increasing angular step size (Fig. 5.7b–d) and is particularly evident in the air. Even with 1° between projections (Fig. 5.7b), it has become difficult to resolve the internal structure seen in Fig. 5.7a.

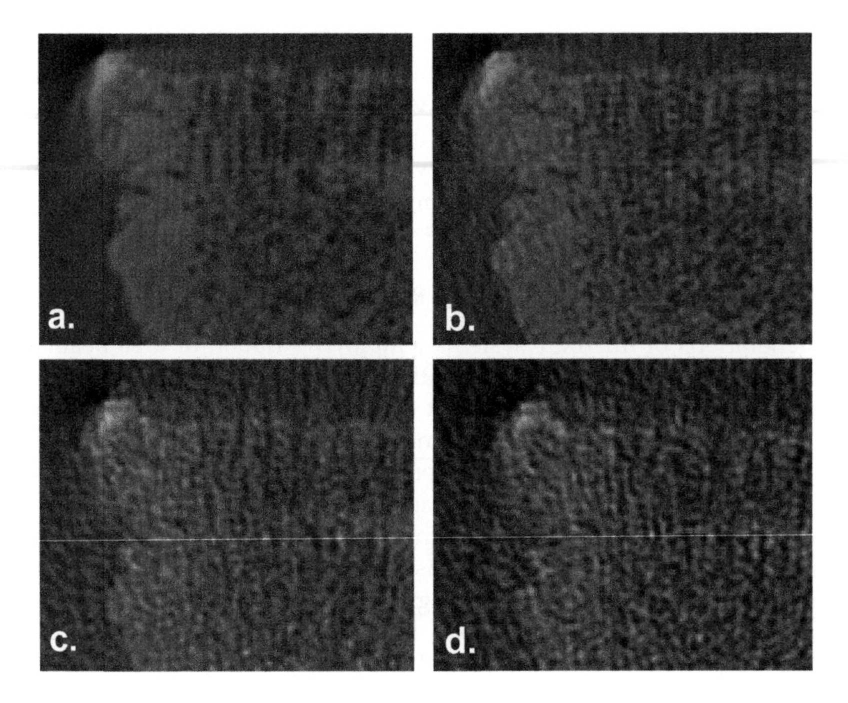

FIGURE 5.7
Illustration of the effect of angular sampling on reconstruction quality. In these small portions of the $(1K)^2$ slice, the lighter the pixel, the more absorbing is the corresponding voxel. (a) 0.25° increment between projections. (b) 1° increment. (c) 2° increment. (d) 4° increment. The same set of projections were used for each reconstruction, that is, all projections for (a), every fourth for (b), etc. Test plate of sea urchin *Lytechinus variegatus* (i.e., a section of the calcite mineralized ellipsoidal endoskeleton enclosing and protecting the urchin's critical organs). The horizontal field of view in each image is 150 voxels. Voxel size equaled 5 μm, and energy was 21 keV. (S.R. Stock, K. I. Ignatiev and F. De Carlo, 2-BM, APS, June 2003.)

5.1.7 Beam Hardening

As mentioned in Section 2.3, use of polychromatic radiation produces an effect called beam hardening: the average photon energy of the beam penetrating the sample increases with increasing sample thickness because the lower energy photons are absorbed at a much higher rate than the higher energy photons. Thus, Eqs. (3.1)–(3.3) are no longer strictly valid as written. Beam hardening combined with scattering leads to cupping in reconstructed slices of a uniform object, that is, a radial gradient in linear attenuation coefficient with abnormally low values at the interior and high values at the periphery. One must always check for this effect when analyzing data from lab microCT systems; only rarely must it be considered with synchrotron systems because these normally employ monochromatic radiation. Figure 5.8 shows a portion of a slice through an aluminum specimen; the green graph shows lower values of the linear attenuation coefficient in the interior compared to the outer portions of the specimen. Dual energy techniques, however, can correct for the effect of collecting data using a range of wavelengths (Engler and Friedman 1990).

5.1.8 Artifacts from High Absorption Features within a Specimen

Nonphysical streaks can radiate from high absorption objects within a lower attenuation matrix (such star artifacts are described in the medical CT literature; for example,

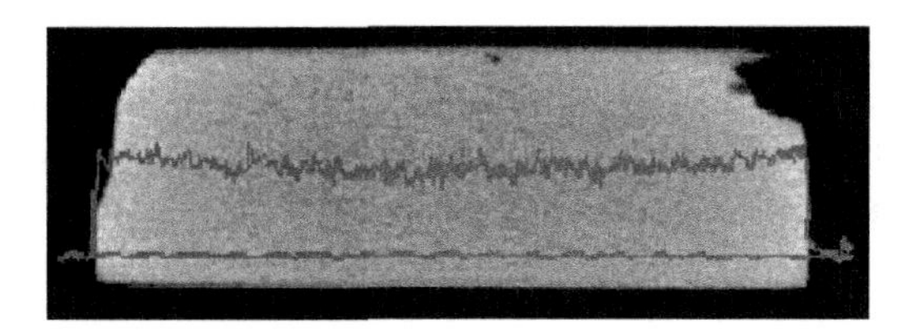

FIGURE 5.8
Beam hardening seen in an aluminum (AA 2090) sample imaged at 70 kVp. The line graph (green) shows the variation of linear attenuation coefficient as a function of position across the 7.45-mm wide specimen. Linear attenuation coefficients vary from 3 cm^{-1}, and significant cupping is seen in the middle of the specimen. Blue and black represent the lowest levels of attenuation. Imaging with a Scanco MicroCT-40 system: 500 projections each with 0.3 s integration time and 512 samples, reconstruction with 12 μm isotropic voxels. (S.R. Stock, K. Ignatiev.)

originating from metal implants in bone). Figure 5.9a shows streaks from wires of a small strain gage attached to a rat tibia. The wires are essentially opaque and affect the values of linear attenuation coefficient elsewhere in the specimen. Streaks can often be seen emanating from and parallel to long, flat specimen faces (Fig. 5.9b), and, if one has the option of designing specimen geometry, such features should be avoided. Streak artifacts generally are attributed to the effect of scatter. One wonders, however, whether the streak originates from the large, very sharp jump in attenuation from inside to outside of the specimen when the projection direction is parallel to the long specimen face; the finite sampling frequencies present in the data may be insufficient to track the jump adequately.

In orthopedics research, metal implants are often present and can cause streak effects. A major objective in orthopedic studies involving implants is measuring the extent to which

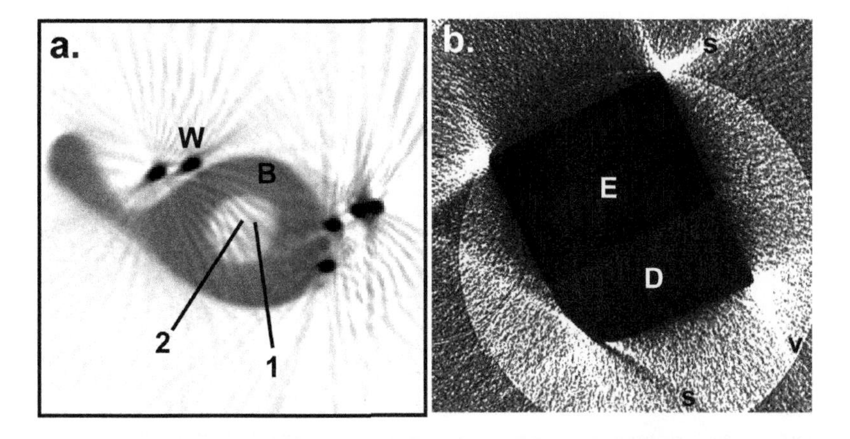

FIGURE 5.9
Streak artifacts. In both images, the darker the pixel, the more absorbing the voxel, and the slices were recorded with a Scanco MicroCT-40 system. (a) Streaks produced by high absorption features (strain gage wires "W"). The linear attenuation coefficients for the identified features are: W 8 cm^{-1}, bone B 2.9 cm^{-1}, high absorption streak 1 1.1 cm^{-1}, and low absorption streak 2 (adjacent to 1) 0.3 cm^{-1}. 70 kVp, 114 μA, 250 projections each with 0.3 s integration and 1,024 samples, reconstruction with 36 μm isotropic voxels, horizontal field of view of 9.2 mm. (b) Streaks "s" emanating from flat specimen surfaces into the surrounding air. The sample is in a small vial "v" and is a section of a bovine incisor containing enamel "E" and dentin "D." The field of view is 3.01 mm, and the histogram has been equalized to show the streaks more clearly. 45 kVp, 177 μA, 1,000 projections each with 0.3 s integration and 2048 samples, reconstruction with 12 μm isotropic voxels. From the data set reported in Vieira et al. (2006).

bone grows in contact with the metal implant. Typically, a strong artifact zone exists around metal implants, which can prevent microCT-based measurement of contact, but recent results suggest that the current generation of lab microCT scanners can overcome this effect and deliver accurate measurements of the amount of contact (Meagher, Parwani et al. 2018).

5.1.9 Artifacts in Truncated Data Sets

Truncated data consist of scans where (a) mass moves into and out of the FOV, (b) projections are missing, perhaps because the beam is blocked along certain projection angles, or (c) the specimen is too thick and absorbing for adequate transmission along certain directions. As mentioned in Chapter 3, geometry may be accurately reproduced in reconstructions from truncated data sets, but linear attenuation coefficient values may be inaccurate. As an example, consider the plate-like specimen of trabecular bone which has one dimension longer than can be fit within the FOV when imaging with high resolution (Xiao, Carlo et al. 2008). The slice in Fig. 5.10a was recorded with lower resolution so that the entire specimen was within the beam over the entire rotation range; the area within the dashed circle was within the FOV when the sample was imaged with higher resolution, that is, with local tomography. A portion of the uncorrected local tomography slice appears in

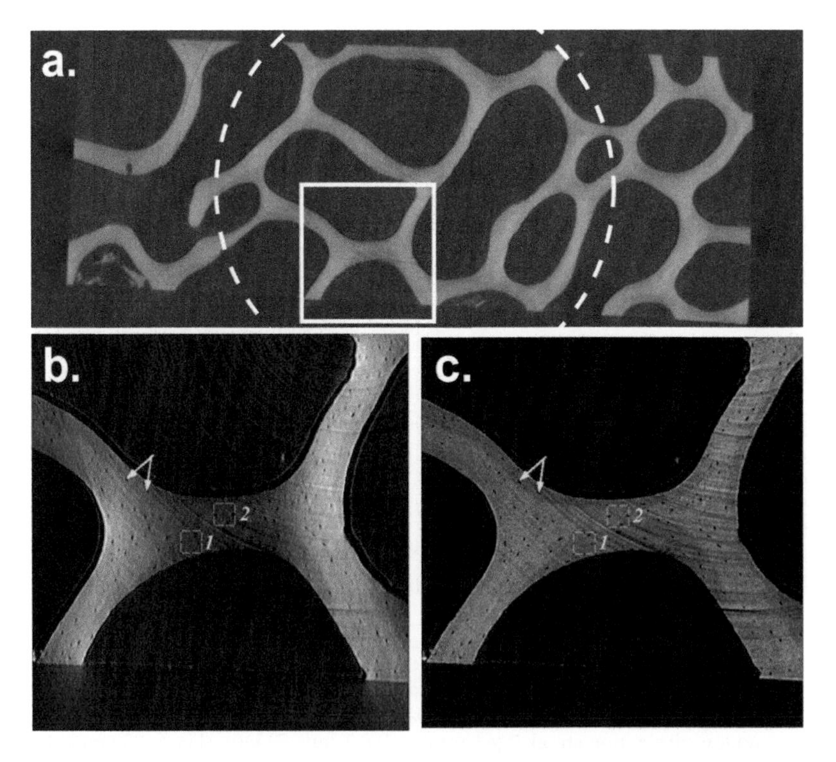

FIGURE 5.10
Local tomography of trabecular bone. (a) Lower-resolution scan of the entire specimen. The dashed circle indicates the FOV scanned at higher resolution. The box shows the area enlarged in (b) and (c). (b) Portion of the uncorrected high-resolution reconstruction. The two arrows point to osteocyte lacunae, and the two small boxes indicate areas compared in the original study. (c) Corrected high-resolution reconstruction. Data were collected with 20 keV photons at 2-BM, APS. Adapted from Xiao et al. (2008).

Fig. 5.10b (the area within the box in Fig. 5.10a). In Fig. 5.10b, the horizontally running trabecula has much lower linear attenuation coefficients than the vertical trabeculae (darker vs light voxels); note that this difference in contrast is not present in Fig. 5.10a and is related to the orientation of the plate. Figure 5.10c shows the same area of the high-resolution local tomography scan after correction for the missing mass.

Local tomography of sea urchin tooth shown in Fig. 3.13 has another type of artifact. The background in the enlargement of Fig. 3.13d is stippled; this does not reflect contrast within the plastic in which the sea urchin jaw is cast but rather angular undersampling. The effect of this stippling is to make it extremely difficult to segment the plates of the tooth.

5.1.10 Phase Contrast Artifacts

Synchrotron microCT data from sources such as APS, ESRF, SLS, and SPring-8 seem to invariably have a strong component of phase contrast in the reconstructions. The "hot edges" seen in radiographs along the projected boundary between materials with different electron density produce edge sharpening in the reconstructions; that is, this contrast can be thought of as a physical analog of the sharpening filters used in reconstruction software. Despite statements to the contrary (Ma, Boughton et al. 2016), these features are not beam hardening. Sometimes these phase effects can lead to anomalously large values of the linear attenuation coefficient in positions within a specimen that are not easily recognizable as being near surfaces. In one study (Stock, Barss et al. 2003a), such unexpected contrast (Fig. 5.11) was not recognized until later (Stock, Ignatiev et al. 2003b). This effect

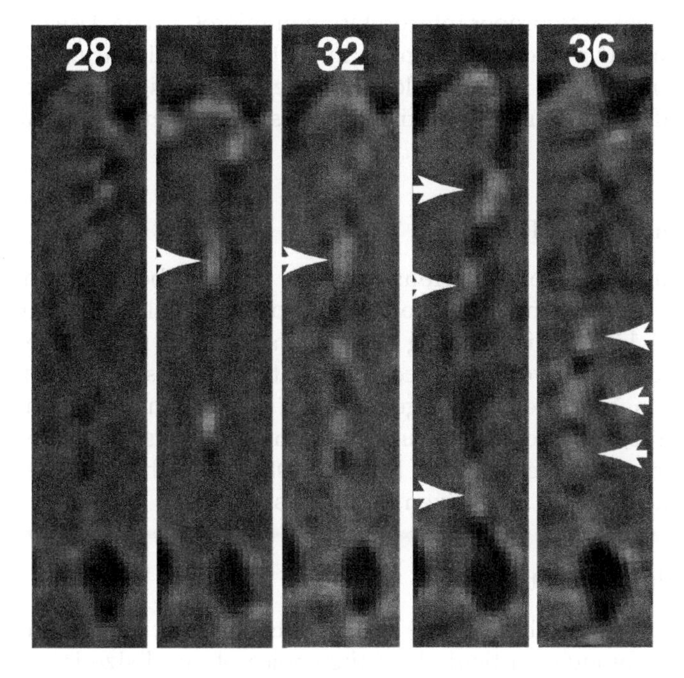

FIGURE 5.11
Anomalous contrast from phase effects seen in channels between calcite carinar process plates in a tooth of the sea urchin *Lytechinus variegatus*. Sections of slices 28, 30, 32, 34, and 36 are shown from left to right, respectively. The white features identified with arrowheads have values of the linear attenuation coefficient three times greater than is possible for calcite. 14 keV, $(1K)^2$ reconstruction, ~1.3 μm isotropic voxels. (Stock, Barss et al. 2003a.)

may or may not be responsible for contrast interpreted as solute segregation (e.g., high local Zn concentration in an Al engineering alloy (Ohgaki, Toda et al. 2006)) or as hypermineralization bands in bone (Stock 2008). Surfaces just below (or above) the slice plane can produce altered contrast at locations, based on the single slice, seemingly unrelated to surfaces. Bands of hypermineralized cartilage were seen in deer antler (Fig. 5.5a, b), and backscattered electron imaging and scanning x-ray excited x-ray fluorescence mapping confirm that highly mineralized regions were present. Considerable care must be taken, therefore, in the interpretation of voxel values in synchrotron microCT. In situations like those mentioned above, viewing a movie paging through a stack of slices can be very helpful.

5.2 Performance: Precision and Accuracy

The accuracy of microCT reconstructions has been considered in many papers. Many direct comparisons (i.e., microCT slices matched with physical sections) have established beyond doubt that microCT provides accurate reconstructions of specimen volumes; further, microCT's limitations are also well established. (That is not to say that microCT data cannot be, and has not been, misinterpreted or the technique misused or misapplied.) Also important is understanding what can be resolved (spatial resolution) and what can be detected (sensitivity limits). Knowledge of the dependence of spatial resolution and contrast sensitivity on experimental parameters as well as the intrinsic microstructural characteristics is central to proper interpretation of microCT data. Use of a physical phantom for calibrating 3D microCT systems is important (Perilli, Baruffaldi et al. 2006, Du, Umoh et al. 2007) and is an often overlooked necessity for high accuracy, high precision work. One manufacturer currently certifies every system using the QRM nano bar phantom (QRM 2019).

5.2.1 Correction for Nonidealities

Section 5.1 discussed not only various artifacts that can be found in CT reconstructions but also several methods for correcting for nonidealities in microCT data collection. Section 5.1.2 mentioned ring correction by knowledge-based smoothing of sinograms. Section 5.1.3 covered how correct rotation centers can be obtained for reconstructions, Section 5.1.4 mentioned geometrical correction for mechanical errors in microCT apparatus, and Section 5.1.9 indicated that errors associated with local tomography can be corrected. Increasingly, ML and DL approaches will be used to correct various errors encountered with given microCT systems.

5.2.2 Partial Volume Effects

Partial volume effects can significantly bias results obtained with microCT or any volumetric imaging technique, particularly if the choice of voxel size is ill-considered relative to the dimensions of the object of interest. Consider the two tubular specimens in Fig. 5.12 that have the same outer diameter (taken as 1.5 mm) but differing wall thicknesses (0.3 mm, left, and 0.55 mm, right). Both are sampled with the same size voxels (0.1 mm × 0.1 mm in the plane of the reconstruction), and, for simplicity, everything normal to the plane is ignored. For the left-hand specimen, 55% of the voxels are partial, but the

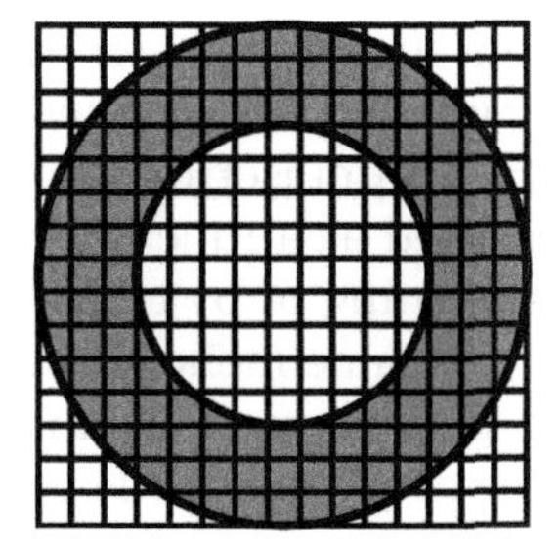
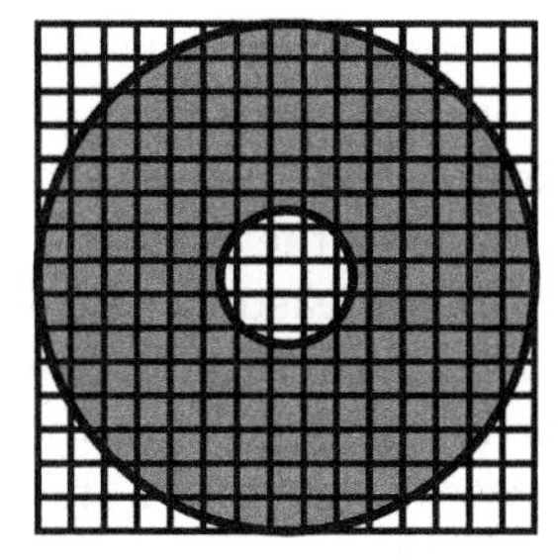

FIGURE 5.12
Two tubular specimens with the same outer diameter but different wall thicknesses sampled with identical voxel sizes.

right-hand specimen contains only 10.5% partial voxels. In a calculation partitioning the slice into solid and empty space using a specific fraction of occupancy, conclusions about the left-hand specimen will be much more sensitive to the selected threshold than the sample shown at the right.

The above example is *not* an academic exercise. Such considerations apply in studies comparing the response to treatment of mice from genetic strains with very different bone phenotypes but with comparable long bone outer diameters, particularly if the imaging is done with voxel dimensions a substantial fraction of the characteristic dimensions of the bone. Further, the clarity of actual scan data sets deteriorated rapidly as voxel size increased, but data sets artificially created from data collected at much higher resolution maintained their clarity to much higher voxel sizes (Cooper, Turinsky et al. 2007).

5.2.3 Detection Limits for High Contrast Features

In many cases, a microCT system may be required to detect the presence of features smaller than the tiniest possible voxel size. If the small features have high enough contrast (e.g., high Z precipitates in a low Z matrix or cracks in a homogenous solid) and are reasonably well dispersed, it is certainly possible to detect these features when they are smaller than the voxel size. Here, cracks in a solid are used as a paradigm for feature detection based on differences in absorption contrast.

Before discussing detection limits established using metal matrix composites, consider the requirements for reliably detecting cracks with very small openings, say 0.5 μm. Assuming that such a crack extends across the entire voxel and that the signal-to-noise ratio in the reconstruction is adequate for detection of one-quarter voxel of empty space, projections of the object must be recorded with 2-μm pixels in order for the sample to be reconstructed with the required voxel size for crack detection. If a 2K × 2K detector array was used, sample diameters up to 4 mm could be studied successfully. It is important to emphasize that detecting a 0.5-μm wide crack in a 4-mm thick specimen requires sensitivity to sample thickness changes of ~10^{-4}. On the one hand, cracks consist of spatially correlated voxels for low absorption, so this helps in detecting such narrow cracks. On the other hand, cracks often follow the interface between phases (i.e., reinforcement and matrix in composites), and this hinders detection.

Crack detectability has been studied in Al/SiC monofilament (uniaxial fiber alignment, ~140-μm fiber diameter with ~30-μm C core) composites using a thin wedge pushed into a sample parallel to the fiber axes (Breunig, Elliott et al. 1992, Breunig, Stock et al. 1993). Two orientations of the wedge were examined, perpendicular to the eight plies of the composite and 45° to the plies. A pencil beam system with a 10-μm diameter collimator, a Pd filter, and an energy-sensitive detector set for AgK_{α} radiation allowed slices perpendicular to the monofilament axes to be reconstructed with 12.5-μm isotropic voxels. Crack opening displacement in the plane of the plies was measured as a function of distance from the tip of the wedge by direct measurement of the opening or by comparing the separation of fibers on either side of the crack with that of the same monofilaments in the first slice beyond the tip of the crack. Results of the two measurement techniques agreed within 2 μm (within about 15% of one voxel). The wedge at 45° to the plies was inserted 2.1 mm into the ~1.5-mm × ~1.5-mm sample, which was not far enough to cause significant fiber fracture, but it did cause significant fiber displacement, fiber–matrix debonding, and ductile rupture of the Al matrix. The wedge perpendicular to the plies caused substantial fiber fracture. Cracks open as little as 1–2 μm, that is, 10% of a voxel's width, could be detected. Opening as a function of distance from the wedge tip was very different for the two geometries and reflected whether or not fiber fracture had occurred. With the 45° wedge, opening decreased fairly uniformly and quite rapidly until about 700 μm from the tip of the wedge, whereas with the 90° wedge, opening was roughly constant and much larger (than that in the 45° sample) between 100 and 1,000 μm from the wedge tip, after which the crack became invisible within 100 μm (Stock 1999).

The relatively well-behaved gradients in crack opening produced in the studies cited above allowed detailed assessment of how well crack opening can be quantified for a given level of noise in the reconstructed images. Besides controlling the amount of opening allowed, the fibers provided a built-in fiducial for measuring crack opening: the change in fiber–fiber separations. In other words, the difference in two monofilaments' separation across the crack at a particular position and that far from the crack tip should give an accurate measure of opening at the position in question. Most samples in which crack opening needs to be measured do not have these fiducials, so that crack opening must be quantified by summing the openings in adjacent partially open voxels. It is difficult to trust the robustness of such a procedure without at least once checking its results against those of a fiducial-based opening measurement; thus, the results of the wedge studies offer important guidance not only for how small an increment of crack opening can be quantified in monolithic samples but also for detection limits (in terms of partial voxels) for other high-contrast features (Stock 1999).

The results of Breunig and coworkers cited above have been confirmed in more recent synchrotron microCT studies of cracked specimens. Sometimes higher sensitivity can be obtained. At synchrotrons such as ESRF and APS, special effort is required to suppress phase contrast, but its presence substantially improves crack visibility beyond that possible with pure absorption contrast. Details are summarized elsewhere (Stock 2008).

5.2.4 Geometry

Assessing the accuracy of microCT reconstructions requires comparison of reconstructions with the results of another independent technique on the same specimen (Stock 2008). Lab microCT agreed with physical measurements of tooth dimensions in one study (Kim, Paik et al. 2007). Lab microCT vs microMRI (magnetic resonance imaging) provided one comparison (Borah, Gross et al. 2001), and synchrotron microCT vs scanning acoustic microscopy of osteonal bone furnished a second example verifying microCT's accuracy (Dalstra, Karaj et al. 2004). MicroCT quantification of porosity in a bone cement was found to be much more repeatable and robust with respect to segmentation thresholds than either radiography or optical microscopy (Cox, Wilcox et al. 2006). Bone formation in polymeric scaffolds was evaluated by proton magnetic resonance microscopy and microCT (Washburn, Weir et al. 2004). Physical sectioning compared to microCT slices has been the subject of still other studies confirming validity of reconstructions: confocal optical microscopy of thin serial sections contrasted with microCT of lung specimens (Kriete, Breithecker et al. 2001); histology vs microCT of cortical bone (Wacheter, Augat et al. 2001); and histomorphometry vs microCT of biopsies of cancellous bone (Cortet, Chappard et al. 2004, Chappard, Retailleau-Gaborit et al. 2005). Calcein labeling is a standard method used to show where new bone is formed, and, in a longitudinal in vivo study of rat tibiae, positions of calcein labeling matched positions where microCT showed new bone had formed (Waarsing, Day et al. 2004). Finite element modeling of microCT-derived bone structures has been validated by testing of rapid prototyping models produced from the microCT data (Su, Campbell et al. 2007).

MicroCT determination of cortical porosity and of mineral levels showed good agreement with the results of axial ultrasound velocity measurements in the human radius (Bossy, Talmant et al. 2004). MicroCT vs radiography of the same portions of human femora showed good agreement for measures of cortical porosity (Cooper, Matyas et al. 2004). Over a significant range of strains ($\sim 1 < \varepsilon < \sim 1.7$), Martin et al. found excellent agreement between the volume fraction of cavities measured with synchrotron microCT and macroscopic measurements of density (Martin, Josserond et al. 2000). The distribution of particle sizes in pumice clasts showed good agreement between synchrotron microCT and the classic crushing + sieving + winnowing method (Gualda and Rivers 2006). Yarn dimensions and spacings in 3D textiles and their variability were identical in measurements performed with lab microCT, surface scanning, and optical microscopy of cross-sections (Desplentere, Lomov et al. 2005).

Sheppard et al. found good agreement between simulations of mercury invasion capillary pressure based on 3D microCT pore quantification in four varied specimens and actual measurements performed on the same specimens, with the main differences being attributable to microporosity below the resolution limit in the reconstructions (Sheppard, Arns et al. 2006). In a specimen of packed, monodisperse beads (3.0-mm nominal diameter), the algorithms of Sheppard et al. determined a mean particle diameter of 2.98 mm (full width at half maximum ~0.06 mm, slightly less than one 63.4-μm voxel) in the reconstruction; in a second specimen of unconsolidated sandstone, the distribution of particle diameters from lab microCT agreed with results of laser light scattering (Sheppard, Arns et al. 2006).

Accuracy and precision of lung tumor volume measurements were determined from respiratory gated in vivo microCT of a mouse model (Cody, Nelson et al. 2005). Lung tumor volumes were both reproducible (2% operator variability) and accurate (6% average error), and tumor number assessed at necropsy correlated significantly with microCT. Relatively poor contrast between soft tissue types (tumors, blood vessels) was typical of absorption microCT, but the authors employed careful differentiation procedures. Spatial resolution was somewhat limited in both microCT (91-μm isotropic voxels) and in optical inspection (0.5-mm detection limit for tumors). Despite these limitations, this study is convincing, in no small part because of the thorough account provided.

Davis discussed expected image quality and accuracy in data obtained with a lab cone beam microCT system (Davis 1999); different specimen geometries were examined for different cone beam angles. Correction for beam hardening in microCT quantification and the effect of beam hardening on resolution have been discussed (Van De Casteele, Dyck et al. 2004). Determination of actual vs nominal resolving power was described elsewhere

(Seifert and Flynn 2002). Generation data for standards used for validation of experimental studies were described in Jiřík et al. (2018).

One investigation of the reproducibility of microCT data collection found the results stable with respect to replication, and displacement of the 6-mm-long ROI by up 4 mm along the axis of the trabecular cored specimen produced little change in microscopic parameters (Nägele, Vogt et al. 2004). Olurin et al. examined the dependence of morphometric indices for closed-cell Al foam and for two thicknesses of Al foil as a function of scan parameters in a lab microCT system and found characteristics such as volume fraction and mean feature thickness did not depend appreciably on voxel size for their specimens and scanner (Olurin, Arnold et al. 2002).

Examination of potential differences in reconstructions produced by different microCT instruments has been the subject of other studies. For example, results of 2nd generation vs 3rd generation synchrotron microCT as well as absorption vs phase microCT have been compared for cat claws (bone as well as tough cornified tissue) (Ham, Barnett et al. 2006). Comparisons of lab and synchrotron microCT are particularly informative for applications such as bone or tooth around metal implants where there is a large difference in absorptivity: One such study examined bone surrounding Ti-implants (Bernhardt, Scharnweber et al. 2004) and a second characterized bone around dental implants (Cattaneo, Dalstra et al. 2004).

Gureyev and coworkers have compared what they term quasi-local tomography with absorption and with phase contrast and found that accurate quantitative data could be obtained through quite small fractions of the specimen volume (Gureyev, Nesterets et al. 2007). A study comparing errors (and their spatial distribution) between reconstructions with local tomography, corrected local tomography, and lower-resolution microCT scans is also of interest (Xiao, Carlo et al. 2007).

Repeated imaging of the same trabecular bone specimen with three different systems (voxel sizes between 14 and 2 μm) showed that the larger size provided adequate parameterization of the trabecular structure (Peyrin, Salome et al. 1998), a result to be expected because trabeculae are typically ~100 μm thick, that is, on the order of seven voxels along the minimum dimension. The effect of different scanning and reconstruction voxel sizes on trabecular bone parameters was examined for one instrument, and, for the extreme case (voxel size of 110 μm vs mean trabecular thickness of 120 μm), differences in specific surface area (i.e., per unit volume of bone) were as large as 100% (Kim, Christopherson et al. 2004). This study suggests that morphometry studies performed on low-resolution pQCT systems should be evaluated very carefully before being accepted (Schmidt, Priemel et al. 2003) for discussion of circumstances where pQCT is accurate; Busignies et al. (2006) for a demonstration that actual low-resolution data is substantially less clear than equivalent resolution data artificially generated from high-resolution data sets; (MacNeil and Boyd 2007) and for a comparison of high-resolution pQCT, available for longitudinal studies of patients, and microCT).

In a study of liquid foam with synchrotron microCT, variation of segmentation threshold by ±3 units (on a 256 level gray scale) altered the volume fraction of liquid phase by on the order of ±2% (Lambert, Cantat et al. 2005). Lab microCT data were collected on trabecular bone biopsies (6-, 23-, and 230-week old porcine vertebrae); the three data sets were investigated systematically using a range of segmentation levels that an observed might select; the segmented scans were converted into FEM and a variation of 0.5% in threshold produced a 5% difference in bone volume fraction and a 9% difference in maximal stiffness for the most sensitive data (6-week sample with lowest bone volume fraction) (Hara, Tanck et al. 2002).

5.2.5 Linear Attenuation Coefficients

In many applications, reconstructed slices are presented in 0–255 contrast scale and analyzed with simple segmentation. High-definition studies require better contrast sensitivity, for example, where one is looking for spatial variation of mineralization levels in bone. One current area of interest in high-contrast sensitivity microCT centers on healthy and impaired bone, and discussion in this section will focus on mineralized tissue. Before turning to mineralized tissue, however, consider an inorganic material system, aluminum-silicon carbide, whose phases differ little in linear attenuation coefficient.

Figure 5.13 shows a portion of a slice of an Al/SiC composite sample imaged at 21 keV with 1.4-μm isotropic voxels. In addition to processing-related pores "p" within the Al matrix, the carbon cores at the center of each fiber are very clear, and two concentric

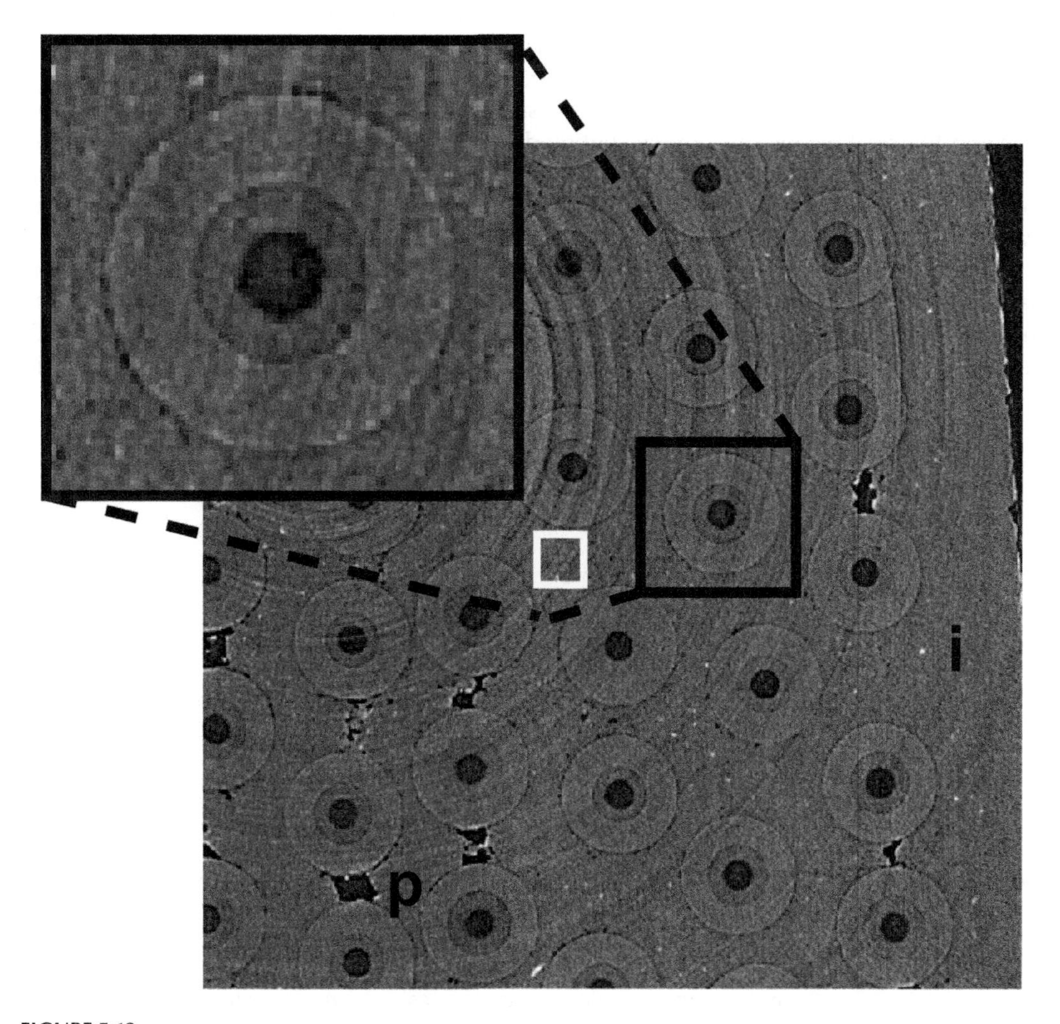

FIGURE 5.13
Slice of an Al/SiC composite showing high absorption impurities "i" and processing-related porosity "p" in the Al matrix. An enlargement (3X) of a single SiC fiber overlaps the upper left of the section of the slice. The small white box shows one area used for determining mean linear attenuation coefficient. 20 keV, 1.4-μm isotropic voxels, 0.25° rotation increment, 0.72-mm horizontal field of view. (S.R. Stock, X. Xiao and F. De Carlo, 2-BM, APS, March, 2006.)

zones of differing contrast are seen in most of the SiC fibers (inset image at upper left of Fig. 5.13). Spatial resolution is high enough to also allow visualization of the thin carbon layer coating the SiC. Within (40 voxel)^2 areas of the Al matrix away from pores or high absorption impurities "i" (e.g., the small white box in the figure) the mean linear attenuation coefficient ± one standard deviation was about 6.4 ± 1.7 cm^{-1}. Within 20×40 voxel area of the outer zone of the SiC fiber (this was the largest rectangular area that would not overlap with the inner zone of SiC), the value averaged about 6.9 ± 1.5 cm^{-1}. These values are similar (after conversion to the same x-ray energy) to those observed earlier (Kinney, Stock et al. 1990). The 7%–8% difference in linear attenuation coefficient provides enough contrast for the SiC fiber visibility, despite the variability in mean linear attenuation of ~25%. Although intrinsic material variability and ring artifacts contribute, the standard deviation values are larger than expected mainly because along the longest specimen direction, transmissivity was a bit greater than 5% in the 12-bit radiographs.

Figure 5.14 shows the radial variation of linear attenuation coefficient for a SiC monofilament fiber in a matrix of Al and identifies the different microstructural zones of the fiber (Kinney, Stock et al. 1990). Radial averaging over many slices was needed to overcome the noise in the data and to reveal the regions of very slightly different composition consistent with reported variations of stoichiometry (Nutt and Wawner 1985, Lerch, Hull et al. 1988). Often, changing the x-ray energy can improve contrast, but in the SiC and Al, linear attenuation coefficients track each other closely for the x-ray energies that can be used for microCT.

Several studies of mineralized tissue have interpreted specific values of linear attenuation coefficient. These include a comparison of quantitative backscattering imaging in the SEM vs microCT of mineralized tissue (Mechanic, Arnaud et al. 1990, Fearne, Elliott et al. 1994, Elliott, Anderson et al. 1997). Scanning acoustic microscopic

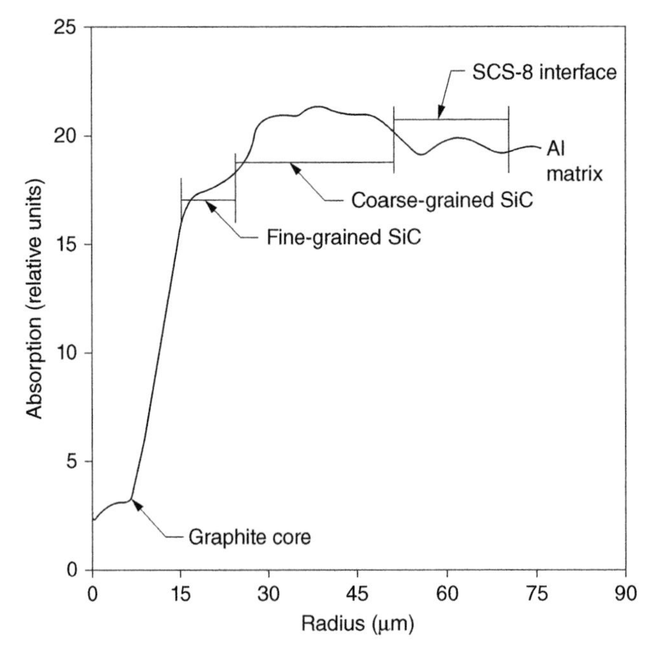

FIGURE 5.14
Radial variation of linear attenuation coefficient in an SiC monofilament in an Al/SiC composite. The differences reflect the different microstructures present. (Kinney, Stock et al. 1990, Breunig 1992.) © Materials Research Society.

maps of specimens have been compared to microCT data (Raum, Cleveland et al. 2006, Raum, Leguerney et al. 2006). Several quantitative studies have appeared on demineralization of tooth enamel (Elliott, Wong et al. 1998, Dowker, Elliott et al. 2004, Delbem, Vieira et al. 2006, Vieira, Delbem et al. 2006). Others have examined linear attenuation coefficients quantitatively in bone (Nuzzo, Lafage-Proust et al. 2002, Borah, Ritman et al. 2005).

Interpretation of linear attenuation coefficient μ is limited by irregular microstructural gradients, by partial volume effects, by counting statistics, and by other sources of noise. In analyzing the mean value $<\mu>$ for a given phase or region, an area typically is defined and the number average for the voxels and its standard deviation σ are computed. Increasing the number of photons sampling the specimen can improve the variance in μ, at least from this source. According to Eq. (3.8), increasing the number of counts transmitted through a specimen by a factor of four will decrease σ (due to counting statistics) by a factor of two. Contributions to the total variance from other sources are, of course, not affected.

Figure 5.15 presents a practical example of the effect of counting statistics on feature visibility in bone that has undergone some remodeling (i.e., replacement of preexisting bone

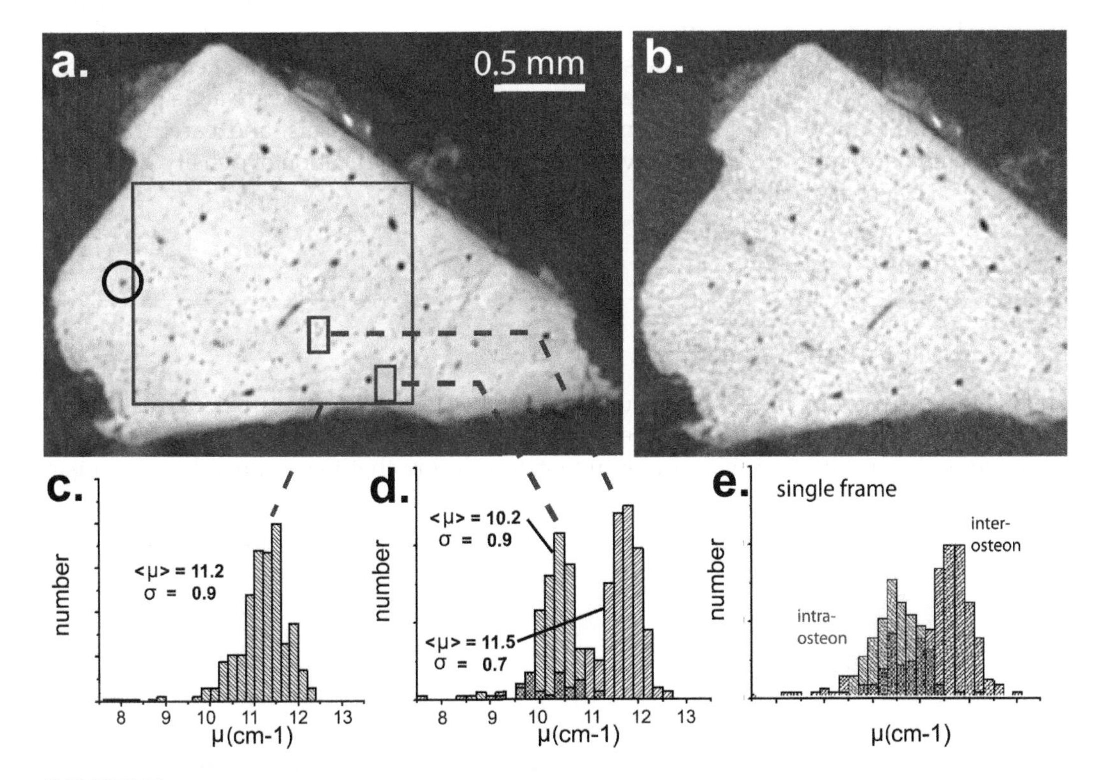

FIGURE 5.15
Effect of counting time on contrast sensitivity in a bone specimen (canine 2–3 year fibula) imaged with (a) 32-frame average and (b) single-frame average. The images are of the same slice and at the same magnification but (b) covers a slightly small horizontal field of view. The mineral level within remodeled osteons (e.g., that inside the black circle in (a) centered on the low absorption Haversian canal) is lower than that in the older bone between the osteons and is more clearly visible in (a) than in (b). (c) Histogram of linear attenuation coefficient values μ (mean $<\mu>$ and standard deviation σ) for the large boxed area in (a). (d) Histograms of the two small areas indicated in (a) (intra-osteonal and inter-osteonal bone) superimposed. (e) Histograms of the areas in (b) matching those in (d) showing the lower contrast sensitivity in the single frame data compared to the frame averaged data. 21 keV, 5-μm isotropic voxels, 0.25° rotation increment, 0.90-mm horizontal field of view. (S.R. Stock and F. De Carlo, 2-BM, APS, November, 2004).

by the action of osteoclasts and osteoblasts; see the figure caption for further details). The reconstructed slice in Fig. 5.15b was collected with a single frame of the CCD and the matching area in Fig. 5.15a with a 32-frame average. The remodeled osteons in Fig. 5.15a appear more clearly against the surrounding matrix of older, interosteonal bone than in Fig. 5.15b. Figure 5.15c shows the histogram of inter- and intra-osteon areas, that is, the overall histogram. Figure 5.15d superimposes histograms of an inter-osteon area and an intra-osteon area for the 32-frame averaged data, and Fig. 5.15e superimposes the same plots for the single-frame data. The histograms are more clearly separated in Fig. 5.15d than in Fig. 5.15e, and this explains the qualitatively better visibility. Consideration of experimental values of $\langle\mu\rangle$ and σ for different areas of the slices reveals quantitatively the extent of improvement. For the slice with the longer counting time (32-frame average), $\langle\mu\rangle \pm \sigma$ was $11.5 \pm 0.7\ cm^{-1}$ for the inter-osteon area and $10.2 \pm 0.9\ cm^{-1}$ for intra-osteon area. For the shorter counting time (single frame), the values were $11.3 \pm 0.8\ cm^{-1}$ for the inter-osteon area and $10.4 \pm 0.8\ cm^{-1}$ for intra-osteon area. With longer counting times, the standard deviations do not improve, certainly not by a factor of anything near $\sqrt{32}$, and the variance in this sample is not dominated by counting statistics but is probably determined by sub-voxel structural variability and other effects.

In data from tube-based microCT systems, beam hardening limits interpretation of gradients of density, particularly when the denser portions of the sample are at its periphery. This was the situation in a study of model pharmaceutical tablets, and careful correction for beam hardening appears to have allowed valid quantification of density gradients (Busignies, Leclerc et al. 2006).

5.3 Contrast Enhancement

High-contrast penetrant liquids can be used to enhance visibility of features such as cracks. Figure 5.16 shows matching slices of the keel of a tooth of the sea urchin *Arbacia punctulata* before and after infiltration with a polytungstate solution that has a much higher linear

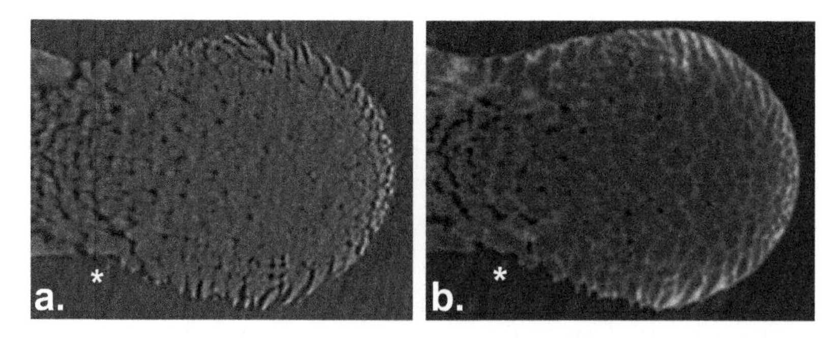

FIGURE 5.16
Contrast enhancement through the use of a high absorption penetrant. Keel of sea urchin tooth (*Arbacia punctulata*) (a) before and (b) after infiltration with a solution of 60% polytungstate (see Stock et al. (2003c) for details) for 3.9 × 103 s. In (b), the solution has filled many of the gaps between prisms in the keel (white polygonal network because the solution is much more absorbing than the calcite mineral) and greatly enhanced the visibility of the individual prisms. Note that the orientation of the two slices could only be closely matched for a small area (just above the asterisk). 21 keV, 1.8-μm isotropic voxels, 0.25° rotation increment, 0.90- mm horizontal field of view. (S.R. Stock, K. I. Ignatiev and F. De Carlo, 2-BM, APS, November, 2003.)

attenuation coefficient than the calcite of the tooth. The keel consists of an array of single crystal prisms running nearly normal to the plane of the slice. At the stage of mineralization shown, the prisms have not yet between cemented together, and the polytungstate has filled the gaps between prisms that were originally filled with soft tissue and fluid. The polytungstate delineated boundaries between prisms, white in Fig. 5.16b, are much clearer than before infiltration.

High atomic number gases such as Xe can also be used to enhance visibility of internal structures. Respiration of Xe, for example, has been used to map areas of active gas transport in lungs (Bayat, Le Duc et al. 2001, Bayat, Porra et al. 2006). Respired Xe enters into and persists within the fatty tissue of the brain and has been used to visualize the soft tissue of the brain.

Materials such as Microfil®, a silicone rubber containing lead chromate, are widely used to perfuse blood vessel networks in animals (Marxen, Thornton et al. 2004); Fig. 8.6b shows an example. Once the rubber hardens, microCT imaging can be used to quantify the network characteristics without further treatment of the tissue. Other options include injecting solution containing materials such as barium sulfate nanoparticles, and a variety of other contrast materials continue to be employed (Disney, Madi et al. 2017, Firouzi, Poursalehi et al. 2017, Kokkonen, Chin et al. 2017, Mok, Leow et al. 2017, Wang, Verboven et al. 2017, Kuva, Sammaljärvi et al. 2018).

Absorption edge difference imaging can increase sensitivity to small concentrations of the element of interest and is absolutely straightforward at most synchrotron imaging beam lines. Applications include transport in low-porosity materials (Altman, Peplinski et al. 2005) and in sands (Wildenschild, Hopmans et al. 2002), mapping of flame retardants (Br, Sb) in polymers (Butler, Ham et al. 2001), mapping Cs adsorption on iron oxide–hydroxide particles (Altman, Rivers et al. 2005), and mapping new bone formation through administration of Pb or Sr labels (Kinney and Ryaby 2001). Tetrachloroethane with 8 vol.% iodobenzene was used in model studies of organic, water-immiscible phase distribution in porous water-filled materials (Schnaar and Brusseau 2005, 2006). Multi-energy data collection and reconstruction algorithms have also received attention for materials where there are no convenient absorption edges (Ham, Willson et al. 2004).

Sensitivity limits to contrast agents have been investigated numerically. Sensitivity to a fixed concentration of KI in water was clearly much better in a coarse sand (mean particle diameter d_{50} = 0.58 mm in a 6 mm diameter sample) than in a fine sand (d_{50} = 0.17 mm in a 1.5-mm diameter sample) because the larger photon flux in the former produces a much higher signal-to-noise ratio (Wildenschild, Hopmans et al. 2002); this example is particularly compelling because the specimens are self-similar; that is, the relative sizes of pore and particle do not vary.

Altman et al. (2005) examined how well porosity could be quantified in various geological specimens; they employed fluid containing high-contrast ions (Cs or I) and compared reconstructions obtained above and below the ion's absorption edge. This difference-imaging approach maximizes sensitivity for the atomic species in question, and Altman et al. established detection limits and uncertainties using a series of solution concentrations in model, large-diameter specimens and examined the effect of pore diameter on sensitivity to Cs in absorption edge difference imaging (Altman, Peplinski et al. 2005).

As mentioned in an earlier chapter and as is covered in examples in later chapters, x-ray phase contrast is being used increasingly with samples where absorption contrast between different components is minimal. Soft tissue specimens, polymeric composites, and Al-SiC composites are examples.

5.4 Data Acquisition Challenges

The large amounts of data collected with lab microCT systems (potentially running 24 h of every day) and with synchrotron microCT (see below) are very challenging to handle. While the specifics vary from system to system and change constantly over time, one gains some idea of where micro- and nanoCT imaging is "going" by comparing past performance estimates with those when this text was being written. Laboratory microCT systems' evolution is covered first. Synchrotron microCT is described based on snapshots of the author's work at 2-BM, APS, in November 2006 and again in March 2019 and serves to illustrate the evolving challenges of exploding data collection rates. The coverage is meant to be illustrative and not exhaustive, and other micro- and nanoCT specialists may have different experiences.

Lab microCT performance is steadily increasing. For a 2001-era Scanco MicroCT-40 (near parallel beam design), data for 2K × 2K reconstructions of 40 slices could be acquired in 15 min if 300-ms integration per projection were used. Reconstructing these slices on the original computer workstation was considerably slower than the acquisition time. A redesigned Scanco MicroCT-40 system (cone beam design) with a larger detector improved data acquisition on a per-slice basis by a factor of 3.5–4 times: 2,000 projections covering 180° with 300-ms integration per projection required 42 min to collect data for 414 slices. The computer workstations supplied with the newer generation MicroCT-40 also cannot reconstruct as rapidly as the data are collected. Newer instruments like the Scanco MicroCT-50 (cone beam design) collecting data at higher definition than that in the above examples might collect FOV of 500 or more slices in about 1 h (detector integration plus various sources of overhead) and produce reconstructed slices of 4.65K × 4.65K. For a single FOV, the projection data can amount to 2–11 GB and the reconstructed slices 11–26 GB. Recording more of sample length, that is, additional FOV, would increase data sizes proportionally.

Comparing a 2-BM, APS, synchrotron microCT imaging session from 2006 with one from 2019 is very instructive. The majority of the description remains for the 2006 session with the 2019 information highlighting the enormous changes in hardware and software capabilities for routine operation. Given that 2-BM is on a bending magnet, its highest resolution mode (full-field, i.e., lens free) produces information-containing voxels of 0.65 μm and data collection rates that are surpassed at other storage rings, but obtaining beam time at these other facilities (and at the APS nanoCT station at 32-ID) is much more difficult than for 2-BM. Thus, 2-BM and other beam lines like it continue to fill an important niche.

In November 2006, the author scanned samples during nine 8-h shifts. The goal was imaging a large number of sea urchin spines for a detailed comparison of "design" variations within one phylogenic family. High throughput was essential, and this dictated that the 2-BM specimen placement robot (De Carlo, Xiao et al. 2006) would be used. One shift was lost to overnight problems with the robot, but otherwise the robot performed smoothly. A 2K × 2K detector and views every 0.25° were used (significant angular undersampling outside the central region of 1K voxels diameter). The detector integration time was ~ 0.2 s/view using the DMM, for a total of ~150 s of actual image acquisition over the ~1,200 s between start and end of data acquisition. The majority of the scan time was occupied mainly by sample motions and handshaking between various hardware components and by data transfer, with a major contribution of nonoptimal tuning of the various network and hardware components (De Carlo 2006), something which has greatly improved since 2006.

There was a significant difference between number of specimen FOVs actually imaged in eight shifts (91, 1.3 specimens/h, ~2,800 s/FOV) and the number expected (192 specimens, 3 specimens/h, 1,200 s/FOV). Unloading the previous batch of sample holders from the robot sample tray and loading the next batch of specimen holders (up to 24 specimens per tray) required about 10 min (the specimens were placed on the holders while the previous tray was being collected), so this was a negligible delay when spread over multiple specimens and multiple trays. Some "wasted" time was inevitable between finishing one tray and starting the next, but this was a very minor component. Entering sample names into the data collection software occupied a surprisingly large amount of time.

In 2006, reconstructing data at 2-BM, APS, was very rapid compared to other aspects of data handling, and, if the users multitasked, they could expect to leave the facility at the end of their shifts with a significant fraction of their data reconstructed and often in hand. Each stack of 2K reconstructed slices from a single FOV amounted to 20–25 GB with an additional 5–10 GB in raw projections, depending on the amount of dynamic compression possible. During the November 2006 run, transferring 1 TB of data (40–50 specimens) from the data analysis cluster (Linux) to an external USB-2 hard drive attached to a PC (Windows) required 40 h. When the drives were attached to a Linux machine and formatted as ext2, the transfer took 15 h (De Carlo 2006).

During writing (March 2019), the author received three shifts of beam time at 2-BM, APS. Data collection was with a 2.56K × 2.56K detector and with 1,400 slices per FOV (maximum extent of the beam at the imaging energy). The beam time nominally started at 8 am, but debugging new data collection scripts meant that data collection started at about 2 pm. Up to 10 FOV were collected on a sample, each FOV required about 5 min to acquire, and the motion between FOV on a sample was a few seconds. A total of 71 FOV were collected before all of the samples were imaged (about 1 am). A total of nine samples were imaged, and a small amount of time was required for sample replacement and realignment between specimens. Reconstructed slices were 25 MB in size, the raw data (projections) totaled 10.2 GB for each FOV, the reconstructed slices for each FOV amounted to 35 GB, and the total storage for the run summed to 3.2 TB.

Data collection rates will continue to increase, and efforts are being organized to cope with this (Parkinson, Gursoy et al. 2019). The GigaFRoST detector at the TOMCAT beam line at SLS (Mokso, Schlepütz et al. 2017), for example, produces data streams at 7.7 GB/s, which can continue over 24 h or longer producing tens of TB per day (Parkinson, Gursoy et al. 2019). One hopes that the overall infrastructure (processors for reconstruction, network, disks, etc.) keeps pace, or at least does not lag further behind. Of course, the biggest bottleneck of all is the lag between bringing the data home and actually doing something with the data (even just paging through slices).

5.5 Specimen Damage

X-radiation is ionizing and has the potential to damage specimens. For the vast majority of specimens, there is no effect. In in vivo imaging of animals, doses are kept quite low, but cellular damage has been noted with synchrotron microCT (Petruzzellis, Pagliarani et al. 2018). In bone samples that were later fractured, synchrotron microCT imaging affected measured toughness (Barth, Launey et al. 2010).

5.6 Speculations

Several trends for future micro- and nanoCT are clear from recently published studies. Consideration of what constitutes an "adequate" study is discussed in the next paragraph. The remainder of the section speculates on directions of future studies. Data representations in the literature have become increasingly sophisticated, and these are discussed in Chapter 6 after some of the quantification methodologies are introduced.

As the number of microCT publications have increased, expectations for the quality of and depth of analysis in published studies have risen. What was a strong PhD thesis in the early- to mid-1990s, became, by the late 1990s (in this author's opinion, as standards vary from discipline to discipline, institution to institution, or country to country), only an adequate MS thesis. The same is true of papers in archival journals. Although long explications of analysis methods and of the principles of microCT data collection remain appropriate for theses, very little of this should appear in journal papers, given previous coverage in the literature. One now expects not only interpretation of geometry defined via single thresholds of structures but also (brief but rigorous) consideration of numerical values of linear attenuation coefficients. If binary segmentation is used for numerical analysis, short but detailed examination of the effect of threshold choice should be incorporated; more complicated segmentation routines require presentation of more details, even if it is in appendices or supplementary material.

Future studies will incorporate more elaborate loading apparatus and environmental chambers (furnaces, cooling stages, high-pressure chambers) and more elaborate and better calibrated monitoring of experimental conditions. Peripherals purpose-built by manufacturers for their commercial microCT systems are available. Emphasis will surely continue on repeated observations of the same specimen: a wider range of in situ and in vivo studies with increased data acquisition rates. A guide to considerations in small animal imaging has appeared (Bouxsein, Boyd et al. 2010). More studies will appear on very fast phenomena using gating to freeze movement such as found in sprays; such gating is quite involved, however, so the number of such studies will remain relatively small, excepting perhaps in the area of in vivo small animal imaging.

Studies looking at evolution in the structure of individual specimens should emphasize incorporation of proper boundary conditions, which in practice means larger volumes of material surrounding the volume of interest. Incorporation of microstructure directly into finite element or other numerical models will be an area that will continue to grow. Reconstructions using 2K × 2K detectors are now standard, and introduction of 8K detector widths (in the plane of reconstruction) will be a direct approach for preserving spatial resolution while examining large diameter samples. One should not forget that increasing image definition by a factor of two requires an equal increase in the number of projections; decreasing the voxel size by a factor of two requires the collection of 2^4 more photons to obtain the same signal-to-noise ratio as in the original reconstruction. Local tomography is now widely used, but there will be limits in specimens with moderate absorption: limits will be defined by the decrease in contrast from the extra absorption of the material outside of the region of interest. More lab and synchrotron microCT systems are in place that can deliver high-energy photons, so the range of sample sizes (and of compositions, i.e., high Z) that can be examined will continue to expand.

MicroCT will be applied more often as part of studies integrating it with other scales of testing and analysis or with other techniques such as x-ray microbeam diffraction mapping. Such multimode studies are described in the following chapters on applications, and one should not forget that methods other than those employing x-rays can be used. One expects more studies will center on key specimens linking the microscale (samples with optimum dimensions for contrast sensitivity) with the macroscopic scale of more normal engineering specimens. Some intermediate-sized specimens may also need to be studied to complete the linkage between different structural scales. Although such studies are not as novel as they were a few years ago, the earlier demonstrations may actually make it easier to organize the resources required for more detailed, multiscale research programs.

The rate at which nanoCT and phase micro/nanoCT studies appear will also continue to increase: not only are more commercial nanoCT systems installed than when the first edition of this book was written but also commercial lab phase contrast microCT systems are now being delivered.

Since the first edition appeared, the instrument makers have considerably expanded their capabilities not just on the hardware side but also with software. The last 5 years before the second edition was written saw many of the small specialist companies acquired by larger firms; the long-term effect of the accompanying change in culture will be interesting. In the first edition, the author predicted a period of consolidation in the area of materials microCT characterization with new applications appearing but truly novel developments or applications being rare. Very rapid data acquisition of dynamic processes is one area that has become very active, and ptychography and other forms of coherent imaging appear poised to "take off" as mainstream methodologies.

References

Altman, S. J., W. J. Peplinski and M. L. Rivers (2005). "Evaluation of synchrotron x-ray computerized microtomography for the visualization of transport processes in low porosity materials." *J Contam Hydrol* **78**: 167–183.

Altman, S. J., M. L. Rivers, M. D. Reno, R. T. Cygan and A. A. Mclain (2005). "Characterization of adsorption sites on aggregate soil samples using synchrotron x-ray computerized microtomography." *Env Sci Technol* **39**: 2679–2685.

Barth, H. D., M. E. Launey, A. A. MacDowell, J. W. Ager and R. O. Ritchie (2010). "On the effect of x-ray irradiation on the deformation and fracture behavior of human cortical bone." *Bone* **46**: 1475–1485.

Bayat, S., G. Le Duc, L. Porra, G. Berruyer, C. Nemoz, S. Monfraix, S. Fiedler, W. Thomlinson, P. Suortti, C. G. Standertskjold-Nordenstam and A. R. A. Sovijarvi (2001). "Quantitative functional lung imaging with synchrotron radiation using inhaled xenon as contrast agent." *Phys Med Biol* **46**: 3287–3299.

Bayat, S., L. Porra, H. Suhonen, C. Nemoz, P. Suortti and A. R. A. Sovijärvi (2006). "Differences in the time course of proximal and distal airway response to inhaled histamine studied by synchrotron radiation CT." *J Appl Physiol* **100**: 1964–1973.

Bentz, D. P., D. A. Quenard, H. M. Kunzel, J. Baruchel, F. Peyrin, N. S. Martys and E. J. Garboczi (2000). "Microstructure and transport properties of porous building materials. II: Three-dimensional x-ray tomographic studies." *Mater Struct* **33**: 147–153.

Bernhardt, R., D. Scharnweber, B. Müller, P. Thurnier, H. Schliephake, P. Wyss, F. Beckmann, J. Goebbels and H. Worch (2004). "Comparison of microfocus and synchrotron x-ray tomography for the analysis of osteointegration around Ti6AlV4 implants." *Euro Cells Mater* **7**: 42–51.

Borah, B., G. J. Gross, T. E. Dufresne, T. S. Smith, M. D. Cockman, P. A. Chmielewski, M. W. Lundy, J. R. Hartke and E. W. Sod (2001). "Three-dimensional microimaging (MRμI and μCT), finite element modeling, and rapid prototyping provide unique insights into bone architecture in osteoporosis." *Anat Rec* **265**: 101–110.

Borah, B., E. L. Ritman, T. E. Dufresne, S. M. Jorgensen, S. Liu, J. Sacha, R. J. Phipps and R. T. Turner (2005). "The effect of risedronate on bone mineralization as measured by microcomputed tomography with synchrotron radiation: Correlation to histomorphometric indices of turnover." *Bone* **37**: 1–9.

Bossy, E., M. Talmant, F. Peyrin, L. Akrout, P. Cloetens and P. Laugier (2004). "An in vitro study of the ultrasonic axial transmission technique at the radius: 1 MHz velocity measurement are sensitive to both mineralization and intracortical porosity." *J Bone Miner Res* **19**: 1548–1556.

Bouxsein, M. L., S. K. Boyd, B. A. Christiansen, R. E. Guldberg, K. J. Jepsen and R. M. ller (2010). "Guidelines for assessment of bone microstructure in rodents using micro-computed tomography." *J Bone Miner Res* **25**: 1468–1486.

Breunig, T. M. (1992). Nondestructive evaluation of damage in SiC/Al metal/matrix composite using X-ray tomographic microscopy. Ph.D. thesis, Georgia Institute of Technology.

Breunig, T. M., J. C. Elliott, S. R. Stock, P. Anderson, G. R. Davis and A. Guvenilir (1992). Quantitative characterization of damage in a composite material using x-ray tomographic microscopy. *X-ray Microscopy III*. A. G. Michette, G. R. Morrison and C. J. Buckley. New York, Springer: 465–468.

Breunig, T. M., S. R. Stock, A. Guvenilir, J. C. Elliott, P. Anderson and G. R. Davis (1993). "Damage in aligned fibre SiC/Al quantified using a laboratory x-ray tomographic microscope" *Composites* **24**: 209–213.

Brunetti, A. and F. De Carlo (2004). A robust procedure for determination of center of rotation in tomography. *Developments in X-ray Tomography IV*. U. Bonse. Bellingham (WA), SPIE. **SPIE Proc Vol 5535**: 652–659.

Busignies, V., B. Leclerc, P. Porion, P. Evesque, G. Couarraze and P. Tchoreloff (2006). "Quantitative measurements of localized density variations in cylindrical tablets using x-ray microtomography." *Euro J Pharm Biopharm* **64**: 38–50.

Butler, L. G., K. Ham, H. Jin and R. L. Kurtz (2001). Tomography at the Louisiana State University CAMD synchrotron: Application to polymer blends. *Developments in X-ray Tomography III*. U. Bonse. Bellingham (WA), SPIE. **SPIE Proc Vol 4503**: 54–61.

Cattaneo, P. M., M. Dalstra, F. Beckmann, T. Donath and B. Melsen (2004). Comparison of conventional and synchrotron-radiation-based microtomography of bone around dental implants. *Developments in X-ray Tomography IV*. U. Bonse. Bellingham (WA), SPIE. **SPIE Proc Vol 5535**: 757–764.

Chappard, D., N. Retailleau-Gaborit, E. Legrand, M. Baslé and M. Audran (2005). "Comparison insight bone measurements by histomorphometry and μCT." *J Bone Miner Res* **20**: 1177–1184.

Cody, D. D., C. L. Nelson, W. M. Bradley, M. Wislez, D. Juroske, R. E. Price, X. Zhou, B. N. Bekele and J. M. Kurie (2005). "Murine lung tumor measurement using respiratory-gated microComputed Tomography." *Invest Radiol* **40**: 263–269.

Cooper, D. M. L., J. R. Matyas, M. A. Katzenberg and B. Hallgrimsson (2004). "Comparison of microcomputed tomographic and microradiaographic measurements of cortical bone porosity." *Calcif Tiss Int* **74**: 437–447.

Cooper, D., A. Turinsky, C. Sensen and B. Hallgrimsson (2007). "Effect of voxel size on 3D microCT analysis of cortical bone porosity." *Calcif Tiss Int* **80**: 211–219.

Cortet, B., D. Chappard, N. Boutry, P. Dubois, A. Cotton and X. Marchandise (2004). "Relationship between computed tomographic image analysis and histomorphometry for microarchitectural characterization of human calcaneus." *Calcif Tiss Int* **75**: 23–31.

Cox, B. D., R. K. Wilcox, M. C. Levesley and R. M. Hall (2006). "Assessment of a three-dimensional measurement technique for the porosity evaluation of PMMA bone cement." *J Mater Sci Mater Med* **17**: 553–557.

Dalstra, M., E. Karaj, F. Beckmann, T. Andersen and P. M. Cattaneo (2004). Osteonal mineralization patterns in cortical bone studied by synchrotron-radiation-based computed microtomography and scanning acoustic microscopy. *Developments in X-ray Tomography IV*. U. Bonse. Bellingham (WA), SPIE. **SPIE Proc Vol 5535**: 143–151.

Davis, G. R. (1999). Image quality and accuracy in x-ray microtomography. *Developments in X-ray Tomography II*. U. Bonse. Bellingham (WA), SPIE. **SPIE Proc Vol 3772**: 147–155.

Davis, G. R. and J. C. Elliott (2006). "Artifacts in x-ray microtomography of materials." *Mater Sci Technol* **22**: 1011–1018.

De Carlo, F. (2006).

De Carlo, F., X. Xiao and B. Tieman (2006). X-ray tomography system, automation and remote access at beamline 2-BM of the Advanced Photon Source. *Developments in X-ray Tomography V*. U. Bonse. Bellingham (WA), SPIE. **SPIE Proc Vol 6318**: 63180K-1–63180K-13.

Deel, E. v., Y. Ridwan, J. N. v. Vliet, S. Belenkov and J. Essers (2016). "In vivo quantitative assessment of myocardial structure, function, perfusion and viability using cardiac micro-computed tomography." *J Vis Exp* **108**: e53603.

Delbem, A. C. B., A. E. M. Vieira, K. T. Sassaki, M. L. Cannon, S. R. Stock, X. Xiao and F. De Carlo (2006). Quantitative analysis of mineral content in enamel using synchrotron microtomography and microhardness analysis. *Developments in X-ray Tomography V*. U. Bonse. Bellingham (WA), SPIE. **SPIE Proc Vol 6318**: 631824-1–631824-5.

Desplentere, F., S. V. Lomov, D. L. Woerdeman, I. Verpoest, M. Wevers and A. Bogdanovich (2005). "MicroCT characterization of variability in 3D textile architecture." *Compos Sci Technol* **65**: 1920–1930.

Disney, C. M., K. Madi, A. J. Bodey, P. D. Lee, J. A. Hoyland and M. J. Sherratt (2017). "Visualising the 3D microstructure of stained and native intervertebral discs using X-ray microtomography." *Sci Rep* **7**: 16279.

Donath, T., F. Beckmann and A. Schreyer (2006). Image metrics for the automated alignment of microtomography data. *Developments in X-ray Tomography V*. U. Bonse. Bellingham (WA), SPIE. **SPIE Proc Vol 6318**: 631818-1–631818-9.

Dowker, S. E. P., J. C. Elliott, G. R. Davis, R. M. Wilson and P. Cloetens (2004). "Synchrotron x-ray microtomographic investigation of mineral concentrations at micrometer scale in sound and carious enamel." *Caries Res* **38**: 514–522.

Du, L. Y., J. Umoh, H. N. Nikolov, S. I. Pollmann, T. Y. Lee and D. W. Holdsworth (2007). "A quality assurance phantom for the performance evaluation of volumetric microCT systems." *Phys Med Biol* **52**: 7087–7108.

Elliott, J. C., P. Anderson, G. R. Davis, F. S. L. Wong, S. E. P. Dowker and N. Kozul (1997). Microtomography in medicine and related fields. *Developments in X-ray Tomography*. U. Bonse. Bellingham (WA), SPIE. **3149**: 2–12.

Elliott, J. C., F. S. L. Wong, P. Anderson, G. R. Davis and S. E. P. Dowker (1998). "Determination of mineral concentration in dental enamel from x-ray attenuation measurements." *Conn Tiss Res* **38**: 61–72.

Engler, P. and W. D. Friedman (1990). "Review of dual-energy computed tomography techniques." *Mater Eval* **48**: 623–629.

Fearne, J. M., J. C. Elliott, F. S. Wong, G. R. Davis, A. Boyde and S. J. Jones (1994). "Deciduous enamel defects in low-birth-weight children: Correlated x-ray microtomographic and backscattered electron imaging study of hypoplasia and hypomineralization." *Anat Embryol* **189**: 375–381.

Feldkamp, L. A., L. C. Davis and J. W. Kress (1984). "Practical cone-beam algorithm." *J Opt Soc Am* **A1**: 612–619.

Firouzi, M., R. Poursalehi, H. H. Delavari, F. Saba and M. A. Oghabian (2017). "Chitosan coated tungsten trioxide nanoparticles as a contrast agent for x-ray computed tomography." *Int J Biol Macromol* **98**: 479–485.

Gualda, G. A. R. and M. Rivers (2006). "Quantitative 3D petrography using x-ray tomography: Application to Bishop Tuff pumice clasts." *J Volcanol Geothermal Res* **154**: 46–62.

Gureyev, T. E., Y. I. Nesterets and S. C. Mayo (2007). "Quantitative quasi-local tomography using absorption and phase contrast." *Opt Comm* **280**: 39–48.

Ham, K., H. A. Barnett, T. Ogunbakin, D. G. Homberger, H. H. Bragulla, K. L. Matthews, C. S. Willson and L. G. Butler (2006). Imaging tissue structures: Assessment of absorption and phase contrast x-ray tomography imaging at 2nd and 3rd generation synchrotrons. *Developments in X-ray Tomography V.* U. Bonse. Bellingham (WA), SPIE. **SPIE Proc Vol 6318**: 631822-1–631822-10.

Ham, K., C. S. Willson, M. L. Rivers, R. L. Kurtz and L. G. Butler (2004). Algorithms for three-dimensional chemical analysis with multi-energy tomographic data. *Developments in X-ray Tomography IV.* U. Bonse. Bellingham (WA), SPIE. **SPIE Proc Vol 5535**: 286–292.

Hara, T., E. Tanck, J. Homminga and R. Huiskes (2002). "The influence of microcomputed tomography threshold variations on the assessment of structural and mechanical trabecular bone properties." *Bone* **31**: 107–109.

Ignatiev, K., S. R. Stock and F. De Carlo (2007).

Jiřík, M., M. Bartoš, P. Tomášek, A. Malečková, T. Kural, J. Horáková, D. Lukáš, T. Suchý, P. Kochová, M. Hubálek Kalbáčová, M. Králíčková and Z. Tonar (2018). "Generating standardized image data for testing and calibrating quantification of volumes, surfaces, lengths, and object counts in fibrous and porous materials using x-ray microtomography." *Microsc Res Tech* **81**: 551–568.

Ketcham, R. A. (2006). New algorithms for ring artifact removal. *Developments in X-ray Tomography V.* U. Bonse. Bellingham (WA), SPIE. **SPIE Proc Vol 6318**: 63180O-1–63180O-7.

Kim, D. G., G. T. Christopherson, X. N. Dong, D. P. Fyhrie and Y. N. Yeni (2004). "The effect of microcomputed tomography scanning and reconstruction voxel size on the accuracy of stereological measurements in human cancellous bone." *Bone* **35**: 1375–1382.

Kim, I., K. S. Paik and S. P. Lee (2007). "Quantitative evaluation of the accuracy of micro-computed tomography in tooth measurement." *Clin Anat* **20**: 27–34.

Kinney, J. H. and J. T. Ryaby (2001). "Resonant markers for noninvasive, three-dimensional dynamic bone histomorphometry with x-ray microtomography." *Rev Sci Instrum* **72**: 1921–1923.

Kinney, J. H., S. R. Stock, M. C. Nichols, U. Bonse, T. M. Breunig, R. A. Saroyan, R. Nusshardt, Q. C. Johnson, F. Busch and S. D. Antolovich (1990). "Nondestructive investigation of damage in composites using x-ray tomographic microscopy." *J Mater Res* **5**: 1123–1129.

Kokkonen, H. T., H. C. Chin, J. Töyräs, J. S. Jurvelin and T. M. Quinn (2017). "Solute transport of negatively charged contrast agents across articular surface of injured cartilage." *Ann Biomed Eng* **45**: 973–981.

Kriete, A., A. Breithecker and W. Rau (2001). 3D imaging of lung tissue by confocal microscopy and microCT. *Three-Dimensional and Multidimensional Microscopy: Image Acquisition and Processing VIII.* J. A. Conchello, C. J. Cogswell and T. Wilson, SPIE. **SPIE Proc Vol 4261**: 40–47.

Kuva, J., J. Sammaljärvi, J. Parkkonen, M. Siitari-Kauppi, M. Lehtonen, T. Turpeinen, J. Timonen and M. Voutilainen (2018). "Imaging connected porosity of crystalline rock by contrast agent-aided x-ray microtomography and scanning electron microscopy." *J Microsc* **270**: 98–109.

Lambert, J., I. Cantat, R. Delannay, A. Renault, F. Graner, J. A. Glazier, I. Veretennikov and P. Cloetens (2005). "Extraction of relevant physical parameters from 3D images of foams obtained by x-ray tomography." *Colloids Surf A* **263**: 295–302.

Lerch, B. A., D. R. Hull and T. A. Leonhardt (1988). As-received microstructures of a SiC/Ti-15-3 composite, NASA Lewis.

Ma, S., O. Boughton, A. Karunaratne, A. Jin, J. Cobb, U. Hansen and R. Abel (2016). "Synchrotron imaging assessment of bone quality." *Clin Rev Bone Miner Metab* **14**: 150–160.

MacNeil, J. A. and S. K. Boyd (2007). "Accuracy of high-resolution peripheral quantitative computed tomography for measurement of bone quality." *Med Eng Phys* **29**: 1096–1105.

Martin, C. F., C. Josserond, L. Salvo, J. J. Blandin, P. Cloetens and E. Boller (2000). "Characterization by x-ray micro-tomography of cavity coalescence during superplastic deformation." *Scripta Mater* **42**: 375–381.

Marxen, M., N. M. Thornton, C.B. Chiarot, G. Klement, J. Koprivnikar, J. G. Sled and R. M. Henkelman (2004). "MicroCT scanner performance and considerations for vascular specimen imaging." *Med Phys* **31**: 305–313.

Meagher, M. J., R. N. Parwani, A. S. Virdi and D. R. Sumner (2018). "Optimizing a micro-computed tomography-based surrogate measurement of bone-implant contact." *J Orthop Res* **36**: 979–986.

Mechanic, G. L., S. B. Arnaud, A. Boyde, T. G. Bromage, P. Buckendahl, J. C. Elliott, E. P. Katz and G. N. Durnova (1990). "Regional distribution of mineral and matrix in the femurs of rats flown on Cosmos 1887 biosatellite." *FASEB J* **4**: 34–40.

Mok, P. L., S. N. Leow, A. E. Koh, H. H. Mohd Nizam, S. L. Ding, C. Luu, R. Ruhaslizan, H. S. Wong, W. H. Halim, M. H. Ng, R. B. Idrus, S. R. Chowdhury, C. M. Bastion, S. K. Subbiah, A. Higuchi, A. A. Alarfaj and K. Y. Then (2017). "Micro-computed tomography detection of gold nanoparticle-labelled mesenchymal stem cells in the rat subretinal layer." *Int J Mol Sci* **18**: 345.

Mokso, R., C. M. Schlepütz, G. Theidel, H. Billich, E. Schmid, T. Celcer, G. Mikuljan, L. Sala, F. Marone, N. Schlumpf and M. Stampanoni (2017). "GigaFRoST: The gigabit fast readout system for tomography." *J Synchrotron Rad* **24**: 1250–1259.

Nägele, V. K., H Vogt, TM Link, R Müller, EM Lochmüller, F Eckstein, (2004). "Technical considerations for microstructural analysis of human trabecular bone from specimens excised from various skeletal sites." *Calcif Tiss Int* **75**: 15–22.

Nghia, T. V., R. C. Atwood and M. Drakopoulos (2015). "Radial lens distortion correction with sub-pixel accuracy for x-ray micro-tomography." *Optics Express* **23**: 32859–32868.

Nutt, S. R. and F. E. Wawner (1985). "Silicon carbon filaments: Microstructure." *J Mater Sci* **20**: 1953–1960.

Nuzzo, S., M. H. Lafage-Proust, E. Martin-Badosa, G. Boivin, T. Thomas, C. Alexandre and F. Peyrin (2002). "Synchrotron radiation microtomography allows the analysis of three-dimensional microarchtiecture and degree of mineralization of human iliac crest biopsy specimens: Effects of etidronate treatment." *J Bone Miner Res* **17**: 1372–1382.

Ohgaki, T., H. Toda, M. Kobayashi, K. Uesugi, M. Niinom, T. Akahori, T. Kobayashi, K. Makii and Y. Aruga (2006). "In situ observations of compressive behaviour of aluminium foams by local tomography using high-resolution tomography." *Phil Mag* **86**: 4417–4438.

Olurin, O. B., M. Arnold, C. Körner and R. F. Singer (2002). "The investigation of morphometric parameters of aluminium foams using micro-computed tomography." *Mater Sci Eng A* **328**: 334–343.

Parkinson, D. Y., D. Gursoy, D. M. Pelt, S. Venkatakrishnan, R. Archibald, K. A. Mohan, T. Bicer, M. Vogelgesang, J. Sethian, N. Wadeson, M. Basham and T. Faragó (2019). *Experimental Analysis Solutions for Leading Experimental Techniques*.

Perilli, E., F. Baruffaldi, M. C. Bisi, L. Cristofolini and A. Cappello (2006). "A physical phantom for the calibration of three-dimensional x-ray microtomography examination." *J Microsc* **222**: 124–134.

Petruzzellis, F., C. Pagliarani, T. Savi, A. Losso, S. Cavalletto, G. Tromba, C. Dullin, A. Bär, A. Ganthaler, A. Miotto, S. Mayr, M. A. Zwieniecki, A. Nardini and F. Secchi (2018). "The pitfalls of in vivo imaging techniques: Evidence for cellular damage caused by synchrotron x-ray computed micro-tomography" *New Phytologist* **220**: 104–110.

Peyrin, F., M. Salome, P. Cloetens, A. M. Laval-Jeantet, E. Ritman and P. Rüegsegger (1998). "MicroCT examinations of trabecular bone samples at different resolutions: 14, 7 and 2 micron level." *Technol Health Care* **6**: 391–401.

QRM. (2019). "QRM nano bar phantom." Retrieved April 4, 2019, from http://www.qrm.de/content/products/microct/microct_barpattern_nano.htm.

Raum, K., R. O. Cleveland, F. Peyrin and P. Laugier (2006). "Derivation of elastic stiffness from site-matched mineral density and acoustic impedance maps." *Phys Med Biol* **512**: 747–758.

Raum, K., I. Leguerney, F. chandelier, M. Talmant, A. Saied, F. Peyrin and P. Laugier (2006). "Site-matched assessment of structural and tissue properties of cortical bone using scanning acoustic microscopy and synchrotron radiation microCT." *Phys Med Biol* **51**: 733–746.

Richter, C. P., H. Young, S. V. Richter, V. Smith-Bronstein, S. R. Stock, X. Xiao, C. Soriano and D. S. Whitlon (2018). "Fluvastatin protects cochleae from damage by high-level noise." *Sci Rep* **8**: 3033.

Rivers, M. L. and Y. Wang (2006). Recent developments in microtomography at GeoSoilEnviroCARS. *Developments in X-ray Tomography V.* U. Bonse. Bellingham (WA), SPIE. **SPIE Proc Vol 6318**: 63180J-1–63180J-5.

Sasov, A. (2001). Comparison of fan-beam, cone-beam and spiral scan reconstruction in x-ray microCT. *Developments in X-ray Tomography III.* U. Bonse. Bellingham (WA), SPIE. **SPIE Proc Vol 4503**: 124–131.

Sasov, A., X. Liu and P. L. Salmon (2008). Compensation of mechanical inaccuracies in microCT and nano-CT. *Developments in X-ray Tomography VI.* S. R. Stock. Bellingham (WA), SPIE. **7078**: 70781C.

Schmidt, C., M. Priemel, T. Kohler, A. Weusten, R. Müller, M. Amling and F. Eckstein (2003). "Precision and accuracy of peripheral quantitative computed tomography (pQCT) in the mouse skeleton compared with histology and microcomputed tomography (μCT)." *J Bone Miner Res* **18**: 1486–1496.

Schnaar, G. and M. L. Brusseau (2005). "Pore-scale characterization of organic immiscible-liquid morphology in natural porous media using synchrotron x-ray microtomography." *Env Sci Technol* **39**: 8403–8410.

Schnaar, G. and M. L. Brusseau (2006). "Characterizing pore-scale dissolution of organic immiscible liquid in natural porous media using synchrotron x-ray microtomography." *Env Sci Technol* **40**: 6622–6629.

Seifert, A. and M. J. Flynn (2002). Resolving power of 3D x-ray microtomography systems. *Medical Imaging 2002: Physics of Medical Imaging.* M. J. Y. L. E. Antonuk. Bellingham (WA), SPIE. **SPIE Proc Vol 4682**: 407–413.

Sheppard, A. P., C. H. Arns, A. Sakellariou, T. J. Senden, R. M. Sok, H. Averdunk, M. Saadatfar, A. Limaye and M. A. Knackstedt (2006). Quantitative properties of complex porous materials calculated from x-ray μCT images. *Developments in X-ray Tomography V.* U. Bonse. Bellingham (WA), SPIE. **SPIE Proc Vol 6318**: 631811-1–631811-5.

Stock, S. R. (1999). "Microtomography of materials." *Int Mater Rev* **44**: 141–164.

Stock, S. R. (2008). "Recent advances in x-ray microtomography applied to materials." *Int Mater Rev* **58**: 129–181.

Stock, S. R., J. Barss, T. Dahl, A. Veis, J. D. Almer and F. De Carlo (2003a). "Synchrotron x-ray studies of the keel of the short-spined sea urchin *Lytechinus variegatus*: Absorption microtomography (microCT) and small beam diffraction mapping." *Calcif Tiss Int* **72**: 555–566.

Stock, S. R., K. I. Ignatiev, T. Dahl, A. Veis and F. D. Carlo (2003b). "Three-dimensional microarchitecture of the plates (primary, secondary and carinar process) in the developing tooth of Lytechinus variegatus revealed by synchrotron x-ray absorption microtomography (microCT)." *J Struct Biol* **144**: 282–300.

Stock, S. R., S. Nagaraja, J. Barss, T. Dahl and A. Veis (2003c). "X-ray microCT study of pyramids of the sea urchin Lytechinus variegatus." *J Struct Biol* **141**: 9–21.

Su, R., G. M. Campbell and S. K. Boyd (2007). "Establishment of an architecture-specific experimental validation approach for finite element modeling of bone by rapid prototyping and high resolution computed tomography." *Med Eng Phys* **29**: 480–490.

Van De Casteele, E., D. V. Dyck, J. Sijbers and E. Raman (2004). The effect of beam hardening on resolution in x-ray microtomography. *Progress in Biomedical Optics and Imaging - Medical Imaging 2004: Imaging Processing.* J. M. Fitzpatrick and M. Sonka. Bellingham (WA), SPIE. **SPIE Proc Vol 5370 III**: 2089–2096.

Vidal, F. P., J. M. Letang, G. Peix and P. Cloetens (2005). "Investigation of artifact sources in synchrotron microtomography via virtual x-ray imaging." *Nucl Instrum Meth B* **234**: 333–348.

Vieira, A. E. M., A. C. B. Delbem, K. T. Sassaki, M. L. Cannon and S. R. Stock (2006). Quantitative analysis of mineral content in enamel using laboratory microtomography and microhardness analysis. *Developments in X-ray Tomography V.* U. Bonse. Bellingham (WA), SPIE. **SPIE Proc Vol 6318**: 631823-1–631823-5.

Waarsing, J. H., J. S. Day, J. C. v. d. Linden, A. G. Ederveen, C. Spanjers, N. D. Clerck, A. Sasov, J. A. N. Verhaar and H. Weinans (2004). "Detecting and tracking local changes in the tibiae of individual rats: A novel method to analyse longitudinal in vivo microCT data." *Bone* **34**: 163–169.

Wacheter, N. J., P. Augat, G. D. Krischak, M. Mentzel, L. Kinzl and L. Claes (2001). "Prediction of cortical bone porosity in vitro by microcomputed tomography." *Calcif Tiss Int* **68**: 38–42.

Wang, Z., P. Verboven and B. Nicolai (2017). "Contrast-enhanced 3D microCT of plant tissues using different impregnation techniques." *Plant Methods* **13**: 105.

Washburn, N. R., M. Weir, P. Anderson and K. Potter (2004). "Bone formation in polymeric scaffolds evaluated by proton magnetic resonance microscopy and x-ray microtomography." *J Biomed Mater Res Pt A* **69**: 738–747.

Weitkamp, T. and P. Bleuet (2004). Automatic geometric calibration for x-ray microtomography based on Fourier and Radon analysis. *Developments in X-ray Tomography IV.* U. Bonse. Bellingham (WA), SPIE. **SPIE Proc Vol 5535**: 623–627.

Wildenschild, D., J. W. Hopmans, C. M. P. Vaz, M. L. Rivers, D. Rickard and B. S. B. Christensen (2002). "Using x-ray tomography in hydrology: Systems, resolutions and limitations." *J Hydrol* **267**: 285–297.

Xiao, X., F. D. Carlo and S. Stock (2007). "Practical error estimation in zoom-in and truncated tomography reconstructions." *Rev Sci Instrum* **78**: 063705-1–063705-7.

Xiao, X., F. D. Carlo and S. R. Stock (2008). X-ray zoom-in tomography of calcified tissue. *Developments in X-ray Tomography VI.* S. Stock. Bellingham (WA), SPIE **7078**: 707810.

6

Experimental Design, Data Analysis, Visualization

This chapter has three main subjects and serves as an introduction for the approaches used in the chapters on applications. The first topic is the design of microCT characterization experiments; although most of the factors involved are considered automatically by experienced tomographers and the presentation of these details will undoubtedly bore those individuals, certain aspects will perhaps not occur to newcomers, and are therefore worth describing. The second is data analysis strategies, that is, how relevant numerical quantities might be extracted for different classes of specimens. The third is data visualization, that is, the presentation of numerical data in a form allowing 2D, 3D, 4D, and 5D interpretations of sample sets. Data representation in 2D and 3D is familiar to most readers; 4D representations include temporal observation of a 3D structure or superposition of a numerical quantity (say magnitude of crack opening) onto a 3D structure (the crack plane); and 5D representations might combine the latter representation (measured quantity folded into the 3D structure) with temporal evolution.

6.1 Experiment Design

Before delving into the details of microCT experimental design, posing several questions will help to focus attention on the big picture and on the choices to be made. Various "guides" to performing microCT have appeared and offer their authors' opinions about how to best apply microCT in the arenas in which these authors work (du Plessis, Broeckhoven et al. 2017, Osborne, Kuntner et al. 2017).

The first two questions concern the imaging conditions and are not independent of specimen considerations that follow:

- What spatial resolution is needed for the application? Specifically, what spatial resolution (voxel size) is optimum, what might be adequate, and what might be marginal?
- What contrast sensitivity is required for the application? This might depend on whether and how the data is to be segmented (is simple binary segmentation sufficient, that is, selection of a threshold voxel value and division of all voxels into two classes, one the material of interest and the other "empty" space) or on whether the voxel values are to be interpreted numerically.

One way of describing the combination of different spatial and contrast resolutions is to use high/medium/low definition to describe the imaging characteristics. High definition refers to imaging with high spatial resolution and high contrast sensitivity. In this context, though, high spatial resolution refers not to absolute voxel dimensions but rather to the number of voxels per unit specimen diameter. Medium definition might refer to high spatial definition combined with moderate contrast sensitivity or the reverse, moderate spatial definition combined with high contrast sensitivity. Questions relating to specimen specifics include the following:

- What is the specimen absorptivity (and what are the available x-ray energies)?
- What specimen constraints exist (minimum representative specimen diameter, minimum specimen volume, what is the required specimen aspect ratio, that is, length to width of the cross-section, what is the acceptable length of time out of freezer or culture chamber)?
- How much hard drive space will be required? How long will it take to acquire data? How slow (fast) is reconstruction? Once analysis routines have been perfected, how long will it take to analyze the required number of specimens?

Questions of the number of specimens required for statistical significance are the province of the specific experiments and are not covered here.

Before turning to the question of extracting numerical data or informative 3D representations from microCT data, the reader may find it is useful to review the design and constraints outlined below for one past study. This study involved the time course of Portland cement degradation by sulfate ions, a significant durability issue and potential safety concern in certain environments.

Sulfate ions, in particular, from the external environment, can produce deleterious phase transformations in Portland cement, a major construction material. One phase that forms is gypsum, and it is associated with the loss of adhesion and strength. A second phase involved in the attack is ettringite; it is associated with expansion and cracking. Sulfate attack of Portland cement remains incompletely understood, and the microCT study aimed at improving this situation by combining 3D measures of damage, including cracking, with 3D x-ray diffraction mapping of the spatial distribution of crystalline reaction products (Stock, Naik et al. 2002, Naik 2003, Jupe, Stock et al. 2004, Naik, Kurtis et al. 2004, Wilkinson, Jupe et al. 2004, Naik, Jupe et al. 2006). Exposure in these accelerated tests was interrupted at different points over 1 year, and the noninvasive sampling allowed the same specimen to be interrogated repeatedly. Some results will be presented in Chapter 10.

Several constraints limited data collection for the sulfate attack study. First, a tube-based microCT system was to be used, and the sulfate damage study had to compete for access with several other projects. Nonetheless, numerous variables (and hence specimens) and four exposure times (up to 52 weeks) were to be studied, including two cement types, multiple sulfate ion concentrations, two cations, and two water-to-cement ratios. Further, the specimens had to be large enough that edge effects would not dominate the observations and that a significant volume in the specimen interior would remain unaffected by the damage. On the other hand, the sample size needed to be small enough that there was adequate x-ray transmissivity at the maximum tube voltage (70 kVp) and that the voxel dimensions were small enough to capture important damage processes.

A cylindrical specimen geometry was selected because it was optimum for microCT and because rotation about the sample axis during x-ray diffraction pattern collection would sample more grains at a constant distance from the surface. A 12-mm diameter and 40-mm length were selected based on the sample transmissivity, on the dimensions of inexpensive (disposable) plastic tubes serving as molds for casting the cement, and on the desire to study effects of sample "corners" as well as to study a length of uniform cylinder away from the cylinder's ends (16 mm of length was one sample diameter or more from the ends). The available time for examining each specimen after each increment of sulfate exposure was about 4 h, and the investigators selected the following for each specimen as compromises between this temporal constraint, sensitivity to small cracks, and coverage of a significant volume:

- 37 μm voxels for $(1K)^2$ reconstruction
- Maximum integration time per projection (0.35 s)
- 390 slices covering 19.5 mm of specimen length (one slice every ~50 μm)

Although details of damage at the 5- to 10-μm level were lost, the microCT voxel size was significantly smaller than the sampling volume for diffraction, an important consideration for the combined interpretation of two types of data.

6.2 Data Analysis

This section introduces some principles of data analysis and some of the methods that have been used to analyze microCT data. The intent is not to go into the depth that one would expect of an image analysis text nor is it to indicate which method(s) is(are) best for given application types. This latter consideration is taken up along with the applications in the following chapters. Before describing the numerical approaches, it is useful to consider the quantities that must be measured and the type and amount of sampling required to return a reliable assessment of these quantities. Specialized software has been developed for different types of specimens; for example, SoilJ for analyzing soils (Koestel 2018), BoneJ for 3D bone morphometry (Doube, Kłosowski et al. 2010), and Blob3D (Ketcham 2005).

Numerical measurements of microstructure are typically used to test hypothetical relationships between macroscopic mechanical properties and the volume fraction of a particular phase, the amount and distribution of fluid or gas transport and the channel dimensions and network characteristics, treatment with a given drug and prevention of degradation of trabecular architecture in osteoporosis models, etc. Microstructural quantities include volume fraction of a phase V_V (in bone quantification software bundled with commercial software systems and in some literature this is written BV/TV, the ratio of bone volume to total volume), surface area per unit volume of the phase of interest S_V, mean volumes of cells, mean diameter of particles d_{50}, mean thickness <Th> of structural elements (or diameter of channels), distribution of Th valves, distribution of particle sizes, structural connectivity, and structure model index (SMI valves indicate whether the objects in a specimen tend to be rod-like, plate-like, or spherical; see Chapter 8). Measures of anisotropy of microstructural features have also received attention, but these tend to be context-specific and will be covered only in conjunction with the associated application.

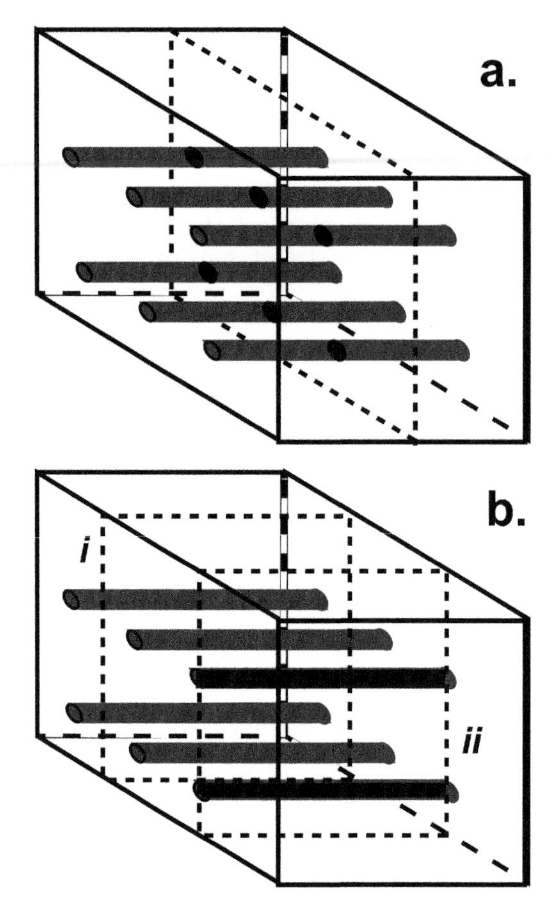

FIGURE 6.1
Illustration of sampling by a single plane. (a) The single plane perpendicular to the axes of the rods captures their cross-sections (black disks), volume fraction, and spacing accurately, but this is quite fortuitous. (b) In a second sampling geometry, plane *"i"* misses the rods entirely and *"ii"* intercepts two rods (black rectangles).

One should also remain aware of the fact that analogous quantities may be measured in different subject areas and called by different names. Quite elaborate analysis routines may be available in the literature that one does not normally encounter, and that is one reason that the applications chapters of this book are organized by structure type rather than traditional discipline.

MicroCT samples volumes, be it as little as the thickness of a single slice (the volume, of course, would be the thickness multiplied by the area of the slices). As illustrated in Fig. 6.1a, a single slice may sometimes give an accurate measure of the microstructure (here the slices are perpendicular to the axes of six circular and parallel rods). The black disks, viewed in perspective, show the intersection of the slice (dashed lines on the cube surfaces) with the rods. A single slice, however, may not represent the actual structure accurately, as Fig. 6.1b shows. The slice plane labeled *"i"* misses the rods entirely and the plane *"ii"* intersects two rods. If the specimen has an isotropic microstructure, one slice may give an accurate value of the volume fraction (a basic stereology result for random structures shows that the area fraction of a phase A_A equals V_V), but, absent the certainty of a truly isotropic microstructure, it is better to sample a volume. Further, measurement of quantities like thicknesses of structures or diameters of particles is almost always biased unless volumetric sampling and analysis are used (the situation in Fig. 6.1a is quite fortunate, and this sampling is rarely attained in practice without 3D data sets).

6.2.1 Segmentation by Voxel Value

In many circumstances, the specimen consists of several discrete phases with distinct absorption or phase contrasts. These individual phases occupy one or more regions within the field of view. Often, only one of these phases is of interest for computational or for visualization purposes and one segments the image (volume) into the phase of interest and everything else. Typically, the segmentation process replaces voxel values of the phase of interest by a binary value of "1" (in physical terms these voxels are treated as solid) and the values of all other voxels with "0" (empty space). Segmentation is often used synonymously with thresholding, and visualizations of volumes employing multiple thresholds and partially transparent solids can be very effective (see Fig. 6.14).

Binary segmentation and selection of a reasonable threshold separating the phase of interest from all other phases is perhaps the most popular approach. Specimens containing high-contrast phases, for example, bone where the bone/marrow contrast is about 10/1 for a lab microCT system operating with an effective x-ray energy of ~25 keV (Rüegsegger, Koller et al. 1996), have histograms with peaks separated by a valley; these materials can be segmented by choosing a threshold within the valley. Such a situation is pictured in Fig. 6.2 for a slice of murine trabecular bone. As has been shown by comparing results on the same specimen reconstructed with different voxel sizes, the specific threshold needed to produce the same volume fraction (a valid constraint given that the data are from the same specimen) will necessarily vary with voxel size (Hangartner 2007). Likewise, valid comparison between high-porosity specimens embedded in plastic and similar unembedded samples (e.g., trabecular bone) requires use of different thresholds (Perilli, Baruffaldi et al. 2007). If data collected for similar samples but with different x-ray energies (kVp for a polychromatic tube source or keV for monochromatic synchrotron x-rays), then different thresholds are required for accurate numerical comparisons and careful validation is essential to ensure that the thresholds are not biasing the analysis.

Confounding effects that necessitate less routine segmentation include large populations of partial voxels resulting, for example, from structures with minimum dimensions on the order of the voxel size (see Figs. 5.12 and 6.3). In addition, situations exist where the material itself contains a spread of linear attenuation coefficients; that is, the material lacks well-defined, well-separated peaks in the histogram. This is the situation shown in Fig. 5.15, and segmentation of remodeled osteons from older bone is problematic at best. Robb and coworkers discuss the range of segmentation approaches used for microCT (Rajagopalan, Lu et al. 2005). The effect of different thresholds and segmentation algorithms were examined recently by others in a study of predictions of permeability in rocks (Latief, Fauzi et al. 2017).

In the absence of a well-defined valley between histogram peaks, one frequent segmentation approach is assignment by inspection: the operator examines a typical slice or slices and selects the threshold which best (to his or her eye) preserves the important fine-scale features of both the solid and the surrounding empty space. Given that this is a highly subjective process, considerable effort has been devoted to the accuracy of conclusions (Hemmatian, Laurent et al. 2017, Rovaris, Queiroz et al. 2018) and the robustness of conclusions derived from small shifts in threshold, the general consensus is that absolute numbers (V_V, Th, etc.) will change somewhat with changing threshold, but, so long as the

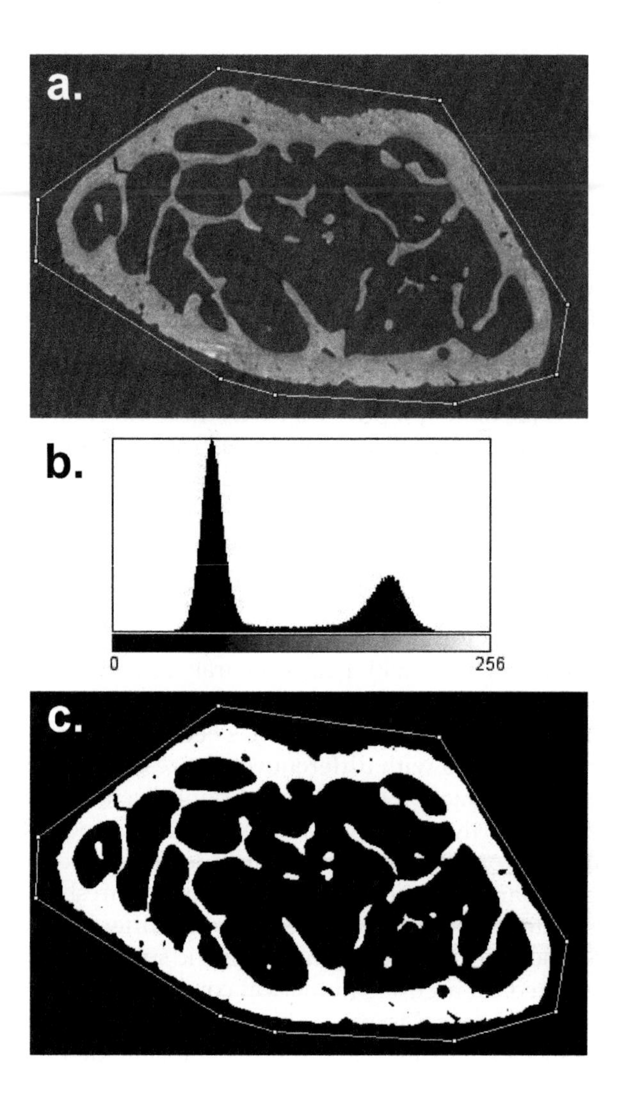

FIGURE 6.2
Binary segmentation applied to a slice of murine trabecular bone. (a) Portion of a synchrotron microCT slice of the murine femur in ethanol, ~2.8-μm isotropic voxels, 512-voxel horizontal field of view, 1K × 1K reconstruction, 17 keV (S.R. Stock, N.M. Rajamannan, F. De Carlo, 2-BM APS, June 2005). (b) Histogram of the polygonal area (>10^5 voxels) within the dashed lines in (a). (c) The voxels with grayscale values 128 and above are shown as white; all other voxels are black.

features being quantified have minimum dimensions greater than perhaps four voxels, comparisons between specimens will be valid (see Section 5.2 on microCT accuracy for more details). The author uses this trial and error method but with the following precaution: the threshold is chosen based on evaluation of a preliminary subset of specimens, a subset explicitly excluded from the actual statistical comparison of different treatment groups (Stock, Ignatiev et al. 2004a).

It is not always possible for a single threshold value to adequately define the phase of interest over the entire specimen. This occurs, for example, when significant beam hardening is present. An alternative to a single, global threshold value is the use of the gradient in grayscale value to define the boundary between phases (Rajon, Pichardo et al. 2006).

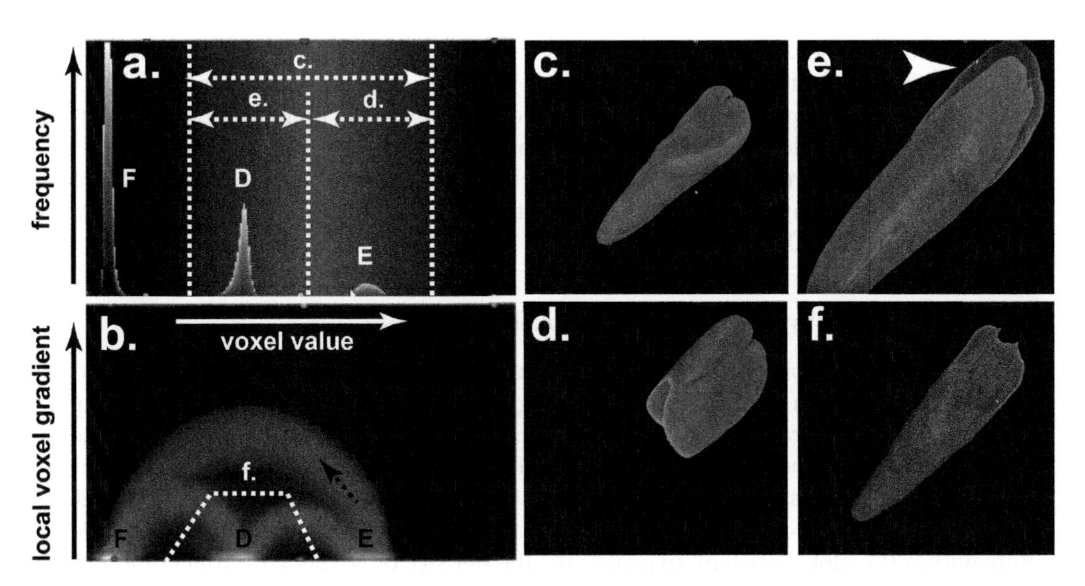

FIGURE 6.3
Illustration of segmentation with 2D histograms (voxel value AND voxel gradient). Note that the renderings differ somewhat in viewing angle and size. (a) 1D histogram (frequency of voxel values, vertical, as a function of voxel value, horizontal) of a tooth imaged with microCT. Labels "F," "D," and "E" denote fluid in which the tooth is immersed, tooth dentin and tooth enamel, respectively. The dotted vertical lines and labels c, d, and e indicate the limits for the voxel values included in the renderings in (c–e). (b) 2D histogram (local voxel gradient, vertical, as a function of voxel value, horizontal) of the same data set used in (a). The grayscale values filling the area of the 2D histogram are the number of voxels with that combination of local gradient and value: the lighter the pixel, the larger the number. The dotted trapezoid indicates the range of gradients and voxel values used to produce (f). The black arrow is explained in the text. (c) Rendering of the entire tooth including both enamel and dentin. (d) Rendering of tooth enamel but not dentin. (e) Rendering of tooth dentin. The halo (arrowhead) consists of partial volumes of enamel. (f) Rendering of tooth dentin using the 2D histogram region shown in (b): the halo of partial volumes of enamel is eliminated. Images courtesy of Graham Davis; see https://www.youtube.com/watch?v=pd9azHTGaB4.

Adaptive or dynamic thresholding, which explicitly accounts for varying background, is another approach (Ramaswamy, Gupta et al. 2004, Brunke, Oldenbach et al. 2005, Feeney, Crawford et al. 2006, Burghardt, Kazakia et al. 2007). Yet another approach is to use two thresholds $T_A < T_B$, with T_A including phase A (e.g., empty space) and none of phase B (e.g., solid) and T_B that includes phase B and none of A; those voxels with values between the thresholds are assigned via indicator kriging to phase A or phase B in such a way as to maximize smoothness of the surface between the phases (Oh and Lindquist 1999, Mendoza, Verboven et al. 2007).

In structures where the solid phase is particularly thin and difficult to resolve reliably, as it often is for walls between cells, the cell boundaries can be determined by a distance transformation plus watershed algorithm (Gonzalez, Woods et al. 2004) (see the following subsections), as, for example, in a study of evolution of a liquid foam described in Chapter 8. In specimen systems where partial volumes and low contrast dominate, selective iterative thresholding may be useful: here an initial threshold is guessed and the results of this binarization are numerically compared to those of different iterations (Montanini 2005). Segmentation by "snake" or active contour models for boundary detection can also be effective (Sheppard, Arns et al. 2006): an initial contour is deformed toward the boundary to be detected so as to minimize a functional designed to have its local minimum at the boundary (Caselles, Kimmel et al. 1997). Spowage et al. described the steps required to identify surface porosity in foams (Spowage, Shacklock et al. 2006). Lindquist explicitly compares three quite different thresholding methods (Lindquist 2001), and the interested reader can take this paper as a starting point for learning more about thresholding. More complex segmentation pipelines have been used as well for automated segmentation, including stingray tissue (Knötel, Seidel et al. 2017) and fracture callus (Bissinger, Götz et al. 2017).

6.2.2 Segmentation by Voxel Value and Voxel Gradient

Voxel values are the most basic component of contrast within images, but human eyes use information other than this when interpreting images. One additional and very important quantity is local gradient in voxel values. Segmentation folding local voxel gradients with voxel values can be very powerful and efficient. This approach can be implemented in the form of a 2D histogram used in the visualization software Drishti (Limaye 2006, 2019)*: the horizontal axis is voxel value, the vertical axis is local voxel gradient, and the points within the plot represent, in grayscale, the density of voxels with that combination of local gradient and value. The example of Fig. 6.3 is of a tooth and illustrates a weakness of 1D segmentation compared to the 2D approach. Figures 6.3a and b show the specimen's 1D and 2D histograms, respectively. The rendering in Fig. 6.3c is of the entire tooth and includes all voxels in the range "c." of Fig. 6.3a. The image in Fig. 6.3d shows tooth enamel only, that is, those voxels in the range "d." Figure 6.3e shows tooth dentin (those voxels in the range "e"); this range includes partial voxels of enamel (plus fluid) and produces the halo indicated by the arrowhead.

Consider what is shown in Fig. 6.3b. There are three strong white domains on the horizontal axis (i.e., where voxel gradients are near zero); these domains labeled "F," "D," and "E" are voxels from the bulk fluid, dentin, and enamel, respectively. In other words, the dentin voxels are surrounded by other dentin voxels and voxel gradients are very small, etc. The light arc between "E" and "F," labeled by the black dotted arrow, represents the set of voxels partially occupied by these two phases. Segmenting the data using the range of voxels and gradients within the trapezoid in Fig. 6.3b eliminates the partial volumes from the rendering (Fig. 6.3f). Sometimes the halo of partial volumes (Fig. 6.3e) is useful but often it is not. In this simple example, one could, with some effort, manually remove the halo from the image, but in other situations (two or more phases with more complicated geometry, perhaps interpenetrating, perhaps with different thicknesses), manual "fixes" are infeasible.

6.2.3 Quantification by the Distance Transform Method

The distance transform method is a powerful albeit simple approach for analyzing structures; here it is defined in terms of the most popular application in microCT, determining an accurate mean "trabecular" thickness or distribution of thicknesses for bone or related structures (Hildebrand and Rüegsegger 1997, Hildebrand, Laib et al. 1999). Analysis proceeds by calculating the metric distance of each solid voxel to the nearest solid-(empty) space surface; that is, this distance is the radius of a sphere centered on this voxel and fitting inside the structure. Redundant (smaller) spheres are eliminated, producing a set of

* Perhaps this segmentation approach is available in other visualization packages, but the author is unaware of this.

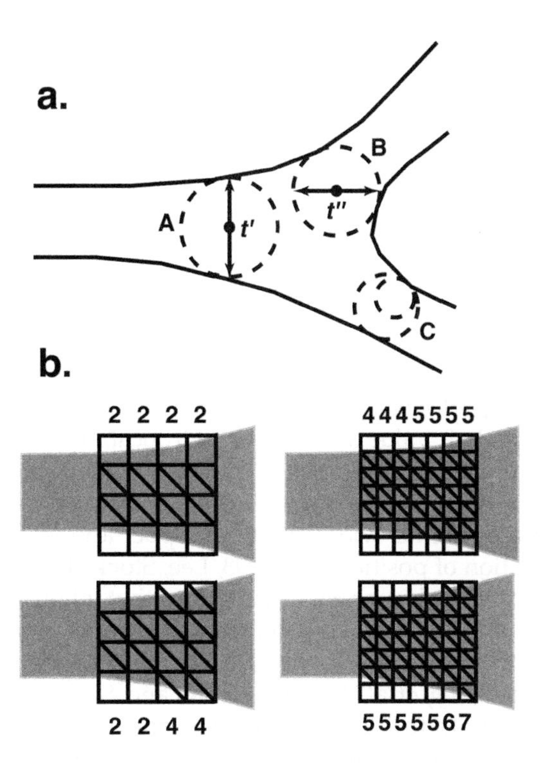

FIGURE 6.4
Two-dimensional illustration of distance transformation determination of plate or strut thickness, a process actually applied in 3D. (a) The dashed circles show that local thicknesses are t′ at A and t″ at B. At C, the largest sphere defines the local thickness. (b) Effect of threshold level and voxel size on the thickness determined. The gray areas represent the same section of "bone," and the large square represents the region of interest (ROI) for analysis. The left column shows the ROI sampled by 4 × 4 voxels, while the right column shows the ROI sampled with (smaller) 7 × 7 voxels. Diagonal slashes show voxels considered to be "bone": in the upper row, the voxels must be totally occupied, while in the lower row, voxels more than one-half occupied are considered solid. Above/below each column of voxels is the number of "bone" voxels in that column. Reprinted from Stock et al. (2004b) with permission.

centers of maximal spheres filling the structure completely (Fig. 6.4). Thicknesses for each portion of the structure are twice the radii, and this allows maps of local thickness value to be produced in 3D renderings as well as the mean thickness or distribution of thicknesses. Mean spacing <Sp> between structural elements is calculated with the same method by simply switching the background and object voxels.

6.2.4 Quantification by Watershed Segmentation

In addition to its use in determining thicknesses, etc., the distance transform can be used for segmentation employing the watershed transformation (Gonzalez, Woods et al. 2004), a methodology based on the idea of catchment basins in geography. A watershed is the ridge dividing areas drained by different rivers or reservoirs, and the approach is to transform the image into a form whose catchment basins are the regions to be identified. An example of such regions is the interior of cells in foams, with the "ridge-lines" being the cell walls. The negative of distance transform yields the ridge lines dividing the regions of the structure, and this can be very effective when a significant fraction of cell walls cannot be detected (Lambert, Cantat et al. 2005). Automated watershed segmentation has also been applied to teeth (Galibourg, Dumoncel et al. 2018).

6.2.5 Quantification by Other Methods

A variety of other segmentation approaches have been used. A number of automated approaches using different techniques have appeared and include: graph-cut and neural network-based segmentation (de Moura Meneses, Bastos Palheta et al. 2018); active contours (Korfiatis, Tassani et al. 2017); fiber segmentation via principal curvature (Kronenberger, Schladitz et al. 2018); multifractal analysis (Torre, Losada et al. 2018); retrieval of subresolution voxels (Smal, Gouze et al. 2018); and ambient occlusion (Titschack, Baum et al. 2018).

One of the shortcomings of voxel value-based thresholding is that voxels are selected primarily on the basis of their value and not on their location. Boundary-following algorithms and related techniques (such as the watershed function) emphasize connected volumes; the boundaries of open channels in SiC/SiC cloth-based composites, for example, were determined; and the channel widths between upper and lower boundary were rapidly computed as a function of position (Lee 1993, Lee, Stock et al. 1998). A related type of analysis is measurement of crack opening by programs that follow the crack from position to position and measure the width as a second step; this works especially well if the crack is first measured under the maximum applied load so that it produces the greatest amount of contrast (Guvenilir, Breunig et al. 1997, Guvenilir and Stock 1998, Guvenilir, Breunig et al. 1999). The difficulty with this approach is that considerable operator intervention is required to fill local gaps (snakes or membranes may be used) or to identify connections that are not obvious locally, that is, structures that may connect after quite some distance out of plane. Percolation operators may help in this regard, that is, allowing virtual particles to bounce around the open structure (random walk) while recording all positions the particle reached. Cluster labeling within segmented data, for example, in transport studies of porous substances (Nakashima, Nakano et al. 2004), is related to this last approach.

Rajagopalan et al. expressed microCT images in terms of its 2D discrete Fourier transform and produced segmentations based on the phase component of the transform (Rajagopalan, Lu et al. 2005). This approach was motivated by the success of image matching methods based on phase images, for example, cross-correlations metrics.

Simultaneous solid-phase and void-phase burns are very useful for analysis of particle characteristics. Starting at an interface, a value is assigned to each voxel that equals the number of voxels it is away from the interface. The local maxima of burn number is a particle center, and all voxels previously identified as solid were assigned to one or another particle with additional steps required to accurately partition contacting grains. Particle volume, surface area, orientation, aspect ratio, and contact statistics flow directly from the assigned particles (Thompson, Willson et al. 2006).

Iterative opening and closing in 3D (n erosions* followed by n dilations with increasing n until no volume remains in the image – note that n erosions remove all structures smaller than $2n$ voxels) is the basis of another method (granulometry) of rapidly computing the distribution of wall thicknesses in a cellular material (Elmoutaouakkil, Fuchs et al. 2003);

* Erosion refers to the operation on a segmented image where the outermost voxel is removed from all around the boundary of the object in question. Dilation refers to the opposite operation, addition of a voxel to all boundary positions. In a single (one-voxel) erosion of a segmented image, all isolated single voxels are removed (in fact all structures with largest dimension three voxels or less). Dilation of the eroded structure restores the structure except where it was completely removed. A moment's reflection reveals that dilation followed by erosion serves to fill in small voids in the segmented objects; see Fig. 6.5.

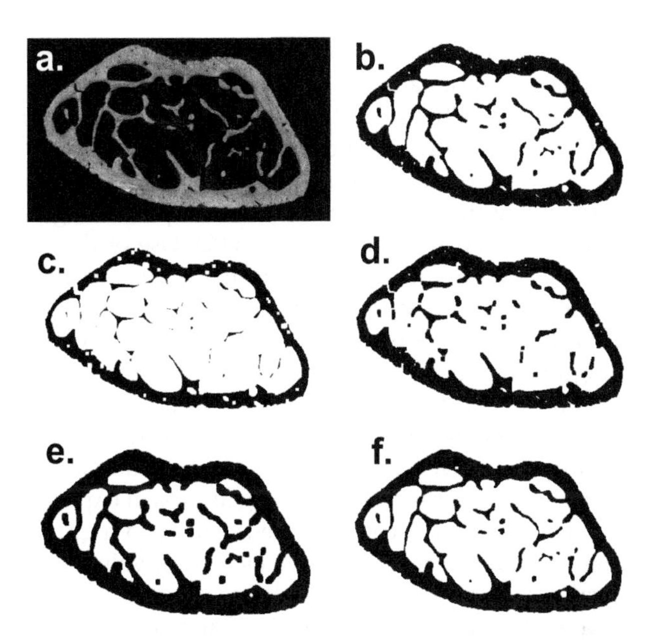

FIGURE 6.5
Illustration of erosion + dilation and dilation + erosion for the slice shown in Fig. 6.2. (a) Grayscale image in 8-bit format (0–255 contrast levels). (b) Image thresholded at 128. (c) Result of two erosions applied to (b). (d) Result of two dilations following the two erosions. Some small thin structures were removed (lower right). (e) Result of two dilations applied to (b). (f) Result of two erosions following two dilations. Comparison of (d) and (f) shows significant differences: for example, the fine porosity in the bone cortex persists in (d) but is eliminated in (f).

the derivative of the remaining volume with respect to structuring element size n gives the (size) distribution of thicknesses (Brunke, Oldenbach et al. 2005). Figure 6.5 illustrates the effect of two erosions followed by two dilations and the effect of two dilations followed by two erosions using the slice shown in Fig. 6.2.

Skeletonization refers to a representation of an object by its centerline, that is, by reducing the object to a one-voxel-wide string or branched string extending to the extreme ends of the original object. This representation of data simplifies analysis of complex arrays of objects, particularly those that are channel- or fiber-like. Skeletonization analysis was used, for example, to study two bonded stainless steel fiber assemblies and to show that the distribution of fiber segment lengths between the two specimens differed as did the distribution of fiber orientations (Tan, Elliott et al. 2006). Such orientation data are often shown on stereographic projections; see Cullity and Stock (2001) or other materials texts for an introduction to this type of plot.

In biology, there are many vessel systems (arteries and veins; kidney microvasculature and glomeruli; lung bronchi, bronchioles, and alveoli; cortical bone Haversian and Volkmann canals) with branching that can be likened to trees. Hundreds of branches are present in a typical organ, and automated analysis routines are essential if an adequate number of replicates are to be analyzed. This last requirement is particularly important in biological studies where interindividual variability is very large. It is important to note the tree-like nature of vessel systems and branches because this differs from the situation pertaining in cellular materials and conditions analysis algorithms. After extraction of the vessel tree, quantitative data can be computed, and numerical analysis of the tree characteristics can be performed. One approach, partition into mother/child/sibling relationships, also known as generational analysis, has been shown to be effective for six or more hierarchical levels (Wan, Ritman et al. 2002); this focuses analysis of functionality and dimensional changes onto equivalent portions of the network. Wan et al. studied the coronary arterial tree of a rat and focused quantification on arterial lumen cross-sectional area, interbranch segment length, branch surface area at equivalent generation, and interbranch and intrabranch levels (Wan, Ritman et al. 2002). Use of the self-similarity of the arterial trees can improve analysis efficiency (Johnson, Karau et al. 1999).

Structural connectivity is important for the mechanical integrity of the array of struts and plates (trabeculae) in cancellous bone and also for pores' function in gas and fluid transport in plants. Significant errors can result if a volume is divided for computational purposes, and this is a particularly important consideration for high-definition synchrotron microCT images of trabecular bones (data sets currently comprising 8 GB or more voxels). Labeling of connected objects on both sides of a common face of two subvolumes and comparing the common border is one promising approach (Apostol and Peyrin 2007). Porosity connectivity in apples was also determined to depend highly on the voxel size of the reconstructed images, and the representative elemental volumes (REV; range of volumes over which a valid statistical average can be computed) were determined for this type of fruit (Mendoza, Verboven et al. 2007). The REV approach was also applied to paper materials studied with synchrotron microCT (Rolland Du Roscoat, Decain et al. 2007). A related method that the authors termed the "mean window technique" was used to determine the properties of a particulate-reinforced metal composite (Borbély, Kenesei et al. 2006). Levitz suggests determining chord distribution functions, linear graphs of retraction (related to skeletonization), and correlated Gaussian fields as useful techniques for analyzing the properties of different porous materials (Levitz 2007).

MicroCT data sets can be used as the basis for incorporating specimen-specific microarchitecture or other microstructure into finite element models. This has been an active area in bone research, discussed in the Chapter 8, and research employing this methodology continues (Fu, Dutt et al. 2006, Gong, Zhang et al. 2007, Su, Campbell et al. 2007), including approaches where local mineral levels are used to define local elastic constants to the individual matrix elements of the modeled solid (Mulder, van Ruijven et al. 2007). Kim and coworkers found that prediction of apparent mechanical properties and structural properties agreed well with experiment regardless of which of the three thresholding methods they used (Kim, Zhang et al. 2007).

In certain circumstances, interpretation of the structures within the volume as an array of idealized solids (spheres, plates, ellipsoids, etc.) can be helpful. It is important, however, to match the idealized form as closely as possible with actual object, say by keeping the volume identical and by minimizing the number of outlier voxels. For example, study of different structural metal foam types revealed differences in cell anisotropy (distribution of cell volumes and of aspect ratios quantified as equivalent ellipsoids; stereographic projections showing distribution of cell axes vs orientation) that correlated with altered mechanical properties (Benaouli, Froyen et al. 2000, 2005).

6.2.6 Image Texture

Images of any type often contain regions with different textures, and, without going into detail, analogy provides a simple illustration of what is meant by texture. Texture refers to the structure contained within a region, and most readers would have little difficulty visualizing smooth, rough, or periodic textures of surfaces (see Fig. 11.19 of

Gonzalez et al. (2004)). MicroCT data sets can be analyzed by using the texture of different regions, but this approach is largely unexplored. One recent study did, however, apply texture analysis to 3D microCT data sets of five porous specimens (Jones, Reztsov et al. 2007). The specimens were mineral carbon forms from different geographical locations with similar topological structure that differed mainly in textural quality. Robust measures of structural texture were extracted from the grayscale images in the form of a set of 96 texture features that comprised the texture vector for a particular sample. The texture vector was then related to the texture space defined by prior measurements on "known" training specimens, that is, probabilities could be computed for the likelihood that a given specimen belonged to each population. One expects that, with further development, this approach will be valuable not only for classification of complex specimens but also for rapid (automated?) identification of key structural differences. Texture features at CT-level resolutions were correlated (by binning microCT data on many cancellous bone specimens) with the underlying microarchitecture in another study (Showalter, Clymer et al. 2006).

A machine vision approach and principal components analysis were applied to phylogenic analysis of sea urchins based on tooth cross-sectional shape (Ziegler, Stock et al. 2012). This study plotted first vs second vs third most important components and found that tooth cross-sections clustered together in a way consistent with phylogenic relationships established by other techniques. Statistical shape analysis was used to analyze bone destruction (Brown, Ross et al. 2017).

6.2.7 Segmentation by Machine Learning/Deep Learning

Machine learning/deep learning was described earlier in Chapter 3 in the context of reconstruction from projections, and ML/DL has been used for some time for image analysis. The example in the previous section on texture analysis is one example. The specific steps used in the example of Fig. 6.6 illustrate the simplicity and power of using ML for segmentation (Marsh 2019). MicroCT images of denim fabric were the input (832 slices). The goal was segmenting fine thread regions separate from the bulk of the coarser fiber material (Fig. 6.6a). The fine thread volumes were manually painted in six training slices (e.g., Fig. 6.6b), and these training data were fed into the neural network 50 times. The training period was so short (12 min) that no effort was made to determine whether a shorter number of iterations would have sufficed. The trained neural network was then used to segment the entire stack of slices (5 min). Figure 6.6c–f shows the resulting renderings. Other examples are machine learning–based in vivo quantification of plant starch reserves (Earles, Knipfer et al. 2018), autocorrelative analysis of large data sets (Kaira, Yang et al. 2018), and analysis of fibrous composites (Madra, Breitkopf et al. 2018).

6.2.8 Interpretation of Voxel Values

The values of the linear attenuation coefficient can be used to analyze changes in microstructure or to quantify volume fraction of unresolved solid in two phase materials. Stock and coworkers used lab microCT to image a demipyramid of the sea urchin *Lytechinus variegatus* at a resolution too coarse for the trabecular structure of stereom to be resolved (Stock, Nagaraja et al. 2003). Because this ossicle consisted of only two phases (high Mg calcite and air), the 3D distribution of linear attenuation coefficient values was interpreted in terms of partial volumes of calcite. A similar approach (analysis of volume fraction in partial voxels) for quantifying crack opening as a function of applied stress and 3D

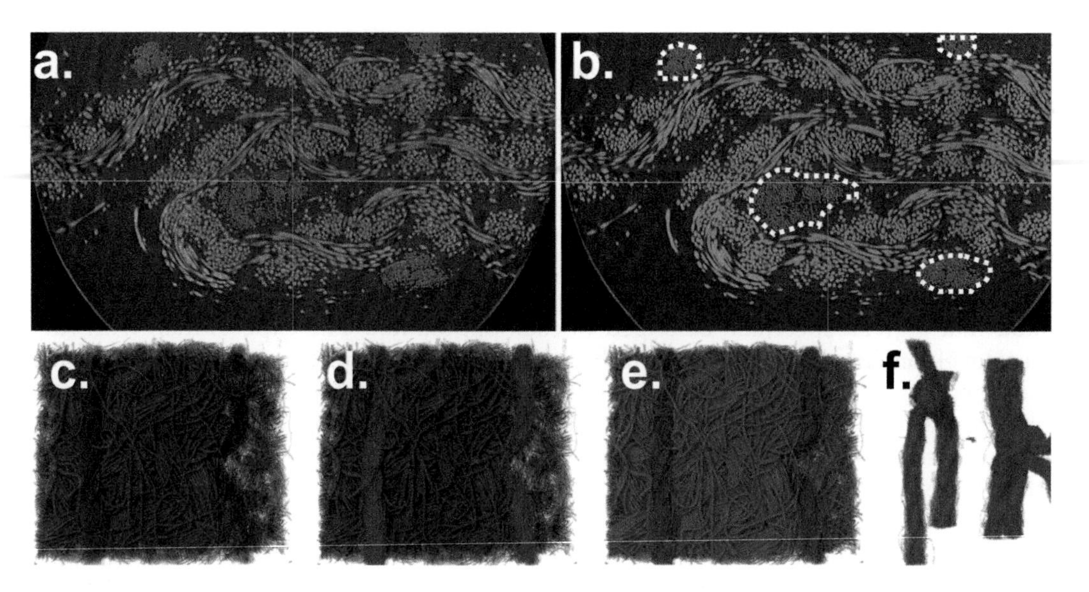

FIGURE 6.6
(color). Machine learning applied to microCT data of a piece of denim fabric. (a) Typical slice of the data set in grayscale contrast. (b) Regions of fine threads painted manually (orange) with the coarser fiber regions in blue. (c) Renderings of the data set before segmentation. (d) Rendering after ML segmentation showing the fine thread volumes in orange and the other fibers in gray. (e) Rendering with fine fibers in orange and coarse fibers in blue. (f) Rendering showing only the fine thread regions (orange). Data collected on a Rigaku nano3dx system with 3.24-μm voxels. Images courtesy of Mike Marsh, Object Research Systems. Data courtesy of Rigaku.

position is described in Chapter 11. Interpretation of voxel values as levels of mineral in bone and importation of these 3D maps as variable Young's moduli in finite element models of trabecular bone is another interesting use of this type of data (Mulder, van Ruijven et al. 2007). A refined estimate of interconnected porosity in rock was based on segmenting not only pores and rock but also voxels with values between the two (Cid, Carrasco-Núñez et al. 2017). Characterization of cavitation in high pressure flow of diesel fuel is another example of this type of approach (Lorenzi, Mitroglou et al. 2017).

6.2.9 Tracking Evolving Structures

There are quite a number of circumstances when prior and altered microstructures must be compared, directly or indirectly, in the same specimen. On the one hand, using the specimen as its own control is an incredibly powerful advantage, given the considerable *a priori* knowledge available. On the other hand, it can be extremely frustrating to accomplish in practice because of difficulties aligning structures and sorting out actual changes from mis- or reorientations. Indirect comparisons are much simpler and include measurements of volume fraction of a phase (bone volume fraction), mean particle sizes (mean trabecular thickness), distributions of particle sizes, etc. There are other analysis options avoiding point-by-point registration, an example of which is covered after registration is considered.

Automatic tracking of the displacement of small readily identifiable features is one method of quantifying deformation fields. In one study, thousands of micropores within the solid were tracked automatically for each increment of deformation, and a microstructure gage formalism was used to calculate maps of the 3D strain tensor components

(Toda, Ohgaki et al. 2006, Toda, Takata et al. 2006). A particle tracking approach has also been used to map displacement fields during mechanical processing (Nielsen, Poulsen et al. 2003, Nielsen, Beckmann et al. 2004, McDonald, Schneider et al. 2006, Zettler, Donath et al. 2006). Particles or micropores can also be used as fiducials to measure changes in crack opening as a function of applied stress (Breunig, Stock et al. 1992, 1993, Toda, Sinclair et al. 2003, 2004). Tracking the motion of specific features also seems to be a viable technique for strain quantification in trabecular bone (Verhulp, van Rietbergen et al. 2004, Liu and Morgan 2007).

Alignment of altered structures is an active area in biomedical research. Displacements in CT images of lungs can be followed through the formalism of image warping (Fan and Chen 1999). Strains in trabeculae from their displacement during loading are a second area of research covered in more detail in Chapter 8. Determination of longitudinal changes in trabeculae by direct comparison of the reconstructed volumes was an ambitious approach employed in an in vivo lab microCT comparison of OVX and control rats (Waarsing, Day et al. 2004). In this study, image registration (translation and rotation between volumetric data sets) was performed to maximize mutual information. The registration algorithm worked well for the controls, for which changes were very small, but the large changes in the OVX animals required a modified approach-based registration of a relatively small number of large invariant structures. For in vivo studies, a special imaging chamber was devised to allow microCT and microPET *inter*modal registration (Chow, Stout et al. 2006). *Intra*modal methodologies for registering longitudinal microCT data sets have also received attention (Boyd, Moser et al. 2006). Also of interest is the volumetric digital image correlation method applied to triaxial compression of rock in order to calculate the 3D strain fields (Lenoir, Bornert et al. 2007).

Registration issues can be avoided to some extent by extraction of the quantity of interest and mapping this quantity in the 3D or 4D (three spatial dimensions plus time) space occupied by the specimen. An example of this is measurement of crack opening as a function of position and its mapping on the 3D crack surface (i.e., the crack's center position) as a fourth dimension, say, as color. Although this sounds quite mysterious at this juncture, actually going through the steps in the analysis in the context of data representations in the following section will, one hopes, illustrate what is meant.

6.3 Data Representation

MicroCT data sets can be used in many ways to explain changes in specimens and to reveal differences between groups of similarly treated specimens. Such illustrations can be as simple as qualitative comparisons of comparable slices or as complex as 3-, 4-, or 5D representations of volumetric data of evolving specimens.

The simplest representation of microCT data is the use of 2D sections through volumetric data. A number of slices images have been presented above. Often, movies paging through the stack of slices can be very effective in establishing an overall impression of the structure or in identifying a particular subset of the volume from detailed interrogation. In certain structures, numerical sections other than the slice plane can be most informative. In some cases, of course, only isolated slices or sets of widely spaced slices are available, and analysis options are limited. Provided the sampling adequately represents the structure, then V_V and the variance of V_V across the specimen can be determined with some confidence. Unless there is *a priori* knowledge about the structure and its orientation with respect to the tomographic rotation axis, apparent thicknesses and other measurements are liable to differ substantially and in an unpredictable way from the actual values.

In volumetric microCT, where one can section the volume numerically along more than one plane, anisotropic structures are often represented by showing three orthogonal sections as in an isometric presentation of the sides of a cube; in metallurgy, combining micrographs from three orthogonal planes has been common for many years. Figure 6.7 shows three orthogonal planes through the reconstructed volume of several different types of sea urchin spines (Stock, Ebert et al. 2006). As described in the figure caption, some of the structure is difficult to appreciate in single sections. Sectioning through stacks of slices along cylindrical surfaces and unwrapping the surface into a flat 2D map (circle cuts) is another way of presenting data simply (Lee 1993).

Another simple method of displaying structure is to apply a threshold to a data and to view the resulting 3D rendering of the voxels more absorbing than the threshold. An

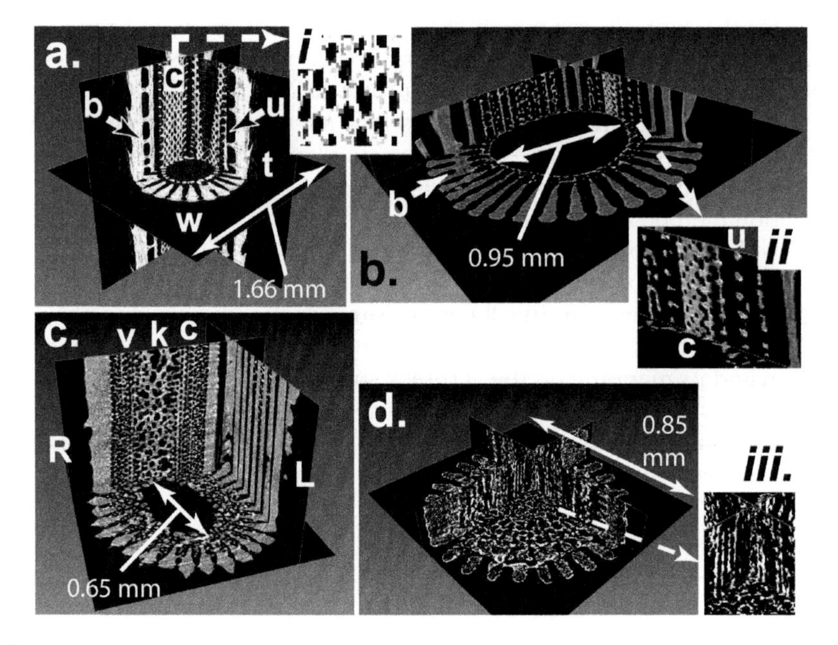

FIGURE 6.7
Three orthogonal sections through reconstructed spines of (a) *Diadema setosum*, (b) *Centrostephanus rodgersii*, (c) *Echinothrix diadema*, and (d) *Lytechninus variegatus* showing three orthogonal slices through the spines. Insets *i*, *ii*, and *iii* are enlargements of the fine stereom indicated by the dashed lines. The wedges "w" of both *C. rodgersii* and *D. setosum* are linked by well-defined bridges "b" and capped by thorns "t" (upward oriented in (a), not visible in (b)). The axial distribution of bridges tends toward regular spacing in (a). In (b), *C. rodgersii* has at least one bridge every 250–300 μm, although the spacing is sometimes much less; bridges link five or more adjacent wedges in the same transverse plane and spiral around the spine axis. In *E. diadema*, however, the wedges are linked by many fine trabeculae "v" (c). In *C. rodgersii* and *D. setosum*, the central cavities are bordered by a thin, nearly circular calcite cylinder that is pierced by a regular array of holes (insets *i* and *ii*). These perforated cylinders connect to the wedges through thin, regularly spaced radial bars "u." In contrast, *E. diadema*'s oval central cavity is bordered by irregular, coarse stereom "k" in which the perforated cylinder is buried (below "c" in (c)). In (d), the regular axial structure of the inner, fine stereom in the *L. variegatus* spine can be appreciated only in 3D (inset *iii*). The data were collected at 2-BM, APS, with a 1K × 1K detector, 0.25° rotation steps and: (a) 18 keV x-rays and reconstructed with 1.66-μm voxels, (b, c) 21 keV x-rays and 5-μm voxels, and (d) 15 keV x-rays and 1.3-μm voxels. Reproduced from Stock et al. (2006).

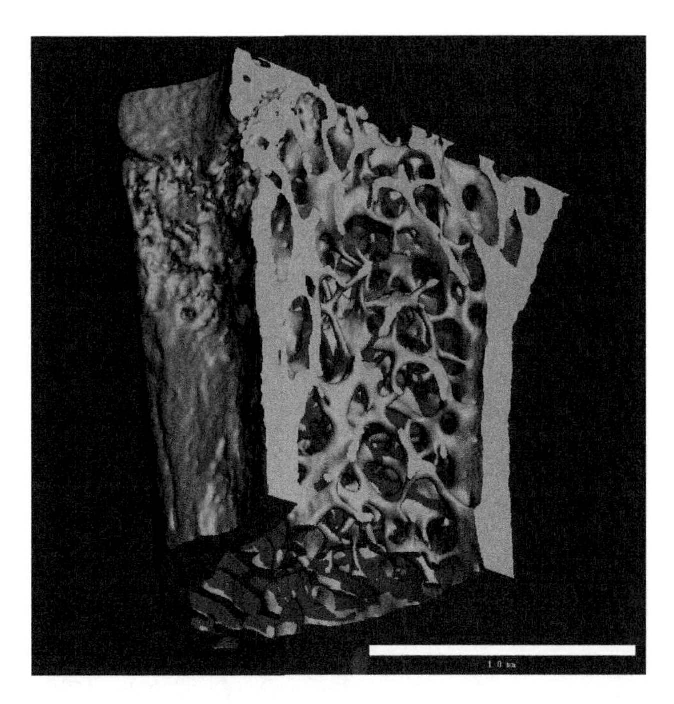

FIGURE 6.8
3D rendering of the trabecular bone in the mouse femur shown in Figs. 6.2 and 6.5. The scale bar at the bottom right is 1 mm long.

example of this sort is used in Fig. 6.8 to show the trabecular structure in a mouse femur. Note that part of the cortex and trabecular bone has been numerically removed to reveal the center. It is always useful to remember that renderings showing the low absorption voxels of an object can be as effective as those showing the high absorption voxels. Figure 6.9 shows the low absorption voxels within a small volume of a rabbit femur; this

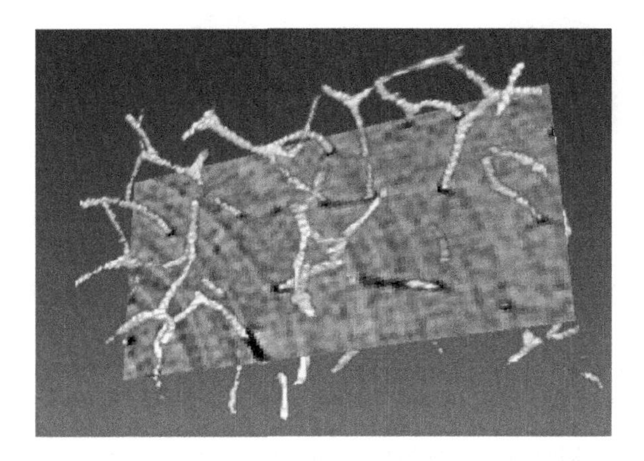

FIGURE 6.9
Small canals in rabbit femoral bone. Low absorption voxels rendered as orange-white, higher absorption voxels rendered transparent. A small section of the one of reconstructed slices within the volume is shown for background as a grayscale image. 21 keV, 1K × 1K reconstruction, 0.25° rotation increment, ~5 μm isotropic voxels. (S.R. Stock, K. Ignatiev, N.M. Rajamannan, F. De Carlo, 2-BM, APS, March 2003.)

data set was noisy and the threshold used truncated some of the channels, making them appear discontinuous in places, but paging through the slices reveals these breaks are not, in fact, present.

Thresholded 3D renderings, however, are not always superior to sets of three orthogonal grayscale sections for visualizing structures. Figure 6.10 compares the two representations of the stereom of a sea urchin demipyramid, a structure with nearly 50 vol.% empty

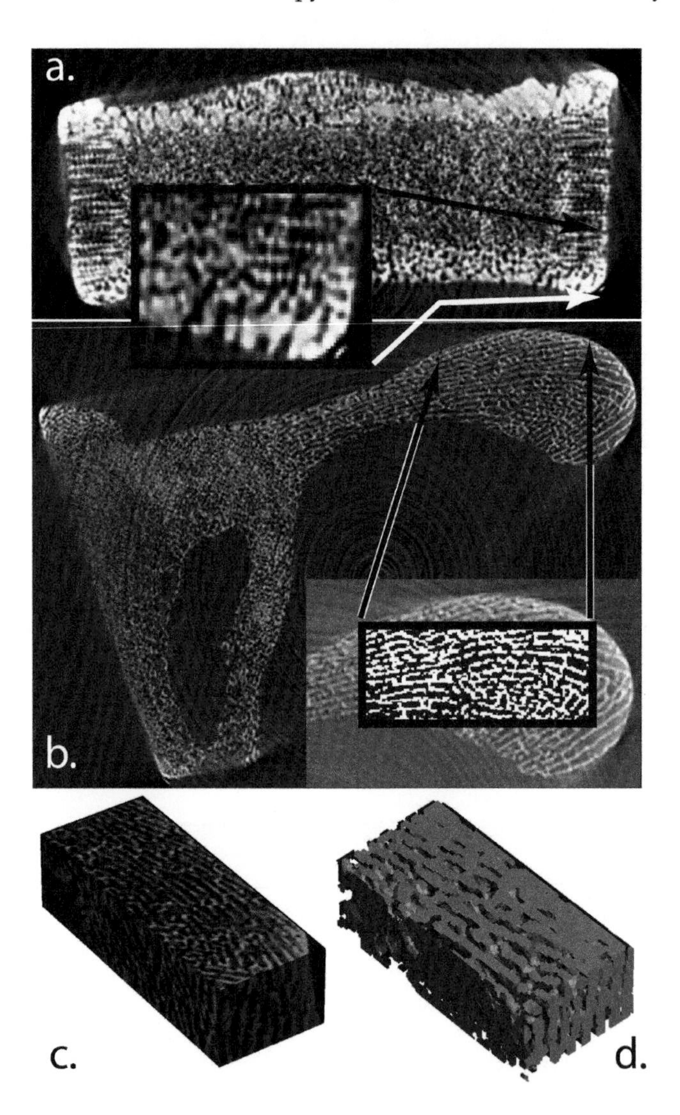

FIGURE 6.10
Synchrotron microCT data for an interambulacral plate of *Lytechinus variegatus* (a) and a demipyramid of *Asthenosoma varium* (b–d). (a) Slice with the plate exterior at the top; the inset box shows the magnified section of the ossicle indicated by the arrows, and the horizontal field of view is 550 voxels (2.75 mm). (b) Slice with an enlarged area inset. The black box defines the ROI for numerical microstructure evaluation; the pixels are binarized to either calcite (white) or void (black). The horizontal field of view is 687 voxels (3.44 mm). (c) Grayscale isometric views and (d) thresholded 3D rendering from within the ROI defined in (b); the volumes are both 37 voxels (0.185 mm) high. In (a–c) the lighter the pixel, the higher the linear attenuation coefficient. In (d) the higher absorption voxels are shown solid and lower values are rendered transparent. 21 keV, 1K × 1K reconstruction, 0.25° rotation increment (Stock, Ignatiev et al. 2004a).

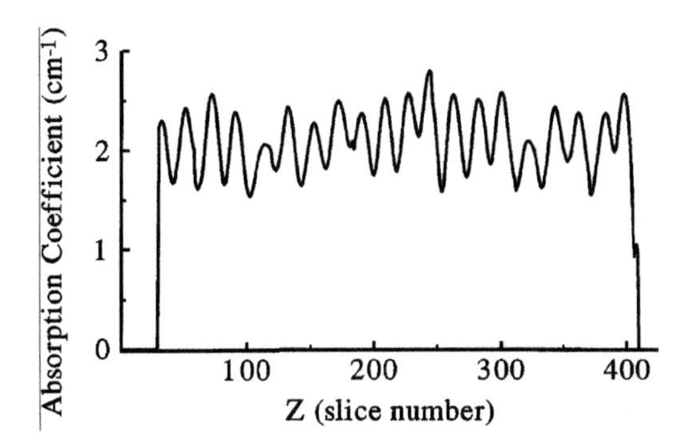

FIGURE 6.11
Mean linear attenuation coefficient per slice for a composite preform of SiC cloths. The 20 peaks correspond to the centers of the 20 cloths stacked perpendicular to the tomography rotation axis (Lee 1993, Lee, Stock et al. 1998).

space. In the author's opinion, Fig. 6.10d, the 3D rendering with a threshold yielding about 50 vol.% solid, is a much poorer visualization of the structure than Fig. 6.10c, the three orthogonal grayscale section representation. It is very difficult to make sense out of very crowded renderings unless one does significant editing and/or color coding of specific structures such as has been used in some recent studies (Donoghue, Bengtson et al. 2006, Parkinson, McDermott et al. 2008).

Representation of microstructural variation using quantities extracted from volumetric microCT data sets requires more work than the approaches described above. It can be as simple as plotting the mean linear attenuation coefficient per slice as a function of position along an important direction (assumed parallel to the tomography rotation axis). Figure 6.11 shows such a plot for a stack of SiC cloths, and the peaks and valleys clearly show the average centers of the cloths (Lee 1993, Lee, Stock et al. 1998). Deformation of two fairly dense Al "foams" produced by powder metallurgy (46% and 37% of full density) was quantified by measuring mean porosity as a function of slice number for compression along the tomographic rotation axis (Wang, Hu et al. 2006), much like what was done the SiC cloths or for liquid foam (Lambert, Cantat et al. 2005). Displacement of local maxima/minima in mean slice porosity could be followed through the different increments of deformation as the porosity diminished by up to a factor of two.

Often, 3D surfaces are represented by the triangular elements of a mesh fitted to the surface. These surface elements can be used numerically for analyses other than visual representations: for each element, its vertices define its 3D position and the normal to the element. Silva and coworkers used this information and the normal directions of neighboring surface elements to quantify bone surface erosion in a rat model of rheumatoid arthritis (Silva, Ruan et al. 2006). This approach should be very valuable in microCT studies in many disciplines.

More complicated analysis is required to extract 3D variations of quantities such as porosity or crack opening, particularly when the quantity needs to be followed as a function of time (for crack opening, as a function of applied stress). It is best to illustrate this approach concretely, and the example used, crack opening as a function of position for different in situ loads (Guvenilir 1995, Guvenilir, Breunig et al. 1997), anticipates more detailed discussion in Chapter 11.

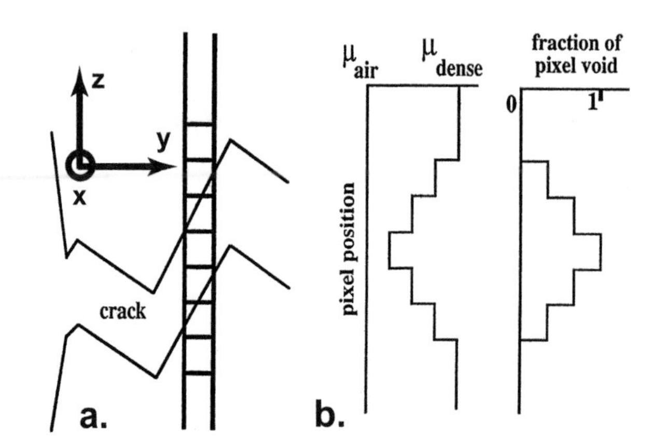

FIGURE 6.12
Illustration of the analysis used to generate the distribution of crack openings shown in Figs. 6.13 and 6.14. (a) A column of voxels (represented by squares) along the loading direction **z** (parallel to the tomography reconstruction axis). Crack openings occupying whole and partial voxels are shown for one (x,y) position. (b) Schematic of values of linear attenuation coefficient along a particular column (left) and corresponding partial voxels of open crack (right). The partial voxels are summed for each column and used in the 2D and 3D representations of the distribution of crack opening.

For rough cracks without significant tortuosity (that is, the cracks do not have many parallel branches but do deviate considerably from a simple surface), crack opening vs position can be determined relatively simply as follows. Assume that the loading direction (**z**) is parallel to the tomography rotation axis and hence perpendicular to the reconstruction plane (containing coordinate axes **x** and **y**, Fig. 6.12). The voxels along each column (x,y) have a range of values, and generally the crack will be at the position with the minimum voxel. In practice one also requires that the identified crack voxel be in the vicinity of the other crack voxels. The column consists of a variable number of partial voxels (Fig. 6.12a), and the difference of the voxels' values from the mean value of the uncracked material represents partial voxels of opening (Fig. 6.12b). These fractions can be added to give the total opening at that position (x,y) and can be projected along **z** onto the x-y plane to give a simple 2D representation of the variation of crack opening. Figure 6.13 shows fatigue crack opening in a cylindrical Al specimen projected onto the x-y plane for four applied loads; the different colors show the local experimental opening (Guvenilir, Breunig et al. 1997). This 2D simplification focuses attention on patterns of opening without the complication of the third spatial dimension. Similar 2D projected views were developed for intercloth channel widths within SiC/SiC cloth-lay-up composites and revealed patterns of opening that depended on the relative displacements of the holes in the woven cloths on either side of the channel (Lee 1993, Lee, Stock et al. 1998). These well-defined patterns (see Chapter 10) and their relationship to hole position were difficult to appreciate in numerical sections or in inverse 3D renderings of the specimen porosity.

The 2D projection of crack openings onto a plane (Fig. 6.13) ignores the important third dimension characteristic of the very rough cracks in this Al alloy. Figure 6.14a shows a mesh map of one of the crack faces, and Fig. 6.14b–d shows fatigue crack opening (in color) at the correct positions on the 3D surface (Guvenilir, Breunig et al. 1997). The contact or lack of contact between peaks and valleys of opposing crack faces is readily apparent and, as is discussed in Chapter 11, changes with applied stress.

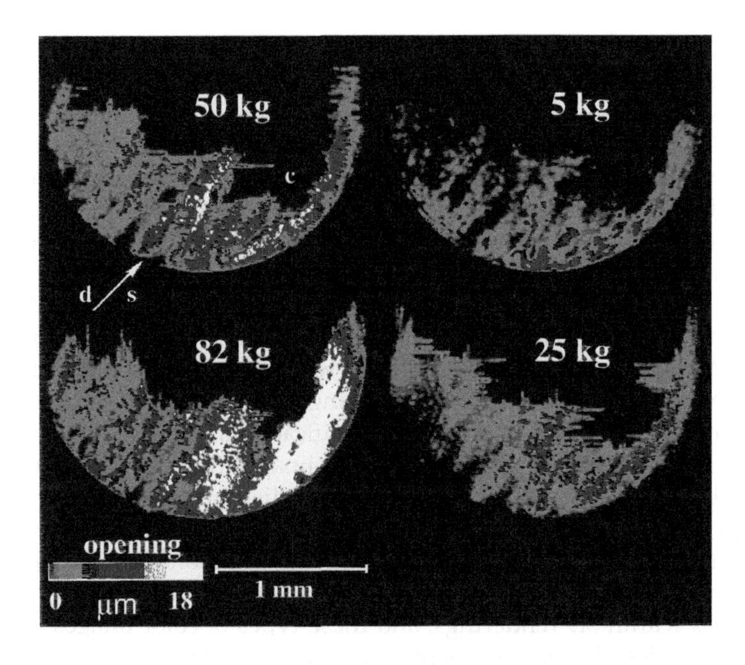

FIGURE 6.13
Crack opening in a sample of an Al alloy projected on the plane normal to the load axis. The color bar (lower left) indicates the total opening at each position, and the symbols are explained in the text in Chapter 11. Maps for four different applied forces (kgf) are shown. The corresponding stress intensities are given in parentheses. (a) 82 kgf (7.1 MPa√m). (b) 50 kgf (4.3 MPa√m). (c) 25 kgf (2.2 MPa√m). (d) 5 kgf (0.4 MPa√m) (Guvenilir, Breunig et al. 1997).

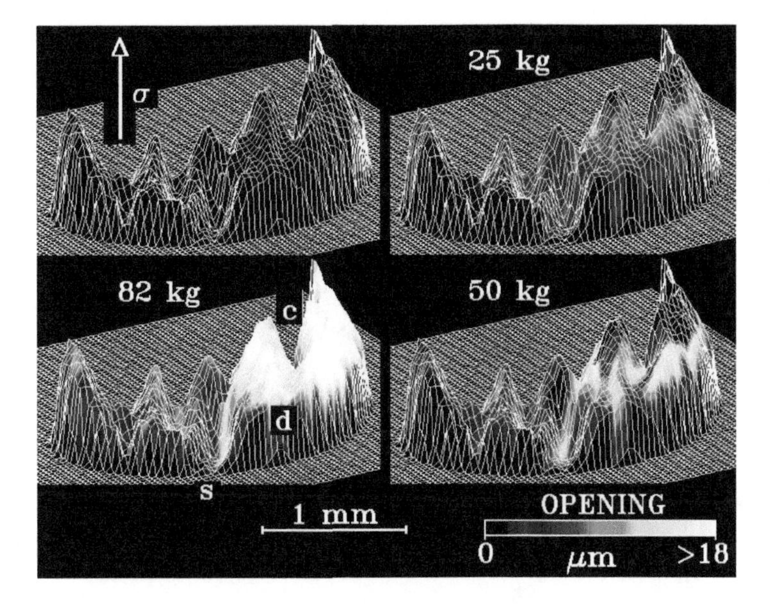

FIGURE 6.14
Crack opening in a sample of an Al alloy in 3D (see Fig. 6.13 for the openings projected onto the plane normal to the load axis). The loading direction is indicated by "σ," the color bar (lower right) indicates the total opening at each position on the 3D mesh, and the other symbols are explained in the text in Chapter 11. Maps for four different applied forces (kgf) are shown. The corresponding stress intensities are given in parentheses. (upper right) Mesh map showing the 3D crack position. (lower left) Crack opening at 82 kgf (7.1 MPa√m). (lower right) Opening at 50 kgf (4.3 MPa√m). (upper right) Opening at 25 kgf (2.2 MPa√m) (Guvenilir, Breunig et al. 1997).

Distance transform data of local structural thicknesses can be readily combined with 3D renderings using local coloring of the solid to emphasize local thickness. This sort of representation is used frequently in biomedical studies (e.g., trabecular bone or blood vessel networks). Figure 6.15 shows the voxels of cracks and voids rendered solid in a cement specimen that had been attacked by sulfate ions. The colors are the local thick thicknesses as determined by the distance transform method, and the results of the study of sulfate attack of cement, mentioned in the experiment design section above, are reviewed in the chapter on environmental interactions.

A very effective use of 3D renderings is to use two different thresholds to represent different volumes of the specimen and to render the more extensive volume semitransparent. Figure 6.16 is of a human heart valve in which significant calcification developed (Rajamannan, Nealis et al. 2005). The 3D rendering shows the soft tissue as semitransparent light blue and the more highly absorbing calcification as white.

Many recent microCT papers incorporate color images (if not in the hardcopy, then in the online version of the journal). As the human eye distinguishes more levels of contrast in color images than in grayscale images, color is sometimes used to increase the dynamic range visible in slices, but this is used less frequently than one would expect. Color has been used primarily to present four- or more dimensional data or to label different discrete subvolumes within 3D renderings, and these types of color images will be increasingly important in descriptions of scientific and engineering studies. More than a few color figures have appeared, however, that could have been equally effective as grayscale images, but this is perhaps an overly pedantic observation.

Supplemental data in particular movies posted on journal websites have become an increasingly important component of publications. Movies paging through a stack of

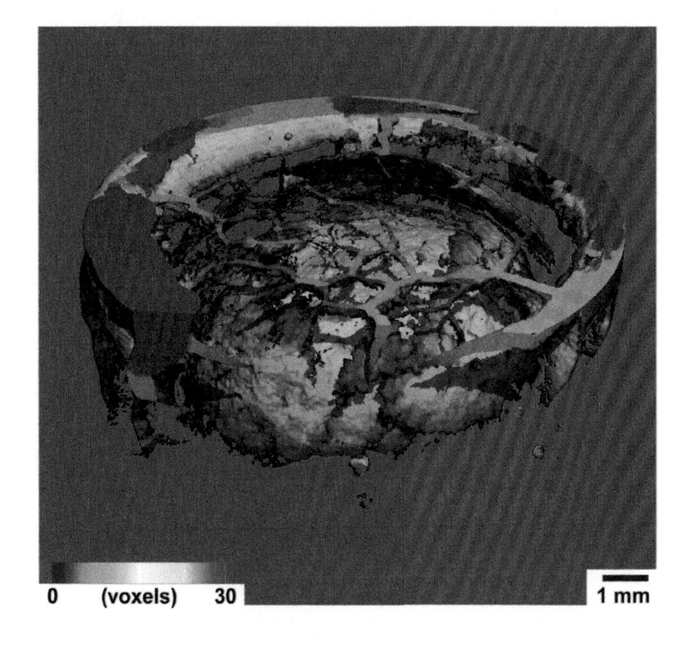

FIGURE 6.15
Three-dimensional rendering of the open voxels within the interior of a Portland cement specimen that had suffered significant damage from sulfate ion attack. The colors represent local thicknesses (in units of voxels, each of which was 37 μm on an edge) determined using the distance transform. Unpublished image from the study is described in Stock et al. (2002), Naik (2003), and Naik et al. (2006).

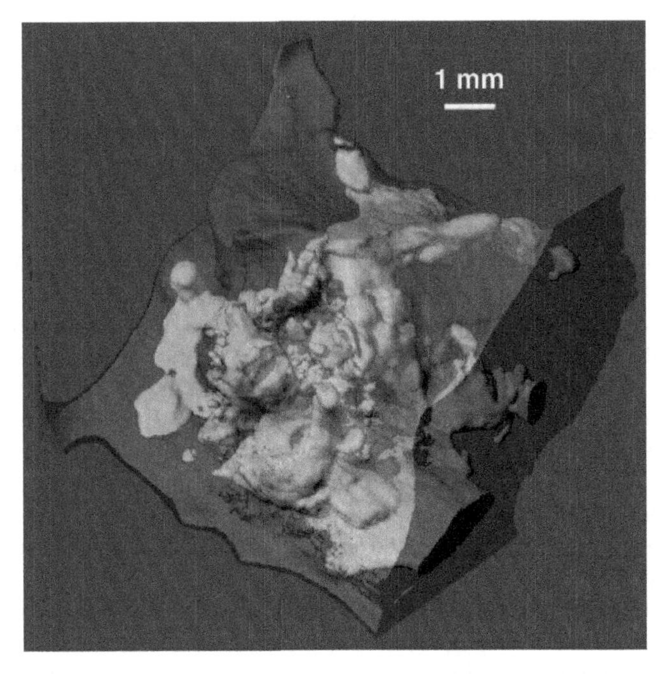

FIGURE 6.16
Rendering of calcified human heart valve employing two thresholds, a lower value for the soft tissue (semi-transparent light blue) and a higher level for the calcification (white). The flat surface represents a numerical section through the volume. (Rajamannan, Nealis et al. 2005) © Lippincott, Williams, and Wilkins, 2005, used with permission.

slices, showing different perspectives on a rendered volume or removing outer layers of an object and exposing interior structure, are popular supplements. Such 3D renderings (spinning or viewing perspective under user control) can be produced without particular difficulty or expense. MicroCT can be used as was used to create physical models with 3D printing and other additive manufacturing methods (e.g., osteoporotic and normal trabecular bone (Borah, Gross et al. 2001)), and this is now very inexpensive. Figure 6.17

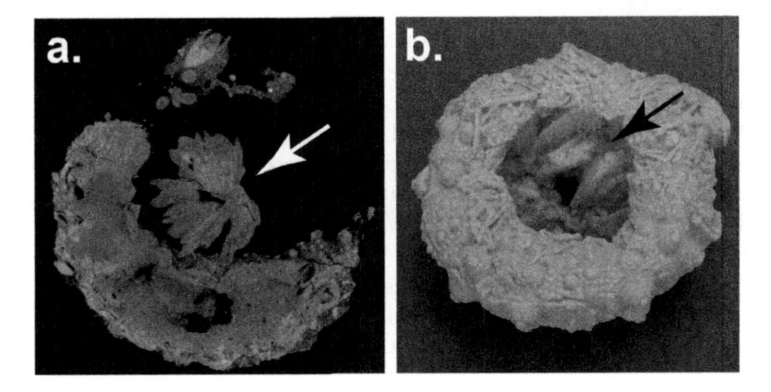

FIGURE 6.17
(a) MicroCT slice of sea urchin fossil. (S.R. Stock and G.R. Davis unpublished data.) (b) Photograph of 3D-printed replica of the fossil. The microCT data were the input for the printer (courtesy A. Limaye). The arrows point to the structures of the sea urchin's jaw.

shows a lab microCT reconstructed slice of a fossil sea urchin and a photo of a 3D-printed replica of the fossil based on the microCT data. For many individuals, there is something profoundly satisfying in handling a physical object. In the case of the visually impaired, this may be the only way they can interact with an object; the Museum of Natural History (London), for example, is investigating use of 3D printing to enlarge microscopic objects (scanned by microCT) for display. In the museum and exhibition realm, interactive augmented reality is being used increasingly; one example is CT data viewed on a tablet and superimposed on the physical object and allowing observation from different viewing positions (Matsuda 2019).

Some groups maintain websites where extensive microCT or CT data sets are posted. For example, the digital morphology website (www.digimorph.org) includes anatomical data on a wide variety of animals, renderings and slices of which can be viewed in a variety of ways. The visible cement website maintained by NIST contains synchrotron microCT data sets and serves as a standard for benchmarking new analysis programs for this class of materials (Bentz, Mizell et al. 2002). TomoBank contains data sets including round-robin samples measured at different synchrotron radiation sources and a collection of phantoms for developers to use to compare performance of different reconstruction approaches (De Carlo, Gürsoy et al. 2018).

As a final note to this chapter, 3D visualization increasingly relies on the use of GPUs to handle the large data sets being studied. By way of context, the author's first 3D rendering of microCT data was produced in 1990 on a dedicated PIXAR machine which was so expensive that only places like weapons labs and movie studios could afford to operate them. Now, high-end GPUs are affordable if not cheap. The challenge going forward will not be to throw more computation power at visualizations but to focus this power to illuminate rather than obscure, to convey information as clearly as in the Minard Map (Tufte 2001).

References

Apostol, L. and F. Peyrin (2007). "Connectivity analysis in very large 3D microtomographic images." *IEEE Trans Nucl Sci* **54**: 167–172.

Benaouli, A. H., L. Froyen and M. Wevers (2000). Micro focus computed tomography of aluminium foams. *X-ray Tomography in Materials Science*. J. Baruchel, J. Y. Buffière, E. Maire, P. Merle and G. Peix. Paris, Hermes Science: 139–154.

Benouali, A. H., L. Froyen, T. Dillard, S. Forest and F. N'Guyen (2005). "Investigation on the influence of cell shape anisotropy on the mechanical performance of closed cell aluminum foams using micro-computed tomography." *J Mater Sci* **40**: 5801–5811.

Bentz, D. P., S. Mizell, S. Satterfield, J. Devaney, W. George, P. Ketcham, J. Graham, J. Porterfield, D. Quenard, F. Vallee, H. Sallee, E. Boller and J. Baruchel (2002). "The visible cement data set." *J Res NIST* **107**: 137–148. See also www.visiblecement.nist.gov.

Bissinger, O., C. Götz, K. D. Wolff, A. Hapfelmeier, P. M. Prodinger and T. Tischer (2017). "Fully automated segmentation of callus by microCT compared to biomechanics." *J Orthop Surg Res* **12**: 108.

Borah, B., G. J. Gross, T. E. Dufresne, T. S. Smith, M. D. Cockman, P. A. Chmielewski, M. W. Lundy, J. R. Hartke and E. W. Sod (2001). "Three-dimensional microimaging (MRμI and μCT), finite element modeling, and rapid prototyping provide unique insights into bone architecture in osteoporosis." *Anat Rec* **265**: 101–110.

Borbély, A., P. Kenesei and H. Biermann (2006). "Estimation of the effective properties of particle-reinforced metal–matrix composites from microtomographic reconstructions." *Acta mater* **54**: 2735–2744.

Boyd, S. K., S. Moser, M. Kuhn, R. J. Klinck, P. L. Krauze, R. Muller and J. A. Gasser (2006). "Evaluation of three-dimensional image registration methodologies for in vivo micro-computed tomography." *J Biomed Eng* **34**: 1587–1599.

Breunig, T. M., S. R. Stock, S. D. Antolovich, J. H. Kinney, W. N. Massey and M. C. Nichols (1992). A framework relating macroscopic measures and physical processes of crack closure of Al-Li Alloy 2090. *Fracture Mechanics: Twenty-Second Symposium (Vol. 1).* H. A. Ernst, A. Saxena and D. L. McDowell. Philadelphia, ASTM. **ASTM STP 1131**: 749–761.

Breunig, T. M., S. R. Stock, A. Guvenilir, J. C. Elliott, P. Anderson and G. R. Davis (1993). "Damage in aligned fibre SiC/Al quantified using a laboratory x-ray tomographic microscope." *Composites* **24**: 209–213.

Brown, J. M., E. Ross, G. Desanti, A. Saghir, A. Clark, C. Buckley, A. Filer, A. Naylor and E. Claridge (2017). "Detection and characterisation of bone destruction in murine rheumatoid arthritis using statistical shape models." *Med Image Anal* **40**: 30–43.

Brunke, O., S. Oldenbach and F. Beckmann (2005). "Quantitative methods for the analysis of synchrotron-µCT datasets of metallic foams." *Eur J Appl Phys* **29**: 73–81.

Burghardt, A. J., G. J. Kazakia and S. Majumdar (2007). "A local adaptive threshold strategy for high resolution peripheral quantitative computed tomography of trabecular bone." *J Biomed Eng* **35**: 1678–1686.

Caselles, V., R. Kimmel and G. Sapiro (1997). "Geodesic active contours." *Int J Computer Vision* **22**: 61–79.

Chow, P. L., D. B. Stout, E. Komisopoulou and A. R. Chatziioannou (2006). "A method of image registration for small animal, multi-modality imaging." *Phys Med Biol* **51**: 379–390.

Cid, H. E., G. Carrasco-Núñez and V. C. Manea (2017). "Improved method for effective rock microporosity estimation using X-ray microtomography." *Micron* **97**: 11–21.

Cullity, B. D. and S. R. Stock (2001). *Elements of X-Ray Diffraction*. Upper Saddle River (NJ), Prentice-Hall.

De Carlo, F., D. Gürsoy, D. J. Ching, K. J. Batenburg, W. Ludwig, L. Mancini, F. Marone, R. Mokso, D. M. Pelt, J. Sijbers and M. Rivers (2018). "TomoBank: A tomographic data repository for computational x-ray science." *Meas Sci Technol* **29**: 034004.

Donoghue, P. C. J., S. Bengtson, X. Dong, N. J. Gostling, T. Huldtgren, J. A. Cunningham, C. Yin, F. P. Z. Yue and M. Stampanoni (2006). "Synchrotron X-ray tomographic microscopy of fossil embryos." *Nature* **442**: 680–683.

Doube, M., M. M. Kłosowski, I. Arganda-Carreras, F. Cordeliéres, R. P. Dougherty, J. Jackson, B. Schmid, J. R. Hutchinson and S. J. Shefelbine (2010). "BoneJ: Free and extensible bone image analysis in ImageJ." *Bone* **47**: 1076–1079.

Earles, J. M., T. Knipfer, A. Tixier, J. Orozco, C. Reyes, M. A. Zwieniecki, C. R. Brodersen and A. J. McElrone (2018). "In vivo quantification of plant starch reserves at micrometer resolution using X-ray microCT imaging and machine learning." *New Phytolog* **218** 1260–1269.

Elmoutaouakkil, A., G. Fuchs, P. Bergounhon, R. Péres and F. Peyrin (2003). "Three-dimensional quantitative analysis of polymer foams from synchrotron radiation x-ray microtomography." *J Phys D* **36**: A37–A43.

Fan, L. and C. W. Chen (1999). Integrated approach to 3D warping and registration from lung images. *Developments in X-Ray Tomography II.* U. Bonse. Bellingham (WA), SPIE. **SPIE Proc Vol 3772**: 24–35.

Feeney, D. S., J. W. Crawford, T. Daniell, P. D. Hallett, N. Nunan, K. Ritz, M. Rivers and I. M. Young (2006). "Three-dimensional microorganization of the soil-root-microbe system." *Microb Ecol* **52**: 151–158.

Fu, X., M. Dutt, A. C. Bentham, B. C. Hancock, R. E. Cameron and J. A. Elliott (2006). "Investigation of particle packing in model pharmaceutical powders using x-ray microtomography and discrete element method." *Powder Technol* **167**: 134–140.

Galibourg, A., J. Dumoncel, N. Telmon, A. Calvet, J. Michetti and D. Maret (2018). "Assessment of automatic segmentation of teeth using a watershed-based method." *Dentomaxill Acial Radiol* **47**: 20170220.

Gong, H., M. Zhang, L. Qin and Y. Hou (2007). "Regional variations in the apparent and tissue-level mechanical parameters of vertebral trabecular bone with aging using micro-finite element analysis." *Annal Biomed Eng* **35**: 1622–1631.

Gonzalez, R. C., R. E. Woods and S. L. Eddins (2004). *Digital Image Processing Using MATLAB®*. Upper Saddle River (NJ), Pearson.

Guvenilir, A. (1995). Investigation into asperity induced closure in an Al-Li alloy using X-ray tomography. Ph.D. thesis, Georgia Institute of Technology.

Guvenilir, A., T. M. Breunig, J. H. Kinney and S. R. Stock (1997). "Direct observation of crack opening as a function of applied load in the interior of a notched tensile sample of Al-Li 2090." *Acta Mater* **45**: 1977–1987.

Guvenilir, A., T. M. Breunig, J. H. Kinney and S. R. Stock (1999). "New direct observations of crack closure processes in Al-Li 2090 T8E41." *Phil Trans Roy Soc (Lond)* **357**: 2755–2775.

Guvenilir, A. and S. R. Stock (1998). "High resolution computed tomography and implications for fatigue crack closure modeling." *Fatigue Fract Eng Mater Struct* **21**: 439–450.

Hangartner, T. N. (2007). "Thresholding technique for accurate analysis of density and geometry in QCT, pQCT and microCT images." *J Musculoskelet Neuronal Interact* **7**: 9–16.

Hemmatian, H., M. R. Laurent, S. Ghazanfari, D. Vanderschueren, A. D. Bakker, J. Klein-Nulend and G. H. van Lenthe (2017). "Accuracy and reproducibility of mouse cortical bone microporosity as quantified by desktop microcomputed tomography." *PLOS One* **12**: e0182996.

Hildebrand, T., A. Laib, R. Müller, J. Dequecker and P. Rüegsegger (1999). "Direct three-dimensional morphometric analysis of human cancellous bone: Microstructural data from spine, femur, iliac crest, and calcaneus." *J Bone Miner Res* **14**: 1167–1174.

Hildebrand, T. and P. Rüegsegger (1997). "A new method for the model independent assessment of thickness in three-dimensional images." *J Microsc* **185**: 67–75.

Johnson, R. H., K. L. Karau, R. C. Molthen and C. A. Dawson (1999). Quantification of pulmonary arterial wall distensibility using parameters extracted from volumetric microCT images. *Developments in X-Ray Tomography II*. U. Bonse. Bellingham (WA), SPIE. **SPIE Proc Vol 3772**: 15–23.

Jones, A. S., A. Reztsov and C. E. Loo (2007). "Application of invariant grey scale features for analysis of porous minerals." *Micron* **38**: 40–48.

Jupe, A. C., S. R. Stock, P. L. Lee, N. N. Naik, K. E. Kurtis and A. P. Wilkinson (2004). "Phase composition depth profiles using spatially resolved energy dispersive x-ray diffraction." *J Appl Cryst* **37**: 967–976.

Kaira, C. S., X. Yang, V. De Andrade, F. De Carlo, W. Scullin, D. Gursoy and N. Chawla (2018). "Automated correlative segmentation of large Transmission X-ray Microscopy (TXM) tomograms using deep learning." *Mater Charact* **142**: 203–210.

Ketcham, R. A. (2005). "Computational methods for quantitative analysis of three-dimensional features in geological specimens." *Geosphere* **1**: 32–41.

Kim, C. H., H. Zhang, G. Mikhail, D. von Stechow, R. Müller, H. S. Kim and X. E. Guo (2007). "Effects of thresholding techniques on μCT-based finite element models of trabecular bone." *J Biomed Eng* **129**: 481–486.

Knötel, D., R. Seidel, S. Prohaska, M. N. Dean and D. Baum (2017). "Automated segmentation of complex patterns in biological tissues: Lessons from stingray tessellated cartilage." *PLOS One* **12**: e0188018.

Koestel, J. (2018). "SoilJ: An ImageJ plugin for the semiautomatic processing of three dimensional x-ray images of soils." *Vadose Zone J* **17**: 170062.

Korfiatis, V. C., S. Tassani and G. K. Matsopoulos (2017). "An Independent Active Contours Segmentation Framework for bone microCT images." *Comput Biol Med* **87**: 358–370.

Kronenberger, M., K. Schladitz, B. Hamann and H. Hagan (2018). "Fiber segmentation in crack regions of steel fiber reinforced concrete using principal curvature." *Image Anal Stereol* **37**: 127–137.

Lambert, J., I. Cantat, R. Delannay, A. Renault, F. Graner, J. A. Glazier, I. Veretennikov and P. Cloetens (2005). "Extraction of relevant physical parameters from 3D images of foams obtained by x-ray tomography." *Colloids Surf A* **263**: 295–302.

Latief, F. D., U. Fauzi, Z. Irayani and G. Dougherty (2017). "The effect of X-ray micro computed tomography image resolution on flow properties of porous rocks." *J Microsc* **266**: 69–88.

Lee, S. B. (1993). Nondestructive examination of chemical vapor infiltration of 0°/90° SiC/Nicalon composites. Ph.D. thesis, Georgia Institute of Technology.

Lee, S. B., S. R. Stock, M. D. Butts, T. L. Starr, T. M. Breunig and J. H. Kinney (1998). "Pore geometry in woven fiber structures: 0°/90° plain-weave cloth lay-up preform." *J Mater Res* **13**: 1209–1217.

Lenoir, N., M. Bornert, J. Desrues, P. Bésuelle and G. Viggiani (2007). "Volumetric digital image correlation applied to X-ray microtomography images from triaxial compression tests on Argillaceous rock." *Strain* **43**: 193–205.

Levitz, P. (2007). "Toolbox for 3D imaging and modeling of porous media: Relationship with transport properties." *Cement Concr Res* **37**: 351–359.

Limaye, A. (2006). Drishti: A volume exploration and presentation tool. *Vis 2006*. Baltimore.

Limaye, A. (2019). "Drishti: A volume exploration and presentation tool." Retrieved March 28, 2019, from http://anusf.anu.edu.au/Vizlab/drishti/index.shtml.

Lindquist, W. (2001). Quantitative analysis of three-dimensional x-ray tomographic images. *Developments in X-Ray Tomography III*. U. Bonse. Bellingham (WA), SPIE. **SPIE Proc Vol 4503**: 103–115.

Liu, L. and E. F. Morgan (2007). "Accuracy and precision of digital volume correlation in quantifying displacements and strains in trabecular bone." *J Biomech* **40**: 3516–3520.

Lorenzi, M., N. Mitroglou, M. Santini and M. Gavaises (2017). "Novel experimental technique for 3D investigation of high-speed cavitating diesel fuel flows by X-ray microcomputed tomography." *Rev Sci Instrum* **88**: 033706.

Madra, A., P. Breitkopf, B. Raghavan and F. Trochu (2018). "Diffuse manifold learning of the geometry of woven reinforcements in composites." *C R Mech* **346**: 532–538.

Marsh, M. (2019). Deep learning - transformative image processing and image segmentation solutions. *Tomography for Scientific Advancement - North America*. Gainesville (FL).

Matsuda, N. (2019). Visualizing an ancient mummy with X-ray imaging and augmented reality. *A Child from Roman Egypt: Historical and Scientific Studies of a Portrait Mummy*. E. Ronkko, T. Terpstra and M. S. Walton. Chicago, The Block Museum of Art: 101–105.

McDonald, S. A., L. C. R. Schneider, A. C. F. Cocks and P. J. Withers (2006). "Particle movement during the deep penetration of a granular material studied by x-ray microtomography." *Scripta Mater* **54**: 191–196.

Mendoza, F., P. Verboven, H. K. Mebatsion, G. Kerckhofs, M. Wevers and B. Nicolai (2007). "Three-dimensional pore space quantification of apple tissue using x-ray computed tomography." *Planta* **226**: 559–570.

Montanini, R. (2005). "Measurement of strain rate sensitivity of aluminium foams for energy dissipation." *Int J Mech Sci* **47**: 26–42.

de Moura Meneses, A. A., D. Bastos Palheta, C. J. Gomes Pinheiro and R. C. Rodrigues Barroso (2018). "Graph cuts and neural networks for segmentation and porosity quantification in synchrotron radiation X-ray microCT of an igneous rock sample." *Appl Rad Isotopes* **133**: 121–132.

Mulder, L., L. J. van Ruijven, J. H. Koolstra and T. M. G. J. van Eijden (2007). "The influence of mineralization on intrabecular stress and strain distribution in developing trabecular bone." *Annal Biomed Eng* **35**: 1668–1677.

Naik, N. (2003). Sulfate attack on Portland cement-based materials: mechanisms of damage and long term performance. Ph.D. thesis, Georgia Institute of Technology.

Naik, N. N., A. C. Jupe, S. R. Stock, A. P. Wilkinson, P. L. Lee and K. E. Kurtis (2006). "Sulfate attack monitored by microCT and EDXRD: Influence of cement type, water-to-cement ratio, and aggregate." *Cement Concr Res* **36**: 144–159.

Naik, N. N., K. E. Kurtis, A. P. Wilkinson, A. C. Jupe and S. R. Stock (2004). Sulfate deterioration of cement-based materials examined by x-ray microtomography. *Developments in X-Ray Tomography IV.* U. Bonse. Bellingham (WA), SPIE. **SPIE Proc Vol 5535**: 442–452.

Nakashima, Y., T. Nakano, K. Nakamura, K. Uesugi, A. Tsuchiyama and S. Ikeda (2004). "Three-dimensional diffusion of non-sorbing species in porous sandstone: Computer simulation based on x-ray microtomography using synchrotron radiation." *J Contam Hydrol* **74**: 253–264.

Nielsen, S. F., F. Beckmann, R. B. Godiksen, K. Haldrup, H. F. Poulsen and J. A. Wert (2004). Measurement of the components of plastic displacement gradients in three dimensions. *Developments in X-Ray Tomography IV.* U. Bonse. Bellingham (WA), SPIE. SPIE Proc Vol **5535**: 485–492.

Nielsen, S. F., H. F. Poulsen, F. Beckmann, C. Thorning and J. Wert (2003). "Measurements of plastic displacement gradient components in three dimensions using marker particles and synchrotron x-ray absorption microtomography." *Acta Mater* **51**: 2407–2415.

Oh, W. and W. B. Lindquist (1999). "Image thresholding by indicator kriging." *IEEE Trans Pattern Anal Mach Intell* **21**: 590–602.

Osborne, D. R., C. Kuntner, S. Berr and D. Stout (2017). "Guidance for efficient small animal imaging quality control." *Mol Imaging Biol* **19**: 485–498.

Parkinson, D. Y., G. McDermott, L. D. Etkin, M. A. Le Gros and C. A. Larabell (2008). "Quantitative 3-D imaging of eukaryotic cells using soft X-ray tomography." *J Struct Biol* **162**: 380–386.

Perilli, E., F. Baruffaldi, M. Visentin, B. Bordini, F. Traina, A. Cappello and M. Viceconti (2007). "MicroCT examination of human bone specimens: Effects of polymethylmethacrylate embedding on structural parameters." *J Microsc* **225**: 192–200.

du Plessis, A., C. Broeckhoven, A. Guelpa and S. G. le Roux (2017). "Laboratory x-ray micro-computed tomography: A user guideline for biological samples." *Gigascience* **6**(6): 1–11.

Rajagopalan, S., L. Lu, M. J. Yaszemski and R. A. Robb (2005). "Optimal segmentation of microcomputed tomographic images of porous tissue-engineering scaffolds." *J Biomed Mater Res* **75A**: 877–887.

Rajamannan, N. M., T. B. Nealis, M. Subramaniam, S. R. Stock, K. I. Ignatiev, T. J. Sebo, J. W. Fredericksen, S. W. Carmichael, T. K. Rosengart, T. C. Orszulak, W. D. Edwards, R. O. Bonow and T. C. Spelsberg (2005). "Calcified rheumatic valve neoangiogenesis is associated with VEGF expression and osteoblast-like bone formation." *Circulation* **111**: 3296–3301.

Rajon, D. A., J. C. Pichardo, J. M. Brindle, K. N. Kielar, D. W. Jokisch, P. W. Patton and W. E. Bolch (2006). "Image segmentation of trabecular spongiosa by inspection of visual gradient magnitude." *Phys Med Biol* **51**: 4447–4467.

Ramaswamy, S., M. Gupta, A. Goel, U. Aaltosalmi, M. Kataja, A. Koponen and B. V. Ramarao (2004). "The 3D structure of fabric and its relationship to liquid and vapor transport." *Colloids Surf A* **241**: 323–333.

Rolland Du Roscoat, S., M. Decain, X. Thibault, C. Geindreau and J. F. Bloch (2007). "Estimation of microstructural properties from synchrotron x-ray microtomography and determination of the REV in paper materials." *Acta Mater* **55**: 2841–2850.

Rovaris, K., P. M. Queiroz, K. F. Vasconcelos, L. D. S. Corpas, B. M. D. Silveira and D. Q. Freitas (2018). "Segmentation methods for micro CT images: A comparative study using human bone samples." *Braz Dent J* **29**(2): 150–153.

Rüegsegger, P., B. Koller and R. Müller (1996). "A microtomographic system for the nondestructive evaluation of bone architecture." *Calcif Tiss Int* **58**: 24–29.

Sheppard, A. P., C. H. Arns, A. Sakellariou, T. J. Senden, R. M. Sok, H. Averdunk, M. Saadatfar, A. Limaye and M. A. Knackstedt (2006). Quantitative properties of complex porous materials calculated from x-ray μCT images. *Developments in X-Ray Tomography V.* U. Bonse. Bellingham (WA), SPIE. **SPIE Proc Vol 6318**: 631811-1–631811-5.

Showalter, C., B. D. Clymer, B. Richmond and K. Powell (2006). "Three-dimensional texture analysis of cancellous bone cores evaluated at clinical CT resolutions." *Osteopor Int* **17**: 259–266.

Silva, M. D., J. Ruan, E. Siebert, A. Savinainen, B. Jaffee, L. Schopf and S. Chandra (2006). "Application of surface roughness analysis on micro-computed tomographic images of bone erosion: Examples using a rodent model of rheumatoid arthritis." *Mole Imaging* **5**: 475–484.

Smal, P., P. Gouze and O. Rodriguez (2018). "An automatic segmentation algorithm for retrieving sub-resolution porosity from X-ray tomography images." *J Petrol Sci Eng* **166**: 198–207.

Spowage, A. C., A. P. Shacklock, A. A. Malcolm, S. L. May, L. Tong and A. R. Kennedy (2006). "Development of characterization methodologies for macroporous materials." *J Porous Mater* **13**: 431–438.

Stock, S. R., T. A. Ebert, K. Ignatiev and F. D. Carlo (2006). Structures, structural hierarchy and function in sea urchin spines. *Developments in X-Ray Tomography V.* U. Bonse. Bellingham (WA), SPIE. **SPIE Proc Vol 6318**: 63180A-1–63180A-4.

Stock, S. R., K. Ignatiev and F. D. Carlo (2004a). Very high resolution synchrotron microCT of sea urchin ossicle structure. *Echinoderms: München.* T. Heinzeller and J. H. Nebelsick. London, Taylor and Francis: 353–358.

Stock, S. R., K. I. Ignatiev, S. A. Foster, L. A. Forman and P. H. Stern (2004b). "MicroCT quantification of in vitro bone resorption of neonatal murine calvaria exposed to IL-1 or PTH." *J Struct Biol* **147** 185–199.

Stock, S. R., S. Nagaraja, J. Barss, T. Dahl and A. Veis (2003). "X-Ray microCT study of pyramids of the sea urchin Lytechinus variegatus." *J Struct Biol* **141**: 9–21.

Stock, S. R., N. N. Naik, A. P. Wilkinson and K. E. Kurtis (2002). "X-ray microtomography (microCT) of the progression of sulfate attack of cement paste." *Cement Concr Res* **32**: 1673–1675.

Su, R., G. M. Campbell and S. K. Boyd (2007). "Establishment of an architecture-specific experimental validation approach for finite element modeling of bone by rapid prototyping and high resolution computed tomography." *Med Eng Phys* **29**: 480–490.

Tan, J. C., J. A. Elliott and T. W. Clyne (2006). "Analysis of tomography images of bonded fibre networks to measure distributions of fiber segment length and fiber orientation." *Adv Eng Mater* **8**: 495–500.

Thompson, K. E., C. S. Willson and K. T. W. Zhang, A. H. Reed, L. Beenken, (2006). "Quantitative computer reconstruction of particulate materials from microtomography images." *Powder Technol* **163**: 169–182.

Titschack, J., D. Baum, K. Matsuyama, K. Boos, C. Färber, W.-A. Kahl, K. Ehrig, D. Meinel, C. Soriano and S. R. Stock (2018). "Ambient occlusion - a powerful algorithm to segment shell and skeletal intrapores in computed tomography data." *Comput Geosci* **115**: 75–87.

Toda, H., T. Ohgaki, K. Uesugi, M. Kobayashi, N. Kuroda, T. Kobayashi, M. Niinomi, T. Akahori, K. Makii and Y. Aruga (2006). "Quantitative assessment of microstructure and its effects on compression behavior of aluminum foams via high-resolution synchrotron x-ray tomography." *Metall Mater Trans A* **37**: 1211–1219.

Toda, H., I. Sinclair, J. Y. Buffiere, E. Maire, T. Connolley, M. Joyce, K. H. Khor and P. Gregson (2003). "Assessment of the fatigue crack closure phenomenon in damage-tolerant aluminium alloy by in-situ high-resolution synchrotron x-ray microtomography." *Phil Mag* **83**: 2429–2448.

Toda, H., I. Sinclair, J. Y. Buffiere, E. Maire, K. H. Khor, P. Gregson and T. Kobayashi (2004). "A 3D measurement procedure for internal local crack driving forces via synchrotron x-ray microtomography." *Acta Mater* **52**: 1305–1317.

Toda, H., M. Takata, T. Ohgaki, M. Kobayashi, T. Kobayashi, K. Uesugi, K. Makii and Y. Aruga (2006). "3-D image-based mechanical simulation of aluminium foams: Effects of internal microstructure." *Adv Eng Mater* **8**: 459–467.

Torre, I. G., J. C. Losada, R. J. Heck and A. M. Tarquis (2018). "Multifractal analysis of 3D images of tillage soil." *Geoderma* **311**: 167–174.

Tufte, E. R. (2001). *The Visual Display of Quantitative Information.* Cheshire (CT), Graphics Press.

Verhulp, E., B. van Rietbergen and R. Huiskes (2004). "A three-dimensional digital image correlation technique for strain measurements in microstructures." *J Biomech* **37**: 1313–1320.

Waarsing, J. H., J. S. Day, J. C. v. d. Linden, A. G. Ederveen, C. Spanjers, N. D. Clerck, A. Sasov, J. A. N. Verhaar and H. Weinans (2004). "Detecting and tracking local changes in the tibiae of individual rats: A novel method to analyse longitudinal in vivo microCT data." *Bone* **34**: 163–169.

Wan, S. Y., E. L. Ritman and W. E. Higgins (2002). "Multi-generational analysis and visualization of the vascular tree in 3D microCT images." *Computers Biol Med* **32**: 55–71.

Wang, M., X. F. Hu and X. P. Wu (2006). "Internal microstructure evolution of aluminum foams under compression." *Mater Res Bull* **41**: 1949–1958.

Wilkinson, A. P., A. C. Jupe, K. E. Kurtis, N. N. Naik, S. R. Stock and P. L. Lee (2004). Spatially resolved energy dispersive x-ray diffraction (EDXRD) as a tool for nondestructively providing phase composition depth profiles on cement and other materials. *Applications of X-rays in Mechanical Engineering 2004*. New York, ASME: 49–52.

Zettler, R., T. Donath, J. F. d. Santos, F. Beckman and D. Lohwasser (2006). "Validation of marker material flow in 4mm thick friction stir welded Al 2024-T351 through computer microtomography and dedicated metallographic techniques." *Adv Eng Mater* **8**: 487–490.

Ziegler, A., S. R. Stock, B. H. Menzec and A. B. Smith (2012). Macro- and microstructural diversity of sea urchin teeth revealed by large-scale micro-computed tomography survey. *Developments in X-Ray Tomography VIII*. S. R. Stock. Bellingham (WA), SPIE. **8506**: 850651.

7

"Simple" Metrology and Microstructure Quantification

The subject of this chapter is metrology and microstructure characterization. Both subjects are covered elsewhere in Chapters 8–12, and the coverage here focuses on "simple" quantification, that is, microCT studies relying on elementary thresholding and on single (one time point) examination of each specimen. More complex interrogations (i.e., studies of cellular solids such as metallic, ceramic, polymeric, or liquid foams or trabecular bone; observations of cracking as a function of applied stress; repeat observations of the same specimen over time) are postponed until the later chapters.

Distribution and morphology of phases is the subject of the first section of this chapter. Pharmaceutical materials and foods are covered first, followed by geological and planetary materials. Examples of studies of monolithic engineered materials (metals, ceramics, polymers) are the next subject. Manufactured composites are the fourth topic; the fifth introduces biological tissues as phases, concentrating on anatomy, and the sixth samples cultural heritage, archeology, and forensic studies.

Metrology refers to the techniques of measurement, and this chapter couples metrology with phylogeny, the study of the evolutionary relations between species, in the second half. The reason for coupling these seemingly dissimilar topics is that quantification and comparison of dimensions or structural motifs underlie the experimental and analysis approaches. The order of the subsections is: industrial metrology, paleontology, cells, plants, invertebrates, and vertebrates.

This chapter may strike the reader as a bit of a potpourri better postponed until the closing pages of the book or as such a miscellany whose topics are best folded among those of the other chapters. Instead, this chapter is collected together because the author sees a common theme of simple analysis approaches answering the questions at hand. It is important to never lose sight of the fact that the simplest approaches can yield important results and that better (more complex analyses) can be the worst enemy of good enough (i.e., a completed, meaningful study answering the questions at hand). The commonality of analysis requirements and approaches is the organizing principle behind the subsequent chapters (at least as far as the author's perception extends), and there is no reason not to start thinking in these terms from the beginning of the coverage of microCT applications.

7.1 Distribution of Phases

Phases can be voids as well as different solid phases. The spatial distribution of and relative amounts of different phases control macroscopic properties to a great extent. For many applications, microCT is the best method of obtaining 3D information or avoiding specimen preparation artifacts. There are many more studies that could have been included, but the reader should be able to gain an appreciation of this area of microCT from the examples below.

7.1.1 Pharmaceuticals and Food

A relatively new application of lab microCT is evaluation of the spatial distribution of phases in pharmaceutical manufactured materials, that is, in solid dosage forms such as tablets and soft-gelatin capsules (Hancock and Mullarney 2005). Thicknesses and interface character in multilayer tablets, microstructure of rapidly-dissolving tablets produced by lyophilization (ice crystallization followed by drying), and particle sizes within controlled-release osmotic tablets are important characteristics directly measurable via microCT. Nondestructive microCT comparison of genuine and counterfeit tablets has the additional advantage of preserving the evidence in patent litigation or other legal proceedings.

Pores in pharmaceutical granules have also been studied with lab microCT (Farber, Tardos et al. 2003), including quantification of localized density variations (Busignies, Leclerc et al. 2006), as has the spatial conformation of components in modified release tablets (Traini, Loreti et al. 2008). Hydrophilic drugs dispersed in polymeric matrices are expected to have decreased release rates as diffusion path length increases, and microCT has been examined as a way of fine tuning the pore structure to achieve more uniform drug release (Wang, Chang et al. 2007). The discrete element method (for simulating structural changes) and microCT have been compared to advance understanding of tablet formation by powder compaction (Fu, Dutt et al. 2006). Modified release through additive manufacturing of tablets, that is, "printlets" (Goyanes, Fina et al. 2017) has been examined as has the effect of the carrier on rapid or controlled release rates (Krupa, Cantin et al. 2017). Magnetic particles are being investigated for controlled drug delivery, for example, to tumors, and microCT has been used to measure these particles' distribution in tissues of animal models (Brunke, Odenbach et al. 2005).

A recent review covers the application of microCT to imaging various types of food (Wang, Herremans et al. 2018). One study focused on the multiple phases in chocolate (Frisullo, Licciardello et al. 2010) and a second on ice cream (Guo, Zeng et al. 2017). Bubbles in wheat dough (Bellido, Scanlon et al. 2006) and in noodle dough (Guillermic, Koksel et al. 2018) were the subject of two other investigations. Teams have also employed microCT to observe banana drying (Léonard, Blacher et al. 2008) and the ripening of mango (Cantre, Herremans et al. 2014). Additional studies on microstructural evolution of food appear at the beginning of Chapter 10.

7.1.2 Geological and Planetary Materials

In one synchrotron microCT study (Gualda and Rivers 2006), quartz, magnetite, sanidine size distributions were measured in pumice clasts (isolated crystals surrounded by a low-density matrix); this investigation was undertaken to address possible limitations of earlier work on the same material. Previous characterization used a crushing + sieving + winnowing procedure to quantify the size distributions (Gualda, Cook et al. 2005), avoiding stereology's well-known limitations in transforming 2D data into true measures of the 3D arrangements in the solid. Such processing, however, tends to cause significant loss of small crystals and frequently fragments larger crystals. This latter artifact is especially significant in that it obscures characterization of fragmentation generated in magmatic processes. As the steps involved in the analysis of the microCT data are characteristic of those often encountered in microCT studies, they are described here in some detail.

In phase quantification studies, the first image analysis step (after reconstruction) is image classification, that is, assigning each voxel in the 3D volume to a given phase. The second step is identifying individual grains, that is, clusters of voxels belonging to a single-phase particle. In the study cited in the previous paragraph (Gualda and Rivers 2006), contrast sensitivity (256 gray levels) was adequate to quantify the volumes and size distributions for particles greater than five voxels diameter in the Bishop Tuff pumice clast; noise limited the investigators' ability to reliably identify smaller particles. Nonetheless, Gualda and Rivers found that the combination of contrast and spatial information allowed distinction between quartz and sanidine, despite the fact that the distribution of linear attenuation coefficients of the minority phase (sanidine) formed an indistinct shoulder of the quartz peak (Gualda and Rivers 2006). The microCT results agreed with earlier results from destructive analysis (Gualda, Cook et al. 2005), namely, that the distribution of quartz particle sizes indicated action of magmatic fragmentation processes but that of magnetite was largely unaffected by the fragmentation process recorded by quartz. The tradeoff between spatial resolution and contrast sensitivity is clearly explained as it affects these results, but the authors do not provide details of the number of x-ray counts recorded in the 2 × 2 binned detector pixels (Gualda and Rivers 2006), and one is unable to assess the numerical extent to which contrast for a given phase is the spread because of counting statistics. It would have been interesting if the investigators had investigated frame averaging for improving contrast sensitivity and increasing the small particle detection limit. Of further interest in this carefully done study is the documentation of exceptional volumes that diverge quite markedly from the rest of the sample: this simple result should serve as a caution to all investigators using microCT of small sections derived from larger objects (Gualda and Rivers 2006).

Bubble (vesicle) characteristics in basalts provide insight into various processes in magma prior to, during, and after eruption; a distribution of bubble sizes, for example, can be due to multiple or continuous nucleation processes or differences in growth rates. Synchrotron microCT of five basalts from different locations showed bubbles were spheroidal and comprised 45 vol.% of lavas to 80% of scoria and were at least 90% interconnected (Song, Jones et al. 2001). Inhomogeneous distribution of bubbles can affect magma fragmentation (Kameda, Ichihara et al. 2017).

Anisotropy measurements of magnetic susceptibility and of elastic constants (from P-wave velocity) were combined with microCT in a study of Callovo-Oxfordian argillite (David, Robion et al. 2007). An unexpected finding was the presence of tubular structures of dense materials (pyrite).

Study of small diameter inclusions and of materials of different voxel or sub-voxel porosity content can be very important and challenging. Microdiamond content and size distribution in kimberlite are important indicators of the likelihood of finding coarser valuable diamonds. MicroCT has been applied to the problem of quantifying diamond content of drill-hole cores, and these authors report tomography research at De Beers (Schena, Favretto et al. 2005). Distribution of pore sizes was examined on oolite (Dos Reis, Nagata et al. 2017). Zones of differing levels or porosity have been studied in mortar and are important in resistance of this construction material to environmental attack (Diamond and Landis 2007). Soil cratering by raindrops (Beczek, Ryzak et al. 2018) and structure of a Martian meteorite (do Nascimento-Dias, de Oliveira et al. 2018) are other studies.

7.1.3 Two or More Phase Metals, Ceramics, and Polymers

Pyun and coworkers used phase microCT to visualize the 3D morphology of polymer blends (Pyun, Bell et al. 2007). The material was polystyrene plus high-density polyethylene; the investigators visualized both phases by segmentation and, when one phase was dissolved, found that the interfacial S_V after various annealing times agreed with the results mercury porosimetry.

Radio-opaque polymers are desirable for dental applications; Anderson and coworkers employed lab microCT to show barium methacrylate monomer did not blend well when diluted in methacrylate, whereas inhomogenieties could not be detected when tin methacrylate was used (Anderson, Ahmed et al. 2006). Void distribution determination in HY-100 steel is one of the few synchrotron microCT studies of very highly attenuating material (Everett, Simmonds et al. 2001). Catalytic conversion of natural gas to (clean) liquid fuels via Fe nanoparticles is of interest for lessening internal combustion-related pollution, and the spatial distribution of these nanoparticles has been studied with fluorescence microCT (Jones, Feng et al. 2005).

Toda and coworkers used nanoCT to image γ' Ag_2Al precipitates in an Al alloy (Toda, Uesugi et al. 2006). The $\{111\}_{Al}$ habit planes of the precipitates were visible in their 3D renderings (Fig. 7.1). Solidification structures in an Al-Si alloy were seen via phase microCT even though the difference in absorption contrast between Al and Si is very small (Fig. 4.10).

7.1.4 Manufactured Composites

In early work on manufactured composites, the pore structure in degassed, and nondegassed reaction-bonded silicon nitride/silicon carbide was studied with 10-μm voxels in the 1 mm × 1 mm cross-section (Stock, Guvenilir et al. 1989), and, not surprisingly, the 15-μm diameter Nicalon fibers could not be resolved. Within a given cross-section of an Al powder processed composite, the content of 12-μm TiB_2 particles was found to deviate substantially from the nominal 20 vol.% of reinforcement: contents as low as 10 vol.% were reported (Mummery, Derby et al. 1995). Images of a SiC monofilament – Si_3N_4 (91 wt.% Si_3N_4, 6 wt.% Y_2O_3, 3 wt.% Al_2O_3) composite have been obtained with microCT (Hirano, Usami et al. 1989); in 111-μm thick slices, the 140-μm diameter fibers and their

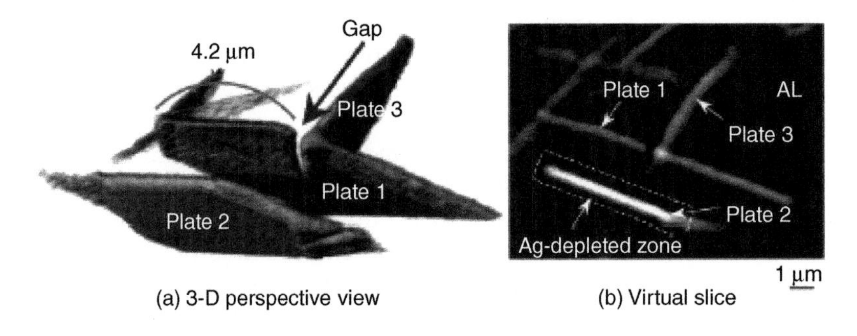

FIGURE 7.1
Plate-like γ' Ag_2Al precipitates imaged with a Fresnel zone plate nanoCT system. (a) 3D view of several precipitates segmented to only show Ag-rich phase. (b) Slice through the plates shown in (a) and showing the Al matrix (gray) and Ag-depleted zones (black) around each plate (white). Imaging at 9.8 keV with ~90-nm isotropic voxels with the precipitates being 2.5–7 voxel thickness (Toda, Uesugi et al. 2006). © 2006, American Institute of Physics.

30-µm diameter cores are quite visible, and the radial variation of linear attenuation coefficient at 24.0 keV for the SiC fibers agrees with others' 21-keV measurements of similar fibers (Kinney, Stock et al. 1990).* Thermomechanical fatigue of ceramic as well as metal matrix composites have also been investigated (Baaklini, Bhatt et al. 1995).

Aluminum matrix composites reinforced by Al_2O_3 whiskers 2–4 µm in diameter and 50–80 µm in length or by Al_2O_3 fibers about 20 µm in diameter have been studied (Bonse, Nusshardt et al. 1991). The sample diameter of the former composite was ~1 mm, while that of the latter, which was not reported, was apparently much less than 1 mm. The whisker-reinforced composite is a material normally used for diesel engine pistons, and its aluminum matrix contains significant levels of Si, Cu, Ni, and Fe. The resulting intermetallic phases were seen to form a 3D network whose mesh size was on the order of 15 µm. These investigations also found that the individual Al_2O_3 fiber images and the distribution of fiber images for the second composite agreed with scanning electron micrographs and demonstrated spatial resolution in the reconstruction of 6 µm (MTF of 80 lines pair/mm at 20% contrast) (Bonse, Nusshardt et al. 1991).

The distribution of Cr particles in alumina, determined via synchrotron microCT, was used as the input for a finite element calculation of residual stresses arising during cooling from the 1,450° C processing temperature (Geandier, Hazotte et al. 2003). Hydrostatic stresses were found in the alumina near the Cr particles. In a separate paper, Geandier and coworkers compared calculations with actual residual stress measurements (Geandier, Weisbecker et al. 2002). Such residual stress measurements are readily performed, nondestructively, using synchrotron high-energy x-ray diffraction (Haeffner, Almer et al. 2005). Similar input was used for finite element modeling (FEM) of dynamic response of porous (and epoxy-infiltrated) shape memory alloy specimens (Qidwai and DeGiorgi 2004).

Calculations of macroscopic properties that have been performed using actual particle spatial and orientation distributions were measured in an Al – 20 vol.% Al_2O_3 composite. Mean field and multiscale modeling and FEM were used to compute various elastic properties, good agreement with experimental moduli measurements was obtained (Borbely, Biermann et al. 2003, Borbély, Kenesei et al. 2006, Kenesei, Biermann et al. 2006, Kenesei, Klohn et al. 2006), and somewhat larger fraction of particles fractured in the interior of the specimen compared to a zone 2–3 particle diameters from the surface. Sanchez et al. quantified graphite volume fraction in Al matrix composites and used the microCT-determined spatial distribution of graphite to calculate flow of Al into the graphite perform and to simulate the spatial distribution of strains from deformation of the solid composite (Sanchez, Narciso et al. 2006). As Heggli et al. discuss in conjunction with their microCT data on a graphite/Al composite, accurate models depend on employing representative volume elements that are sufficiently large to be representative of the material on a macroscopic scale (but are small enough to be tractable numerically), and these authors concluded that reasonably accurate predictions of resistivity could be obtained using ensemble averaging over a sufficient number of small models (Heggli, Etter et al. 2005). Titanium dioxide–polymer composites (for bioimplants) were compacted with radial density gradients observed (based on linear attenuation coefficients from averaged voxel values), but it is unclear whether the reported variation was not, in fact, an artifact of beam hardening, Ti being quite absorbing (De Santis, Catauro et al. 2007).

The distribution of different carbon-based phases in a composite was studied with phase microCT (Coindreau, Vignoles et al. 2003). The refractive index decrement of resin,

* Figures 5.10 and 5.11 and associated text as well as discussion in Chapter 11 cover these other results on SiC fibers in Al matrices.

FIGURE 7.2
Three orthogonal sections through a C/C composite from phase microCT. The volume enclosed by the planes is 0.0033 mm^3, and the voxel size equaled 0.745 μm. Phase contrast shows the C fibers (dark gray in the image) and the carbon matrix deposited on the fibers (light gray) (Martín-Herrero and Germain 2007).

carbon fibers, and deposited carbon were clearly different and allowed clear segmentation of the different phases. Absorption contrast would have revealed only porosity. Carbon fibers in a carbon matrix composite have been the subject of another phase microCT study (Martín-Herrero and Germain 2007). Phase contrast provides enough contrast to differentiate the fibers from the matrix (Fig. 7.2), something that would never be seen with absorption contrast.

Quite a few other microCT composite studies have appeared and are covered elsewhere in this volume.

7.1.5 Biological Tissues as Phases (Anatomy)

Different biological tissue types can be regarded as phases in the materials sense, and the number of studies employing microCT of tissue specimens, scaffolds for implants, etc. dwarfs those of engineering materials. Mineralized tissues such as bone (apatite, calcium phosphate, plus collagen) or echinoderm ossicles (calcite, calcium carbonate) can take the form of a cellular solid (i.e., plates and struts surrounded by soft tissue), and it is sensible to review these studies in the chapter on cellular solids. Likewise, studies of blood vessel and airway networks are covered in the porous solids chapter. MicroCT studies of dense mineralized tissues (cortical bone, sea urchin teeth) are distributed within other subsections, but pathological calcifications have also been studied.

Several examples of pathological calcification have already been presented. Figure 4.6 shows a synchrotron and matched lab microCT slice of a mouse model of osteoarthritis; the highly porous bone on the right side resulted from the treatment. Calcifications as part of juvenile dermatomyositis have been studied (Stock, Ignatiev et al. 2004a, Stock, Rajamannan et al. 2004). Kidney stones have also been the subject of studies, both their microstructure and their location within the kidney (Stock, Rajamannan et al. 2004). Williams and coworkers studied calcium oxalate/apatite calculi attached to renal papilla (Williams Jr., Matlaga et al. 2006). Figure 6.14 shows a semitransparent 3D rendering of a human heart valve that became heavily calcified (Rajamannan, Nealis et al. 2005). Quantitative comparisons based on lab microCT data have shown statistically significant differences in the volume of calcification between control and disease-affected heart valves (Rajamannan, Subramanium et al. 2005). Other microCT studies of calcification gone wrong abound.

MicroCT of soft tissue is mostly done with phase contrast in order to differentiate between different soft tissue types, although this is not always necessary for brain tissue (Müller, Germann et al. 2006) or for fixed lung sections (Shimizu, Ikezoe et al. 2000). Tumor tissue has been well-differentiated from healthy tissue (Momose, Takeda et al. 1998a, 1998b, 1999, Momose and Hirano 1999), and different normal tissue structures (e.g., mammary ducts) can also be made out (Takeda, Momose et al. 1998). An example where in vivo absorption microCT was used to study soft tissue features (lung tumors in a mouse model) is typical of such studies: *a priori* information was necessary to differentiate tumors from large blood vessels (Weber, Peterson et al. 2004). Respiratory gating was used in this study of the accuracy of microCT characterization, and further details are covered in the section on accuracy. Injection of contrast agents that concentrate in the soft tissue type of interest has been used to good effect to image murine liver tumors, but success depends on adequate contrast enhancement (Weber, Peterson et al. 2004), a process that may be difficult to control.

Eyes and ears have been the subject of a number of microCT studies. Porcine (Leszczyński, Sojka-Leszczyńska et al. 2018) and rabbit (Ivanishko, Bravin et al. 2017) eyes have been studied. Demineralization and light staining with Os and imaging with phase-contrast synchrotron microCT allows nerve structure quantification in normal vs hearing-damaged cochleae (Richter, Young et al. 2018, Tan, Jahan et al. 2018). Finite element modeling of the human middle ear is an interesting study (De Greef, Pires et al. 2017) as is the study of beam path of infrared neural stimulation in the guinea pig cohlea (Moreno, Rajguru et al. 2011).

Not all medical x-ray applications require full 3D information, and considerations of dose limitation and/or specimen geometry often dictate that simple radiography is employed. It is worth a brief mention of the few phase radiographic imaging studies, then, because this modality may become more important clinically in the future. Digital phase mammography has received attention because sensitivity per unit dose to tumor cells or tumor precursors such as microcalcifications is much greater than with conventional mammography (Arfelli, Assante et al. 1998, Kotre and Birch 1999, Yu, Takeda et al. 1999). Diffraction-enhanced radiography of cartilage in disarticulated as well as in intact joints is quite promising (Mollenhauer, Aurich et al. 2002, Li, Zhong et al. 2003, Muehleman, Chapman et al. 2003, Muehleman, Majumdar et al. 2004), although the technical challenges of covering the FOV for human joints such as the ankle are considerable. Studies of the temporal mandibular joint and the mandibular symphysis have provided interesting information on use-induced changes (Nicholson, Stock et al. 2006, Ravosa,

Kloop et al. 2007, Ravosa, Kunwar et al. 2007, Ravosa, Stock et al. 2007, Mirahmadi, Koolstra et al. 2017).

Spatial patterning and sex differences are other areas that have been studied by microCT. The 3D local orientation of myocytes in human cardiac tissue is one example (Varray, Mirea et al. 2017) and the discrimination of brain neurites of normal patients or those diagnosed with schizophrenia (Mizutani, Saiga et al. 2019). Patterns of trabecular structure were studied in the proximal femur for growing humans (Milovanovic, Djonic et al. 2017) as were sex differences in human mandibular condyles (Coogan, Kim et al. 2018).

7.1.6 Cultural Heritage, Archeology, and Forensics

Analysis of manufacture and prior conservation of a prayer bead is a fascinating cultural heritage study (Ellis, Suda et al. 2017) illustrating the power of microCT for studying specimens that must absolutely not be altered by examination. Many microCT studies of pottery shards have appeared, and are not reviewed here except to note observation of domesticated rice (*Oryza sativa*) within pottery shards from early Neolithic sites (Barron, Turner et al. 2017). Methods for preserving waterlogged archeological wood have been evaluated with microCT (Bugani, Modugno et al. 2009), and wood preserved through corrosion of metal objects has been examined (Haneca, Deforce et al. 2012). Some artifacts such as manuscript scrolls have degraded so much over centuries that they cannot be unwrapped so that their contents can be read. MicroCT has been used to read the text of such scrolls (Rosin, Lai et al. 2018). Even more impressive to the author is use of microCT to read degraded movie footage and to play back both the video and audio streams (Liu, Rosin et al. 2016).

In forensics, microCT has been applied to the quantitative analysis of knife cuts in bone (Norman, Watson et al. 2018) and in the study of a Bronze Age human femur with an embedded arrowhead (Flohr, Brinker et al. 2015). Molecular human identification has conventionally focused on DNA sampling from dense, weight-bearing cortical bone tissue, but studies of trabecular bone have found greater yields of DNA from trabecular bone, and one study found that osteocyte lacunae density (site of DNA-containing material in bone) was higher in trabecular compared to cortical bone (Andronowski, Mundorff et al. 2017). Quantitative analysis of changes in trabecular spacing has been observed in skull bones in the near postmortem period (Le Garff, Mesli et al. 2018). Precision of microCT results on bone have also been quantified in a forensics setting (Le Garff, Mesli et al. 2017). Age quantification during the blow fly intra-puparial period has been assayed using microCT (Martín-Vega, Simonsen et al. 2017).

Bone recovered from archeological digs can provide important information about the human population's activity levels. The proximal femora and humeri, for example, were examined from five modern human populations spanning hunter-gatherer to industrial era societies, and microCT-based 3D histomorphometry revealed interesting changes (Doershuk, Saers et al. 2019). Considerable caution must be exercised when examining archeological bones because diagenetic changes can be substantial and quite variable, even within burials from the same time period and site. Figure 7.3 shows enlarged areas of reconstructed slices of three human second metacarpal bones; the first three panels show archeological bones with different diagenetic effects that would affect numerical measurements, and the fourth panel shows a bone that was not subjected to alternation.

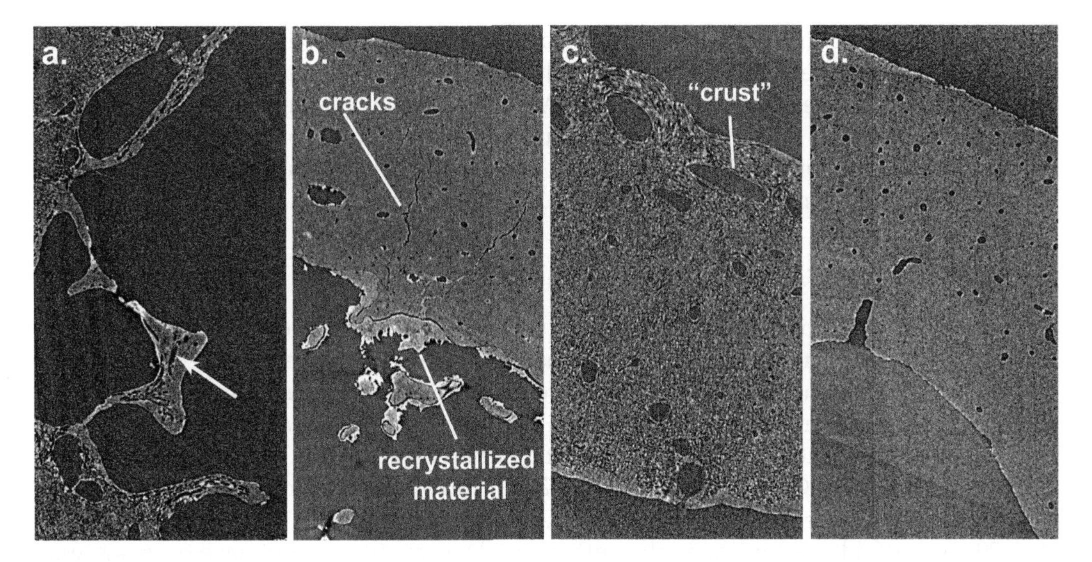

FIGURE 7.3
Portions of reconstructed slices of human second metacarpal bones. (a–c) Archeological bones showing different types of diagenesis that could affect numerical analyses and (d) a bone that has not been altered. (a) Voids attributed to bacterial breakdown of the bone (arrow); these are particularly prominent in trabeculae. (b) Postmortem microcracks within the bone and cortical structure appearing similar to those in (d). The material coating structures within the marrow cavity and with more attenuating voxels (lighter gray) is recrystallized material. Position resolved transmission x-ray diffraction shows only carbonated apatite (cAp) is present in any abundance, leading to the conclusion that the deposits labeled recrystallized material are cAp and have higher density than the cortical bone. (c) The stippled contrast in the cortical bone suggests that fine scale porosity (diameter < 10 μm) has been produced by diagenesis. Thin crusts of more attenuating material appear on the surfaces of the Haversian canals. (d) Bone unaffected by diagenesis. The data were collected with a white beam, so contrast sensitivity is too low to reveal differences in mineral level between newer osteons and the surrounding older bone (see Fig. 5.15). Filtered white beam, 2K × 2K reconstructions, 2.6-μm isotropic voxels, 0.12° rotation steps. (S.R. Stock, S. Mays, C. Soriano, 2-BM, APS, June 2016.)

7.2 Metrology and Phylogeny

MicroCT is frequently used to measure internal or external dimensions and shapes both in manufactured and biological objects. The abilities to view structures in 3D and to take internal measurements along the proper 3D directions are essential.

7.2.1 Industrial Metrology

Industry has long used large specialized CT systems for inspection of high value (large) components. At general (nonmedical) tomography meetings, one hears increasingly about use of widespread use of microCT to monitor manufacturing processes, but the citable information is difficult to come by.

High-definition inspection of fuel injector components has been investigated in the automobile industry (Bauer, Bessler et al. 2004). Stop action, high temporal resolution microCT of evolution of fuel spray (5.1 μs temporal and 150 μm spatial resolution) has been achieved using the pulsed nature of the APS storage ring (Liu, Liu et al. 2004, Im, Fezzaa et al. 2007). These experiments relied on the reproducibility of the spray from injection to injection.

Voids in manufactured components can be harmless or can be critical flaws requiring the component be scrapped. MicroCT has been used to examine the influence of melt strategies on defect population in Ti–6Al–4V components (Tammas-Williams, Zhao et al. 2015) on flaws within additive manufactured titanium component (duPlessisa, leRouxa et al. 2015). Voids in Portland cement concrete have been characterized in the context of pavement material (Lu, Peterson et al. 2018).

7.2.2 Additive Manufacturing

With the advent of affordable systems and of a wide variety of numerical templates, additive manufacturing has become very popular, and microCT has been used to assess the results of "3D-printing." Synchrotron microCT and micromechanical modeling have been used to characterize surfaces produced by metal additive manufacturing (Kantzos, Cunningham et al. 2018). Young's modulus of bamboo parenchyma has been modeled via 3D printing and microCT (Dixon, Muth et al. 2018), and 3D-printed microvascular network design and polymer self-healing were also examined with microCT (Postiglione, Alberini et al. 2017). Scaffolds containing high synthetic hydroxyapatite content have been successfully used in a spine fusion model, and bone ingrowth quantified using microCT (Jakus, Rutz et al. 2016).

7.2.3 Paleontology

Tafforeau and coworkers reviewed application of synchrotron microCT for nondestructive 3D studies of paleontological specimens (Tafforeau, Boistel et al. 2006). MicroCT of structures in fossil fish (Dominguez, Jacobson et al. 2002) and in ossicles of fossil sea urchins (Stock and Veis 2003) have been reported. The internal structure of fossil embryos from the Lower Cambrian were preserved by early stages of diagenesis, and the 3D structures of blastomeres (cells formed by the early division of the fertilized egg) revealed important information about the initial diversification of metazoan animals (Donoghue, Bengtson et al. 2006). Figure 7.4 shows the preserved blastomere structure in cleavage stage embryos using a combination of 3D renderings and virtual sectioning planes. Mazurier et al. focused on fossilized trabecular bone (Mazurier, Volpato et al. 2006).

High-resolution CT has found considerable application in anthropology and primatology. In primatology, a new species of great ape from ~10 Ma ago was established and, through analysis of its dentition and dentinoenamel junction (DEJ), related to the gorilla clade (Suwa, Kono et al. 2007). Enamel thickness is an important diagnostic characteristic for hominoids, and, given that small samples sizes (fragments, small number of specimens) are the rule rather than the exception, enamel and dentin volume quantification with microCT (i.e., establishing these phylogenic characters numerically) is particularly important (Gantt, Kappleman et al. 2006). McErlain and coworkers found that microCT could reveal dental anatomy (attrition of enamel, internal 3D caries structure) in a 500-year-old tooth (McErlain, Chhem et al. 2004). MicroCT of the cranial vault of an Oligocene anthropoid (~29–30 Ma) revealed previous estimates of the endocranial volume were too large, and this study suggested greater caution is needed in phylogenic analyses of these and related animals (Simons, Seiffert et al. 2007). MicroCT was found to be a very useful diagnostic tool for studying a wide range of pathologies in a collection of historical skulls (Rühli, Kuhn et al. 2007). Lab microCT was used to study the symphyseal ontogeny (mandible) of a sub-fossil lemur (Ravosa, Stock et al. 2007). Synchrotron microCT has been used to study, with very high spatial resolution, the dentin, enamel, and DEJ of

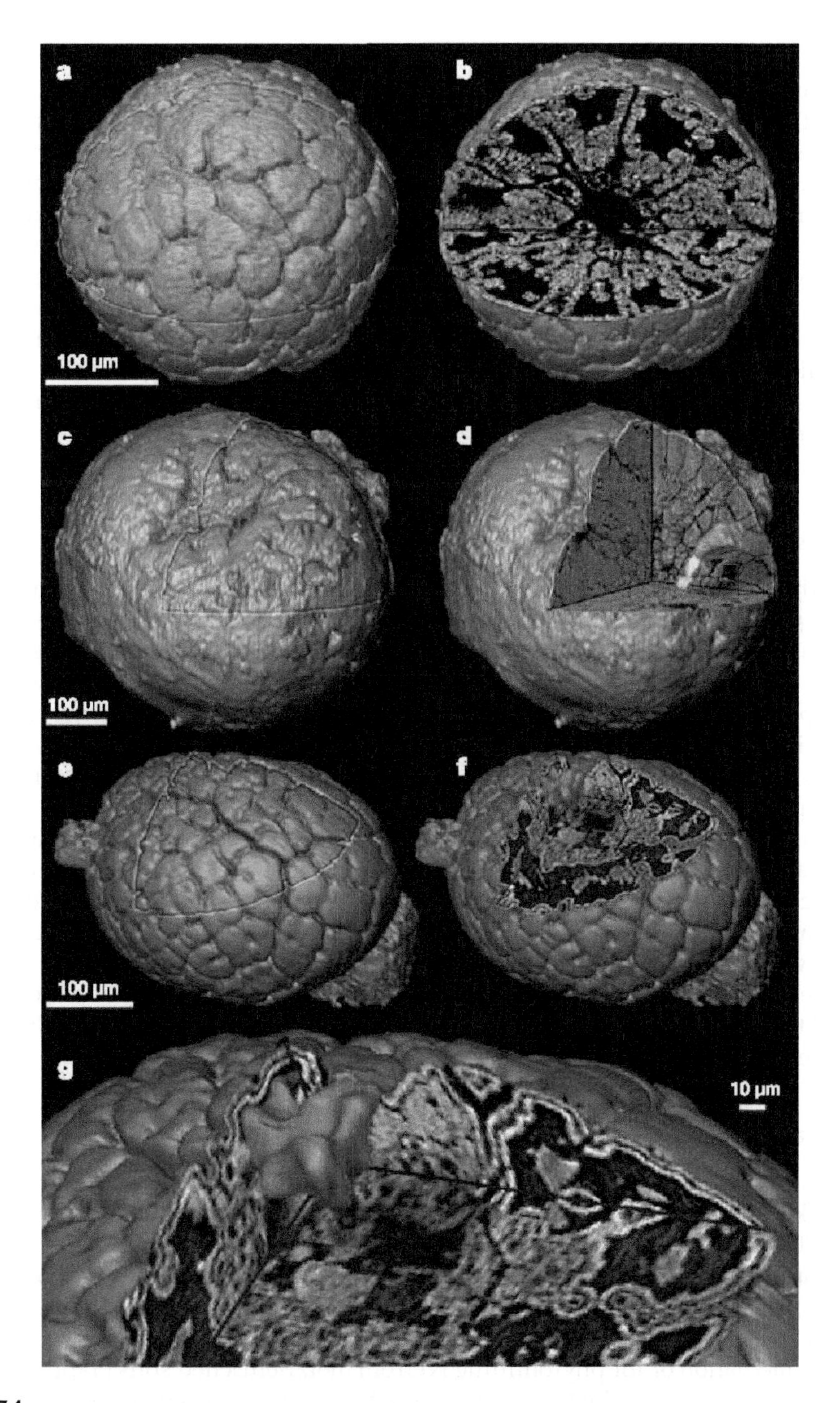

FIGURE 7.4
Fossil cleavage-stage embryos from the Lower Cambrian. (a,b) Divisions between adjacent blastomeres are variably preserved on this embryo's surface and within. (c,d) Divisions between nearly all blastomeres are preserved to their full extent (the light orange and yellow renderings show a column of blastomeres). (e–g) Divisions between blastomeres are not preserved, and the orange structure is a rendering of a cavity within the diagenetic infilling (Donoghue, Bengtson et al. 2006).

Homo neanderthalensis (Macchiarelli, Bondioli et al. 2006) and early *Homo sapiens* (Smith, Tafforeau et al. 2007).

MicroCT has been used in a variety of other phylogenic and paleontological applications. One study reappraised the envenoming capacity of a proto-mammal (Benoit, Norton et al. 2017). MicroCT was used to reanalyze fossil whip spider to determine its phylogenic place relative extant relatives (Garwood, Dunlop et al. 2017) and to analyze the fossil of a bird-like amphibious dinosaur (Cau, Beyrand et al. 2017). Even though most of a well preserved centipede was visible in amber, portions key to assigning it to a new species (or not) could not be seen visually, and synchrotron microCT was required to identify these structures (Edgecombe, Vahtera et al. 2012). Brain evolution in squirrels was interpreted using a microCT-produced virtual endocast (Bertrand, Amador-Mughal et al. 2017), and thorax musculature of Odonata was used via microCT as an important phylogenetic character system, with a matrix of 298 characters with 697 character states (Buesse, Heckmann et al. 2018).

7.2.4 Cells

Imaging the structures within individual cells is extremely challenging because of the great care that must be taken to keep them in their native or near-native state. The cells must remain hydrated while being fixed in place, and the best contrast between organic structures (carbon-containing) and between the encompassing water is in the water window (Chapter 2). This type of x-ray micro- and nanoCT is therefore the province of specialized facilities such as beam line 2 of the ALS.

Contrast between different organelles is great enough to differentiate the various structures in typical studies (Larabell and Le Gros 2004, Parkinson, McDermott et al. 2008). The 3D rendering reproduced in Fig. 7.5 shows the organelles (the nuclei appear in orange, for example), and dimensions, volumes, and numbers of organelles were tabulated from the volumetric data (Parkinson, McDermott et al. 2008). Mechanisms of rapid killing of bacteria were examined in one study (Chongsiriwatana, Lin et al. 2017). Sickling of red blood cells was observed in another study (Darrow, Zhang et al. 2016). The ambitious 4D nucleome project seeks to map the structure and dynamics of the human and mouse genomes in space and time with the goal of gaining deeper mechanistic insights into how the nucleus is organized and functions and employs microCT (Dekker, Belmont et al. 2017).

7.2.5 Flora

Structures of terrestrial and marine plants have also been measured with microCT. In grain kernels, for example, infestations and fissures could be readily detected (Dogan 2007). Localization of iron in seeds of *Arabidopsis* was studied with fluorescence microCT and related to the presence of a specific transporter VIT-1 (Kim, Punshon et al. 2006). Fluorescence microCT was also used to study metal storage mechanisms in the metal hyperaccumulator *Alyssum murale*, a plant developed as a commercial crop for phytoremediation or for phytomining Ni from metal-enriched soils (Tappero, Peltier et al. 2007). In the stalk of the (marine) horsetail, microCT showed silica that occurs in a thin continuous layer of the entire outer epidermis and is highly concentrated in the knob regions of the long epidermis cells (Sapei, Gierlinger et al. 2007).

Conifer tissues have been the subject of microCT studies, fossils in amber (Moreau, Néraudeau et al. 2017), and anatomy of two coexisting arid-zone coniferous trees investigating desiccation tolerance (Sevanto, Ryan et al. 2018). Phenotyping through analysis

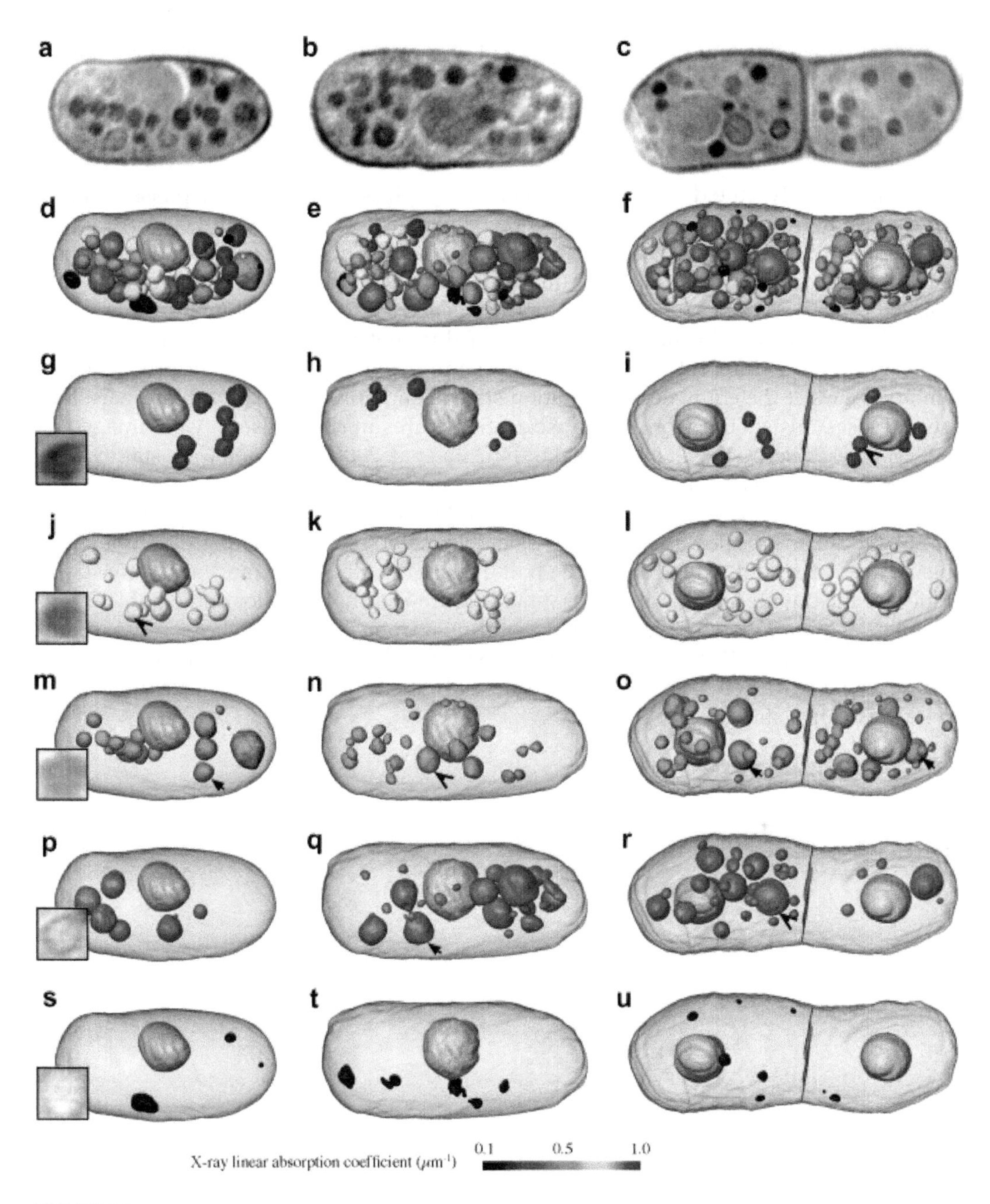

FIGURE 7.5
Cells of *Schizosaccharomyces pombe* imaged by soft x-ray tomography. (a–c) One voxel thick (20 nm) slices through the reconstructed volumes of four cells, with the darker gray indicating higher absorption. (d–f) Colored surfaces representing the boundaries of organelles and the plasma membrane. The colors correspond to the average linear attenuation coefficients inside the organelles (see the color bar at the bottom of the figure). Note that the nuclei should actually be colored blue but are shown as orange to differentiate them from the other structures. The length of the cell in (a) is 5 μm. (g–u) The same surfaces as in (d–f) but with each color isolated from the others. At the left side of each row there is a grayscale slice image of one of the organelles shown in that row; the arrows point to mitochondria and the wedges to other organelles (Parkinson, McDermott et al. 2008).

of seeds (Zhang, Du et al. 2018) and analysis of evolution using seeds (Spencer, Garwood et al. 2017) are examples of other studies.

7.2.6 Insecta, Mollusca, and Echinodermata

Insects are ideally suited for imaging with x-ray phase contrast, and phase-contrast microCT for this application has been recently reviewed (Betz, Wegst et al. 2007). Although considerable understanding has been gained on phenomenon such as insect respiration using real-time phase radiography to quantify tracheal volume changes (Westneat, Betz et al. 2003, Socha, Westneat et al. 2007), microCT is needed to gain a complete understanding of this 3D process. Recording a number of different radiographs from each projection (sampling the different respiratory stages) and using post-experiment processing to combine views at the equivalent respiration state is an approach used with animals, and one supposes this should work with insects. Structures in small insects were imaged in an SEM-based microCT system (Tanisako, Tsuruta et al. 2006). Neurons were stained in a study of *Drosophila* central nervous system using microCT (Mizutani, Takeuchi et al. 2007).

Silkworms eat mulberry leaves that contain crystals of calcium oxalate, and have evolved specialized organs for processing this unpalatable material. These Malpighian tubules lie outside the worm's gut, extend about one-half the length of the worm, and, at certain stages of life, are filled with calcium oxalate crystals. Figure 7.6a shows a slice of the gut "G" and a number of Malpighian tubules "M" surrounding it from a 13-day-old larva preserved in concentrated ethanol. Phase-enhanced edge contrast allows other soft tissue structure to be observed. Figure 7.6b shows a 3D rendering of high absorption voxels (calcium oxalate crystals) within a stack of ~2K slices.

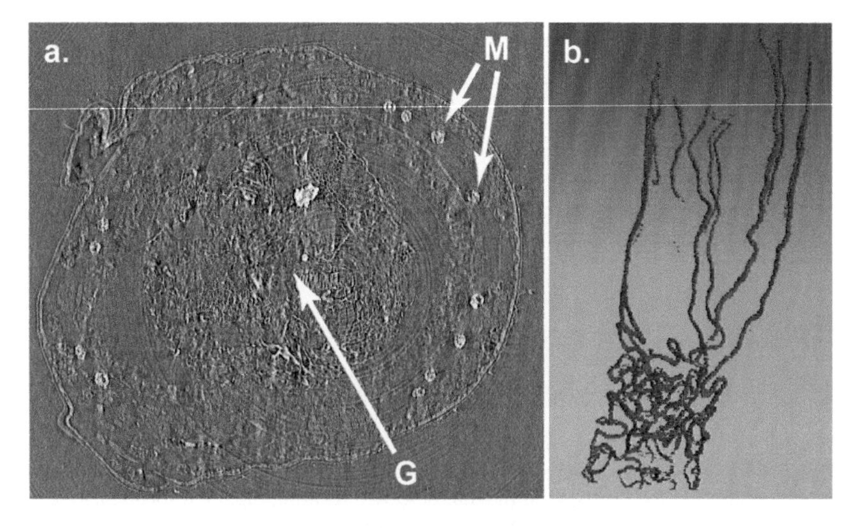

FIGURE 7.6
Silkworm slice and 3D rendering of calcium oxalate–filled Malpighian tubes. (a) Slice, 480 voxels horizontal field of view. Gut "G" and Malpighian tubules "M" filled with high absorption calcium oxalate crystals (compared to the rest of the larva and surrounding fluid) are labeled. (b) 3D rendering showing the high absorption voxels as pink, that is, the Malpighian tubes. The larva's rectum is at the bottom of the ~2.5-mm-long rendering, and the wavy structure of the tubules is natural. 14.6 keV, 2K × 2K reconstruction, 1.4-µm voxels, 0.125° rotation increment. (S.R. Stock, M.A. Webb, A.J. Wyman, K.I. Ignatiev and F. De Carlo, 2-BM, APS, December 2007).

One recent study produced a 3D analysis of a type specimen (Akkari, Ganske et al. 2018). Taxonomy of ants has received considerable attention (Agavekar, Hita Garcia et al. 2017, Garcia, Fischer et al. 2017, Staab, Garcia et al. 2018).

Changes in snail shells during growth were studied through in vivo microCT (Postnov, Zarowski et al. 2006). Synchrotron microCT and SEM were compared in another study of the formation of embryonic and juvenile snail shells (Marxen, Prymak et al. 2008).

Sea urchins form complex geometric structures, generally of single-crystal calcite, and, in their hierarchy of structures, appear to employ as wide a range of strengthening strategies as have been employed in human engineered composites. Three main types of calcite structures are produced by sea urchins: porous stereom, a structure whose topology resembles that of trabecular bone and which is discussed in Chapter 8 and also shown in Fig. 5.7, a dense structure of plates and prisms comprising the teeth, and open, highly regular microarchitecture of certain spines (Fig. 6.9 and the following paragraph). The literature on the microarchitecture of sea urchin mineralized tissue is quite extensive and scattered, but, aside from very brief comments relating to microCT, this interesting subject is left unexamined.

Sea urchin spines protect the animal's body from predators and high-energy, rough surf environments, and the spines of different phylogenic families have different characteristic architectures as well as a wide range of lengths and diameters. In addition to stereom, spines can contain radial wedges or a dense cortical shell. A wide range of sea urchins has adopted the strategy of employing long, hollow spines capable of rapid reorientation, and the regularity of some of these structures are remarkable. Reports of structural analyses have appeared recently (Stock, Ebert et al. 2006, Stock, De Carlo et al. 2007), and numerical analysis of the structures revealed by microCT may shed light onto why certain characteristic structures have persisted over millions of years, providing insight into functional advantages conferred by certain structures. Figure 6.9 shows the complex 3D spine structure of one type of sea urchin: the radial wedges and hollow center concentrate mass where it will be most effective in resisting bending, and the bridges between wedges and the perforated central cylinder link the wedges. This structure is a single crystal, and the exquisite biological control of mineralization geometry and crystallography at ocean temperatures is even more remarkable because the calcite's high Mg content grows at equilibrium only about several hundred degrees Celsius. Several studies suggest that spatial constraint of the mineralization space controls single-crystal calcite growth in sea urchin ossicles (Park and Meldrum 2002, Aizenburg, Muller et al. 2003). Materials processing employing this strategy and suitable macromolecular crystallization adjuncts appear to be a very attractive route to improved cellular solids.

Sea urchin teeth differ from the urchins' other skeletal elements in that the tooth does not consist of stereom (see the section on cellular materials) nor of wedges and bridges, but rather of a dense, single-crystal calcite structure suited to rasping food from the surfaces of rocks. As sea urchin teeth contain their entire developmental history, they have received considerable attention as a biomineralization model. At the tooth's aboral end, deep within the urchin, little mineral is present, but, by the aboral end, that is, the cutting edge, mineral has filled virtually all of the tooth's volume. Within the tooth, myriad reinforcement strategies are employed to make a functional structure from calcite, an otherwise wretched structural material (Stock, Ignatiev et al. 2004b, Stock 2014).

Sea urchin teeth have cross-sections shaped like "U" or "T" (e.g., slice shown in Fig. 5.16 shows the leg of a "T"-shaped tooth. Note that sand dollars, closely related to the regular sea urchins discussed here, have teeth with diamond cross-sections), but brevity requires the discussion be limited to the "T"-shaped or camarodont teeth. The leg of the "T" is called the keel, and the bar across the top is termed the flange. The five teeth mounted in the five pyramids (each consisting of two demipyramids, see Chapter 8) of the oral apparatus have their flanges tangent to the oral cavity and their keels running radially from the flange to the axis of the jaw structure. Each tooth functions as a T-girder, a structure providing considerable resistance to bending per unit mass, with the flange loaded in compression and the keel loaded in tension.

The microstructure of the mature flange seems well adapted to compression loading and the keel to tensile loading. The flange consists of a stack of plates paralleling the upper and lower surfaces of the bar of the "T" and close to parallel to the tooth's axis, and the interior of the flange contains thin needles running along the tooth's axis and extending into the keel, where they widen into much larger diameter prisms. The flange plates and needles are compressed end-on during eating, a geometry allowing high resistance to wear and to catastrophic failure. The keel, loaded in tension when the flange is compressed on hard surfaces, is essentially a composite structure reinforced by fibers (prisms) aligned along the tensile loading axis. In addition to the prisms within the center of the keel, teeth of sea urchins such as *Lytechinus variegatus* also have carinar process plates running along the flanks of the keel. This is different from the stirodont tooth structure pictured in Fig. 5.16 where the carinar process plates are absent. Synchrotron microCT observations of carinar process plate orientation in *L. variegatus* suggested that these plates (on the sides of the keel) serve to prevent deflection (and fracture) of the keel along a secondary bending axis (Fig. 7.7); that is, a situation that might be encountered when only one side of the flange is in contact with a hard substrate (Stock, Ignatiev et al. 2003). Different camarodonts have different arrangements of carinar process plates (Fig. 7.8), although all of those studied to date have plates arranged to resist the bending hypothesized for *L. variegatus* (Stock, Ignatiev et al. 2004a, Stock, Xiao et al. 2013). Appreciation of the possible function of the carinar process plates occurred only after (microCT-derived) 3D data sets were available.

7.2.7 Vertebrates

Studies of fish skeletons as biomineralization models have appeared (Neues, Arnold et al. 2006, Neues, Goerlich et al. 2007), the latter study being of particular interest because a normal wild type and mutant are compared. MicroCT-based phenomics in the zebrafish skeleton has been reported (Hur, Gistelinck et al. 2017). MicroCT has been used in a study of the sonic organ of the singing midshipman fish, although MRI was the imaging main modality (Forbes, Morris et al. 2006); it would be interesting to study this organ in vivo using x-ray phase imaging.

Pneumatization, that is, extension of pulmonary air sacs into marrow cavities, is an important adaptation of birds. Forelimb anatomy of a sparrowhawk has been examined in 3D (Bribiesca-Contreras and Sellers 2017). MicroCT was used to study cortical and trabecular bone in pneumatic and apneumatic thoracic vertebrae (two different species of ducks) and found a significant difference in microarchitecture (Fajardo, Hernandez et al. 2007). Cranial shape evolution in adaptive radiations of birds has been examined with microCT (Tokita, Yano et al. 2017).

MicroCT of trabecular structures has been used as the basis for improved dosimetry estimates for different bone sites (Kramer, Khoury et al. 2006, 2007). MicroCT was used as the input for quantifying skeletal chord lengths for estimating doses received by the cells in the bone marrow during nuclear medicine therapy (Shah, Rajon et al. 2005). Those interested in this subject are directed to the references of these papers for others' application of microCT to dosimetry.

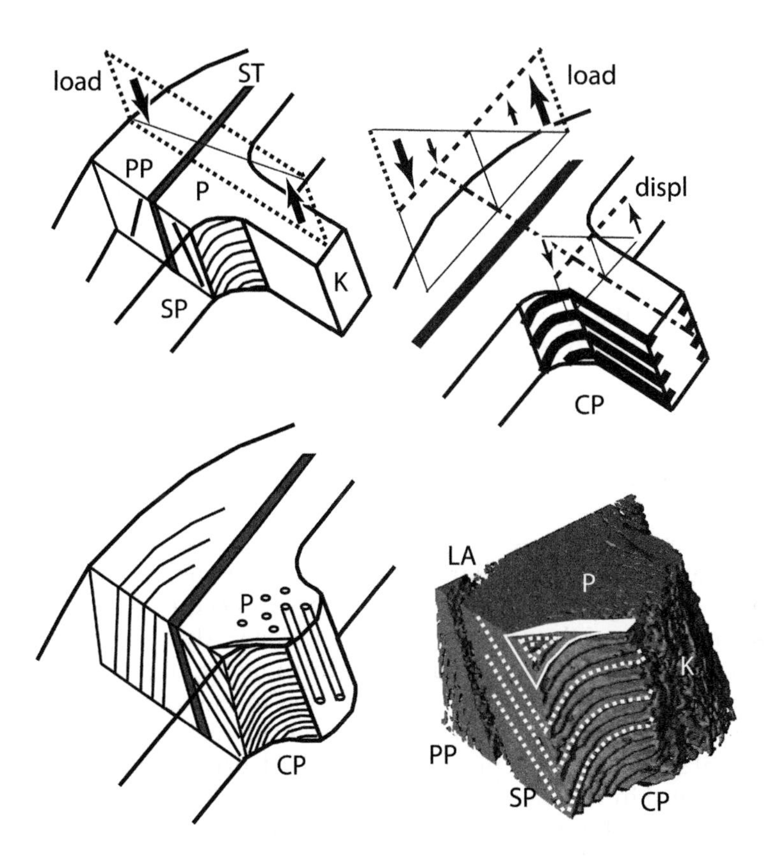

FIGURE 7.7
Function of *Lytechinus variegatus* (Echinoida, Toxopneustida) carinar process plates in resisting transverse tooth bending. In all drawings and the rendering, the cutting edge of the tooth is up, and CP denotes the carinar process plates, K the keel, P the prisms, and LA the low absorption portion of the tooth seen in slices. (Upper left) Primary bending to which the tooth is exposed (compression in the flange and tension in the keel). The stone part ST (center of the flange), the primary plates PP, and secondary plates SP are shown schematically. (Upper right) Possible displacements in the keel in response to uneven flange loading during scraping on an uneven surface, that is, transverse bending about a second axis. (Lower left) Geometry of the reinforcing elements of the structure. (Lower right) MicroCT rendering of a portion of the tooth in the same orientation as the schematics and illustrating the relative orientations and locations of plates and prisms. Reprinted from Stock et al. (2003), © 2004, with permission from Elsevier.

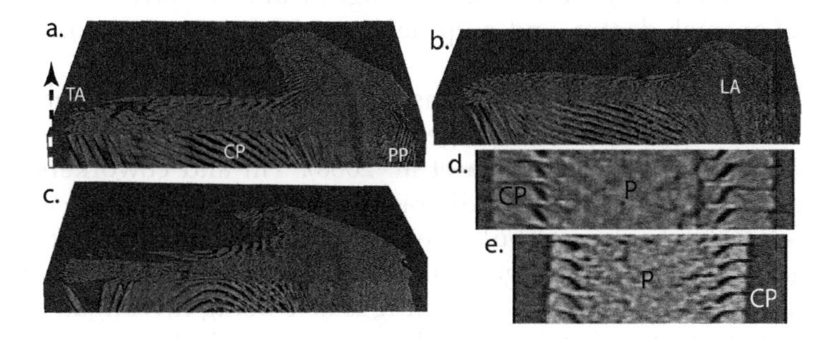

FIGURE 7.8
Carinar process plates in different camarodont teeth shown in 3D orthogonal sections and planes transverse to two keels. Each image extends 50 voxels (~85 μm) vertically (i.e., along tooth axis TA). LA, PP, CP, P are the low absorption area of the keel, primary plates, carinar process plates, prisms, respectively. (a,c) *Strongylocentrotus franciscanus* (Echinoida, Strongylocentrotus). (b,d) *Hetereocentrotus trigonarius* (Echinoida, Echinometrida). (c) *Paracentrotus lividus* (Echinoida, Echinida) (Stock, Ignatiev et al. 2004b).

Comparison of trabecular structure across different mouse genetic strains (Martín-Badosa, Amblard et al. 2003) and of femoral heads of two different primates with very different sizes and loading environments (MacLatchy and Müller 2002) were other studies with a metrology component. Skeletal microCT atlases of different mouse strains and skeletal structures also have appeared (Jacob and Chole 2006, Perlyn, DeLeon et al. 2006, Chan, Kovacevic et al. 2007, Olafsdottir, Darvann et al. 2007, Maga, Tustison et al. 2017, Richbourg, Martin et al. 2017). Comparison of genetic strain-average structure with aligned microCT and MRI 3D data sets illustrate how the atlases can be used to correlate features of the central nervous system with those of the surrounding bones (Chan, Kovacevic et al. 2007). Birth defects in mice were the focus of another study (Liu, Kim et al. 2017).

Examples of population studies in humans are quantification of root surface area of permanent teeth in a Chinese population (Gu, Zhu et al. 2017). Differences in meniscus dimensions were observed between sexes using microCT (Mickiewicz, Walczak et al. 2018).

Nielsen et al. provided a three-dimensional rendering of the root canal in an endodontically prepared human maxillary molar, and the tissue loss (i.e., increase in volume of the root canal space) due to the endodontic procedure was clear even in the 127-μm × 127-μm voxels of the reconstructions (Nielsen, Alyassih et al. 1997). Dowker et al. obtained similar images of human upper third molar, upper central incisor, and upper lateral incisor with and without files inserted in the root canal; these data consisted of isotropic ~39-μm voxels and showed considerable detail, such as unfilled space in the canal sealant and the presence of dentin debris (Dowker, Davis et al. 1997). These data are useful in illustrating the variety and complexity of root canals during computer-assisted dental training. Another library of microCT images for teaching 3D dental structures also has been reported (Seifert, Flynn et al. 2004). It is interesting to contemplate whether x-ray microCT of engineering materials can provide analogous computer-assisted learning opportunities for undergraduate and graduate engineering or science students. In 2017, perhaps the most active area of microCT publication was on root canals.

Peters et al. used lab microCT to analyze root canal geometry by adapting tools developed for histomorphometry of trabecular bone (Peters, Laib et al. 2000). Macromorphology of human tooth roots (Plotino, Grande et al. 2006) and distance transform illustration of local diameter of the tooth pulp (Dougherty and Kunzelmann 2007) are also of interest. Figure 7.9 shows images from the latter study: 3D tri-level segmentation of enamel, dentin, and pulp (left) and 3D local pulp widths determined by the distance transform method.

MicroCT of the auditory apparatus and associated blood vessels has received attention (Shibata, Matsumoto et al. 1999, Vogel, Beckmann et al. 2001, Muller 2004, Gea, Decraemer et al. 2005, Müller, Lareida et al. 2006). Yin and coworkers have examined guinea pig cochleae with both diffraction-enhanced microCT and histology (Yin, Zhao et al. 2007). Scala vestibuli pressure and 3D stapes velocities were analyzed for the gerbil using microCT-generated 3D solid structures (Deacraemer, de La Rochefoucauldi et al. 2007). MicroCT scanning is also being investigated as a tool for evaluating the surgical positioning of cochlear implant electrodes and the damage produced by the electrode placement (Postnov, Zarowski et al. 2006); note that samples were cut from the surrounding temporal bone and were not imaged in human patients in vivo.

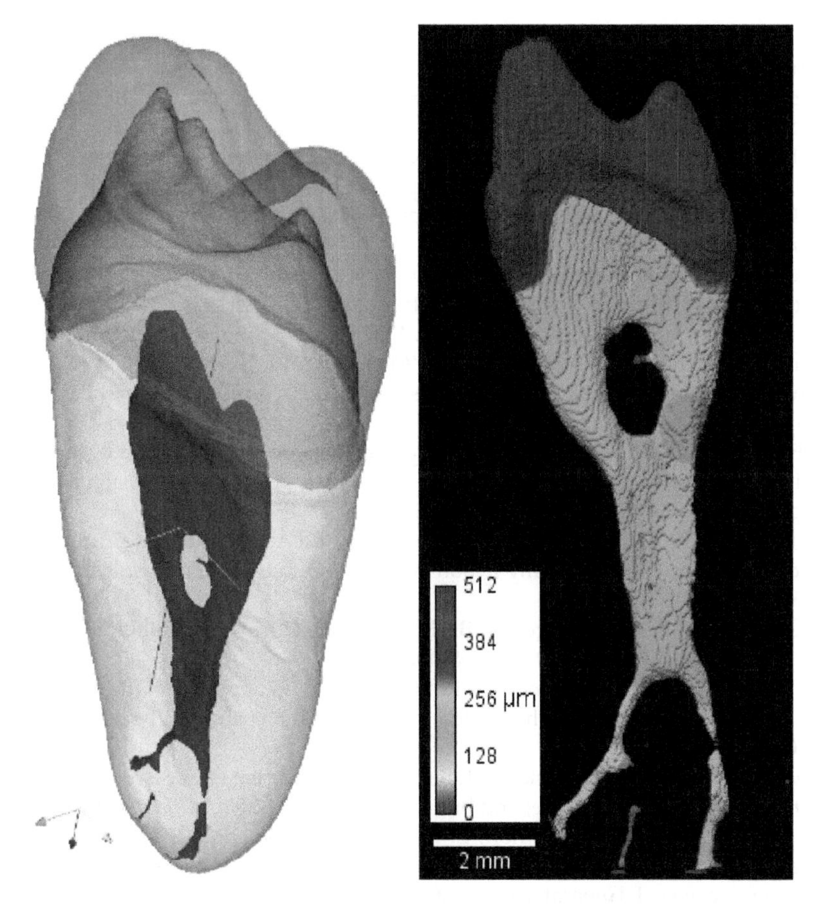

FIGURE 7.9
(left) 3D rendering of a tooth segmented into enamel (cyan), dentin (yellow), and pulp (red). (right) Local thicknesses (color) of the pulp of the tooth in (a) determined with the distance transform method. 324 × 254 × 669 voxels. (Dougherty and Kunzelmann 2007) Copyright © 2007 Microscopy Society of America. Reprinted courtesy of the Microscopy Society and Cambridge University Press.

References

Agavekar, G., F. Hita Garcia and E. P. Economo (2017). "Taxonomic overview of the hyperdiverse ant genus Tetramorium Mayr (Hymenoptera, Formicidae) in India with descriptions and X-ray microtomography of two new species from the Andaman Islands." *PeerJ* **5**: e3800.

Aizenburg, J., D. A. Muller, J. L. Grazul and D. R. Hartmann (2003). "Direct fabrication of large micropatterned single crystals." *Science* **299**: 1205–1208.

Akkari, N., A. S. Ganske, A. Komerički and B. Metscher (2018). "New avatars for Myriapods: Complete 3D morphology of type specimens transcends conventional species description (Myriapoda, Chilopoda)." *PLOS One* **13**: e0200158.

Anderson, P., Y. Ahmed, M. P. Patel, G. R. Davis and M. Braden (2006). "X-ray microtomographic studies of novel radio-opaque polymeric materials for dental applications." *Mater Sci Technol* **22**: 1094–1097.

Andronowski, J. M., A. Z. Mundorff, I. V. Pratt, J. M. Davoren and D. M. Cooper (2017). "Evaluating differential nuclear DNA yield rates and osteocyte numbers among human bone tissue types: A synchrotron radiation microCT approach." *Forensic Sci Int Genet* **28**: 211–218.

Arfelli, F., M. Assante, V. Bonvicini, A. Bravin, G. Canatore, E. Castelli, L. D. Palma, M. D. Michiel, R. Longo, A. Olivo, S. Pani, D. Pontoni, P. Poropat, M. Prest, A. Rashevsky, G. Tromba, A. Vacchi, E. Vallazza and F. Zanconati (1998). "Low dose phase contrast x-ray medical imaging." *Phys Med Biol* **43**: 2845–2852.

Baaklini, G. Y., R. T. Bhatt, A. J. Eckel, P. Engler, M. G. Castelli and R. W. Rauser (1995). "X-ray microtomography of ceramic and metal matrix composites." *Mater Eval* **53**: 1040–1044.

Barron, A., M. Turner, L. Beeching, P. Bellwood, P. Piper, E. Grono, R. Jones, M. Oxenham, N. K. T. Kien, T. Senden and T. Denham (2017). "MicroCT reveals domesticated rice (*Oryza sativa*) within pottery sherds from early Neolithic sites (4150–3265 cal BP) in Southeast Asia." *Sci Rep* **7**: 7410.

Bauer, W., F. T. Bessler, E. Zabler and R. B. Bergmann (2004). Computer tomography for nondestructive testing in the automotive industry. *Developments in X-Ray Tomography IV.* U. Bonse. Bellingham (WA), SPIE. **SPIE Proc Vol 5535**: 464–472.

Beczek, M., M. Ryzak, K. Lamorski, A. Sochan, R. Mazur and A. Bieganowski (2018). "Application of X-ray computed microtomography to soil craters formed by raindrop splash." *Geomorphology* **303**: 357–361.

Bellido, G. G., M. G. Scanlon, J. H. Page and B. Hallgrimsson (2006). "The bubble size distribution in wheat flour dough." *Food Res Int* **39**: 1058–1066.

Benoit, J., L. A. Norton, P. R. Manger and B. S. Rubidge (2017). "Reappraisal of the envenoming capacity of *Euchambersia mirabilis* (Therapsida, Therocephalia) using µCT-scanning techniques." *PLOS One* **12**: e0172047.

Bertrand, O. C., F. Amador-Mughal and M. T. Silcox (2017). "Virtual endocast of the early Oligocene *Cedromus wilsoni* (Cedromurinae) and brain evolution in squirrels." *J Anat* **230**: 128–151.

Betz, O., U. Wegst, D. Weide, M. Heethoff, L. Helfen, W. K. Lee and P. Cloetens (2007). "Imaging applications of synchrotron phase-contrast microtomography in biological morphology and biomaterials science. I. General aspects of the technique and its advantages in the analysis of millimetre-sized arthropod structure." *J Microsc* **227**: 51–71.

Bonse, U., R. Nusshardt, F. Busch, R. Pahl, J. H. Kinney, Q. C. Johnson, R. A. Saroyan and M. C. Nichols (1991). "X-ray tomographic microscopy of fibre-reinforced materials." *J Mater Sci* **26**: 4076–4085.

Borbely, A., H. Biermann, O. Hartmann and J. Buffiere (2003). "The influence of the free surface on the fracture of alumina particles in an Al-Al_2O_3 metal-matrix composite" *Comput Mater Sci* **26**: 183–188.

Borbély, A., P. Kenesei and H. Biermann (2006). "Estimation of the effective properties of particle-reinforced metal–matrix composites from microtomographic reconstructions." *Acta mater* **54**: 2735–2744.

Bribiesca-Contreras, F. and W. I. Sellers (2017). "Three-dimensional visualisation of the internal anatomy of the sparrowhawk (Accipiter nisus) forelimb using contrast-enhanced micro-computed tomography." *PeerJ* **5**: e3039.

Brunke, O., S. Odenbach, C. Fritsche, I. Hilger and W. A. Kaiser (2005). "Determination of magnetic particle distribution in biomedical applications by x-ray microtomography." *J Mag Mater* **289**: 428–430.

Buesse, S., S. Heckmann, T. Hoernschemeyer and S. M. Bybee (2018). "The phylogenetic relevance of thoracic musculature: A case study including a description of the thorax anatomy of Zygoptera (Insecta: Odonata) larvae." *Syst Entomol* **43**: 31–42.

Bugani, S., F. Modugno, J. J. Łucejko, G. Giachi, S. Cagno, P. Cloetens, K. Janssens and L. Morselli (2009). "Study on the impregnation of archaeological waterlogged wood with consolidation treatments using synchrotron radiation microtomography." *Anal Bioanal Chem* **395**: 1977–1985.

Busignies, V., B. Leclerc, P. Porion, P. Evesque, G. Couarraze and P. Tchoreloff (2006). "Quantitative measurements of localized density variations in cylindrical tablets using x-ray microtomography." *Euro J Pharm Biopharm* **64**: 38–50.

Cantre, D., E. Herremans, P. Verboven, J. Ampofo-Asiama and B. Nicolaï (2014). "Characterization of the 3-D microstructure of mango (*Mangifera indica* L. cv. Carabao) during ripening using X-ray computed microtomography." *Innov Food Sci Emerg Technol* **24**: 28–39.

Cau, A., V. Beyrand, D. F. A. E. Voeten, V. Fernandez, P. Tafforeau, K. Stein, R. Barsbold, K. Tsogtbaatar, P. J. Currie and P. Godefroit (2017). "Synchrotron scanning reveals amphibious ecomorphology in a new clade of bird-like dinosaurs." *Nature* **552**: 395–399.

Chan, E., N. Kovacevic, S. K. Y. Ho, R. M. Henkelman and J. T. Henderson (2007). "Development of a high resolution three-dimensional surgical atlas of the murine head for strains 129S1/SvImJ and C57Bl/6J using magnetic resonance imaging and micro-computed tomography." *Neuroscience* **144**: 604–615.

Chongsiriwatana, N. P., J. S. Lin, R. Kapoor, M. Wetzler, J. C. Rea, M. K. Didwania, C. H. Contag and A. Barron (2017). "Intracellular biomass flocculation as a key mechanism of rapid bacterial killing by cationic, amphipathic antimicrobial peptides and peptoids." *Sci Rep* **7**: 16718.

Coindreau, O., G. Vignoles and P. Cloetens (2003). "Direct 3D microscale imaging of carbon-carbon composites with computed holotomography." *Nucl Instrum Meth B* **200**: 308–314.

Coogan, J. S., D. G. Kim, T. L. Bredbenner and D. P. Nicolella (2018). "Determination of sex differences of human cadaveric mandibular condyles using statistical shape and trait modeling." *Bone* **106**: 35–41.

Darrow, M., Y. Zhang, B. P. Cinquin, E. A. Smith, R. M. Boudreau, R. H. Rochat, M. F. Schmid, Y. Xia, C. A. Larabell and W. Chiu (2016). "Visualizing red blood cell sickling and the effects of inhibition of sphingosine kinase 1 using soft x-ray tomography." *J Cell Sci* **129**: 3511–3517.

David, C., P. Robion and B. Menéndez (2007). "Anisotropy of elastic, magnetic and microstructural properties of the Callovo-Oxfordian argillite." *Phys Chem Earth* **32**: 145–153.

De Greef, D., F. Pires and J. J. Dirckx (2017). "Effects of model definitions and parameter values in finite element modeling of human middle ear mechanics." *Hear Res* **344**: 195–206.

De Santis, R., M. Catauro, L. Di Silvio, L. Manto, M. G. Raucci, L. Ambrosio and L. Nicolais (2007). "Effects of polymer amount and processing conditions on the in vitro behaviour of hybrid titanium dioxide/polycaprolactone composites." *Biomaterials* **28**: 2801–2809.

Deacraemer, W. F., O. de La Rochefoucauldi, W. Dong, S. M. Khanna, J. J. J. Dirckx and E. S. Olson (2007). "Scala vestibuli pressure and three-dimensional stapes velocity measured in direct succession in gerbil." *J Acoust Soc Am* **121**: 2774–2791.

Dekker, J., A. S. Belmont, M. Guttman, V. O. Leshyk, J. T. Lis, S. Lomvardas, L. A. Mirny, C. C. O'Shea, P. J. Park, B. Ren, J. C. Politz, J. Shendure, S. Zhong and D. N. network (2017). "The 4D nucleome project." *Nature* **549**(7671): 219–226.

Diamond, S. and E. Landis (2007). "Microstructural features of a mortar as seen by computed microtomography." *Mater Struct* **40**: 989–993.

Dixon, P. G., J. T. Muth, X. Xiao, M. A. Skylar-Scott, J. A. Lewis and L. J. Gibson (2018). "3D printed structures for modeling the Young's modulus of bamboo parenchyma." *Acta Biomater* **68**: 90–98.

Doershuk, L. J., J. P. P. Saers, C. N. Shaw, T. Jashashvili, K. J. Carlson, J. T. Stock and T. M. Ryan (2019). "Complex variation of trabecular bone structure in the proximal humerus and femur of five modern human populations." *Am J Phys Anthropol* **168**: 104–118.

Dogan, H. (2007). "Nondestructive imaging of agricultural products using x-ray microtomography." *Microsc Microanal* **13**(Suppl 2): 1316CD.

Dominguez, P., A. G. Jacobson and R. P. S. Jefferies (2002). "Paired gill slits in a fossil with a calcite skeleton." *Nature* **417**: 841–844.

Donoghue, P. C. J., S. Bengtson, X. Dong, N. J. Gostling, T. Huldtgren, J. A. Cunningham, C. Yin, F. P. Z. Yue and M. Stampanoni (2006). "Synchrotron X-ray tomographic microscopy of fossil embryos." *Nature* **442**: 680–683.

Dos Reis, P. J., R. Nagata and C.R. Appoloni (2017). "Pore size distributions convolution for microtomographic images applied to Shark Bay's oolite." *Micron* **98**: 49–54.

Dougherty, R. P. and K. H. Kunzelmann (2007). "Computing local thickness of 3D structures with ImageJ." *Microsc Microanal* **13**(Suppl 2): 1678CD.

Dowker, S. E. P., G. R. Davis and J. C. Elliott (1997). "X-ray microtomography: nondestructive three-dimensional imaging for in vitro endodontic studies." *Oral Surg Oral Med Oral Pathol Oral Radiol Endod* **83**: 510–516.

duPlessisa, A., S. G. leRouxa, J. Els, G. Booysen and D. C. Blaine (2015). "Application of microCT to the non-destructive testing of an additive manufactured titanium component." *Case Stud Nondest Test Eval* **4**: 1–7.

Edgecombe, G. D., V. Vahtera, S. R. Stock, A. Kallonen, X. Xiao, A. Rack and G. Gribet (2012). "A scolopocryptopid centipede (Chilopoda: Scolopendromorpha) from Mexican amber: Synchrotron microtomography and phylogenetic placement using a combined morphological and molecular dataset." *Zool J Linn Soc* **166**: 768–786.

Ellis, L., A. Suda, R. M. Martin, E. Moffatt, J. Poulin and A. J. Nelson (2017). The virtual deconstruction of a prayer bead in the Thomson collection at the Art gallery of Ontario with MicroCT scanning and advanced 3D analysis software. *Prayer Nuts, Private Devotion, and Early Modern Art Collecting*. E. Wetter and F. Scholten. Riggisberg, Abegg-Stiftung.

Everett, R. K., K. E. Simmonds and A. B. Geltmacher (2001). "Spatial distribution of voids in HY-100 steel by x-ray tomography." *Scripta Mater* **44**: 165–169.

Fajardo, R. J., E. Hernandez and P. M. O'Connor (2007). "Post cranial skeletal pneumaticity: A case study in the use of quantitative microCT to assess vertebral structure in birds." *J Anat* **211**: 138–147.

Farber, L., G. Tardos and J. N. Michaels (2003). "Use of x-ray tomography to study the porosity and morphology of granules." *Powder Technol* **132**: 57–63.

Flohr, S., U. Brinker, A. Schramm, U. Kierdorf, A. Staude, J. Piek, D. Jantzen, K. Hauensteinf and J. Orschiedt (2015). "Flint arrowhead embedded in a human humerus from the Bronze age site in the Tollense valley, Germany – a high-resolution microCT study to distinguish antemortem from perimortem projectile trauma to bone." *Int J Paleopath* **9**: 76–81.

Forbes, J. G., H. D. Morris and K. Wang (2006). "Multimodal imaging of the sonic organ of *Porichthys notatus*, the singing midshipman fish." *Mag Res Imaging* **24**: 321–331.

Frisullo, P., F. Licciardello, G. Muratore and M. A. Del Nobile (2010). "Microstructural characterization of multiphase chocolate using X-ray microtomography." *J Food Sci* **75**: E469–E476.

Fu, X., M. Dutt, A. C. Bentham, B. C. Hancock, R. E. Cameron and J. A. Elliott (2006). "Investigation of particle packing in model pharmaceutical powders using x-ray microtomography and discrete element method." *Powder Technol* **167**: 134–140.

Gantt, D. G., J. Kappleman, R. A. Ketcham, M. E. Alder and T. H. Deahl (2006). "Three-dimensional reconstruction of enamel thickness and volume in humans and hominoids." *Eur J Oral Sci* **114**(suppl 1): 360–364.

Garcia, F. H., G. Fischer, C. Liu, T. L. Audisio and E. P. Economo (2017). "Next-generation morphological character discovery and evaluation: an X-ray microCT enhanced revision of the ant genus Zasphinctus Wheeler (Hymenoptera, Formicidae, Dorylinae) in the Afrotropics." *Zookeys* **693**: 33–93.

Garwood, R. J., J. A. Dunlop, B. J. Knecht and T. A. Hegna (2017). "The phylogeny of fossil whip spiders." *BMC Evol Biol* **17**: 105.

Gea, S. L. R., W. F. Decraemer and J. J. J. Dirckx (2005). "Region of interest microCT of the middle ear: a practical approach." *J X-ray Sci Technol* **13**: 137–147.

Geandier, G., A. Hazotte, S. Denis, A. Mocellin and E. Maire (2003). "Microstructural analysis of alumina chromium composites by x-ray tomography and 3-D finite element simulation of thermal stresses." *Scripta Mater* **48**: 1219–1224.

Geandier, G., P. Weisbecker, S. Denis, A. Hazotte, A. Mocellin, J. L. Lebrun and E. Elkaim (2002). "X-ray diffraction analysis of residual stresses in alumina-chromium composites and comparison with numerical simulations." *Mater Sci Forum* **404–407**: 547–552.

Goyanes, A., F. Fina, A. Martorana, D. Sedough, S. Gaisford and A. W. Basit (2017). "Development of modified release 3D printed tablets (printlets) with pharmaceutical excipients using additive manufacturing." *Int J Pharm* **527**: 21–30.

Gu, Y., Q. Zhu, Y. Tang, Y. Zhang and X. Feng (2017). "Measurement of root surface area of permanent teeth in a Chinese population." *Arch Oral Biol* **81**: 26–30.

Gualda, G. A. R., D. L. Cook, R. Chopra, L. Qin, A. T. Anderson Jr and M. Rivers (2005). "Fragmentation, nucleation and migration of crystals and bubbles in the bishop tuff rhyolitic magma." *Trans Roy Soc Edin Earth Sci* **95**: 375–390.

Gualda, G. A. R. and M. Rivers (2006). "Quantitative 3D petrography using x-ray tomography: Application to Bishop Tuff pumice clasts." *J Volcanol Geothermal Res* **154**: 46–62.

Guillermic, R.-M., F. Koksel, X. Sun, D. W. Hatcher, M. T. Nickerson, G. S. Belev, M. A. Webb, J. H. Page and M. G. Scanlon (2018). "Bubbles in noodle dough: Characterization by X-ray microtomography." *Food Res Int* **105**: 548–555.

Guo, E., G. Zeng, D. Kazantsev, P. Rockett, J. Bent, M. Kirkland, G. Van Dalen, D. S. Eastwood, D. St John and P. D. Lee (2017). "Synchrotron X-ray tomographic quantification of microstructural evolution in ice cream: A multi-phase soft solid." *RSC Adv* **7**: 15561–15573.

Haeffner, D. R., J. D. Almer and U. Lienert (2005). "The use of high energy x-rays from the Advanced Photon Source to study stresses in materials." *Mater Sci Eng A* **399**: 120–127.

Hancock, B. C. and M. P. Mullarney (2005). "X-ray microtomography of solid dosage forms." *Pharm Technol* **29**: 92–100.

Haneca, K., K. Deforce, M. N. Boone, D. v. Loo, M. Dierick, J. v. Acker and J. v. d. Bulcke (2012). "X-ray sub-micron tomography as a tool for the study of archaeological wood preserved through the corrosion of metal objects." *Archaeometry* **54**: 893–905.

Heggli, M., T. Etter, P. Wyss, P. J. Uggowitzer and A. A. Gusev (2005). "Approaching representative volume element size in interpenetrating phase composites." *Adv Eng Mater* **7**: 225–229.

Hirano, T., K. Usami and K. Sakamoto (1989). "High resolution monochromatic tomography with x-ray sensing pickup tube." *Rev Sci Instrum* **60**: 2482–2485.

Hur, M., C. A. Gistelinck, P. Huber, J. Lee, M. H. Thompson, A. T. Monstad-Rios, C. J. Watson, S. K. McMenamin, A. Willaert, D. M. Parichy, P. Coucke and R. Y. Kwon (2017). "MicroCT-based phenomics in the zebrafish skeleton reveals virtues of deep phenotyping in a distributed organ system." *Elife* **6**: e26014.

Im, K. S., K. Fezzaa, Y. J. Wang, X. Liu, J. Wang and M. C. Lai (2007). "Particle tracking velocimetry using fast x-ray phase-contrast imaging." *Appl Phys Lett* **90**: 091919.

Ivanishko, Y., A. Bravin, S. Kovalev, P. Lisutina, M. Lotoshnikov, A. Mittone, S. Tkachev and M. Tkacheva (2017). "Feasibility study of the 3D visualization at high resolution of intra-cranial rabbit eyes with X-ray CT phase-contrast imaging." *Invest Ophthalmol Vis Sci* **58**: 5941–5948.

Jacob, A. and R. A. Chole (2006). "Survey anatomy of the paranasal sinuses in the normal mouse." *Laryngoscope* **116**: 558–563.

Jakus, A. E., A. L. Rutz, S. W. Jordan, A. Kannan, S. M. Mitchell, C. Yun, K. D. Koube, S. C. Yoo, H. E. Whiteley, C.-P. Richter, R. D. Galiano, W. K. Hsu, S. R. Stock, E. L. Hsu and R. N. Shah (2016). "Hyperelastic 'bone': A highly versatile, growth factor–free, osteoregenerative, scalable, and surgically friendly biomaterial." *Sci Transl Med* **358**: 358ra127.

Jones, K. W., H. Feng, A. Lanzirotti and D. Mahajan (2005). "Synchrotron x-ray microprobe and computed microtomography for characterization of nanocatalysts." *Nucl Instrum Meth B* **241**: 331–334.

Kameda, M., M. Ichihara, S. Maruyama, N. Kurokawa, Y. Aoki, S. Okumura and K. Uesugi (2017). "Advancement of magma fragmentation by inhomogeneous bubble distribution." *Sci Rep* **7**: 16755.

Kantzos, C. A., R. W. Cunningham, V. Tari and A. D. Rollett (2018). "Characterization of metal additive manufacturing surfaces using synchrotron X-ray CT and micromechanical modeling." *Compu Mech* **61**: 575–580.

Kenesei, P., H. Biermann and A. Borbely (2006). "Estimation of elastic properties of particle reinforced metal-matrix composites based on tomographic images." *Adv Eng Mater* **8**: 500–506.

Kenesei, P., A. Klohn, H. Biermann and A. Borbely (2006). "Mean field and multiscale modeling of a particle reinforced metal-matrix composite based on microtomographic investigations." *Adv Eng Mater* **8**: 506–510.

Kim, S. A., T. Punshon, A. Lanzirotti, L. Li, J. M. Alonso, J. R. Ecker, J. Kaplan and M. L. Guerinot (2006). "Localization of iron in Arabidopsis seed requires the vacuolar membrane transporter VIT1." *Science* **314**: 1295–1298.

Kinney, J. H., S. R. Stock, M. C. Nichols, U. Bonse, T. M. Breunig, R. A. Saroyan, R. Nusshardt, Q. C. Johnson, F. Busch and S. D. Antolovich (1990). "Nondestructive investigation of damage in composites using x-ray tomographic microscopy." *J Mater Res* **5**: 1123–1129.

Kotre, C. J. and I. P. Birch (1999). "Phase contrast enhancement of x-ray mammography: A design study." *Phys Med Biol* **44**: 2853–2866.

Kramer, R., H. J. Khoury, J. W. Vieira and I. Kawrakow (2006). "Skeletal dosimetry in the MAX06 and the FAX06 phantoms for external exposure to photons based on vertebral 3D-microCT images." *Phys Med Biol* **51**: 6265–6289.

Kramer, R., H. J. Khoury, J. W. Vieira and I. Kawrakow (2007). "Skeletal dosimetry for external exposure to photons based on μCT images of spongiosa from different bone sites." *Phys Med Biol* **52**: 6697–6716.

Krupa, A., O. Cantin, B. Strach, E. Wyska, Z. Tabor, J. Siepmann, A. Wróbel and R. Jachowicz (2017). "In vitro and in vivo behavior of ground tadalafil hot-melt extrudates: How the carrier material can effectively assure rapid or controlled drug release." *Int J Pharm* **528**: 498–510.

Larabell, C. A. and M. A. Le Gros (2004). "X-ray tomography generates 3-D reconstructions of the yeast, *Saccharomyces cerevisiae*, at 60-nm resolution." *Mole Biol Cell* **15**: 957–962.

Le Garff, E., V. Mesli, Y. Delannoy, T. Colard, J. De Jonckheere, X. Demondion and V. Hédouin (2017). "The precision of micro-tomography in bone taphonomic experiments and the importance of registration." *Forensic Sci Int* **273**: 161–167.

Le Garff, E., V. Mesli, E. Marchand, H. Behal, X. Demondion, A. Becart and V. Hedouin (2018). "Is bone analysis with μCT useful for short postmortem interval estimation?" *Int J Legal Med* **132**: 269–277.

Léonard, A., S. Blacher, C. Nimmol and S. Devahastin (2008). "Effect of far-infrared radiation assisted drying on microstructure of banana slices: An illustrative use of X-ray microtomography in microstructural evaluation of a food product." *J Food Eng* **85**: 154–162.

Leszczyński, B., P. Sojka-Leszczyńska, D. Wojtysiak, A. Wróbel and R. Pędrys (2018). "Visualization of porcine eye anatomy by X-ray microtomography." *Exp Eye Res* **167**: 51–55.

Li, J., Z. Zhong, R. Lidtke, K. E. Kuettner, C. Peterfy, E. Aliyeva and C. Muehleman (2003). "Radiography of soft tissue of the foot and ankle with diffraction enhanced imaging." *J Anat* **202**: 463–470.

Liu, X., A. J. Kim, W. Reynolds, Y. Wu and C. W. Lo (2017). "Phenotyping cardiac and structural birth defects in fetal and newborn mice." *Birth Defects Res* **109**: 778–790.

Liu, X., J. Liu, X. Li, S. K. Cheong, D. Shu, J. Wang, M. W. Tate, A. Ercan, D. R. Schuette, M. J. Renzi, A. Woll and S. M. Gruner (2004). Development of ultrafast computed tomography of highly transient fuel sprays. *Developments in X-Ray Tomography IV.* U. Bonse. Bellingham (WA), SPIE. **SPIE Proc Vol 5535**: 21–28.

Liu, C., P. L. Rosin, Y.-K. Lai, G. R. Davis, D. Mills and C. Norton (2016). "Recovering historical film footage by processing microtomographic images." *Digital Heritage Progress in Cultural Heritage: Documentation, Preservation, and Protection*. M. Ioannides, E. Fink, A. Moropoulou, M. Hagedorn-Saupe, A. Fresa, G. Liestøl, V. Rajcic and P. Grussenmeyer Berlin, Springer: 219–231.

Lu, H., K. Peterson and O. Chernoloz (2018). "Measurement of entrained air-void parameters in Portland cement concrete using micro X-ray computed tomography." *Int J Pavement Eng* **19**: 109–121.

Macchiarelli, R., L. Bondioli, A. Debenath, A. Mazurier, J. F. Tournepiche, W. Birch and C. Dean (2006). "How Neaderthal molar teeth grew." *Nature* **444**: 748–751.

MacLatchy, L. and R. Müller (2002). "A comparison of the femoral head and neck trabecular architecture of Galago and Perodicticus using microcomputed tomography." *J Human Evol* **43**: 89–105.

Maga, A. M., N. J. Tustison and B. B. Avants (2017). "A population level atlas of *Mus musculus* craniofacial skeleton and automated image-based shape analysis." *J Anat* **231**: 433–443.

Martín-Badosa, E., D. Amblard, S. Nuzzo, A. Elmoutaouakkilo, L. Vico and F. Peyrin (2003). "Excised bone structures in mice: imaging at three-dimensional synchrotron microCT." *Radiology* **229**: 921–928.

Martín-Herrero, J. and C. Germain (2007). "Microstructure reconstruction of fibrous C/C composites from X-ray microtomography." *Carbon* **45**: 1242–1253.

Martín-Vega, D., T. J. Simonsen, M. Wicklein and M. J. R. Hall (2017). "Age estimation during the blow fly intra-puparial period: a qualitative and quantitative approach using micro-computed tomography." *Int J Legal Med* **131**: 1429–1448.

Marxen, J. C., O. Prymak, F. Beckmann, F. Neues and M. Epple (2008). "Embryonic shell formation in the snail *Biomphalaria glabrata*: A comparison between scanning electron microscopy (SEM) and synchrotron radiation microcomputer tomography." *J Mollus Stud* **74**: 19–25.

Mazurier, A., V. Volpato and R. Macchiarelli (2006). "Improved noninvasive microstructural analysis of fossil tissues by means of SR-microtomography." *Appl Phys A* **83**: 229–233.

McErlain, D. D., R. K. Chhem, R. N. Bohay and D. W. Holdsworth (2004). "Micro-computed tomography of a 500-year-old tooth: Technical note." *Can Assoc Radiol J* **55**: 242–245.

Mickiewicz, P., M. Walczak, M. Łaszczyca, D. Kusz and Z. Wróbel (2018). "Differences between sexes in the standard and advanced dimensioning of lateral meniscal allografts." *Knee* **25**: 8–14.

Milovanovic, P., D. Djonic, M. Hahn, M. Amling, B. Busse and M. Djuric (2017). "Region-dependent patterns of trabecular bone growth in the human proximal femur: A study of 3D bone microarchitecture from early postnatal to late childhood period." *Am J Phys Anthropol* **164**: 281–291.

Mirahmadi, F., J. H. Koolstra, F. Lobbezoo, G. H. van Lenthe and V. Everts (2017). "Ex vivo thickness measurement of cartilage covering the temporomandibular joint." *J Biomech* **52**: 165–168.

Mizutani, R., R. Saiga, A. Takeuchi, K. Uesugi, Y. Terada, Y. Suzuki, V. De Andrade, F. De Carlo, S. Takekoshi, C. Inomoto, N. Nakamura, I. Kushima, S. Iritani, N. Ozaki, S. Ide, K. Ikeda, K. Oshima, M. Itokawa and M. Arai (2019). "Three-dimensional alteration of neurites in schizophrenia." *Transl Psychiatry* **9**: 85.

Mizutani, R., A. Takeuchi, T. Hara, K. Uesugi and Y. Suzuki (2007). "Computed tomography imaging of the neuronal structure of *Drosophila* brain." *J Synchrotron Rad* **14**: 282–287.

Mollenhauer, J., M. Aurich, Z. Zhong, C. Muehleman, A. A. Cole, M. Hasnah, O. Oltulu, K. E. Kuettner, A. Margulis and L. D. Chapman (2002). "Diffraction enhanced x-ray imaging of articular cartilage." *Osteoarthr Cartilage* **10**: 163–171.

Momose, A. and K. Hirano (1999). "The possibility of phase contrast X-ray microtomography." *Jpn J Appl Phys* **38 Suppl 1**: 625–629.

Momose, A., T. Takeda, Y. Itai, J. Tu and K. Hirano (1999). Recent observations with phase-contrast computed tomography. *Developments in X-Ray Tomography II*. U. Bonse. Bellingham (WA), SPIE. **SPIE Proc Vol 3772**: 188–195.

Momose, A., T. Takeda, Y. Itai, A. Yoneyama and K. Hirano (1998a). Perspective for medical applications of phase contrast x-ray imaging. *Medical Applications of Synchrotron Radiation*. C. U. M. Ando. Tokyo, Springer: 54–61.

Momose, A., T. Takeda, Y. Itai, A. Yoneyama and K. Hirano (1998b). "Phase contrast tomographic imaging using an x-ray interferometer." *J Synchrotron Rad* **5**: 309–314.

Moreau, J. D., D. Néraudeau, V. Perrichot and P. Tafforeau (2017). "100-million-year-old conifer tissues from the mid-Cretaceous amber of Charente (western France) revealed by synchrotron microtomography." *Ann Bot* **119**: 117–128.

Moreno, L. E., S. M. Rajguru, A. I. Matic, N. Yerram, A. M. Robinson, M. Hwang, S. Stock and C.-P. Richter (2011). "Infrared neural stimulation: Beam path in the guinea pig cochlea." *Hear Res* **282**: 289–302.

Muehleman, C., L. D. Chapman, K. E. Kuettner, J. Rieff, J. A. Mollenhauer, K. Massuda and Z. Zhong (2003). "Radiography of rabbit articular cartilage with diffraction enhanced imaging." *Anat Rec* **272A**: 392–397.

Muehleman, C., S. Majumdar, A. S. Issever, F. Arfelli, R. H. Menk, L. Rigon, G. Heitner, B. Reime, J. Metge, A. Wagner, K. E. Kuettner and J. Mollenhauer (2004). "X-ray detection of structural orientation in human articular cartilage." *Osteoarthr Cartilage* **12**: 97–105.

Muller, R. (2004). "A numerical study of the role of the tragus in the big brown bat." *J Acous Soc Am* **116**: 3701–3712.

Müller, B., M. Germann, D. Jeanmonod and A. Morel (2006). Three-dimensional assessment of brain tissue morphology. *Developments in X-Ray Tomography V.* U. Bonse. Bellingham (WA), SPIE. **SPIE Proc Vol 6318**: 631803-1–631803-8.

Müller, B., A. Lareida, F. Beckmann, G. M. Diakov, F. Kral, F. Schwarm, R. Stoffner, A. R. Gunkel, R. Glueckert, A. Schrott-Fischer, J. Fischer, A. Andronache and W. Freysinger (2006). Anatomy of the murine and human cochlea visualized at the cellular level by synchrotron radiation based microcomputed tomography. *Developments in X-Ray Tomography V.* U. Bonse. Bellingham (WA), SPIE. **SPIE Proc Vol 6318**: 631805-1–631805-9.

Mummery, P. M., B. Derby, P. Anderson, G. R. Davis and J. C. Elliott (1995). "X-ray microtomographic studies of metal matrix composites using laboratory x-ray sources." *J Microsc* **177**: 399–406.

do Nascimento-Dias, B. L., D. F. de Oliveira, A. S. Machado, O. M. O. Araújo, R. T. Lopes and M. J. dos Anjos (2018). "Utilization of nondestructive techniques for analysis of the Martian meteorite NWA 6963 and its implications for astrobiology" *X-ray Spectromet* **47**: 86–91.

Neues, F., W. H. Arnold, J. Fischer, F. Beckmann, P. Gaengler and M. Epple (2006). "The skeleton and pharyngeal teeth of zebrafish (Danio rerio) as a model of biomineralization in vertebrates." *Mat-wiss Werkstofftech* **37**: 426–431.

Neues, F., R. Goerlich, J. Renn, F. Beckmann and M. Epple (2007). "Skeletal deformations in medaka (Oryzias latipes) visualized by synchrotron radiation micro-computer tomography (SRμCT)." *J Struct Biol* **160**: 236–240.

Nicholson, E. K., S. R. Stock, M. W. Hamrick and M. J. Ravosa (2006). "Biomineralization and adaptive plasticity of the temporomandibular joint in myostatin knockout mice." *Arch Oral Biol* **51**: 37–49.

Nielsen, R. B., A. M. Alyassih, D. D. Peters, D. L. Carnes and J. Lancester (1997). "Microcomputed tomography: An advanced system for detailed endodontic research." *Oral Surg Oral Med Oral Pathol Oral Radiol Endod* **83**: 510–516.

Norman, D. G., D. G. Watson, B. Burnett, P. M. Fenne and M. A. Williams (2018). "The cutting edge - microCT for quantitative toolmark analysis of sharp force trauma to bone." *Forensic Sci Int* **283**: 156–172.

Olafsdottir, H., T. A. Darvann, N. V. Hermann, E. Oubel, B. K. Ersboll, A. F. Frangi, P. Larsen, C. A. Perlyn, G. M. Morriss-Kay and S. Kreiborg (2007). "Computational mouse atlases and their application to automatic assessment of craniofacial dysmorphology caused by the Crouzon mutation $Fgfr2^{C342Y}$." *J Anat* **211**: 37–52.

Park, R. J. and F. C. Meldrum (2002). "Synthesis of single crystals of calcite with complex morphologies." *Adv Mater* **14**: 1167–1169.

Parkinson, D. Y., G. McDermott, L. D. Etkin, M. A. Le Gros and C. A. Larabell (2008). "Quantitative 3-D imaging of eukaryotic cells using soft X-ray tomography." *J Struct Biol* **162**: 380–386.

Perlyn, C. A., V. B. DeLeon, C. Babbs, D. glover, L. Burell, T. Darvann, S. Kreiborg and G. Morriss-Kay (2006). "The cranialfacial phenotype of the crouzon mouse: Analysis of a model for syndromic craniosynostosis using three-dimensional microCT." *Cleft Palate Craniofac J* **43**: 740–748.

Peters, O. A., A. Laib, P. Rüegsegger and F. Barbakow (2000). "Three-dimensional analysis of root canal geometry by high resolution computed tomography." *J Dent Res* **79**: 1405–1409.

Plotino, G., N. M. Grande, R. Pecci, R. Bedini, C. H. Pameijer and F. Somma (2006). "Three-dimensional imaging using microcomputed tomography for studying tooth macromorphology." *J Am Dent Assoc* **137**: 1555–1561.

Postiglione, G., M. Alberini, S. Leigh, M. Levi and S. Turri (2017). "Effect of 3D-printed microvascular network design on the self-healing behavior of cross-linked polymers." *ACS Appl Mater Interf* **9**: 14371–14378.

Postnov, A., A. Zarowski, N. de Clerck, F. Vanpoucke, F. E. Offeciers, D. Van Dyck and S. Peeters (2006). "High resolution microCT scanning as an innovatory tool for evaluation of the surgical positioning of cochlear implant electrodes." *Acta Oto-Laryngologica* **126**: 467–474.

Pyun, A., J. R. Bell, K. H. Won, B. M. Weon, S. K. Seol, J. H. Je and C. W. Macosko (2007). "Synchrotron X-ray microtomography for 3D imaging of polymer blends." *Macromolecules* **40**: 2029–2035.
Qidwai, M. A. and V. G. DeGiorgi (2004). "Numerical assessment of the dynamic behavior of hybrid shape memory alloy composite." *Smart Mater Struct* **13**: 134–145.
Rajamannan, N. M., T. B. Nealis, M. Subramaniam, S. R. Stock, K. I. Ignatiev, T. J. Sebo, J. W. Fredericksen, S. W. Carmichael, T. K. Rosengart, T. C. Orszulak, W. D. Edwards, R. O. Bonow and T. C. Spelsberg (2005). "Calcified rheumatic valve neoangiogenesis is associated with VEGF expression and osteoblast-like bone formation." *Circulation* **111**: 3296–3301.
Rajamannan, N. M., M. Subramanium, S. Stock, F. Caira and T. C. Spelsberg (2005). "Atorvastatin inhibits hypercholesterolemia-induced calcification in the aortic valves via the Lrp5 receptor pathway." *Circulation* **112** [suppl I]: I-229–I-234.
Ravosa, M. J., E. P. Kloop, J. Pinchoff, S. R. Stock and M. Hamrick (2007). "Plasticity of mandibular biomineralization in myostatin-deficient mice." *J Morphol* **268**: 275–282.
Ravosa, M. J., R. Kunwar, S. R. Stock and M. S. Stack (2007). "Pushing the limit: masticatory stress and adaptive plasticity in mammalian craniomandibular joints." *J Exp Biol* **210**: 628–641.
Ravosa, M. J., S. R. Stock, E. L. Simons and R. Kunwar (2007). "MicroCT analysis of symphyseal ontogeny in a subfossil lemur (Archaeolemur)." *Int J Primatol* **28**: 1385–1396.
Richbourg, H. A., M. J. Martin, E. R. Schachner and M. A. McNulty (2017). "Anatomical variation of the tarsus in common inbred mouse strains." *Anat Rec* **300**: 450–459.
Richter, C. P., H. Young, S. V. Richter, V. Smith-Bronstein, S. R. Stock, X. Xiao, C. Soriano and D. S. Whitlon (2018). "Fluvastatin protects cochleae from damage by high-level noise." *Sci Rep* **8**: 3033.
Rosin, P. L., Y. K. Lai, C. Liu, G. R. Davis, D. Mills, G. Tuson and Y. Russell (2018). "Virtual recovery of content from X-ray micro-tomography scans of damaged historic scrolls." *Sci Rep* **8**: 11901.
Rühli, F. J., G. Kuhn, R. Evison, R. Müller and M. Schultz (2007). "Diagnostic value of microCT in comparison with histology in the qualitative assessment of historical human skull bone pathologies." *Am J Phys Anthropol* **133**: 1099–1111.
Sanchez, S. A., J. Narciso, F. Rodriguez-Reinoso, D. Bernard, I. G. Watson, P. D. Lee and R. J. Dashwood (2006). "Characterization of lightweight graphite based composites using x-ray microtomography." *Adv Eng Mater* **8**: 491–495.
Sapei, L., N. Gierlinger, J. Hartmann, R. Nöske, P. Strauch and O. Paris (2007). "Structural and analytical studies of silica accumulations in Equisetum hyemale." *Anal Bioanal Chem* **389**: 1249–1257.
Schena, G., S. Favretto, L. Santoro, A. Pasini, M. Bettuzzi, F. Casali and L. Mancini (2005). "Detecting microdiamonds in kimberlite drill-hole cores by computed tomography." *Int J Miner Process* **75**: 173–188.
Seifert, A., M. J. Flynn, K. Montgomery and P. Brown (2004). Visualization of x-ray microtomography data for a human tooth atlas. *Medical Imaging 2004: Visualization, Image-Guided Procedures, and Display, 2004*. R. Galloway, Jr. Bellingham (WA), SPIE. **SPIE Proc Vol 5367**: 747–757.
Sevanto, S., M. Ryan, L. T. Dickman, D. Derome, A. Patera, T. Defraeye, R. E. Pangle, P. J. Hudson and W. T. Pockman (2018). "Is desiccation tolerance and avoidance reflected in xylem and phloem anatomy of two coexisting arid-zone coniferous trees?" *Plant Cell Environ* **41**: 1551–1564.
Shah, A. P., D. A. Rajon, D. W. Jokisch, P. W. Patton and W. E. Bolch (2005). "A comparison of skeletal chord length distributions in the adult male." *Health Phys* **89**: 199–215.
Shibata, T., S. Matsumoto and T. Nagano (1999). "Tomograms of the arterial system of the human fetal auditory apparatus obtained by very high resolution microfocus x-ray CT and 3D reconstruction." *Acta Anat Nippon* **74**: 545–553.
Shimizu, K., J. Ikezoe, H. Ikura, H. Ebara, T. Nagareda, N. Yagi, K. Umetani, K. Uesugi, K. Okada, A. Sugita and M. Tanaka (2000). Synchrotron radiation microtomography of the lung specimens. *Medical Imaging 2000: Physics of Medical Imaging*. J. T. Dobbins III and J. M. Boone. Bellingham (WA), SPIE. **SPIE Proc Vol 3977**: 196–204.
Simons, E. L., E. R. Seiffert, T. M. Ryan and Y. Attia (2007). "A remarkable female cranium of the early Oligocene anthropoid *Aegyptopithecus zeuxis* (Catarrhini, Propliopithecidae)." *PNAS* **104**: 8731–8736.

Smith, T. M., P. Tafforeau, D. J. Reid, R. Grun, S. Eggins, M. Boutakiout and J. J. Hublin (2007). "Earliest evidence of modern human life history in North African early *Homo sapiens*." *PNAS* **104**: 6128–6133.

Socha, J. J., M. W. Westneat, J. F. Harrison, J. S. Waters and W. K. Lee (2007). "Real-time phase-contrast x-ray imaging: A new technique for the study of animal form and function." *BMC Biol* **5**: 6.

Song, S. R., K. W. Jones, W. B. Lindquist, B. A. Dowd and D. L. Sahagian (2001). "Synchrotron x-ray computed tomography: studies on vesiculated rocks." *Bull Volcanol* **63**: 252–263.

Spencer, A. R. T., R. J. Garwood, A. R. Rees, R. J. Raine, G. W. Rothwell, N. T. J. Hollingworth and J. Hilton (2017). "New insights into Mesozoic cycad evolution: An exploration of anatomically preserved Cycadaceae seeds from the Jurassic Oxford Clay biota." *Peer J* **5**: e3723.

Staab, M., F. H. Garcia, C. Liu, Z. H. Xu and E. P. Economo (2018). "Systematics of the ant genus Proceratium Roger (Hymenoptera, Formicidae, Proceratiinae) in China – with descriptions of three new species based on microCT enhanced next-generation-morphology." *Zookeys* **770**: 137–192.

Stock, S. R. (2014). "Sea urchins have teeth? A review of their microstructure, biomineralization, development and mechanical properties." *Conn Tiss Res* **55**: 41–51.

Stock, S. R., F. De Carlo, X. Xiao and T. A. Ebert (2007). Bridges between radial wedges (septs) in two diadematid spine types. *12th International Echinoderm Conference*. Durham, Echinoderms.

Stock, S. R., T. A. Ebert, K. Ignatiev and F. D. Carlo (2006). Structures, structural hierarchy and function in sea urchin spines. *Developments in X-Ray Tomography V*. U. Bonse. Bellingham (WA), SPIE. **SPIE Proc Vol 6318**: 63180A-1–63180A-4.

Stock, S. R., A. Guvenilir, T. L. Starr, J. C. Elliott, P. Anderson, S. D. Dover and D. K. Bowen (1989). "Microtomography of silicon nitride/silicon carbide composites." *Ceram Trans* **5**: 161–170.

Stock, S. R., K. Ignatiev, F. De Carlo, J. D. Almer and A. Veis (2004a). Multiple mode x-ray study of 3-D tooth microstructure across different sea urchin families. *Proceedings 8th International Conference on Chemistry and Biology of Mineralized Tissue*. W. J. Landis and J. Sodek. Toronto, University of Toronto Press: 107–110.

Stock, S. R., K. Ignatiev, P. L. Lee, K. Abbott and L. M. Pachman (2004b). "Pathological calcification in juvenile dermatomyositis (JDM): microCT and synchrotron x-ray diffraction reveal hydroxyapatite with varied microstructures." *Conn Tiss Res* **45**: 248–256.

Stock, S. R., K. I. Ignatiev, T. Dahl, A. Veis and F. D. Carlo (2003). "Three-dimensional microarchitecture of the plates (primary, secondary and carinar process) in the developing tooth of Lytechinus variegatus revealed by synchrotron x-ray absorption microtomography (microCT)." *J Struct Biol* **144**: 282–300.

Stock, S. R., N. M. Rajamannan, E. R. Brooks, C. B. Langman and L. M. Pachman (2004). Pathological calcifications studies with microCT. *Developments in X-Ray Tomography IV*. U. Bonse. Bellingham (WA), SPIE. **SPIE Proc Vol 5535**: 424–431.

Stock, S. R. and A. Veis (2003). Preliminary microfocus x-ray computed tomography survey of echinoid fossil microstructure. *Applications of X-Ray Computed Tomography in the Goesciences*. F. Mees, R. Swennen, M. van Geet and P. Jacobs. London, Geological Society of London. Special Publication **215**: 225–235.

Stock, S. R., X. Xiao, S. R. Stock and A. Ziegler (2013). "Quantification of carinar process plate orientation in camarodont sea urchin teeth." *Cah Biol Mar* **54**: 735–741.

Suwa, G., R. T. Kono, S. Katoh, B. Asfaw and Y. Beyene (2007). "A new species of great ape from the late Miocene epoch in Ethiopia." *Nature* **448**: 921–924.

Tafforeau, P., R. Boistel, E. Boller, A. Bravin, M. Brunet, Y. Chaimanee, P. Cloetens, M. Feist, J. Hoszowska, J. J. Jaeger, R. F. Kay, V. Lazzari, L. Marivaux, A. Nel, C. Nemoz, X. Thibault, P. Vignaud, S. Zabler, H. Riesemeier, P. Fratzl and P. Zaslansky (2006). "Applications of X-ray synchrotron microtomography for non-destructive 3D studies of paleontological specimens." *Appl Phys A* **83**: 195–202.

Takeda, T., A. Momose, E. Ueno and Y. Itai (1998). "Phase contrast x-ray CT image of breast tumor." *J Synchrotron Rad* **5**: 1133–1135.

Tammas-Williams, S., H. Zhao, F. Léonard, F. Derguti, I. Todd and P. B. Prangnell (2015). "XCT analysis of the influence of melt strategies on defect population in Ti–6Al–4V components manufactured by selective electron beam melting." *Mater Charact* **102**: 47–61.

Tan, X., I. Jahan, Y. Xu, S. R. Stock, C. C. Kwan, C. Soriano, X. Xiao, B. Fritzsch, J. García-Añoveros and C.-P. Richter (2018). "Auditory neural activity in congenital deaf mice induced by infrared neural stimulation." *Sci Rep* **8**: 388.

Tanisako, A., S. Tsuruta, A. Hori, A. Okumura, C. Miyata, C. Kuzuryu, T. Obi and H. Yoshimura (2006). MicroCT of small insects by projection x-ray microscopy. *Developments in X-Ray Tomography V.* U. Bonse. Bellingham (WA), SPIE. **SPIE Proc Vol 6318**: 63180B-1–63180B-8.

Tappero, R., E. Peltier, M. Gräfe, K. Heidel, M. Ginder-Vogel, K. J. T. Livi, M. L. Rivers, M. A. Marcus, R. L. Chaney and D. L. Sparks (2007). "Hyperaccumulator *Alyssum murale* relies on a different metal storage mechanism for cobalt than for nickel." *New Phyt* **175**: 641–654.

Toda, H., K. Uesugi, A. Takeuchi, K. Minami, M. Kobayashi and T. Kobayashi (2006). "Three-dimensional observation of nanoscopic precipitates in an aluminum alloy by microtomography with Fresnel zone plate optics." *Appl Phys Lett* **89**: 143112.

Tokita, M., W. Yano, H. F. James and A. Abzhanov (2017). "Cranial shape evolution in adaptive radiations of birds: Comparative morphometrics of Darwin's finches and Hawaiian honeycreepers." *Philos Trans R Soc Lond B Biol Sci* **372**: 20150481.

Traini, D., G. Loreti, A. S. Jones and P. M. Young (2008). "X-ray computed microtomography for the study of modified release systems." *Microsc Anal* **22**: 13–15.

Varray, F., I. Mirea, M. Langer, F. Peyrin, L. Fanton and I. E. Magnin (2017). "Extraction of the 3D local orientation of myocytes in human cardiac tissue using X-ray phase-contrast microtomography and multi-scale analysis." *Med Image Anal* **38**: 117–132.

Vogel, U., F. Beckmann, T. Zahnert and U. Bonse (2001). Microtomography of the human middle and inner ear. *Developments in X-Ray Tomography III.* U. Bonse. Bellingham (WA), SPIE. **SPIE Proc Vol 4503**: 146–155.

Wang, Y., H. I. Chang, D. F. Wertheim, A. S. Jones, C. Jackson and A. G. A. Coombes (2007). "Characterisation of the macroporosity of polycaprolactone-based biocomposites and release kinetics for drug delivery." *Biomater* **28**: 4619–4627.

Wang, Z., E. Herremans, S. Janssen, D. Cantre, P. Verboven and B. Nicolaï (2018). "Visualizing 3D food microstructure using tomographic methods: advantages and disadvantages." *Annu Rev Food Sci Technol* **9**: 323–343.

Weber, S. M., K. A. Peterson, B. Durkee, C. Qi, M. Longino, T. Warner, Jr. F. T. Lee and J. P. Weichert (2004). "Imaging of murine liver tumor using microCT with a hepatocyte-selective contrast agent: Accuracy is dependent on adequate contrast enhancement." *J Surg Res* **119**: 41–45.

Westneat, M. W., O. Betz, R. W. Blob, K. Fezzaa, W. J. Cooper and W. K. Lee (2003). "Tracheal respiration in insects visualized with synchrotron x-ray imaging." *Science* **299**: 558–560.

Williams Jr., J. C., B. R. Matlaga, S. C. Kim, M. E. Jackson, A. J. Sommer, J. A. McAteer, J. E. Lingeman and A. P. Evan (2006). "Calcium oxalate calculi found attached to the renal papilla: Preliminary evidence for early mechanisms in stone formation." *J Endourol* **20**: 885–890.

Yin, H. X., T. Zhao, B. Lo, H. Shu, Z. F. Huang, X. L. Gao, P. P. Zhu, Z. Y. Wu and S. Q. Luo (2007). "Visualization of guinea pig cochleae with computed tomography of diffraction enhanced imaging and comparison with histology." *J X-ray Sci Technol* **15**: 73–84.

Yu, Q., T. Takeda, K. Umetani, E. Ueno, Y. Itai, Y. Hiranaka and T. Akatsuka (1999). "First experiment by two-dimensional digital mammography with synchrotron radiation." *J Synchrotron Rad* **6**: 1148–1152.

Zhang, Y., J. Du, L. Ma, X. Pan, J. Wang and X. Guo (2018). "Three-dimensional segmentation, reconstruction and phenotyping analysis of maize kernel based on microCT images." *Fresenius Environ Bull* **27**: 3965–3969.

8

Cellular or Trabecular Solids

The subject of this chapter is cellular solids, that is, materials with significant volume fractions of empty, fluid-filled, or soft-tissue-filled space. This usage of "cellular" is, of course, quite different from that a microbiologist might intuitively envision. The common features between engineered materials such as foams and natural cellular materials such as wood or cancellous bone require similar analyses approaches. MicroCT studies of these materials are therefore collected together in this chapter.

The following section gives background on cellular materials. Section 8.2 describes microCT studies of static cellular structures, and Section 8.3 expands coverage to studies of temporally evolving, nonbiological cellular materials. Section 8.4 describes some of the many microCT results on mineralized tissues such as cancellous (trabecular) bone, and Section 8.5 highlights a few studies on scaffolds being developed for cell ingrowth; that is, as biomedical implants.

8.1 Cellular Solids

Following the definition of Gibson and Ashby, a cellular solid consists of an interconnected network of solid struts (rods) or plates that form the edges and faces of cells, respectively (Gibson and Ashby 1999). In other words, faces separate two cells, and edges are common to three or more subvolumes (cells) of the larger structure. In 3D, the cells are polyhedra filling space and such materials are often termed foams. If the cells connect through open faces (the only material being struts at the cells edges), the material is termed as open cell foam; if the faces are solid, sealing off adjacent cells, it is a closed cell foam. Many cellular materials are produced by plants and animals, including wood, cork, and bone. Engineered (by humans) cellular solids employ all classes of materials, including composites, and are used for thermal insulation, packaging (energy absorption), structures (for high specific strengths), buoyancy (marine), and scaffolding for cell growth. The microstructural characteristics of cellular solids are difficult to quantify, except with noninvasive, 3D methods such as microCT. In what follows, for purposes of simplicity, the solid phase will be discussed as if it occupies a relatively small fraction of the total volume.

As noted by Maire et al., cellular solids are often very challenging to analyze with tomographic techniques, particularly with respect to the different levels within the hierarchy of structural scales influencing the materials performance, specifically the scale of the constitutive material and the scale of cellular microstructure (Maire, Fazekas et al. 2003b). The interplay between voxel size, contrast sensitivity, and field of view (FOV) is particularly prominent in materials such as foams. If, on the one hand, the distribution of cell sizes is important, then large voxel sizes may be required for the FOV to span the half-dozen or more cells required to represent the structure adequately, and an instrument optimized for these parameters (perhaps an industrial or medical peripheral CT system) might prove more efficient than a microCT system. If, on the other hand, features within the cell walls are of central importance, then microCT or even nanoCT on small sections of the material is required. Both scales sometimes can be studied productively on the same instrument, and several investigations have employed two extreme resolution/FOV combinations on a single instrument (Salvo, Belestin et al. 2004), two or more systems (Peyrin, Salome et al. 1998, Brunke, Oldenbach et al. 2004), or local tomography techniques.

A structure like a cellular solid, that is, a complex mixture of empty space and solid, requires several parameters in order to describe its microstructure, specifically how much material is present and how the material is distributed spatially. These microstructural characteristics have largely been defined in studies of cancellous bone, and unbiased methods of measuring these quantities are products of these studies (Odgaard 2001). The amount of material is given by the volume fraction of solid V_V. In certain applications, the mean cell size and the distribution of cell sizes of a foam must be specified in order for macroscopic properties to be predicted. Also important is the 3D distribution of solid material, the components of which can be in the form of plates and struts (rods); these are characterized by quantities such as the surface area per unit volume S_V and mean thickness <Th> of the structural elements. Accurate values of S_V and <Th> cannot be derived from isolated slices unless one can assume the individual structural elements are all plates or all rods: this is one circumstance where a data set of contiguous slices is essential. As discussed in detail elsewhere (Odgaard 1997), one cannot simply make measurements of apparent thickness in the individual slices of a volume without risk of introducing significant, unpredictable bias into the data.

Additional quantities of importance include the connectivity density (Conn.D), the structural anisotropy, and the structure model index (SMI). Connectivity reports the number of redundant trabeculae in the structure, that is, trabeculae that can be cut without increasing the number of separate parts of the structure, and Conn.D is calculated by dividing the connectivity by the examination volume. A more complete discussion of Euler numbers, connectivity, and edge effects is beyond the scope of this review (Odgaard 2001), as these topics are not emphasized in the examples below. Anisotropy can be defined by main directions (perpendiculars to symmetry planes in the structure) and by numbers quantifying the concentration of directions around the main direction (Odgaard 2001). The fabric tensor compactly describes orthotropic architectural anisotropy via a 3×3 matrix of eigenvectors giving main directions and eigenvalues the degree of concentration around main directions (Cowin 1985). Alternatively, the degree of anisotropy DA can be computed using the mean intercept length (MIL) method (Odgaard 1997, 2001). The SMI relates the convexity of the structure to a model type and allows one to determine, for example, whether a given structure is more rod-like or more plate-like (Hildebrand and Rüegsegger 1997). An array of ideal (flat) plates has $SMI = 0$, a set of ideal cylindrical rods has $SMI = 3$, and a set of spheres has $SMI = 4$. If sufficient "air bubbles" are present within the structure, $SMI < 0$.

8.2 Static Cellular Structures

Before covering evolving cellular structures and biological/biomedical structures in detail, it is useful to consider some of the wide variety of cellular materials to which microCT has been applied. Although there are a few reports of plant-related cellular structures, some of which are mentioned in the following paragraph, the bulk of this section is comprised of inorganic and engineered polymeric materials.

Seeds often have a cellular structure, and, based on analysis of the cellular structure, phase contrast microCT has been used to study the phylogeny of fossil seeds from the Cretaceous (Friis, Crane et al. 2007). The cortical flesh of the apple fruit can be regarded as a cellular material, and microCT quantified differences in pore structure between two cultivars, despite considerable variability within the pore spaces of each cultivar (Mendoza, Verboven et al. 2007). MicroCT has shown that differences in porosity in banana slices result from different drying techniques (Léonard, Blacher et al. 2008).

Porosity and grain structure in wood have been studied (Illman and Dowd 1999, Steppe, Cnudde et al. 2004, Vetter, Cnudde et al. 2006). More recent phase contrast microCT of spruce wood quantified total porosity, tracheid diameter, cell wall thickness, and pit diameter and suggested that subvoxel-sized features could be determined indirectly using watershed segmentation (Trtik, Dual et al. 2007). Figure 8.1 shows three orthogonal planes through a reconstructed volume of spruce wood; a wide range of pore dimensions are present and the axes of these cylindrical structures are primarily along the plant's axis (vertical in Fig. 8.1) but also in other directions. In other words, many different levels of structural hierarchy contribute to the twin tasks of biological viability and of weight-bearing plus resistance to bending deformation. More details are found elsewhere (Wilkes 2008).

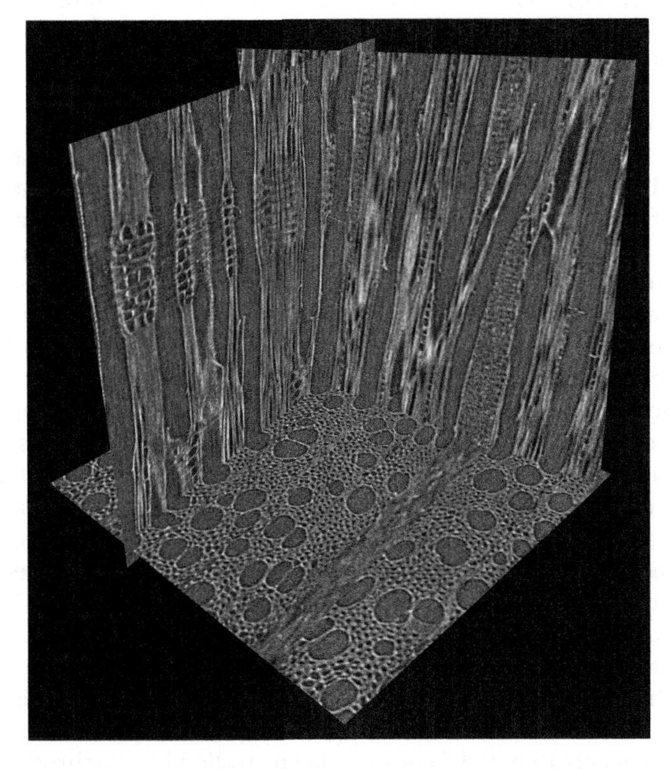

FIGURE 8.1
Synchrotron microCT of wood. Three orthogonal sections are shown with the plant's axis (i.e., the tree trunk axis) vertical. The lighter the pixel, the more absorbing is the corresponding voxel. (Unpublished synchrotron microCT data, T.E. Wilkes, K.T. Faber, F. De Carlo, S.R. Stock, 2-BM, APS, June 2007). See also Wilkes (2008).

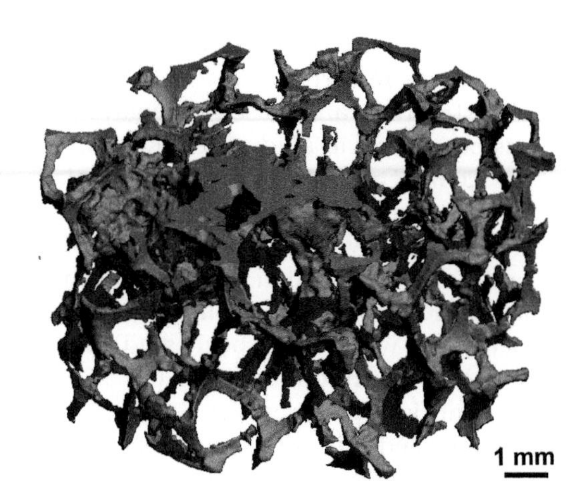

FIGURE 8.2
3D rendering of an open cell Al foam showing only the solid. (Unpublished lab microCT data, Y. Matsumoto, A.H. Brothers, D.C. Dunand and S.R. Stock, October 2005.) See also Matsumoto et al. (2007).

Figure 8.2 shows a 3D rendering of an open cell Al foam, and microCT is very valuable for studying processes such as chemical milling of metal foams (Matsumoto, Brothers et al. 2007). The elements in Fig. 8.2 are mostly struts and not plates. In comparison with trabecular bone (e.g., Fig. 6.6), the structure appears much more angular and transitions seem to be more abrupt. Several different composite foams based on natural macromolecules have been characterized by microCT (Darder, Matos et al. 2017, Ghafar, Parikka et al. 2017, Colonetti, Sanches et al. 2018).

Open-cell aluminum foams have been used as substrates/materials sources for zeolite catalyst growth, and lab microCT was used to characterize the starting cell diameters and strut thicknesses as well as to show that the zeolite film had homogeneous thickness (Scheffler, Herrmann et al. 2004). Pd-Ag/SiO_2 xerogel catalysts supported on Al_2O_3 foams were also characterized by microCT (Blacher, Léonard et al. 2004). Polyurethane foams containing residues of petroleum industry catalysts were studied as recoverable pH-sensitive sorbents for aqueous pesticides (Almeida, Ayres et al. 2018), and acid–base polymeric foams for oil recovery (Cherukupally, Acosta et al. 2017). Synchrotron microCT was used to characterize the internal cellular structure within the walls of hollow fibers used for filtration (Remigy and Meireles 2006) and the complex 3D porosity in a polymeric microfiltration membrane (Remigy, Meireles et al. 2007), a study that would have been difficult to perform with other methods. Synchrotron microCT of replication processed open-celled aluminum foams was used to measure the internal architecture of these microcellular materials (Goodall, Marmottant et al. 2007). Cell shape did not affect uniaxial mechanical properties as long as the cells were roughly equiaxed; spherical and angular particle-derived foam propertied differed otherwise. Permeability in the Darcy regime of fluid flow, however, was relatively insensitive to details of pore shape.

Bubble size distributions in simulated or actual volcanic silicate foams have been quantified with synchrotron microCT, and 2D and 3D measures of the distributions were compared and related to models of bubble growth (Robert, Baker et al. 2004). Pore initiation by blowing agents in metal foams has been studied by synchrotron microCT, and pores in prealloyed Al powders nucleated around the blowing agent particles, whereas they tended to nucleate around Si particles in Al-Si powder blends (Helfen, Baumbach et al. 2005). In a more well-developed, closed-cell foam produced by the blowing agent route, a large number of 3D parameters were quantified for pores (volume fraction, equivalent radius, surface area, sphericity), for the distribution of cell wall material (via granulometry) between cell faces and edges, and for Ti particle geometric parameters and size distribution (Brunke, Oldenbach et al. 2005). The 3D spatial distribution of Ti particles was quantified and related to foam morphology in a simple representation, projection of the mass within each slice along the length of the cylindrical sample that sufficed to show the principal foaming direction (Brunke, Oldenbach et al. 2005). Different foaming times also altered the distribution of cell wall material (Brunke, Oldenbach et al. 2004).

X-ray tomography (using a fairly large voxel size, 90 μm, dictated by the required FOV for a material with cells up to 5-mm diameter) of different structural metal foam types revealed differences in cell anisotropy (distribution of cell volumes and of aspect ratios quantified as equivalent ellipsoids (Benaouli, Froyen et al. 2000, 2005); stereographic projections showing distribution of cell axes vs orientation) that correlated with altered mechanical properties (Benouali, Froyen et al. 2005). Defects in the cell walls (corrugations, holes, and cracks), however, substantially reduced Young's modulus and strength from theoretical values (Benouali, Froyen et al. 2005). Another lab microCT report of a metal foam (Maire and Buffière 2000) was followed by detailed FEM of foam deformation (Maire, Fazekas et al. 2003b), see below.

Synchrotron microCT of polyvinylchloride foam (an example where cell diameters were smaller than 0.2 mm) showed that a higher processing temperature produced higher volume fraction of porosity, larger equivalent cell diameters, decreased wall thicknesses, and decreased degree of anisotropy (Elmoutaouakkil, Fuchs et al. 2003). Interestingly, selection of small subvolumes produced the same results as for larger volumes of interest. Polymeric foams for hydrogen storage applications have also been examined with microCT.

Glass foams can be produced from silicate wastes (power generation waste plus bottle fragments plus a SiC foaming agent) and used for thermal or acoustic insulation, etc., a process of considerable interest given the amount of this waste that has been accumulating for years. Lab microCT has been used to relate foam structure and properties with processing temperature (Wu, Boccaccini et al. 2006). Wu and coworkers studied lightweight glass-ceramic foams derived from silicate wastes by varying processing temperature and amount of foaming agent (Wu, Boccaccini et al. 2007). Good agreement was obtained between experimental conductivity data and simulations based on microCT-derived finite element analysis, and the foams behaved mechanically in a manner typical of brittle foams.

8.3 Temporally Evolving, Nonmineralized Tissue Cellular Structures

The 3D evolution of liquid foams over tens of hours has been studied by synchrotron microCT (Lambert, Cantat et al. 2005), and the progress of drainage and bubble coarsening posed particular problems in this study. While the minimum time between each set of projections for a single reconstruction was 8 min (10 mm FOV, 10 μm voxels, 50 ms exposure per projection, and 900 radiographs per data set), nearly instantaneous changes such as bubbles bursting or moving produced occasional reconstruction artifacts. Despite an inability to image a significant fraction of thin film area, segmentation of the image into a set of bubbles could be done automatically, using a distance transformation plus watershed algorithm. Even with boundary bubbles removed, the subvolume contained

~2×10^3 bubbles at the start of the experiment, an order of magnitude increase over the number of bubbles that could be analyzed by other techniques such as MRI or optical tomography. A large portion of the distribution of bubble volumes could be characterized as exponential at a given time. The authors of this study also examined the number of faces per bubble, a quantity important in coarsening kinetics (Lambert, Cantat et al. 2005).

Rapid synchrotron microCT (<30 s per reconstructed volume) has also been applied to bubble formation in bread, that is, during the process of rising of the dough where the volume fraction of porosity increases from ~0.1 to ~0.7 (Babin, Valle et al. 2006). As dough rises over a period of 2–3 h with relatively stable cell walls, unlike soap bubbles, the changes during each data collection increment were negligible. The first stage consisted of nucleation and growth of many similarly sized spherical bubbles. Bubble interaction and growth by coalescence characterized the second stage, and the third stage was dominated by coalescence-driven formation of large, irregularly shaped voids. Tracking individual bubbles during the first stage of rising allowed the kinetics of bubble growth to be compared to models. Baking of dough was also examined (Babin, Valle et al. 2006). Drying of materials such as pomelo peels can produce spatially varying internal strains, and microCT and digital volume correlation have been used to study this process (Wang, Pan et al. 2018).

Mechanical deformation of cellular materials is generally performed incrementally either in an external testing apparatus or in an in situ load frame specially constructed for microCT. While numerous apparatuses (for applying loads in situ during microCT) have been built, their designs differ little from those developed in the first half of the 1990s and described in the chapter on microCT of deformed or cracked specimens (Breunig, Stock et al. 1992, 1993, Breunig, Nichols et al. 1994, Hirano, Usami et al. 1995).

With microCT characterization of foams, there is concern that the results of interrupted testing differ from those obtained during continuous loading. The lab microCT studies of Nazarian et al., however, found the results of stepwise compression of several types of cellular materials agreed well with data from conventional continuous testing (Nazarian and Muller 2004, Nazarian, Stauber et al. 2005).

MicroCT observation of deformation of cellular solids is frequently compared to FEM of the deformation based on the initial (microCT-determined) structure. In principle, this allows the step-by-step FEM predictions of deformation process to be compared with experiment and assumptions in the FEM to be refined. The steps involved in meshing microCT for FEM are covered elsewhere (Keaveny 2001, van Rietbergen 2001, van Rietbergen and Huiskes 2001, Maire, Fazekas et al. 2003b, Youssef, Maire et al. 2005, Plougonven, Bernard et al. 2006), and the reader is advised to search the bone literature for further discussion of approaches for cellular material.

Several microCT studies of elastomeric foams have appeared and include both open- and closed-cell materials (Kinney, Marshall et al. 2001, Elliott, Windle et al. 2002, Youssef, Maire et al. 2005, Plougonven, Bernard et al. 2006). These studies were aimed at improving understanding of deformation processes underlying the three stages of the typical elastomeric foam stress–strain curve (linear elastic stage I at lowest stresses transitioning into a plateau region; stage II, where little increase in stress produces large increases in strain and finishing in a densification regime; stage III, of rapidly rising stress). Kinney and coworkers studied 1-mm thick, 15-mm diameter silica-reinforced (~25 wt%) polysiloxane foam pads (Kinney, Marshall et al. 2001). Synchrotron microCT of this closed-cell material was performed at four deformation levels from 0% to 35% strain, and analysis concentrated on the fraction of surface connected pores and finite element modeling of the structure. Because the pore diameters and mean cell wall thicknesses were ~1/3 and ~1/10 the specimen thickness, it is difficult to know whether the results can be applied to other geometries.

Plougonven et al. studied a polypropylene foam subjected to alternating cycles of interrupted shocks followed by synchrotron microCT imaging (Plougonven, Bernard et al. 2006, Viot, Bernard et al. 2007); this simulates energy absorption behavior during crashes. The investigators concentrated on behavior at the mesoscopic scale of grains (1–3 mm diameter within the 10-mm diameter specimen), and the hierarchy of structural scales of pores in the foam complicated the analysis. Phase contrast effects were suppressed before filtering suppressed noise, and a distance transform operation defined the cell walls. The relative densities of the different grains were then computed as a function of deformation. A similar approach combined with numerical modeling was used to study shock loading of an Al foam (Bourne, Bennett et al. 2008).

Shape memory polymer foams can be compacted as much as 80% and still experience full strain recovery over multiple cycles. MicroCT of one such foam revealed bending buckling and cell collapse with increasing compression, consistent with results of numerical simulations (Di Prima, Lesniewski et al. 2007).

In an open-celled polyurethane foam, Elliott and coworkers imaged the structure at 14 strains with the observations clustered at the critical regions of the stress–strain curve (i.e., the transition between linear elastic and plateau regimes, the transition between plateau and densification regions, and in the final stages of densification) (Elliott, Windle et al. 2002). Their focus was on strut deformation processes, and they employed local tomography to observe the central ~7 mm of a 25-mm × 25-mm high specimen. The initial stages of compression (below 4% strain, in the linear elastic regime) were taken up by small amounts of bending in struts that were both longer than average and inclined to the compression axis. More severe bending and reorientation of struts within a localized deformation band accommodated strains up to ~23%. The densification regime began after multiple collapse bands formed and impinged on one another. Node and strut models of the deformation were applied, but further improvements were found to be necessary (Elliott, Windle et al. 2002). Brydon et al. considered, in much greater detail, numerical modeling of densification of open-celled carbon foams, whose structure had been measured by synchrotron microCT (Brydon, Bardenhagen et al. 2005).

The microCT study of a closed-cell polyurethane foam stands in contrast with that described in the previous paragraph (Youssef, Maire et al. 2005). Synchrotron microCT (local reconstruction) was also performed on the central portion of a larger specimen (central ~2 mm of a 6-mm diameter specimen) at four strains (maximum of 20%), and the investigators performed a thorough sensitivity analysis of FEM parameters and used FEM to fill in the gaps between microCT measurements. Direct comparison of FEM-derived strain distributions in the actual foam structure showed cell wall rupture occurred at local strain concentrators.

Metal foams often fail at significant lower stresses than expected from theory, and this has motivated microCT imaging of foams after discrete increments of compression (Bart-Smith, Bastawros et al. 1998, Kádár, Maire et al. 2004, Montanini 2005, Toda, Ohgaki et al. 2006a, Toda, Takata et al. 2006b, Wang, Hu et al. 2006). Even simple analysis methods and modest spatial resolution can provide informative results. Deformation of two fairly dense Al "foams" produced by powder metallurgy (46% and 37% of full density), for example, was quantified by measuring mean porosity as a function of slice number (compression along the tomographic rotation axis), much like what was done earlier for density gradients in chemical vapor infiltration (Lee 1993) or more recently for liquid foam (Lambert, Cantat et al. 2005). Displacement of local maxima/minima in mean slice porosity could be followed through the different increments of deformation as the porosity diminished by up to a factor of two. Correlation coefficients for cross-sections between two deformation states were never lower than 0.55, and the data confirm the phenomenon that the higher the cross-section porosity, the greater the deformation of that cross-section. Similar approaches were used by others for an Al foam (Desischer, Kottar et al. 2000) and for compression of a glass wool (Badel, Letang et al. 2003).

Medical CT (in plane voxel sizes a significant fraction of a millimeter and out-of-plane voxels even larger) can reveal some details of wall deformation of closed-cell Al foams and can identify sites of weakness (Bart-Smith, Bastawros et al. 1998), but higher spatial resolution is required to produce data that can be incorporated usefully into numerical or analytical models. MicroCT of indentation tested, low-density Al foams (between 0.06% and 0.17% of full density) revealed the deformation zone was confined to a spherical cap immediately under the indenter with only occasional buckling of cell walls farther than one cell diameter from the zone directly under the indenter. Finite element modeling based on the reconstruction agreed with analytical expressions for the size of the deformation zone (Kádár, Maire et al. 2004).

Significant deformation occurs during the manufacture of open-cell Ni foams used as battery electrodes, and synchrotron microCT of these foams has proven very useful in characterizing tension and compression damage (Dillard, N'Guyen et al. 2005). In tension, bending, stretching, and alignment of struts were observed, and the particulars depended on the initial anisotropy of the foam. Compression led to strain localization due to buckling of struts, and this differed from the large rotations observed under compression in polyurethane (Elliott, Windle et al. 2002).

Repeated observations (after different compression increments) with synchrotron microCT revealed damage accumulation in the cell walls and plateau borders of an Al foam (Ohgaki, Toda et al. 2006, Toda, Ohgaki et al. 2006a, Toda, Takata et al. 2006b); local tomography was required to provide adequate spatial resolution within the volume of interest while allowing the specimen tested to have a large enough cross-section to be representative of the foam. Displacement of thousands of micropores within the solid was tracked automatically for each increment of deformation, and a microstructure gage formalism was used to calculate maps of the 3D strain tensor components. Characteristics of pores and the microcracks that nucleated from them were quantified, and the distribution of strain was discussed in terms of these features. The microCT-derived 3D structure was imported into FEM (Toda, Ohgaki et al. 2006a, Toda, Takata et al. 2006b). Similar data are required in applications in medicine, for example, displacements in CT images of lungs through image warping (Fan and Chen 1999).

Compression of a low-porosity, closed-cell amorphous metal foam was observed with synchrotron microCT at 14 strains during loading and three stains during unloading (Demetriou, Hanan et al. 2007). Yielding proceeded by percolation of an elastic buckling instability. Plots of individual cell displacements vs cell coordinate along the loading direction showed the width of the collapse zone.

Cellular structures can also be embedded in another solid phase. Madi and coworkers used synchrotron microCT to study a refractory consisting of zirconia grains coated by a softer amorphous (glassy) phase that formed the walls of the cellular system (Madi, Forest et al. 2007). Finite element modeling of the resulting structure was used to investigate the influence of the amount of glassy phase on stiffness.

Lab microCT of two syntactic metal foam types (preforms of hollow ceramic spheres infiltrated by liquid metal of two different Al alloys) was used to characterize deformation modes and to interpret features in stress–strain curves (Balch,

O'Dwyer et al. 2005). The foam with commercial purity Al showed barreling over a large part of its area with uniform, rather modest plastic deformation of the matrix coupled with sphere fracture and a thin crush band. The foam with 7075–T6 matrix had damage concentrated in two intersecting and much thicker crush bands oriented ~45° to the compression axis; sharp stress drops after the peak stress resulted from the formation of the crush bands.

Adrien et al. compared syntactic foams containing hollow glass microspheres and one of three polymeric matrices (relatively stiff epoxy, relatively compliant polypropylene, and polyurethane) using synchrotron microCT with 0.7-μm voxels (Adrien, Maire et al. 2007). The thickness of the spheres was rather constant but their diameters were strongly distributed. During stepwise confined compression in the dry state, strong adherence was observed between the glass spheres and polyurethane, but it was weaker between the other two materials and the spheres. In the foams with more compliant matrices, damage was more homogeneous and mainly affected the larger spheres, while in the stiffer material (epoxy), damage was localized in bands.

Examining the behavior of the foam and of the foam–sheet interface in Al-sandwich material required microCT at two very different scales (Salvo, Belestin et al. 2004). Nonetheless, microCT allowed strains to be calculated based on measurements at four deformation states. The interface appeared to be sound, and the foam in the sandwich behaved like similar free-standing foams examined by this group (Maire, Elmoutaouakkil et al. 2003a, Maire, Fazekas et al. 2003b).

8.4 Mineralized Tissue

Mineralized tissue is an important naturally-occurring biomaterial that has been studied extensively with microCT. Studies as diverse as skeletal banding in coral (Lewis, Lough et al. 2017) and automated assessment of bone changes in osteoarthritis (Das Neves Borges, Vincent et al. 2017) have appeared. Indeed, a number of commercial lab microCT systems were optimized for studying the mineralized tissue of widest clinical interest, bone. Several books provide useful background on bone and its microarchitecture (Martin, Burr et al. 1998, Cowin 2001, McCreadie, Goulet et al. 2001, Currey 2002).

This section begins with the microCT studies of porous mineralized tissue structures other than the bone. Vertebrate bone and tooth are based on nanocrystallites of carbonated apatite (cAp), and sea urchins and other echinoderms use calcite. Biopsies of human tissue or whole limbs of small animal models continue to be a major fraction of the microCT literature; it is no longer practical to discuss all of these studies, or for that matter, to even list them, and more than a few examples follow in the second through fourth sections. Section 8.4.2 describes the motivations for studying cancellous bone and covers the older studies first in order to give the reader a historical perspective, admittedly at the risk of fragmenting certain topics. Section 8.4.3 provides a sampling of more recent studies on bone growth and aging, specifically on structural changes in cancellous bone produced by aging or by treatment of conditions such as osteoporosis. Section 8.4.4 considers mechanical testing and numerical modeling of mechanical properties. The reader should note that discussion of "dense" cAp-based mineralized tissue (cortical bone and tooth) is postponed until Chapters 10 and 11 in order to retain focus on cellular materials. The reader also should note that, in this section "cellular" is taken to mean structures analogous to the foams described above and not structures containing biological cells (i.e., containing nuclei and other organelles).

8.4.1 Echinoderm Stereom

Most of the microCT characterization of the fenestrated tissue (stereom) of animals of this phylum has been concentrated on sea urchins. Papers have appeared analyzing stereom of starfish (Blowes, Egertová et al. 2017) and of brittle stars (Titschack, Baum et al. 2018).

The calcite stereom of sea urchins (mentioned briefly in Section 7.2.6) has many different fabrics (Smith 1980), that is, dimensions and arrangements of struts and plates, and synchrotron microCT has high enough resolving power to study many types of stereom. Figure 6.8 shows a slice of one of the many plates comprising the test (protective globe enclosing the sea urchin's internal organs) and of a demipyramid (part of the jaw structure) (Stock, Ignatiev et al. 2004a). Within the region indicated in Fig. 6.8b, BV/TV = 0.487, <Th> = 19.5 µm, <Sp> = 20.7 µm, and SMI = 0.22, that is, the structure consisted primarily of plates. Determination of specific stereom fabric type (e.g., galleried, labyrinthic, perforate) from segmented volumes will require further development of analysis tools.

MicroCT imaging of cellular mineralized tissues is useful even when the microstructure cannot be resolved. The finest stereom found in echinoderm ossicles can have submicrometer dimensions, for example, and nanoCT would be required. Interpretation of linear attenuation coefficients as partial voxels of calcite in a demipyramid of a sea urchin was discussed in Section 6.2.8.

8.4.2 Cancellous Bone – Motivations for Study and the Older Literature

Aging populations are characteristic of countries in the developed world, and osteoporosis, a disease of increased bone fragility seen in both elderly females and males, is of great concern because of the resulting morbidity and mortality related to bone fractures. The classical view is that decreased bone mineral density (BMD) explains fracture susceptibility. This paradigm fails in a significant fraction of patients, and the first refinement to this model is the proviso that decreased bone competence may also be the product of inferior bone microarchitecture, that is, impaired networks of struts and plates of the trabecular bone, a major constituent of vertebrae and femoral head, both major risk sites for osteoporosis-related fractures. Improving treatment of osteoporosis in the aged has motivated, at least in part, microCT studies of bone.

Quite some time ago, it was realized that microCT was an ideal method for characterizing bone microarchitecture. Early microCT studies of trabecular bone, that is, those in the years before commercial microCT systems became widespread, had to establish the reliability of the technique. These studies focused on imaging trabeculae (Layton, Goldstein et al. 1988, Feldkamp, Goldstein et al. 1989, Dedrick, Goulet et al. 1991, Dedrick, Goldstein et al. 1993, Goldstein, Goulet et al. 1993, Bonse, Busch et al. 1994, Elliott, Anderson et al. 1994, Bonse and Busch 1996, Hildebrand, Laib et al. 1997, Kinney, Haupt et al. 1997, Kirby, Davis et al. 1997, Ladd and Kinney 1997, Müller and Hayes 1997, Peyrin, Salomé et al. 1997) and on establishing robust methodologies for threshold definition and for topological analysis (Engelke, Umgießer et al. 1997) – loss of trabecular connectivity was hypothesized as a major portion of the microarchitectural portion of bone quality deterioration.

Earlier serial sectioning of cancellous bone and unbiased estimates of connectivity did not reveal a simple relationship between connectivity and bone volume fraction (Odgaard and Gunderson 1993). Using microCT, Feldkamp and coworkers investigated anisotropy and connectivity within 8-mm cubes of human cancellous bones (Layton, Goldstein et al. 1988). Goulet et al. suggest that mechanical influences seem to alter the amount of bone by changing trabecular thickness, while hormonal and/or chemical influences act to affect the number of trabeculae (Goulet, Goldstein et al. 1994). It is also almost certain that the correlations between experimentally measured macroscopic properties and modeling based on the actual microstructure would not have been demonstrated without x-ray microCT. In theory, one could measure elastic response and then use exhaustive serial sectioning to define the trabecular network for use in 3D modeling; in practice, the labor required prohibits this approach.

Quantitative (conventional) x-ray CT and dual energy absorptivity (i.e., quantitative radiography) have demonstrated that density measurements of human femora explained no more than 30%–40% of observed variance in modulus and 50%–60% of the variance in ultimate stress, and that the orientation of cancellous cubes in the principal compressive trabeculae region was a significant contributor to mechanical properties independent of the bone density (Cody, McCubbrey et al. 1996), a direct indication that trabecular microstructure is important in bone strength. Changes in trabecular morphology (thickness, relative density, bone fraction, and separation) were studied in a guinea pig model (Layton, Goldstein et al. 1988, Dedrick, Goulet et al. 1991) and in a canine model of osteoarthritis (Kuhn, Goulet et al. 1990); of particular interest in the canine study was that osteoarthritis was induced in one knee of the dog while the other knee was left unaffected and served as a control; earlier microCT work showed less than 5% differences in mean bone morphology variables between legs. Kinney and coworkers imaged trabecular bone architecture of rat tibia *in vivo* before and 5 weeks after ovariectomy-induced osteoporosis and found a major loss in trabecular bone following ovariectomy (OVX) (Kinney, Lane et al. 1992, 1995). More important was the change from an interconnected plate- and strut-like structure to one that is mostly disconnected, dangling trabecular elements (a morphology not observed in non-OVX rats). Finally, a strong linear relationship was observed between connectivity and volume of trabecular bone.

Bonse et al. showed volume renderings of cancellous bone biopsies from a single patient over 15 years; the 3D renderings clearly showed the large loss of bone as a result of chronic hemodialysis (Bonse, Busch et al. 1996). Numerical values for the volume fraction of bone and the ratio of the bone surface area to bone volume for five biopsies suggested that the main change was in the number and connectivity of trabeculae and not in their thickness. Others have studied trabecular changes in pre- and postmenopausal women and numerically simulated changes in density and architecture beginning with the normal, premenopausal structure (Müller and Hayes 1997).

Images of microcallus formations within trabecular bone have been interpreted to show the healing process after microfracture (Bonse and Busch 1996). Guldberg and coworkers inserted a small chamber into dogs' femora and tibia and observed tissue repair with microCT and other methods after the 7-mm internal diameter chambers were removed from the animals (Guldberg, Caldwell et al. 1997). Repair response with and without mechanical stimulation was studied, and after 12 weeks the bone volume fraction was 75% larger in the loaded group relative to the unloaded control group. Trabecular connectivity was increased for the loaded bone relative to the control groups, the trabeculae were much thicker in the loaded biopsies, but the trabecular spacing was the same as was the number of trabeculae. The results support the hypothesis that newly formed bone is deposited in mechanically appropriate location, and Guldberg et al. suggest that the limitation of strain magnitude is important in the process of tissue adaptation.

8.4.3 Cancellous Bone – Growth and Aging

Most studies of cancellous bone have been on human tissue or several popular animal models. Because high definition microCT data sets covering trabecular networks can be extremely large, special approaches are required for connectivity analyses (Apostol and Peyrin 2007). Although one might regard the need for such expedients as transitory, given the rapid increase in RAM and other aspects of computational power, and soon to go the way of foreground/background computing techniques, this is probably not the case: the push for ever higher definition, that is, more voxels per slice, and the increasing data collection rates suggest pressure on analysis algorithms will continue.

Before addressing the main subject of this section, brief mention might be useful of microCT studies of other animals' trabecular bone, studies that do not fit into the flow of the other text. MicroCT of the trabecular portion of deer antler revealed a porosity profile along the antler diameter, with a granulometry-determined pore size distribution dependent on the sample's original site (Leonard, Guiot et al. 2007). Skeletal pneumaticity, produced by the extension of pulmonary air sacs into marrow spaces, is a hallmark adaptation of some birds, and microCT was used to quantify differences between thoracic vertebrae in one species of duck that possess a pneumatic structure and a second with apneumatic vertebrae (Fajardo, Hernandez et al. 2007). One expects differences in bone microarchitecture for animals that are phylogenically close relatives but which have different modes of locomotion, and microCT of trabecular bone in the femoral head and neck for two such primate genera showed no difference in cancellous bone volume density but did show differences in trabecular orientation (MacLatchy and Müller 2002).

Bone undergoes large changes during development and growth, and a few such microCT studies are mentioned in this paragraph and in scattered places throughout subsequent paragraphs. Synchrotron microCT and synchrotron x-ray diffraction were used to study the microarchitectural and physical changes of human vertebrae during fetal growth (Nuzzo, Meneghini et al. 2003); the trabecular bone network was denser than in adult vertebrae, perhaps in compensation for the immature cortical shell in the fetal bone. Tibia and spine from IGF-1 deficient and wild-type fetal mice (18th day of gestation) were imaged with synchrotron microCT; considerable variation in mineral level and bone structure was observed along the length of the tibiae, and, interestingly, von Kossa staining, a standard histological procedure for detecting mineral in tissues, revealed no mineral content in the IGF-1 -/- spinal ossification center, while synchrotron microCT clearly indicated the presence of highly attenuating components (Burghardt, Wang et al. 2007).

An important maturation stage in the long bones of many animals is the fusion of the growth plate, and, until microCT was applied to the problem, it was unclear whether epiphyseal growth plates remained open throughout life in rats, a popular skeletal (trabecular) model for aging humans. Rats are not good models for Haversian bone remodeling because this process does not occur, and, if their bones never ceased to extend, however slowly, this would further limit the usefulness of the rat model. The microCT data of Martin et al. suggest that growth plate fusion does take place in rats and the fraction closed follows a sigmoidal pattern (Martin, Ritman et al. 2003).

Bone adapts during aging, and microCT has been employed to investigate sex- and site-specific differences in trabecular bone in humans. For the femoral neck, in vivo densitometry of nearly 700 women and direct microCT measurements of 13 pairs of postmortem specimens revealed that strength in old age was largely achieved during growth by differences (from individual to individual) in the distribution rather than the amount of bone in a given femoral neck cross-section (Zebaze, Jones et al. 2007). Transmenopausal changes in trabecular bone were assessed in 38 paired transilial biopsies (pre- and post-menopause specimens from the same individual), variables characterizing bone structure significantly decreased from values before menopause (BV/TV, trabecular number), and increased SMI suggested that trabecular microarchitecture was transforming from a plate-like to a rod-like structure (Akhter, Lappe et al. 2007). In vitro microCT of bone structure in subjects older than 52 years revealed significant sex differences in microarchitectural parameters at the distal radius, femoral neck, and femoral trochanter but not the iliac crest, calcaneus, and lumbar vertebral body (Eckstein, Matsuura et al. 2007).

Baseline studies of sex and age effects on trabecular bone have also been the subject of studies in animal models. Paired biopsies of goat iliac crest (pre- and post-OVX) demonstrated significant deterioration of the bone microarchitecture 6 months after OVX (Siu, Qin et al. 2004). As rats and mice lack Haversian remodeling, marmosets, which do exhibit this type of remodeling, were investigated as a model of age-related changes to bone which was found to resemble human bone (Bagi, Volberg et al. 2007). MicroCT was used to assess age-related changes in bone architecture of male and female C57BL/6J mice (eight time points between 1 and 20 months of age), and deterioration in vertebral and femoral microarchitecture began early, continued throughout life, was more pronounced at the femoral metaphysis than in the vertebrae and was greater in females than males (Glatt, Canalis et al. 2007). Sex and LRP-5 gene-related changes in trabecular bone were investigated using microCT, and significant differences were noted at multiple sites (Dubrow, Hruby et al. 2007).

There have been many studies of response of bone in animal models to different challenges to bone integrity and to drugs administered to combat this degradation. The experimental details of one study on a mouse model of osteoporosis (but not its results) are described in the following paragraph as an exemplar of how microCT is being employed. Note because one animal is used per time point, the design differs from longitudinal studies involving in vivo observation.

Bouxsein et al. compared OVX-induced bone loss between five inbred mouse strains to establish whether genetic regulation of bone loss occurred under the influence of estrogen deficiency (Bouxsein, Myers et al. 2005). Between six and nine animals for each strain and for OVX and sham-OVX (control) were studied, and lab microCT with 12-μm isotropic voxels was performed on dissected bones: right femur, right tibia, and fifth lumbar vertebra. Note that sham operations are necessary for the controls because the stress of the surgery could affect bone retention. For vertebrae, 250–300 transverse slices were collected to cover the bone; the volume from the growth plate to 1.8 mm distal was scanned in the tibia; diaphysis characteristics were measured in the femur. Binary segmentation was used along the scanner's software to calculate assumption-free bone characteristics such as bone volume fraction, trabecular thickness, and SMI. Analysis of variance was used to determine whether observed differences were statistically significant.

MicroCT has examined trabecular bone loss accompanying steroid treatment for inflammation (McLaughlin, Mackintosh et al. 2002), sciatic neurectomy (Ito, Nishida et al. 2002a), and gastrectomy (Stenstrom, Olander et al. 2000). The long-term effect of bisphosphonates on bone (perforations, microcracking) also has been examined (Jin, Cobb et al. 2017, Ma, Goh et al. 2017). MicroCT demonstrated that acute alcohol exposure (Obermeyer, Yonick et al. 2012), secondhand smoke (Santiago, Zamarioli et al. 2017), and severe hemorrhagic shock (Bundkirchen, Macke et al. 2017) impaired fracture healing. Patterns of bone resorption and the effect of dose were examined for calvarial culture and IL-1 and PTH treatment (Stock, Ignatiev et al. 2004b). Effects of disease processes on bone (infections, tumors, arthritis) have also been examined with microCT (Balto, Müller et al. 2000, Kurth and Müller 2001, Pettit, Ji et al. 2001, Boyd, Müller et al. 2002, Patel, Issever et al. 2003, Morenko, Bove et al. 2004, Sone, Tamada et al. 2004, Hu, Gerseny et al. 2011). Degradation of the 3D microarchitecture of subchondral trabecular human bone has been studied in osteoarthritis and compared these areas without cartilage with cores from neighboring sites that retained their cartilage (Chappard, Peyrin et al. 2006).

Bone loss at menopause has been studied extensively; for example, using OVX in animal models to mimic estrogen loss (Lane, Thompson et al. 1998, Ito, Nishida et al. 2002b). A study of five inbred mouse strains found OVX-related skeletal response varied in a site- and compartment-specific fashion among the different strains, supporting the hypothesis that bone loss during and after menopause is partly genetically regulated (Bouxsein, Myers et al. 2005). Trabecular bone in senescent marmosets and its response to alendronate therapy resembled human bone (Bagi, Volberg et al. 2007). Postmenopausal osteoporosis is often treated with bisphosphonates, and several studies of bisphosphonates' effects on bone loss have been reported for animal models (Nuzzo, Lafage-Proust et al. 2002, Dufresne, Chmielewski et al. 2003, Ito, Azuma et al. 2003, Borah, Ritman et al. 2005, Yoon, Kim et al. 2017). Parathyroid hormone (PTH) (Dempster, Cosman et al. 2001, Lane, Yao et al. 2003b, Gittens, Wohl et al. 2004, Lotinum, Evans et al. 2004) or fibroblast growth factor (FGF) (Lane, Kumer et al. 2003a, Lane, Yao et al. 2003b) treatment of osteoporosis models have also been investigated.

MicroCT has been employed to examine several other strategies for improving bone fragility. Ex vivo data on vertebrae of rats administered strontium ranelate revealed improved mechanical properties (increased plastic energy to failure) of the vertebrae, increased elastic modulus in individual trabeculae (via nanoindentation), and improved microarchitecture (via microCT) compared to controls (Ammann, Badoud et al. 2007). Extremely low-level, high-frequency oscillatory motion was examined in chronically unloaded mouse limbs; compared to normal age-matched controls, oscillations attenuated the decline in trabecular microarchitecture–related mechanical properties as assessed by finite element modeling (Ozcivici, Garman et al. 2007).

Alveolar bone is highly porous and supports teeth through the connection of the periodontal ligaments. Dalstra and coworkers used synchrotron microCT to image human, simian, and porcine jaw segments and were able to visualize blood vessels and fibers of the periodontal ligaments (Dalstra, Cattaneo et al. 2006, 2007). Differences in the distribution of linear attenuation coefficient values in the close vicinity of the ligaments suggested remodeling activity at those locations.

Bone structure is expected to change in response to in vivo loading. Changes in bone loading history are easily imposed in rat and mouse models and form the basis for a number of microCT studies (Ishijima, Tsuji et al. 2002, Amblard, Lafage-Proust et al. 2003, David, Laroche et al. 2003, Joo, Sone et al. 2003, Squire, Donahue et al. 2004). Craniomandibular joints are expected to change in response to different forces encountered in eating (i.e., to exhibit adaptive plasticity) and to attachment of different sized muscles. MicroCT was used to examine such changes in rabbits (animals eating a hard diet vs those on a soft diet), in myostatin-deficient mice vs normal mice (the larger-than-normal muscle mass of the former has led to informal label "muscle mice"), and in subfossils of the extinct primate *Archaeolemur* (Nicholson, Stock et al. 2006, Ravosa, Kloop et al. 2006a, Ravosa, Kunwar et al. 2006b, 2007a, Ravosa, Stock et al. 2007b). Tanaka et al. used lab microCT and several calibration standards to compare mineral levels in the mandibles of rats fed hard vs soft diets; significant differences were observed between groups as well as at different locations (Tanaka, Sano et al. 2007).

Mulder and coworkers used microCT to study altering architecture and mineralization during development of trabecular bone in the mandibular condyle (Mulder, van Ruijven et al. 2007b); the inhomogeneous finite element models (tissue moduli scaled to the local degree of mineralization) derived from the data identified important structural anisotropy. In one related paper, microCT of pig mandibular condyles revealed mean degree of mineralization and intratrabecular differences in mineralization between surfaces and cores of trabecular elements increased during development from 8 weeks prepartum to 108 weeks postpartum (Willems, Mulder et al. 2007). Figure 8.3 shows 3D microCT renderings of the condyle of 43-week-old pig; the more highly magnified images of several

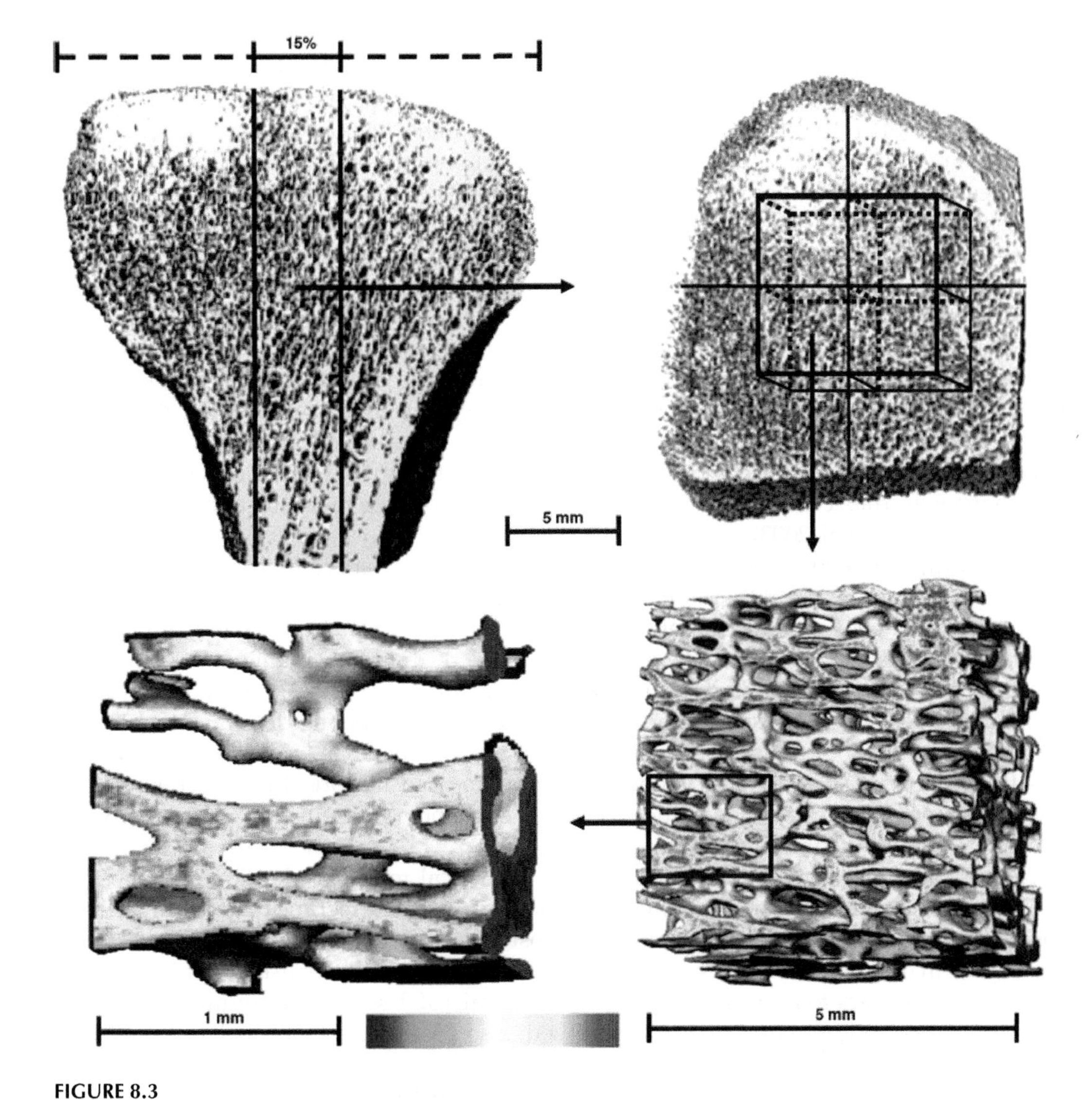

FIGURE 8.3
Rendering of the left condyle of a 43-week-old pig. The upper right image is a frontal cross-section. The upper left image is the medial view of the sagittal slice (posterior at the left) showing selection of four subvolumes. The lower left subvolume (posterioinferior) is shown magnified at the lower right, and a small number of trabeculae are shown still further magnified at the lower left. The color in the magnified images at the bottom shows increasing degree of mineralization from blue to red (Willems, Mulder et al. 2007).

trabeculae show, via color, the intratrabecular distribution of mineral. In another related paper on pig mandibular condyles, tissue stiffness (via nanoindentation) and mineral level (microCT) were low at trabecular surfaces and higher in the cores, directly demonstrating the link between developing stiffness and degree of mineralization (Mulder, Koolstra et al. 2007a). A fourth paper in this series found that remodeling in human mandibular condylar trabecular bone was larger than in the adjacent cortical bone (Renders, Mulder et al. 2007).

In many animals, mineral content varies considerably from older, mature areas (highest mineral content) to newly remodeled osteons (lower mineral content). The sensitivity limits of microCT make it ill-suited for quantifying small differences in composition, but a number of studies have shown that the mineralization levels in bone can be mapped. Synchrotron microCT of low mineralized vs high mineralized volumes in trabecular as a function of bisphosphonate treatment was the subject of one study (Borah, Ritman et al. 2005). Microradiography of thin sections of bone has long been used to show remodeling, and determination of the degree of bone mineralization via this method has been compared to that with synchrotron microCT (Nuzzo, Lafage-Proust et al. 2002). Synchrotron microCT has also been used to study the degree of mineralization in human normal vertebra, osteoblastic metastases and sites with degenerative osteosclerosis (Sone, Tamada et al. 2004), and in different mouse genetic strains (Martín-Badosa, Amblard et al. 2003, Bayat, Apostol et al. 2005).

Several in vivo and in situ microCT characterization studies have appeared. Multiple observations of rat bone by synchrotron microCT are one example (Lane, Thompson et al. 1998, Lane, Yao et al. 2003), and this approach has been extended to mice (Bayat, Apostol et al. 2005). The reader should consult the latter study for detailed discussion of the relationship between x-ray dose and signal-to-noise ratio in the reconstructions. Observation of changes in trabeculae in the murine hind-limb unloading model is another example (David, Laroche et al. 2003).

Determination of longitudinal changes in trabeculae by direct comparison of the reconstructed volumes was a rather ambitious approach employed in an in vivo lab microCT comparison of OVX and control rats (Waarsing, Day et al. 2004). In this study, image registration was difficult in the OVX animals because of the large changes, but clear patterns of bone resorption and apposition were documented for the OVX rats, and new bone formation was shown to correspond with areas of calcein labeling. It is also possible to administer Pb and Sr to living animals in order to label the bone in much the same way as fluorochromes such as tetracycline or calcein are used in optical microscopy of thin sections; imaging to either side of the absorption edge increases sensitivity in synchrotron microCT studies (Kinney and Ryaby 2001).

Ruimerman and coworkers proposed a 3D numerical model of developing bone that takes into account different mechano-biological pathways affecting bone (Ruimerman, Hilbers et al. 2005): osteocyte stimulation of osteoblast bone formation, an effect of elevated strain in the bone matrix; microcrack, and disuse promotion of osteoclast resorption. The model appears to reproduce features seen in microCT scans of growing pigs and, if osteoclast resorption frequencies are increased, also produces structures like those seen in osteoporotic bone.

A prominent feature of clinical radiographs is the texture produced by the (partially) overlapping trabeculae at the ends of long bones and within vertebrae. An obvious question is whether analysis of the texture can provide useful information about the trabecular microarchitecture, and this has, in fact, received attention. Luo et al. used synchrotron microCT to show that plain radiographs contain architectural information directly related to the underlying 3D structure; they suggested that a well-controlled sequence of radiographs might allow monitoring of trabecular changes in vivo and identifying individuals at increased risk of osteoporotic fracture (Luo, Kinney et al. 1999). Later synchrotron microCT-based analysis by others examined which 2D texture parameters correlated best with the underlying 3D structure (Apostol, Peyrin et al. 2004). This approach may prove useful for analysis of other cellular solids, for example, metal foams undergoing high strain rate deformation (with complete tomographic reconstructions only possible of the initial and final states), but much more detailed descriptions of the radiograph's texture need to be developed (instead of one or a few global measures of texture, a dozen or more parameters might be appropriate), and *a priori* knowledge could be utilized (for this example, the known initial structure and corresponding radiograph, the final structure and radiograph and the fact that the ith radiograph resulted from the $(i-1)^{th}$ structure and evolved into the $(i+1)^{th}$ structure). Other related papers have appeared (Showalter, Clymer et al. 2006, Jennane, Harba et al. 2007).

In addition to absorption microCT, trabecular bone has also been imaged with scattered x-radiation (Kleuker and Schulze 1997). Recently, Stock and coworkers performed x-ray scattering-based reconstruction of the diaphyses of a rabbit femur (Stock, De Carlo et al. 2008); other studies are cited in Chapter 12 as are combination of x-ray scattering-based quantification of internal strain states accompanying loading (Almer and Stock 2005, 2007) with tomographic reconstruction.

8.4.4 Cancellous Bone – Deformation, Damage, and Modeling

The relationship between elastic modulus and microarchitecture of trabecular bone is the first subject of this section. Damage and fracture studies are described afterwards.

In one early study, numerous 8-mm cubes of human cancellous bone (four cadavers without known bone disorders, various bones, and cube orientations) were imaged with 50-µm isotropic voxels prior to mechanical determination of modulus and ultimate strength (Goldstein, Goulet et al. 1993, Goulet, Goldstein et al. 1994). Bone volume fractions ranged from 6% to 36%, trabeculae thickness from 0.10 to 0.19 mm, and spacings from 0.32 to 1.67 mm. In normal bone, strong correlations were found between the independent structural measures of bone volume fraction, trabecular plate number, and connectivity, and strong relationships were found for modulus and ultimate strength, accounting for 80% to 90% of the variance in these properties. Other works on modeling elastic moduli, based on three-dimensional networks defined by microCT, appear elsewhere (Hildebrand, Laib et al. 1997, Kinney, Haupt et al. 1997, Ladd and Kinney 1997).

Elastic moduli from indentation measurements have been determined for PTH- and FGF-treated bones and structures compared with microCT (Lane, Yao et al. 2003). Bone modeling and simulation were the subject of another study (Müller 1999). MicroCT-derived bone microstructures are routinely imported into finite element models, and the response to mechanical loading was investigated numerically (Kinney, Haupt et al. 2000, Niebur, Feldstein et al. 2000, Borah, Gross et al. 2001, Homminga, Huiskes et al. 2001, van Rietbergen 2001, van Rietbergen and Huiskes 2001, Boyd, Müller et al. 2002, Hara, Tanck et al. 2002, Pistoia, Rietbergen et al. 2003, Stölken and Kinney 2003, van Eijden, Ruijven et al. 2004).

Smoothing methods have been examined for automatic finite element model generation from microCT volumetric data (Boyd and Müller 2006); this is important in terms of model efficiency and accuracy. In another microCT study, effects of three types of thresholding techniques applied to trabecular bone did not change predictions of apparent mechanical properties and structural properties, both of which agreed well with experimental measurements (Kim, Zhang et al. 2007).

Selective laser sintering of polymer powders has been used to reproduce trabecular bone structures derived from microCT data sets (Cosmi and Dreossi 2007, Su, Campbell et al. 2007). Numerical modeling of the moduli of observed structure was then compared to moduli measured experimentally for the fabricated specimens in both these studies.

Most mechanical tests of trabecular bone are on excised specimens where the peripheral trabeculae have lost some of their load-bearing capability due to loss of connectivity. This introduces a bias in the measured value of the elastic modulus that must be corrected if models are to correctly represent the actual load carrying capacity of the structure. MicroCT-based finite element models of aged human vertebral trabecular revealed that the widths over which the peripheral trabeculae were mostly unloaded were between 0.19 and 0.58 mm in ~8-mm diameter cores (Ün, Bevill et al. 2006). Note that trabecular thickness averages about 0.14 mm in humans (McCreadie, Goulet et al. 2001) and is slightly greater than 0.10 mm in mice (McLaughlin, Mackintosh et al. 2002). Actual experiments were found to underestimate the true, "side"-artifact free Young's modulus by 27% on the average, and by over 50% in certain experiments (Ün, Bevill et al. 2006).

As mentioned above, trabecular bone specimens can have primarily plate-like elements or mainly rod-like elements. Based on their synchrotron microCT data, Follet et al. found one finite element model worked better predicting the compressive Young's modulus for specimens containing rod-like trabecular arrays (those with very low bone volume fraction), and a second was superior with plate-like trabecular samples (Follet, Peyrin et al. 2007).

A digital correlation technique was developed for strain measurement in trabecular bone; typical values for the standard deviation of strain approached 0.01 (Verhulp, van Rietbergen et al. 2004), indicating that, at that level of development, the technique was limited to strain measurements beyond the yield strain. Digital correlation methods have advanced significantly since then (2004). Trabecular bone can be expected to differ from site to site and between different mammals; several different digital volume correlation methods and microCT were used to compare real and simulated displacement fields for the bovine distal femur, bovine proximal tibia, rabbit distal femur, rabbit proximal tibia, rabbit vertebra, and human vertebra; across all bone types, differences in strains ranged between 345 and 794 $\mu\varepsilon$ (Liu and Morgan 2007). Regional variations in apparent and tissue-level mechanical parameters were studied in 90 trabecular bone samples from six human L4 vertebrae combining micro-finite element analysis and microCT; significant differences were observed (Gong, Zhang et al. 2007), a very interesting result indeed.

Imaging of entire vertebrae allowed, via importation into a finite element model, assessment of the contribution of cortical and trabecular structure to strength; this study also compared actual and numerical estimates of yield stress for OVX vs control vs OVX plus alfacalcidol rats (Ito, Nishida et al. 2002a). In a similar study, Eswaran and coworkers used microCT-derived finite element models of human vertebrae to determine that the biomechanical role of the thin cortical shell can be substantial and vary with position: 45% of the load carried at the mid transverse section vs 15% close to the endplates (Eswaran, Gupta et al. 2006). Strength in bisphosphonate-treated bone has been characterized along with bone microarchitecture in OVX rats (Ito, Azuma et al. 2003). Healthy and osteoporotic human proximal femora were compared using microCT microstructure and 3D micro- and continuum-finite element models of loading during a fall (Verhulp, van Rietbergen et al. 2006). Figure 8.4 compares maximum principal stress on the midplane of the healthy and diseased bones at the three different finite element resolutions, and the different

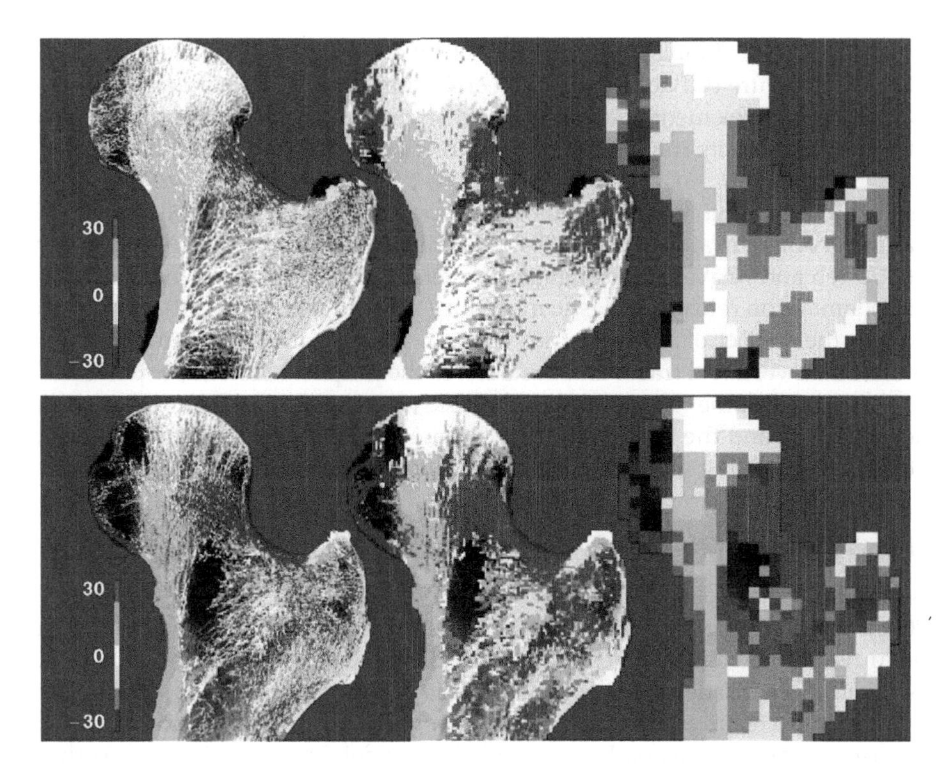

FIGURE 8.4
Finite element models at three resolutions for a healthy proximal femur (top row) and an osteoporotic femur (bottom row). Maximum principal stresses (MPa, in color) from the 3D models are shown for the bones' mid-planes; the stress values in the binned models (center and right) were divided by the local volume fraction of bone in order to estimate the average tissue-level stresses and to provide a better comparison with the highest resolution model. The left column is the model based on 80-μm isotropic voxels, the elements in the middle column were enlarged eight times (0.64-mm edges), and the elements in the right column were enlarged 38 times (3.04-mm edges). The head of the femur is at the upper left, the applied force of 1,000 N is distributed over the left side of the head, the nodes at the surface of the trochanter (far right side of the femur in the figure) were fixed in the horizontal direction, and the nodes at the distal end of the femur (bottom of the figure) were restricted to vertical motion. These loading conditions are those used in standard models of falling forces (Verhulp, van Rietbergen et al. 2006).

resolutions produced comparable results. Importation into numerical models is particularly valuable in highly variable biological systems because the same starting structure can be perturbed virtually and the response determined numerically.

Mulder and coworkers focused on the distribution of strains within individual trabeculae as they developed in fetal and newborn porcine mandibular condylar bone (Mulder, van Ruijven et al. 2007c). MicroCT measurements of the spatial distribution of mineral level within the trabecula (higher in the center lower toward the surfaces) were the input for the finite element model, and the inhomogeneous mineral distribution and finite element results suggested to the authors that the mineral distribution contributed to development of a structure that is able to resist increasing loads without an increase in average deformation.

The position of teeth within the mandible can affect fractures at the mandibular angle; in particular, impacted third molars are associated clinically with increased risk of fracture at this site. MicroCT and finite element analysis were used to investigate the effect of partially impacted third molars on clinically observed fractures (Takada, Abe et al. 2006). The 3D microstructure did not differ between human mandibles with and without third molars, but the finite element analysis showed stress concentration around the root apex of the third molar and its transmission in a direction matching clinical findings of angle fractures.

Damage accumulation in cancellous bone specimens has been followed during in situ loading (Nazarian and Muller 2004, Thurner, Wyss et al. 2004, Nazarian, Stauber et al. 2005). Yeh and Keaveny investigated the relative effects of trabecular microfracture and microdamage on reloading elastic modulus (Yeh and Keaveny 2001). In situ straining of bone was the subject of another study (Bleuet, Roux et al. 2004). Compressive fatigue testing of human vertebral trabecular bone led to the proposal of a simple power law relationship between volume fraction of bone, fabric eigenvalue, applied stress, and number of cycles to failure and the suggestion that fatigue life of such trabecular systems can be predicted for this loading mode (Rapillard, Charlebois et al. 2006).

8.4.5 Mineralized Cartilage

Sharks and rays have skeletons consisting of mineralized cartilage (Porter, Beltrán et al. 2006), and age at death has been investigated through microCT imaging of banding in skeletal tissues (Geraghty, Jones et al. 2012). The microstructural elements of elasmobranch mineralized tissues are mineralized tesserae, and microCT indicates that some tesseral anatomical features are common among species of all major elasmobranch groups despite large variation in tesseral shape and size (Seidel, Lyons et al. 2016). At the macro level, microCT showed the pectoral fins of one skate were comprised of radially oriented fin rays, formed by staggered mineralized skeletal elements stacked end-to-end while at the micro level, the midsection of each radial element was composed of three mineralized components, which consist of discrete segments (tesserae) that are mineralized cartilage and embedded in unmineralized cartilage (Huang, Hongjamrassilp et al. 2017).

8.5 Implants and Tissue Scaffolds

This section describes another very active research area with numerous microCT studies. Section 8.5.1 describes solid implants and bone formation; although the implants are not cellular materials, the bone formed fits into this category. Section 8.5.2 describes some of the scaffold processing studies. Section 8.5.3 covers bone ingrowth into scaffolds.

8.5.1 Implants

Bone formation around implants and the resulting structural integrity (or lack thereof) is a biomedical engineering topic of increasing importance as the number of hip (and knee) replacements increases. Examples of microCT studies include the 3D analysis of bone formation around Ti implants (Bernhardt, Scharnweber et al. 2006), characterization of Mg bone implant degradation (Witte, Fischer et al. 2006a, 2006b), study of bone around dental implants (Cattaneo, Dalstra et al. 2004), and investigation of human tooth–alveolar bone complex (Dalstra, Cattaneo et al. 2006). One complication in many such imaging studies is that the high absorptivity of many implants such as stainless steel or Ti washes out contrast in bone. In the author's experience, this can cripple analysis if contrast is confined to linear, 256-level gray scale; such an 8-bit approach seems to be favored in many different analyses, so this is not an academic concern. Use of nonlinear contrast scales or high-end clipping can be effective; contrast enhancements for bone structures adjacent to implants are described elsewhere (Tesei, Casseler et al. 2005, DaPonte, Clark et al. 2006). In synchrotron microCT, phase contrast at edges of different low absorption tissue types may also aid segmentation in the presence of a very absorbing material.

Itokawa and coworkers examined a novel composite material, PMMA-containing hydroxyapatite particles, for closing cranial defects (Itokawa, Hiraide et al. 2007). After 12 months implantation in beagles, microCT revealed that bone was preferentially attached to hydroxyapatite particles intersecting the composite's surface. In another microCT study, BMP-2-coated particles were reported to improve bone repair (Orth, Kruse et al. 2017). Others investigated alendronate linked to polymer microparticles (Wang, Rajalakshmanan et al. 2017). Treatment of scaffolds with platelets has been investigated as a method of improving bone healing (Zhang, Niu et al. 2017).

Four-millimeter mid-diaphyseal defects in murine femoral were grafted with live autografts or processed allografts and allowed to heal for 6, 9, 12, or 18 weeks (Reynolds, Hock et al. 2007). Significant statistical correlations between combinations of microCT-derived structural parameters (graft and callus volume and cross-sectional polar moment of inertia) with measured ultimate torque and torsional rigidity.

Measurement of preexisting bone geometry by CT and design of matching replacement implants via rapid prototyping techniques such as selective laser sintering are a developing area of clinical interest. MicroCT of a minipig model found significant bone growth interior and exterior to an implanted polycaprolactone condylar ramus unit at 1- and 3-month time points (Smith, Flanagan et al. 2007).

8.5.2 Scaffold Structures and Processing

Scaffolds for bone replacement via osteointegration have been actively researched (Hollister, Chu et al. 2001), and microCT is a natural tool for studying these structures that generally mimic trabecular bone. Biocompatible materials including scaffolds were discussed in Müller et al. (2001), and other reports of scaffold structures and function include (Thurner, Karamuk et al. 2001, Thurner, Müller et al. 2003, Donath, Beckmann et al. 2004, Thurner, Wyss et al. 2004, Irsen, Leukers et al. 2006). In another study, porosity characteristics were quantified in six very different graft materials; an order of magnitude spread in specific internal surface areas was observed, and the porosity was interconnecting in all but one material (Vanis, Rheinbach et al. 2006). Effect of pore structure on bone ingrowth was investigated in a Ti-based scaffold (Kapat, Srivas et al. 2017).

Porous metal or ceramic composite scaffolds have been used for bone regeneration and investigated with microCT. One example is in vitro and in vivo evaluation of MgF_2-coated Mg alloy porous scaffolds (Yu, Zhao et al. 2017). Repair of rabbit segmental bone defects by peptide-treated porous tantalum scaffolds is a second example (Wang, Li et al. 2017). Ceramic or ceramic-polymer composites scaffolds were investigated by several groups (Shao, Ke et al. 2017, Shim, Won et al. 2017).

Starch/polycaprolactone biodegradable scaffolds were produced by a 3D-plotting technology, and coverage of the scaffolds by bone-like apatite layers was investigated by microCT (Oliveira, Malafaya et al. 2007). Deposition under static, agitation, and circulating flow perfusion conditions were compared, and the last was the most effective at producing well-defined apatite layers in the inner parts of the scaffolds. Phase contrast microCT was used to image bioresorbable polymeric scaffolds (Tu, Hu et al. 2017).

Partap and coworkers described controlled porosity alginate hydrogels synthesized by simultaneous micelle templating and internal gelation (Partap, Muthutantri et al. 2007). MicroCT revealed relatively monodisperse pore sizes, high total pore volume, and high degrees of porosity. Higher surfactant concentrations produced smaller pores with lower pore volume, and the surfactant could be completely washed from the scaffolds, something that is essential for use in tissue engineering. Freeze-casting of PLGA/cAp composite scaffolds was another processing route investigated (Schardosim, Soulié et al. 2017).

Fourier phase segmentation methods were used to study porosity in biodegradable poly (propylene fumarate) scaffolds formed by the solvent casting particulate (sodium chloride) leaching process (Rajagopalan, Lu et al. 2005). A similar NaCl dissolution processing route was used with a composite of polylactic acid and soluble calcium phosphate glass; a three-factor (NaCl particle size, glass particle size, glass volume fraction), two-level design of experiments approach employed, and microCT characterization and finite element modeling of the structure were compared with stiffness measurements from compression tests (Charles-Harris, del Valle et al. 2007).

Jones and coworkers examined foaming sol–gel-derived bioactive glass scaffolds, focusing on the pores and interconnects in the structure (Jones, Poologasundarampillai et al. 2007). They extracted small volumes containing single pores, rendered only the pores, and labeled the portions of the surface comprising interconnects to adjacent pores. Their maps of predicted flow paths (Fig. 8.5) and analyses of predicted permeability as a function of the size of the representative elemental volume are of particular interest.

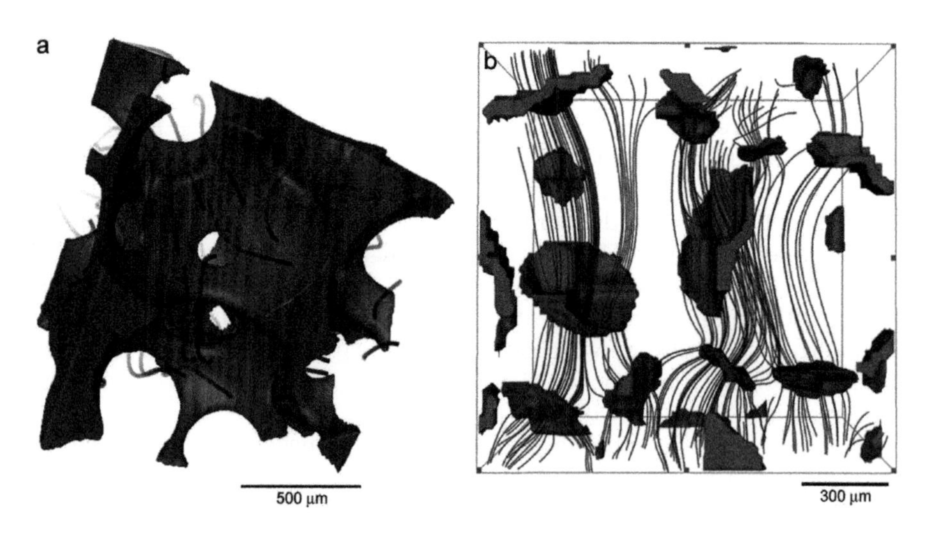

FIGURE 8.5
(a) Small subvolume of a scaffold rendered as a solid object with flow illustrated by streak lines, ribbons representing the path a mass-less particle would follow. (b) A larger volume with the solid removed and only the pore interconnects shown (i.e., the opposite of how the interconnects were shown in (a)). The streak lines converge at the interconnects and illustrate the importance of interconnect size on flow (Jones, Poologasundarampillai et al. 2007).

8.5.3 Bone Growth into Scaffolds

Cancedda and coauthors recently reviewed bulk and surface studies on trabecular bone, scaffolds, and tissue-engineered bones (Cancedda, Cedola et al. 2007). The combination of microCT imaging and microbeam x-ray scattering provides very useful information, and this paper is covered in more detail in Chapter 12.

Bone ingrowth into scaffolds occurs via several steps. Before bone mineral can be formed, the matrix material must be in place; and cell seeding must precede deposition of the matrix. Hofmann et al. used the NaCl particle dissolution route to fabricate porous silk scaffolds with designed gradients of pores (mimicking what is seen in vertebrae) and compared static vs dynamic mesenchymal stem cell seeding in the scaffolds (Hofmann, Hagenmüller et al. 2007). The fluid-dynamic microenvironment in tissue construct was examined in another microCT study (Cioffi, Boschetti et al. 2006).

Quantification of bone on hydroxyapatite scaffolds (Mastrogiacomo, Komlev et al. 2004) and 3D study of bone ingrowth into calcium phosphate biomaterials (Weiss, Obadia et al. 2003) are studies of interest. Komlev et al. examined highly porous hydroxyapatite scaffolds before implantation and after they were seeded in vitro with bone marrow stromal cells and implanted for 8, 16, or 24 weeks in immunodeficient mice and measured volume fraction, average thickness, and distribution of newly formed bone (Komlev, Peyrin et al. 2006). Komlev and coworkers found new bone thickness increased from week 8 to week 16; new bone thickness did not increase from week 16 to 24, but mineralization of the bone matrix continued during this time.

Bone formation in polymeric scaffolds has been evaluated by proton magnetic resonance microscopy and microCT (Washburn, Weir et al. 2004). Other studies include pore interconnectivity in bioactive glass foams (Atwood, Jones et al. 2004), structure and properties of clinical coralline implants (Knackstedt, Arns et al. 2006), scaffolds of cAp-collagen sponges (Itoh, Shimazu et al. 2004), and polymer composite scaffolds (Mathieu, Mueller et al. 2006).

Jones et al. compared bone ingrowth in hydroxyapatite scaffolds with a cellular structure with those with a more strut-like structure (Jones, Arns et al. 2007). Porosity, pore size distribution, pore constriction sizes, and pore topology were measured for the original scaffold structure, and the distribution of bone ingrowth after 4 or 12 weeks implantation was quantified. For the early time period, growth was from the periphery of the scaffold with a constant decrease in bone mineralization into the scaffold volume. Bone ingrowth was strongly enhanced for pore diameters greater than 100 μm. Bone ingrowth was also characterized for polymer–hydroxyapatie composites used in a rat model of posteriolateral spine fusion (Jakus, Rutz et al. 2016), and a typical slice from one of these scaffolds is shown in Fig. 8.6 along with a 3D rendering showing blood vessel infiltration into another scaffold.

The efficacy of engineered scaffolds for bone formation is typically evaluated in nonweight- bearing locations or in the presence of stress-shielding devices (bone plates or external fixation), primarily because these scaffolds are not designed to carry weight. Chu and coworkers designed a weight-carrying, biodegradable, tubular porous scaffold for repair of segmental bone defects in rat femora (Chu, Warden et al. 2007). Scaffolds containing BMP-2 maintained bone length throughout and allowed bone bridging; those without collapsed after 15 weeks and failed to induce bridging.

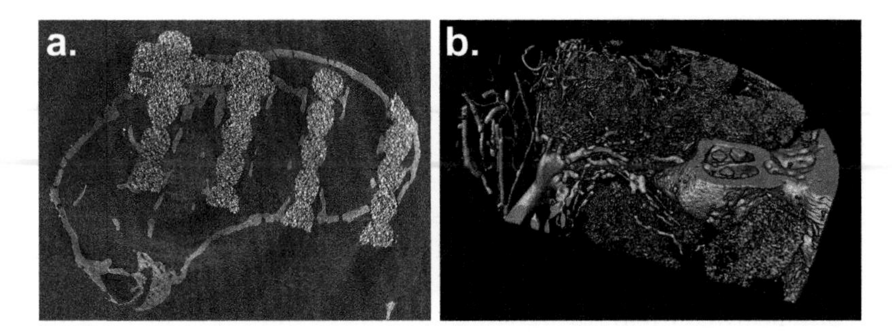

FIGURE 8.6
Synchrotron microCT data on 3D-printed hydroxyapatite-polymer scaffolds. (a) Reconstructed slice showing the scaffold and bone. The punctate appearing white-gray objects are the scaffold struts containing hydroxyapatite particles, and a shell of bone (light gray) has nearly completely covered the scaffold and has grown deeply into the macropores between the columns of struts. The horizontal FOV is 3.975 mm. Imaged at 2-BM, APS, with 25-keV photons, 0.12° rotation steps between projections over 180°, and reconstructed with 2.65-μm isotropic voxels. For more details, see Jakus et al. (2016). (b) 3D rendering of a slightly scaffold material (compared to that in panel (a), black-white punctate features) with blood vessels filled with the contrast agent MicroFil and rendered red. A small volume of bone is visible and is rendered cream. Unpublished data, 2017, Hallman, Hsu, and Stock from 2-BM, APS, with 24.9-keV photons, 0.12° rotation steps between projections over 180°, and reconstructed with 1.30-μm isotropic voxels.

References

Adrien, J., E. Maire, N. Gimenez and V. Sauvant-Moynot (2007). "Experimental study of the compression behaviour of syntactic foams by in situ X-ray tomography." *Acta Mater* **55**: 1667–1679.

Akhter, M. P., J. M. Lappe, K. M. Davies and R. R. Recker (2007). "Transmenopausal changes in the trabecular bone structure." *Bone* **41**: 111–116.

Almeida, M. L. B., E. Ayres, F. C. C. Moura and R. L. Oréfice (2018). "Polyurethane foams containing residues of petroleum industry catalysts as recoverable pH-sensitive sorbents for aqueous pesticides." *J Hazard Mater* **346**: 285–295.

Almer, J. D. and S. R. Stock (2005). "Internal strains and stresses measured in cortical bone via high-energy x-ray diffraction." *J Struct Biol* **152**: 14–27.

Almer, J. D. and S. R. Stock (2007). "Micromechanical response of mineral and collagen phases in bone." *J Struct Biol* **157**: 365–370.

Amblard, D., M. H. Lafage-Proust, A. Laib, T. Thomas, P. Rüegsegger, C. Alexandre and L. Vico (2003). "Tail suspension induces bone loss in skeletally mature mice in the C57BL/6J strain but not in the C3H/HeJ strain." *J Bone Miner Res* **18**: 561–569.

Ammann, P., I. Badoud, S. Barraud, R. Dayer and R. Rizzoli (2007). "Strontium ranelate treatment improves trabecular and cortical intrinsic bone tissue quality, a determinant of bone strength." *J Bone Miner Res* **22**: 1419–1425.

Apostol, L. and F. Peyrin (2007). "Connectivity analysis in very large 3D microtomographic images." *IEEE Trans Nucl Sci* **54**: 167–172.

Apostol, L., F. Peyrin, S. Yot, O. Basset, C. Odet, J. Tabary, J. M. Dinten, E. Boller, V. Boudousq and P. O. Kotzki (2004). A procedure for the evaluation of 2D radiographic texture analysis to assess 3D bone micro-architecture. *Medical Imaging 2004: Image Processing*. J. M. Fitzpatrick and M. Sonka. Bellingham (WA), SPIE. **SPIE Proc Vol 5370**: 195–206.

Atwood, R. C., J. R. Jones, P. D. Lee and L. L. Hench (2004). "Analysis of pore interconnectivity in bioactive glass foams using x-ray microtomography." *Scripta Mater* **51**: 1029–1033.

Babin, P., G. D. Valle, H. Chiron, P. Cloetens, J. Hoszowska, P. Pernot, A. L. Réguerre, L. Salvo and R. Dendievel (2006). "Fast x-ray tomography analysis of bubble growth and foam setting during breadmaking." *J Cereal Sci* **43**: 393–397.

Badel, E., J. M. Letang, G. Peix and D. Babot (2003). "Quantitative microtomography: Measurement of density distribution in glass wool and local evolution during a one-dimensional compressive load." *Meas Sci Technol* **14**: 410–420.

Bagi, C. M., M. Volberg, M. Moalli, V. Shen, E. Olson, N. Hanson, E. Berryman and C. J. Andresen (2007). "Age-related changes in marmoset trabecular and cortical bone and response to alendronate therapy resemble human bone physiology and architecture." *Anat Rec* **290**: 1005–1016.

Balch, D. K., J. G. O'Dwyer, G. R. Davis, C. M. Cady, G. T. Gray. III and D. C. Dunand (2005). "Plasticity and damage in aluminum syntactic foams deformed under dynamic and quasi-static conditions." *Mater Sci Eng A* **391**: 408–417.

Balto, K., R. Müller, D. C. Carrington, J. Dobeck and P. Stashenko (2000). "Quantification of periapical bone destruction in mice by microcomputed tomography." *J Dent Res* **79**: 35–40.

Bart-Smith, N., A. F. Bastawros, D. R. Mumm, A. G. Evans, D. J. Sypeck and H. N. G. Wadley (1998). "Compressive deformation and yielding mechanisms in cellular Al alloys determined using x-ray tomography and surface strain mapping." *Acta Mater* **46**: 3583–3592.

Bayat, S., L. Apostol, E. Boller, T. Borchard and F. Peyrin (2005). "In vivo imaging of bone microarchitecture in mice with 3D synchrotron radiation microtomography." *Nucl Instrum Meth A* **548**: 247–252.

Benaouli, A. H., L. Froyen and M. Wevers (2000). Micro focus computed tomography of aluminium foams. *X-Ray Tomography in Materials Science*. J. Baruchel, J. Y. Buffière, E. Maire, P. Merle and G. Peix. Paris, Hermes Science: 139–154.

Benouali, A. H., L. Froyen, T. Dillard, S. Forest and F. N'Guyen (2005). "Investigation on the influence of cell shape anisotropy on the mechanical performance of closed cell aluminum foams using micro-computed tomography." *J Mater Sci* **40**: 5801–5811.

Bernhardt, R., D. Scharnweber, B. Müller, F. Beckmann, J. Goebbels, J. Jansen, H. Schliephake and H. Worch (2006). 3D analysis of bone formation around Ti implants using microcomputed tomography (μCT). *Developments in X-Ray Tomography V*. U. Bonse. Bellingham (WA), SPIE. **SPIE Proc Vol 6318**: 631807-1–631807-10.

Blacher, S., A. Léonard, B. Heinrichs, N. Tcherkassova, F. Ferauche, M. Crine, P. Marchot, E. Loukine and J. P. Pirard (2004). "Image analysis of x-ray microtomograms of Pd-Ag/SiO2 xerogels catalysts supported on Al2O3 foams." *Colloids Surf A* **241**: 201–206.

Bleuet, P., J. P. Roux, Y. Dabin and G. Boivin (2004). In situ microtomography study of human bones under strain with synchrotron radiation. *Developments in X-Ray Tomography IV*. U. Bonse. Bellingham (WA), SPIE. **SPIE Proc Vol 5535**: 129–136.

Blowes, L. M., M. Egertová, Y. Liu, G. R. Davis, N. J. Terrill, H. S. Gupta and M. R. Elphick (2017). "Body wall structure in the starfish *Asterias rubens*." *J Anat* **231**: 325–341.

Bonse, U. and F. Busch (1996). "X-ray computed microtomography (μCT) using synchrotron radiation." *Prog Biophys Molec Biol* **65**: 133–169.

Bonse, U., F. Busch, O. Gunnewig, F. Beckmann, G. Delling, M. Hahn and A. Kvick (1996). "Microtomography (μCT) applied to structure analysis of human bone biopsies." *ESRF Newsletter* (March): 21–23.

Bonse, U., F. Busch, O. Gunnewig, F. Beckmann, R. Pahl, G. Delling, M. Hahn and W. Graeff (1994). "3-D computed x-ray tomography of human cancellous bone at 8 microns spatial and 10-4 energy resolution." *Bone Miner* **25**: 25–38.

Borah, B., G. J. Gross, T. E. Dufresne, T. S. Smith, M. D. Cockman, P. A. Chmielewski, M. W. Lundy, J. R. Hartke and E. W. Sod (2001). "Three-dimensional microimaging (MRμI and μCT), finite element modeling, and rapid prototyping provide unique insights into bone architecture in osteoporosis." *Anat Rec* **265**: 101–110.

Borah, B., E. L. Ritman, T. E. Dufresne, S. M. Jorgensen, S. Liu, J. Sacha, R. J. Phipps and R. T. Turner (2005). "The effect of risedronate on bone mineralization as measured by microcomputed tomography with synchrotron radiation: correlation to histomorphometric indices of turnover." *Bone* **37**: 1–9.

Bourne, N. K., K. Bennett, A. M. Milne, S. A. MacDonald, J. J. Harrigan and J. C. F. Millett (2008). "The shock response of aluminium foams." *Scripta Mater* **58**: 154–157.

Bouxsein, M. L., K. S. Myers, K. L. Schultz, L. R. Donahue, C. J. Rosen and W. G. Beamer (2005). "Ovariectomy-induced bone loss varies among inbred strains of mice." *J Bone Min Res* **20**: 1085–1092.

Boyd, S. K. and R. Müller (2006). "Smooth surface meshing for automated finite element model generation from 3D image data." *J Biomech* **39**: 1287–1295.

Boyd, S. K., R. Müller and R. F. Zernicke (2002). "Mechanical and architectural bone adaptation in early stage experimental osteoarthritis." *J Bone Miner Res* **17**: 687–694.

Breunig, T. M., M. C. Nichols, J. S. Gruver, J. H. Kinney and D. L. Haupt (1994). "Servo-mechanical load frame for in situ, non-invasive, imaging of damage development." *Ceram Eng Sci Proc* **15**: 410–417.

Breunig, T. M., S. R. Stock, S. D. Antolovich, J. H. Kinney, W. N. Massey and M. C. Nichols (1992). A framework relating macroscopic measures and physical processes of crack closure of Al-Li Alloy 2090. *Fracture Mechanics: Twenty-Second Symposium (Vol. 1)*. H. A. Ernst, A. Saxena and D. L. McDowell. Philadelphia, ASTM. **ASTM STP 1131**: 749–761.

Breunig, T. M., S. R. Stock and R. C. Brown (1993). "Simple load frame for in situ computed tomography and x-ray tomographic microscopy." *Mater Eval* **51**: 596–600.

Brunke, O., S. Oldenbach and F. Beckmann (2004). Structural characterization of aluminium foams by means microcomputed tomography. *Developments in X-Ray Tomography IV*. U. Bonse. Bellingham (WA), SPIE. **SPIE Proc Vol 5535**: 453–463.

Brunke, O., S. Oldenbach and F. Beckmann (2005). "Quantitative methods for the analysis of synchrotron-μCT datasets of metallic foams." *Eur J Appl Phys* **29**: 73–81.

Brydon, A. D., S. G. Bardenhagen, E. A. Miller and G. T. Seidler (2005). "Simulation of the densification of real open-celled foam microstructures." *J Mech Phys Sol* **53**: 2638–2660.

Bundkirchen, K., C. Macke, J. Reifenrath, L. M. Schäck, S. Noack, B. Relja, P. Naber, B. Welke, M. Fehr, C. Krettek and C. Neunaber (2017). "Severe hemorrhagic shock leads to a delayed fracture healing and decreased bone callus strength in a mouse model." *Clin Orthop Relat Res* **475**: 2783–2794.

Burghardt, A. J., Y. Wang, H. Elalieh, X. Thibault, D. Bikle, F. Peyrin and S. Majumdar (2007). "Evaluation of fetal bone structure and mineralization in IGF-I deficient mice using synchrotron radiation microtomography and Fourier transform infrared spectroscopy." *Bone* **40**: 160–168.

Cancedda, R., A. Cedola, A. Giuliani, V. Komlev, S. Lagomarsino, M. Mastrogiacomo, F. Peyrin and F. Rustichelli (2007). "Bulk and interface investigations of scaffolds and tissue-engineered bones by x-ray microtomography and x-ray microdiffraction." *Biomaterials* **28**: 2505–2524.

Cattaneo, P. M., M. Dalstra, F. Beckmann, T. Donath and B. Melsen (2004). Comparison of conventional and synchrotron-radiation-based microtomography of bone around dental implants. *Developments in X-Ray Tomography IV*. U. Bonse. Bellingham (WA), SPIE. **SPIE Proc Vol 5535**: 757–764.

Chappard, C., F. Peyrin, A. Bonnassie, G. Lemineur, B. Brunet-Imbault, E. Lespessailles and C. L. Benhamou (2006). "Subchondral bone micro-architectural alterations in osteoarthritis: A synchrotron micro-computed tomography study." *Osteoarth Cartilage* **14**: 215–223.

Charles-Harris, M., S. del Valle, E. Hentges, P. Bleuet, D. Lacroix and J. A. Planell (2007). "Mechanical and structural characterisation of completely degradable polylactic acid/calcium phosphate glass scaffolds." *Biomaterials* **28**: 4429–4438.

Cherukupally, P., E. J. Acosta, J. P. Hinestroza, A. M. Bilton and C. B. Park (2017). "Acid-base polymeric foams for the adsorption of micro-oil droplets from industrial effluents." *Environ Sci Technol* **51**: 8552–8560.

Chu, T. M. G., S. J. Warden, C. H. Turner and R. L. Stewart (2007). "Segmental bone regeneration using a load-bearing biodegradable carrier of bone morphogenetic protein-2." *Biomaterials* **28**: 459–467.

Cioffi, M., F. Boschetti, M. T. Raimondi and G. Dubini (2006). "Modeling evaluation of the fluid-dynamic microenvironment in tissue-engineered construct: a microCT based model." *Biotech Bioeng* **93**: 500–510.

Cody, D. D., D. A. McCubbrey, G. W. Divine, G. J. Gross and S. A. Goldstein (1996). "Predictive value of proximal femoral bone densitometry in determining local orthogonal material properties." *J Biomech* **29**: 753–761.

Colonetti, V. C., M. F. Sanches, V. C. de Souza, C. P. Fernandes, D. Hotza and M. G. N. Quadri (2018). "Cellular ceramics obtained by a combination of direct foaming of soybean oil emulsified alumina suspensions with gel consolidation using gelatin." *Ceram Int* **44**: 2436–2445.

Cosmi, F. and D. Dreossi (2007). "Numerical and experimental structural analysis of trabecular architectures." *Meccan* **42**: 85–93.

Cowin, S. C. (1985). "The relationship between the elasticity tensor and the fabric tensor." *Mech Mater* **4**: 137–147.

Cowin, S. C., Ed. (2001). *Bone Mechanics Handbook*. Boca Raton, CRC Press.

Currey, J. D. (2002). *Bones - Structure and Mechanics*. Princeton (NJ), Princeton University Press.

Dalstra, M., P. M. Cattaneo and F. Beckmann (2007). "Synchrotron radiation-based microtomography of alveolar support tissues." *Orthod Craniofacial Res* **9**: 199–205.

Dalstra, M., P. M. Cattaneo, F. Beckmann, M. T. Sakima, G. Lemor, M. G. Laursen and B. Melsen (2006). Microtomography of the human tooth-alveolar bone complex. *Developments in X-Ray Tomography V.* U. Bonse. Bellingham (WA), SPIE. **SPIE Proc Vol 6318**: 631804-1–631804-9.

DaPonte, J. S., M. Clark, P. Nelson, T. Sadowski and E. Wood (2006). Quantitative confirmation of visual improvements to microCT bone density images. *Visual Information Processing XV.* Z. Rahman, S. E. Reichenbach and M. A. Neifeld. Bellingham (WA), SPIE. **SPIE Proc Vol 6246**: 62460D-1–62460D-9.

Darder, M., C. R. S. Matos, P. Aranda, R. F. Gouveia and E. Ruiz-Hitzky (2017). "Bionanocomposite foams based on the assembly of starch and alginate with sepiolite fibrous clay." *Carbohydr Polym* **157**: 1933–1939.

Das Neves Borges, P., T. L. Vincent and M. Marenzana (2017). "Automated assessment of bone changes in cross-sectional microCT studies of murine experimental osteoarthritis." *PLOS One* **12**: 0174294.

David, V., N. Laroche, B. Boudignon, M. H. Lafage-Proust, C. Alexandre, P. Rüegsegger and L. Vico (2003). "Noninvasive in vivo monitoring of bone architecture alterations in hindlimb-unloaded female rats using novel three-dimensional microcomputed tomography." *J Bone Miner Res* **18**: 1622–1631.

Dedrick, D. K., S. A. Goldstein, K. D. Brandt, B. L. O'Connor, R. W. Goulet and M. Albrecht (1993). "A longitudinal study of subchondral plate and trabecular bone in cruciate-deficient dogs with osteoarthritis followed up for 54 months." *Arthritis Rheum* **36**: 1460–1467.

Dedrick, D. K., R. Goulet, L. Huston, S. A. Goldstein and G. G. Bole (1991). "Early bone changes in experimental osteoarthritis using microscopic computed tomography." *J Rheumatol Suppl* **27**: 44–45.

Demetriou, M. D., J. C. Hanan, C. Veazey, M. Di Michiel, N. Lenoir, E. Üstündag and W. L. Johnson (2007). "Yielding of metallic glass foam by percolation of an elastic buckling instability." *Adv Mater* **19**: 1957–1962.

Dempster, D. W., F. Cosman, E. Kurland, H. Zhou, J. Nieves, L. Woelfert, E. Shane, K. Plaveti, R. Müller, J. Bilezikian and R. Lindsay (2001). "Effects of daily treatment with parathyroid hormone on bone microarchitecture and turnover in patients with osteoporosis: a paired biopsy study." *J Bone Miner Res* **16**: 1846–1853.

Desischer, H. P., A. Kottar and B. Foroughi (2000). Determination of local mass density distribution. *X-Ray Tomography in Materials Science*. J. Baruchel, J. Y. Buffière, E. Maire, P. Merle and G. Peix. Paris, Hermes Science: 165–176.

Di Prima, M. A., M. Lesniewski, K. Gall, D. L. McDowell, T. Sanderson and D. Campbell (2007). "Thermomechanical behavior of epoxy shape memory polymer foams." *Smart Mater Struct* **16**: 2330–2340.

Dillard, T., F. N'Guyen, E. Maire, L. Salvo, S. Forest, Y. Bienvenu, J. D. Bartout, M. Croset, R. Dendievel and P. Cloetens (2005). "3D quantitative image analysis of open-cell nickel foams under tension and compression loading using x-ray microtomography." *Phil Mag* **85**: 2147–2175.

Donath, T., F. Beckmann, R. G. J. C. Heijkants, O. Brunke and A. Schreyer (2004). Characterization of polyurethane scaffolds using synchrotron-radiation-based computed microtomography. *Developments in X-Ray Tomography IV.* U. Bonse. Bellingham (WA), SPIE. **SPIE Proc Vol 5535**: 775–782.

Dubrow, S. A., P. M. Hruby and M. P. Akhter (2007). "Gender specific LRP5 influences on trabecular bone structure and strength." *J Musculoskelet Neuronal Interact* **7**: 166–173.

Dufresne, T. E., P. A. Chmielewski, M. D. Manhart, T. D. Johnson and B. Borah (2003). "Risedronate preserves bone architecture in early postmenopausal women in 1 year as measured by three-dimensional microcomputed tomography." *Calcif Tiss Int* **73**: 423–432.

Eckstein, F., M. Matsuura, V. Kuhn, M. Priemel, R. Müller, T. M. Link and E. M. Lochmüller (2007). "Sex differences of human trabecular bone microstructure in aging are site-dependent." *J Bone Min Res* **22**: 817–824.

van Eijden, T. M. G. J., L. J. Ruijven and E. B. W. Giesen (2004). "Bone tissue stiffness in the mandibular condyle is dependent on the direction and density of the cancellous structure." *Calcif Tiss Int* **75**: 502–508.

Elliott, J. A., A. H. Windle, J. R. Hobdell, G. Eeckhaut, R. J. Oldman, W. Ludwig, E. Boller, P. Cloetens and J. Baruchel (2002). "In-situ deformation of an open-cell flexible polyurethane foam characterised by 3D computed microtomography." *J Mater Sci* **37**: 1547–1555.

Elliott, J. C., P. Anderson, X. J. Gao, F. S. L. Wong, G. R. Davis and S. E. P. Dowker (1994). "Application of scanning microradiography and x-ray microtomography to studies of bone and teeth." *J X-ray Sci Technol* **4**: 102–117.

Elmoutaouakkil, A., G. Fuchs, P. Bergounhon, R. Péres and F. Peyrin (2003). "Three-dimensional quantitative analysis of polymer foams from synchrotron radiation x-ray microtomography." *J Phys D* **36**: A37–A43.

Engelke, K., G. Umgießer, S. Prevrhal and W. Kalendar (1997). "Three-dimensional analysis of trabecular bone structure: The need for spongiosa standard models." *Developments in X-Ray Tomography.* U. Bonse. Bellingham (WA), SPIE. **3149**: 53–61.

Eswaran, S. K., A. Gupta, M. F. Adams and T. M. Keaveny (2006). "Cortical and trabecular load sharing in the human vertebral body." *J Bone Min Res* **21**: 307–314.

Fajardo, R. J., E. Hernandez and P. M. O'Connor (2007). "Post cranial skeletal pneumaticity: A case study in the use of quantitative microCT to assess vertebral structure in birds." *J Anat* **211**: 138–147.

Fan, L. and C. W. Chen (1999). "Integrated approach to 3D warping and registration from lung images." *Developments in X-Ray Tomography II.* U. Bonse. Bellingham (WA), SPIE. **SPIE Proc Vol 3772**: 24–35.

Feldkamp, L. A., S. A. Goldstein, A. M. Parfitt, G. Jesion and M. Kleerekopes (1989). "The direct examination of three-dimensional bone architecture in vitro by computed tomography." *J Bone Miner Res* **4**: 3–11.

Follet, H., F. Peyrin, E. Vidal-Salle, A. Bonnassie, C. Rumelhart and P. J. Meunier (2007). "Intrinsic mechanical properties of trabecular calcaneus determined by finite-element models using 3D synchrotron microtomography." *J Biomech* **40**: 2174–2183.

Friis, E. M., P. R. Crane, K. R. Pedersen, S. Bengtson, P. C. J. Donoghue, G. W. Grimm and M. Stampanoni (2007). "Phase contrast x-ray microtomography links Cretaceous seeds with Gnetales and Bennettitales." *Nature* **450**: 549–552.

Geraghty, P. T., A. S. Jones, J. Stewart and W. G. Macbeth (2012). "Micro-computed tomography: An alternative method for shark ageing." *J Fish Biol* **80**: 1292–1299.

Ghafar, A., K. Parikka, D. Haberthür, M. Tenkanen, K. S. Mikkonen and J. P. Suuronen (2017). "Synchrotron reveals the fine three-dimensional porosity of composite polysaccharide aerogels." *Mater (Basel)* **10**: 871.

Gibson, L. J. and M. F. Ashby (1999). *Cellular Solids: Structure and Properties, 2nd Ed.* Cambridge, Cambridge University Press.

Gittens, S. A., G. R. Wohl, R. F. Zernicke, J. R. Matyas, P. Morley and H. Uludag (2004). "Systemic bone formation with weekly PTH administration in ovariectomized rats." *J Pharm Pharmaceut Sci* **7**: 27–37.

Glatt, V., E. Canalis, L. Stadmeyer and M. L. Bouxsein (2007). "Age-related changes in trabecular architecture differ in female and male C57BL/6J mice." *J Bone Min Res* **22**: 1197–1207.

Goldstein, S. A., R. Goulet and D. McCubbrey (1993). "Measurement and significance of three-dimensional architecture to the mechanical integrity of trabecular bone." *Calcif Tissue Int* **53**(Suppl 1): S127–S133.

Gong, H., M. Zhang, L. Qin and Y. Hou (2007). "Regional variations in the apparent and tissue-level mechanical parameters of vertebral trabecular bone with aging using micro-finite element analysis." *Annal Biomed Eng* **35**: 1622–1631.

Goodall, R., A. Marmottant, L. Salvo and A. Mortensen (2007). "Spherical pore replicated microcellular aluminium: Processing and influence on properties." *Mater Sci Eng A* **465**: 124–135.

Goulet, R. W., S. A. Goldstein, M. J. Ciarelli, J. L. Kuhn, M. B. Brown and L. A. Feldkamp (1994). "The relationship between the structural and orthogonal compressive properties of trabecular bone." *J Biomech* **27**: 375–389.

Guldberg, R. E., N. J. Caldwell, X. E. Guo, R. W. Goulet, S. J. Hollister and S. A. Goldstein (1997). "Mechanical stimulation of tissue repair in the repair in the hydraulic bone chamber." *J Bone Miner Res* **12**: 1295–1302.

Hara, T., E. Tanck, J. Homminga and R. Huiskes (2002). "The influence of microcomputed tomography threshold variations on the assessment of structural and mechanical trabecular bone properties." *Bone* **31**: 107–109.

Helfen, L., T. Baumbach, P. Pernot, P. Cloetens, H. Stanzick, K. Schladitz and J. Banhart (2005). "Investigation of pore initiation in metal foams by synchrotron-radiation tomography." *Appl Phys Lett* **86**: 231907-1–231907-3.

Hildebrand, T., A. Laib, D. Ulrich, A. Kohlbrenner and P. Rüegsegger (1997). "Bone structure as revealed by microtomography." *Developments in X-Ray Tomography.* U. Bonse. Bellingham (WA), SPIE. **3149**: 34–43.

Hildebrand, T. and P. Rüegsegger (1997). "Quantification of bone microarchitecture with the structure model index." *Comp Meth Biomed Eng* **1**: 15–23.

Hirano, T., K. Usami, Y. Tanaka and C. Masuda (1995). "In situ x-ray CT under tensile loading using synchrotron radiation." *J Mater Res* **10**: 381–385.

Hofmann, S., H. Hagenmüller, A. M. Koch, R. Müller, G. Vunjak-Novakovic, D. L. Kaplan, H. P. Merkle and L. Meinel (2007). "Control of in vitro tissue-engineered bone-like structures using human mesenchymal stem cells and porous silk scaffolds." *Biomaterials* **28**: 1152–1162.

Hollister, S. J., T. M. G. Chu, J. W. Halloran and S. E. Feinberg (2001). "Design and manufacture of bone replacement scaffolds." *Bone Mechanics Handbook, 2nd Ed.* S. C. Cowin. Boca Raton CRC Press: 14-11–14-19.

Homminga, J., R. Huiskes, B. v. Rietbergen, P. Rüegsegger and H. Weinans (2001). "Introduction and evaluation of a gray-value voxel conversion technique." *J Biomech* **34**: 513–517.

Hu, Z., H. Gerseny, Z. Zhang, Y.-J. Chen, A. Berg, Z. Zhang, S. Stock and P. Seth (2011). "Oncolytic adenovirus expressing soluble TGFβ receptor II-Fc-mediated inhibition of established bone metastases: A safe and effective systemic therapeutic approach for breast cancer." *Molecular Therapy* **19**: 1609–1618.

Huang, W., W. Hongjamrassilp, J. Y. Jung, P. A. Hastings, V. A. Lubarda and J. McKittrick (2017). "Structure and mechanical implications of the pectoral fin skeleton in the longnose skate (Chondrichthyes, Batoidea)." *Acta Biomater* **51**: 393–407.

Illman, B. and B. Dowd (1999). High resolution microtomography for density and spatial information about wood structures. *Developments in X-Ray Tomography II.* U. Bonse. Bellingham (WA), SPIE. **SPIE Proc Vol 3772**: 198–204.

Irsen, S. H., B. Leukers, B. Bruckschen, C. Tille, H. Seitz, F. Beckmann and B. Müller (2006). Image-based analysis of the internal microstructure of bone replacement scaffolds fabricated by 3D printing. *Developments in X-Ray Tomography V.* U. Bonse. Bellingham (WA), SPIE. **SPIE Proc Vol 6318**: 631809-1–631809-10.

Ishijima, M., K. Tsuji, S. R. Rittling, T. Yamashita, H. Kurosawa, D. T. Denhardt, A. Nifuji and M. Noda (2002). "Resistance to unloading-induced three-dimensional bone loss in osteopontin-deficient mice." *J Bone Miner Res* **17**: 661–667.

Ito, M., Y. Azuma, H. Takagi, T. Kamimura, K. Komoriya, T. Ohta and H. Kawaguchi (2003). "Preventive effects of sequential treatment with alendronate and 1 alpha-hydroxyvitamin D3 on bone mass and strength in ovariectomized rats." *Bone* **33**: 90–99.

Ito, M., A. Nishida, A. Koga, S. Ikeda, A. Shiraishi, M. Uetani, K. Hayashi and T. Nakamura (2002a). "Contribution of trabecular and cortical components to the mechanical properties of bone and their regulating parameters." *Bone* **31**: 351–358.

Ito, M., A. Nishida, T. Nakamura, M. Uetani and K. Hayashi (2002b). "Differences of three-dimensional trabecular microstructure in osteopenic rat models caused by ovariectomy and neurectomy." *Bone* **30**: 594–598.

Itoh, M., A. Shimazu, I. Hirata, Y. Yoshida, H. Shintani and M. Okazaki (2004). "Characterization of CO_3Ap-collagen sponges using x-ray high resolution microtomography." *Biomaterials* **25**: 2577–2583.

Itokawa, H., T. Hiraide, M. Moriya, M. Fujimoto, G. Nagashima, R. Suzuki and T. Fujimoto (2007). "A 12 month in vivo study on the response of bone to a hydroxyapatite–polymethylmethacrylate cranioplasty composite." *Biomaterials* **28**: 4922–4927.

Jakus, A. E., A. L. Rutz, S. W. Jordan, A. Kannan, S. M. Mitchell, C. Yun, K. D. Koube, S. C. Yoo, H. E. Whiteley, C.-P. Richter, R. D. Galiano, W. K. Hsu, S. R. Stock, E. L. Hsu and R. N. Shah (2016). "Hyperelastic 'bone': A highly versatile, growth factor–free, osteoregenerative, scalable, and surgically friendly biomaterial." *Sci Transl Med* **358**: 358ra127.

Jennane, R., R. Harba, G. Lemineur, S. Bretteil, A. Estrade and C. L. Benhamou (2007). "Estimation of the 3D self-similarity parameter of trabecular bone from its 2D projection." *Med Image Anal* **11**: 91–98.

Jin, A., J. Cobb, U. Hansen, R. Bhattacharya, C. Reinhard, N. Vo, R. Atwood, J. Li, A. Karunaratne, C. Wiles and R. Abel (2017). "The effect of long-term bisphosphonate therapy on trabecular bone strength and microcrack density." *Bone Joint Res* **6**: 602–609.

Jones, A. C., C. H. Arns, A. P. Sheppard, D. W. Hutmacher, B. K. Milthorpe and M. A. Knackstedt (2007). "Assessment of bone ingrowth into porous biomaterials using microCT." *Biomaterials* **28**: 2491–2504.

Jones, J. R., G. Poologasundarampillai, R. C. Atwood, D. Bernard and P. D. Lee (2007). "Non-destructive quantitative 3D analysis for the optimisation of tissue scaffolds." *Biomaterials* **28**: 1404–1413.

Joo, Y. I., T. Sone, M. Fukunaga, S. G. Lim and S. Onodera (2003). "Effects of endurance exercise on three-dimensional trabecular bone microarchitecture in young growing rats." *Bone* **33**: 485–493.

Kádár, C., E. Maire, A. Borbély, G. Peix, J. Lendvai and Z. Rajkovits (2004). "X-ray tomography and finite element simulation of the indentation behavior of metal foams." *Mater Sci Eng A* **387–389**: 321–325.

Kapat, K., P. K. Srivas, A. P. Rameshbabu, P. P. Maity, S. Jana, J. Dutta, P. Majumdar, D. Chakrabarti and S. Dhara (2017). "Influence of porosity and pore-size distribution in Ti(6)Al(4) V foam on physicomechanical properties, osteogenesis, and quantitative validation of bone ingrowth by micro-computed tomography." *ACS Appl Mater Interf* **9**(45): 39235–39248.

Keaveny, T. M. (2001). Strength of trabecular bone. *Bone Mechanics Handbook, 2nd Ed.* S. C. Cowin. Boca Raton, CRC Press: 16-11–16-42.

Kim, C. H., H. Zhang, G. Mikhail, D. von Stechow, R. Müller, H. S. Kim and X. E. Guo (2007). "Effects of thresholding techniques on μCT-based finite element models of trabecular bone." *J Biomed Eng* **129**: 481–486.

Kinney, J. H., D. L. Haupt, M. Balooch, A. J. C. Ladd, J. T. Ryaby and N. E. Lane (2000). "Three-dimensional morphometry of the L6 vertebra in the ovariectomized rat model of osteoporosis: Biomechanical implications." *J Bone Miner Res* **15**: 1981–1991.

Kinney, J. H., D. L. Haupt and A. J. C. Ladd (1997). "Applications of synchrotron microtomography in osteoporosis research." *Developments in X-Ray Tomography*. U. Bonse. Bellingham (WA), SPIE. **3149**: 64–68.

Kinney, J. H., N. Lane, S. Majumdar, S. J. Marshall and G. W. J. Marshall (1992). "Noninvasive three-dimensional histomorphology using x-ray tomographic microscopy." *J Bone Miner Res* 7(Suppl 1): S136.

Kinney, J. H., N. E. Lane and D. L. Haupt (1995). "In vivo, three-dimensional microscopy of trabecular bone." *J Bone Min Res* **10**: 264–270.

Kinney, J. H., G. W. Marshall, S. J. Marshall and D. L. Haupt (2001). "Three-dimensional imaging of large compressive deformations in elastomeric foams." *J Appl Polym Sci* **80**: 1746–1755.

Kinney, J. H. and J. T. Ryaby (2001). "Resonant markers for noninvasive, three-dimensional dynamic bone histomorphometry with x-ray microtomography." *Rev Sci Instrum* **72**: 1921–1923.

Kirby, B. J., J. R. Davis, J. A. Grant and M. J. Morgan (1997). "Monochromatic microtomographic imaging of osteoporotic bone." *Phys Med Biol* **42**: 1375–1385.

Kleuker, U. and C. Schulze (1997). A novel approach to the Rayleigh-to-Compton method: Wavelength dispersive tomography. *Developments in X-Ray Tomography*. U. Bonse. Bellingham (WA), SPIE. **3149**: 177–185.

Knackstedt, M. A., C. H. Arns, T. J. Senden and K. Gross (2006). "Structure and properties of clinical coralline implants measured via 3D imaging and analysis." *Biomaterials* **27**: 2776–2786.

Komlev, V. S., F. Peyrin, M. Mastrogiacomo, A. Cedola, A. Papadimitropoulos, F. Rustichelli and R. Cancedda (2006). "Kinetics of in vivo bone deposition by bone marrow stromal cells into porous calcium phosphate scaffolds: an X-ray computed microtomography study." *Tiss Eng* **12**: 3449–3458.

Kuhn, J. L., R. W. Goulet, M. Pappas and S. A. Goldstein (1990). "Morphometric and anisotropic symmetries of the canine distal femur." *J Orthop Res* **8**: 776–780.

Kurth, A. A. and R. Müller (2001). "The effect of an osteolytic tumor on the three-dimensional trabecular bone morphology in an animal model." *Skeletal Radiol* **30**: 94–98.

Ladd, A. J. C. and J. H. Kinney (1997). "Elastic constants of cellular structures." *Physica A* **240**: 349–360.

Lambert, J., I. Cantat, R. Delannay, A. Renault, F. Graner, J. A. Glazier, I. Veretennikov and P. Cloetens (2005). "Extraction of relevant physical parameters from 3D images of foams obtained by x-ray tomography." *Colloids Surf A* **263**: 295–302.

Lane, N. E., J. Kumer, W. Yao, T. Breunig, T. Wronski, G. Modin and J. H. Kinney (2003a). "Basic fibroblast growth factor forms new trabeculae that physically connect with pre-existing trabeculae, and this new bone is maintained with an anti-resorptive agent and enhanced with an anabolic agent in an osteopenic rat model." *Osteoporos Int* **14**: 374–382.

Lane, N. E., J. M. Thompson, D. Haupt, D. B. Kimmel, G. Modin and J. H. Kinney (1998). "Acute changes in trabecular bone connectivity and osteoclast activity in the ovariectomized rat in vivo." *J Bone Miner Res* **13**: 229–236.

Lane, N. E., W. Yao, J. H. Kinney, G. Modin, M. Balooch and T. J. Wronski (2003b). "Both hPTH(1-34) and bFGF increase trabecular bone mass in osteopenic rats but they have different effects on trabecular bone architecture." *J Bone Miner Res* **18**: 2105–2115.

Layton, M. W., S. A. Goldstein, R. W. Goulet, L. A. Feldkamp, D. J. Kubinski and G. G. Bole (1988). "Examination of subchondral bone architecture in experimental osteoarthritis by microscopic computed axial tomography." *Arthritis Rheum* **31**: 1400–1405.

Lee, S. B. (1993). Nondestructive examination of chemical vapor infiltration of 0°/90° SiC/Nicalon composites. Ph.D. thesis, Georgia Institute of Technology.

Leonard, A., L. P. Guiot, J. P. Pirard, M. Crine, M. Balligand and S. Blacher (2007). "Non-destructive characterization of deer (*Cervus elaphus*) antlers by X-ray microtomography coupled with image analysis." *J Microsc* **225**: 258–263.

Léonard, A., S. Blacher, C. Nimmol and S. Devahastin (2008). "Effect of far-infrared radiation assisted drying on microstructure of banana slices: An illustrative use of X-ray microtomography in microstructural evaluation of a food product." *J Food Eng* **85**: 154–162.
Lewis, B., J. M. Lough, M. C. Nash and G. Diaz-Pulido (2017). "Presence of skeletal banding in a reef-building tropical crustose coralline alga." *PLOS One* **12**: 0185124.
Liu, L. and E. F. Morgan (2007). "Accuracy and precision of digital volume correlation in quantifying displacements and strains in trabecular bone." *J Biomech* **40**: 3516–3520.
Lotinum, S., G. L. Evans, J. T. Bronk, M. E. Bolander, T. J. Wronski, E. L. Ritman and R. T. Turner (2004). "Continuous parathyroid hormone induces cortical porosity in the rat: Effects on bone turnover and mechanical properties." *J Bone Miner Res* **19**: 1165–1171.
Luo, G., J. H. Kinney, J. L. Kaufman, D. Haupt, A. Chiabrera and R. S. Siffert (1999). "Relationship between plain radiographic patterns and three-dimensional trabecular architecture in the human calcaneus." *Osteoporos Int* **9**: 339–345.
Ma, S., E. L. Goh, A. Jin, R. Bhattacharya, O. R. Boughton, B. Patel, A. Karunaratne, N. T. Vo, R. Atwood, J. P. Cobb, U. Hansen and R. L. Abel (2017). "Long-term effects of bisphosphonate therapy: perforations, microcracks and mechanical properties." *Sci Rep* **7**: 43399.
MacLatchy, L. and R. Müller (2002). "A comparison of the femoral head and neck trabecular architecture of Galago and Perodicticus using microcomputed tomography." *J Human Evol* **43**: 89–105.
Madi, K., S. Forest, M. Boussuge, S. Gailliègue, E. Lataste, J. Y. Buffière, D. Bernard and D. Jeulin (2007). "Finite element simulations of the deformation of fused-cast refractories based on X-ray computed tomography." *Comput Mater Sci* **39**: 224–229.
Maire, E. and J. Buffière (2000). X-ray tomography of aluminium foams and Ti/SiC composites. *X-Ray Tomography in Materials Science*. J. Baruchel, J. Y. Buffière, E. Maire, P. Merle and G. Peix. Paris, Hermes Science: 115–126.
Maire, E., A. Elmoutaouakkil, A. Fazekas and L. Salvo (2003a). "In situ x-ray tomography measurements of deformation in cellular solids." *MRS Bull* **28**: 284–289.
Maire, E., A. Fazekas, L. Salvo, R. Dendievel, S. Youssef, P. Cloetens and J. M. Letang (2003b). "X-ray tomography applied to the characterization of cellular materials. Related finite element modeling problems." *Compos Sci Technol* **63**: 2431–2443.
Martin, R. B., D. B. Burr and N. A. Sharkey (1998). *Skeletal Tissue Mechanics*. New York, Springer.
Martin, E. A., E. L. Ritman and R. T. Turner (2003). "Time course of epiphyseal growth plate fusion in rat tibiae." *Bone* **32**: 261–267.
Martín-Badosa, E., D. Amblard, S. Nuzzo, A. Elmoutaouakkilo, L. Vico and F. Peyrin (2003). "Excised bone structures in mice: imaging at three-dimensional synchrotron microCT." *Radiology* **229**: 921–928.
Mastrogiacomo, M., V. S. Komlev, M. Hausard, F. Peyrin, F. Turquier, S. Casari, A. Cedola, F. Rustichelli and R. Cancedda (2004). "Synchrotron radiation microtomography of bone engineered from bone marrow stromal cells." *Tiss Eng* **10**: 1767–1774.
Mathieu, L. M., T. L. Mueller, P. E. Bourban, D. P. Pioletti, R. Müller and J. A. E. Månson (2006). "Architecture and properties of anisotropic polymer composite scaffolds for bone tissue engineering." *Biomaterials* **27**: 905–916.
Matsumoto, Y., A. H. Brothers, S. R. Stock and D. C. Dunand (2007). "Uniform and graded chemical milling of aluminum foams." *Mater Sci Eng A* **447**: 150–157.
McCreadie, B. R., R. W. Goulet, L. A. Feldkamp and S. A. Goldstein (2001). Hierarchical structure of bone and micro-computed tomography. *Noninvasive Assessment of Trabecular Bone Architecture and the Competence of Bone*. S. Majumdar and B. K. Bay. New York, Kluwer: 67–83.
McLaughlin, F., J. Mackintosh, B. P. Hayes, A. McLaren, I. J. Uings, P. Salmon, J. Humphreys, E. Meldrum and S. N. Farrow (2002). "Glucocorticoid-induced osteopenia in the mouse as assessed by histomorphometry, microcomputed tomography and biochemical markers." *Bone* **30**: 924–930.
Mendoza, F., P. Verboven, H. K. Mebatsion, G. Kerckhofs, M. Wevers and B. Nicolai (2007). "Three-dimensional pore space quantification of apple tissue using x-ray computed tomography." *Planta* **226**: 559–570.

Montanini, R. (2005). "Measurement of strain rate sensitivity of aluminium foams for energy dissipation." *Int J Mech Sci* **47**: 26–42.
Morenko, B. J., S. E. Bove, L. Chen, R. E. Guzman, P. Juneau, T. M. A. Bocan, G. K. Peter, R. Arora and K. S. Kilgore (2004). "In vivo micro computed tomography of subchondral bone in the rat after intra-articular administration of monosodium iodoacetate." *Contemp Top Lab Anim Sci* **43**: 39–43.
Mulder, L., J. H. Koolstra, J. M. J. den Toonder and T. M. G. J. van Eijden (2007a). "Intratrabecular distribution of tissue stiffness and mineralization in developing trabecular bone." *Bone* **41**: 256–265.
Mulder, L., L. J. van Ruijven, J. H. Koolstra and T. M. G. J. van Eijden (2007b). "Biomechanical consequences of developmental changes in trabecular architecture and mineralization of the pig mandibular condyle." *J Biomech* **40**: 1575–1582.
Mulder, L., L. J. van Ruijven, J. H. Koolstra and T. M. G. J. van Eijden (2007c). "The influence of mineralization on intrabecular stress and strain distribution in developing trabecular bone." *Annal Biomed Eng* **35**: 1668–1677.
Müller, R. (1999). Microtomographic imaging in the process of bone modeling and simulation. *Developments in X-Ray Tomography II*. U. Bonse. Bellingham (WA), SPIE. **SPIE Proc Vol 3772**: 63–76.
Müller, R. and W. C. Hayes (1997). Biomechanical competence of microstructural bone in the progress of adaptive bone remodeling. *Developments in X-Ray Tomography*. U. Bonse. Bellingham (WA), SPIE. **3149**: 69–81.
Müller, B., P. Thurnier, F. Beckmann, T. Weitkamp, C. Rau, R. Bernhardt, E. Karamuk, L. Eckert, J. Brandt, S. Buchloh, E. Wintermantel, D. Scharnweber and H. Worch (2001). Nondestructive three-dimensional evaluation of biocompatible materials by microtomography using synchrotron radiation. *Developments in X-Ray Tomography III*. U. Bonse. Bellingham (WA), SPIE. **SPIE Proc Vol 4503**: 178–188.
Nazarian, A. and R. Muller (2004). "Time-lapsed microstructural imaging of bone failure behavior." *J Biomech* **37**: 55–65.
Nazarian, A., M. Stauber and R. Müller (2005). "Design and implementation of a novel mechanical testing system for cellular solids." *J Biomed Mater Res* **73B**: 400–411.
Nicholson, E. K., S. R. Stock, M. W. Hamrick and M. J. Ravosa (2006). "Biomineralization and adaptive plasticity of the temporomandibular joint in myostatin knockout mice." *Arch Oral Biol* **51**: 37–49.
Niebur, G. L., M. J. Feldstein, J. C. Yuen, T. J. Chen and T. M. Keaveny (2000). "High-resolution finite element models with tissue strength asymmetry accurately predict failure of trabecular bone." *J Biomech* **33**: 1575–1583.
Nuzzo, S., M. H. Lafage-Proust, E. Martin-Badosa, G. Boivin, T. Thomas, C. Alexandre and F. Peyrin (2002). "Synchrotron radiation microtomography allows the analysis of three-dimensional microarchtiecture and degree of mineralization of human iliac crest biopsy specimens: Effects of etidronate treatment." *J Bone Miner Res* **17**: 1372–1382.
Nuzzo, S., C. Meneghini, P. Braillon, R. Bouvier, S. Mobilio and F. Peyrin (2003). "Microarchitectural and physical changes during fetal growth in human vertebral bone." *J Bone Miner Res* **18**: 760–768.
Obermeyer, T., D. Yonick, K. Lauing, S. Kristen, S. Stock, R. Nauer, P. Strotman, R. Shankar, R. Gamelli, M. Stover and J. J. Callaci (2012). "Mesenchymal stem cells facilitate fracture repair in an alcohol-induced impaired healing model." *J Orthop Trauma* **26**: 712–718.
Odgaard, A. (1997). "Three-dimensional methods for quantification of cancellous bone architecture." *Bone* **20**: 315–328.
Odgaard, A. (2001). "Quantification of cancellous bone architecture." *Bone Mechanics Handbook, 2nd Ed*. S. C. Cowin. Boca Raton, CRC Press: 14-11–14-19.
Odgaard, A. and H. J. G. Gunderson (1993). "Quantification of connectivity in cancellous bone, with special emphasis on 3-D reconstructions." *Bone* **14**: 173–182.

Ohgaki, T., H. Toda, M. Kobayashi, K. Uesugi, M. Niinom, T. Akahori, T. Kobayashi, K. Makii and Y. Aruga (2006). "In situ observations of compressive behaviour of aluminium foams by local tomography using high-resolution tomography." *Phil Mag* **86**: 4417–4438.

Oliveira, A. L., P. B. Malafaya, S. A. Costa, R. A. Sousa and R. L. Reis (2007). "Micro-computed tomography as a potential tool to assess the effect of dynamic coating routes on the formation of biomimetic apatite layers on 3D-plotted biodegradable polymeric scaffolds." *J Mater Sci Mater Med* **18**: 211–223.

Orth, M., N. J. Kruse, B. J. Braun, C. Scheuer, J. H. Holstein, A. Khalil, X. Yu, W. L. Murphy, T. Pohlemann, M. W. Laschke and M. D. Menger (2017). "BMP-2-coated mineral coated microparticles improve bone repair in atrophic non-unions." *Eur Cell Mater* **33**: 1–12.

Ozcivici, E., R. Garman and S. Judex (2007). "High-frequency oscillatory motions enhance the simulated mechanical properties of non-weight bearing trabecular bone." *J Biomech* **40**: 3404–3411.

Partap, S., A. Muthutantri, I. U. Rehman, G. R. Davis and J. A. Darr (2007). "Preparation and characterisation of controlled porosity alginate hydrogels made via a simultaneous micelle templating and internal gelation process." *J Mater Sci* **42**: 3502–3507.

Patel, V., A. S. Issever, A. Burghardt, A. Laib, M. Ries and S. Majumdar (2003). "MicroCT evaluation of normal and osteoarthritic bone structure in human knee specimens." *J Orthop Res* **21**: 6–13.

Pettit, A. R., H. Ji, D. v. Stechow, R. Müller, S. R. Goldring, Y. Choi, C. Benoist and E. M. Gravallese (2001). "TRANCE/RANKL knockout mice are protected from bone erosion in a serum transfer model of arthritis." *Am J Pathol* **159**: 1689–1699.

Peyrin, F., M. Salome, P. Cloetens, A. M. Laval-Jeantet, E. Ritman and P. Rüegsegger (1998). "MicroCT examinations of trabecular bone samples at different resolutions: 14, 7 and 2 micron level." *Technol Health Care* **6**: 391–401.

Peyrin, F., M. Salomé, F. Denis, P. Braillon, A. M. Laval-Jeanlet and P. Cloetens (1997). 3D imaging of fetus vertebra by synchrotron radiation microtomography. *Developments in X-ray tomography*. U. Bonse. Bellingham (WA), SPIE. **3149**: 44–52.

Pistoia, W., B. v. Rietbergen and P. Rüegsegger (2003). "Mechanical consequences of different scenarios for simulated bone atrophy and recovery in the distal radius." *Bone* **33**: 937–945.

Plougonven, E., D. Bernard and P. Viot (2006). Quantitative analysis of the deformation of polypropylene foam under dynamic loading. *Developments in X-Ray Tomography V*. U. Bonse. Bellingham (WA), SPIE. **SPIE Proc Vol 6318**: 631813-1–631813-10.

Porter, M. E., J. L. Beltrán, T. J. Koob and A. P. Summers (2006). "Material properties and biochemical composition of mineralized vertebral cartilage in seven elasmobranch species (Chondrichthyes)." *J Exp Biol* **209**: 2920–2928.

Rajagopalan, S., L. Lu, M. J. Yaszemski and R. A. Robb (2005). "Optimal segmentation of microcomputed tomographic images of porous tissue-engineering scaffolds." *J Biomed Mater Res* **75A**: 877–887.

Rapillard, L., M. Charlebois and P. K. Zysset (2006). "Compressive fatigue behavior of human vertebral trabecular bone." *J Biomech* **39**: 2133–2139.

Ravosa, M. J., E. P. Kloop, J. Pinchoff, S. R. Stock and M. Hamrick (2006a). "Plasticity of mandibular biomineralization in myostatin-deficient mice." *J Morphol* **268**: 275–282.

Ravosa, M. J., R. Kunwar, E. K. Nicholson, E. B. Klopp, J. Pinchoff, S. R. Stock and M. S. Stack (2006b). Adaptive plasticity in mammalian masticatory joints. *Developments in X-Ray Tomography V*. U. Bonse. Bellingham (WA), SPIE. **SPIE Proc Vol 6318**: 63180D-1–63180D-9.

Ravosa, M. J., R. Kunwar, S. R. Stock and M. S. Stack (2007a). "Pushing the limit: Masticatory stress and adaptive plasticity in mammalian craniomandibular joints." *J Exp Biol* **210**: 628–641.

Ravosa, M. J., S. R. Stock, E. L. Simons and R. Kunwar (2007b). "MicroCT analysis of symphyseal ontogeny in a subfossil lemur (Archaeolemur)." *Int J Primatology* **28**: 1385–1396.

Remigy, J. C. and M. Meireles (2006). "Assessment of pore geometry and 3-D architecture of filtration membranes by synchrotron radiation computed microtomography." *Desalination* **199**: 501–503.

Remigy, J. C., M. Meireles and X. Thibault (2007). "Morphological characterization of a polymeric microfiltration membrane by synchrotron radiation computed microtomography." *J Membrane Sci* **305**: 27–35.

Renders, G. A. P., L. Mulder, L. J. van Ruijven and T. M. G. J. van Eijden (2007). "Porosity of human mandibular condylar bone." *J Anat* **210**: 239–248.

Reynolds, D. G., C. Hock, S. Shaikh, J. Jacobson, X. Zhang, P. T. Rubery, C. A. Beck, R. J. O'Keefe, A. L. Lerner, E. M. Schwarz and H. A. Awad (2007). "Micro-computed tomography prediction of biomechanical strength in murine structural bone grafts." **40**: 3178–3186.

van Rietbergen, B. (2001). Micro-FE analyses of bone - state of the art. *Noninvasive Assessment of Trabecular Bone Architecture and the Competence of Bone*. B. K. Bay, S. Majumdar. New York, Kluwer/Plenum. **Adv Exp Med Biol Vol 496**: 21–30.

van Rietbergen, B. and R. Huiskes (2001). Elastic constants of cancellous bone. *Bone Mechanics Handbook, 2nd Ed.* S. C. Cowin. Boca Raton CRC Press: 15-11–15-24.

Robert, G., D. R. Baker, M. L. Rivers, E. Allard and J. Larocque (2004). Comparison of the bubble size distribution in silicate foams using 2-dimensional images and 3-dimensional x-ray microtomography. *Developments in X-Ray Tomography IV*. U. Bonse. Bellingham (WA), SPIE. **SPIE Proc Vol 5535**: 505–513.

Ruimerman, R., P. Hilbers, B. van Rietbergen and R. Huiskes (2005). "A theoretical framework for strain-related trabecular bone maintenance and adaptation." *J Biomech* **38**: 931–941.

Salvo, L., P. Belestin, E. Maire, M. Jacquesson, C. Vecchionacci, E. Boller, M. Bornert and P. Doumalin (2004). "Structure and mechanical properties of AFS sandwiches studied by in-situ compression tests in x-ray microtomography." *Adv Eng Mater* **6**: 411–415.

Santiago, H. A., A. Zamarioli, M. D. Sousa Neto and J. B. Volpon (2017). "Exposure to secondhand smoke impairs fracture healing in rats." *Clin Orthop Relat Res* **475**: 894–902.

Schardosim, M., J. Soulié, D. Poquillon, S. Cazalbou, B. Duployer, C. Tenailleau, C. Rey, R. Hübler and C. Combes (2017). "Freeze-casting for PLGA/carbonated apatite composite scaffolds: Structure and properties." *Mater Sci Eng C Mater Biol Appl* **77**: 731–738.

Scheffler, F., R. Herrmann, W. Schwieger and M. Scheffler (2004). "Preparation and properties of an electrically heatable aluminium foam/zeolite composite." *Micropor Mesopor Mater* **67**: 53–59.

Seidel, R., K. Lyons, M. Blumer, P. Zaslansky, P. Fratzl, J. C. Weaver and M. N. Dean (2016). "Ultrastructural and developmental features of the tessellated endoskeleton of elasmobranchs (sharks and rays)." *J Anat* **229**: 681–702.

Shao, H., X. Ke, A. Liu, M. Sun, Y. He, X. Yang, J. Fu, Y. Liu, L. Zhang, G. Yang, S. Xu and Z. Gou (2017). "Bone regeneration in 3D printing bioactive ceramic scaffolds with improved tissue/material interface pore architecture in thin-wall bone defect." *Biofabrication* **9**: 025003.

Shim, J. H., J. Y. Won, J. H. Park, J. H. Bae, G. Ahn, C. H. Kim, D. H. Lim, D. W. Cho, W. S. Yun, E. B. Bae, C. M. Jeong and J. B. Huh (2017). "Effects of 3D-Printed Polycaprolactone/β-tricalcium phosphate membranes on guided bone regeneration." *Int J Mol Sci* **18**: 899.

Showalter, C., B. D. Clymer, B. Richmond and K. Powell (2006). "Three-dimensional texture analysis of cancellous bone cores evaluated at clinical CT resolutions." *Osteopor Int* **17**: 259–266.

Siu, W. S., L. Qin, W. H. Cheung and K. S. Leung (2004). "A study of trabecular bones in ovariectomized goats with micro-computed tomography and peripheral quantitative computed tomography." *Bone* **35**: 21–26.

Smith, A. B. (1980). *Stereom Microstructure of the Echinoid Test. Special Papers in Palaeontology No. 25.* London, The Palaeontology Assoc.

Smith, M. H., C. L. Flanagan, J. M. Kemppainen, J. A. Sack, H. Chung, S. Das, S. J. Hollister and S. E. Feinberg (2007). "Computed tomography-based tissue-engineered scaffolds in craniomaxillofacial surgery." *Int J Med Robotics Comput Assist Surg* **3**: 207–216.

Sone, T., T. Tamada, Y. Jo, H. Miyoshi and M. Fukunaga (2004). "Analysis of three-dimensional microarchitecture and degree of mineralization in bone metastases from prostate cancer using synchrotron microcomputed tomography." *Bone* **35**: 432–438.

Squire, M., L. R. Donahue, C. Rubin and S. Judex (2004). "Genetic variations that regulate bone morphology in the male mouse skeleton do not define its susceptibility to mechanical unloading." *Bone* **35**: 1353–1360.

Stenstrom, M., B. Olander, D. Lehto-Axtelius, J. E. Madsen, L. Nordsletten and G. A. Carlsson (2000). "Bone mineral density and bone structure parameters as predictors of bone strength: an analysis using computerized microtomography and gastrectomy-induced osteopenia in the rat." *J Biomech* **33**: 289–297.

Steppe, K., V. Cnudde, C. Girard, R. Lemeur, J. P. Cnudde and P. Jacobs (2004). "Use of x-ray computed microtomography for non-invasive determination of wood anatomical characteristics." *J Struct Biol* **148**: 11–21.

Stock, S. R., F. De Carlo and J. D. Almer (2008). "High energy x-ray scattering tomography applied to bone." *J Struct Biol* **161**: 144–150.

Stock, S. R., K. Ignatiev and F. D. Carlo (2004a). Very high resolution synchrotron microCT of sea urchin ossicle structure. *Echinoderms: München*. T. Heinzeller and J. H. Nebelsick. London, Taylor and Francis: 353–358.

Stock, S. R., K. I. Ignatiev, S. A. Foster, L. A. Forman and P. H. Stern (2004b). "MicroCT quantification of in vitro bone resorption of neonatal murine calvaria exposed to IL-1 or PTH." *J Struct Biol* **147**: 185–199.

Stölken, J. S. and J. H. Kinney (2003). "On the importance of geometric nonlinearity in finite element simulations of trabecular bone failure." *Bone* **33**: 494–504.

Su, R., G. M. Campbell and S. K. Boyd (2007). "Establishment of an architecture-specific experimental validation approach for finite element modeling of bone by rapid prototyping and high resolution computed tomography." *Med Eng Phys* **29**: 480–490.

Takada, H., S. Abe, Y. Tamatsu, S. Mitarashi, H. Saka and Y. Ide (2006). "Three-dimensional bone microstructures of the mandibular angle using microCT and finite element analysis: Relationship between partially impacted mandibular third molars and angle fractures." *Dent Traumatol* **22**: 18–24.

Tanaka, E., R. Sano, N. Kawai, G. E. J. Langenbach, P. Brugman, K. Tanne and T. M. G. J. van Eijden (2007). "Effect of food consistency on the degree of mineralization in the rat mandible." *Annal Biomed Eng* **35**: 1617–1621.

Tesei, L., F. Casseler, D. Dreossi, L. Mancini, G. Tromba and F. Zanini (2005). "Contrast-enhanced x-ray microtomography of the bone structure adjacent to oral implants." *Nucl Instrum Meth A* **548**: 257–263.

Thurner, P., E. Karamuk and B. Müller (2001). "3D characterization of fibroblast cultures on PRT textiles." *Euro Cells Mater* **2**(Suppl 1): 57–58.

Thurner, P., B. Müller, F. Beckmann, T. Weitkamp, C. Rau, R. Müller, J. A. Hubbell and U. Sennhauser (2003). "Tomography studies of human foreskin fibroblasts on polymer yarns." *Nucl Instrum Meth B* **200**: 397–405.

Thurner, P. J., P. Wyss, R. Voide, M. Stauber, B. Müller, M. Stampanoni, J. A. Hubbell, R. Müller and U. Sennhauser (2004). Functional microimaging of soft and hard tissue using synchrotron light. *Developments in X-Ray Tomography IV*. U. Bonse. Bellingham (WA), SPIE. **SPIE Proc Vol 5535**: 112–128.

Titschack, J., D. Baum, K. Matsuyama, K. Boos, C. Färber, W.-A. Kahl, K. Ehrig, D. Meinel, C. Soriano and S. R. Stock (2018). "Ambient occlusion - a powerful algorithm to segment shell and skeletal intrapores in computed tomography data." *Comput Geosci* **115**: 75–87.

Toda, H., T. Ohgaki, K. Uesugi, M. Kobayashi, N. Kuroda, T. Kobayashi, M. Niinomi, T. Akahori, K. Makii and Y. Aruga (2006a). "Quantitative assessment of microstructure and its effects on compression behavior of aluminum foams via high-resolution synchrotron x-ray tomography." *Metall Mater Trans A* **37**: 1211–1219.

Toda, H., M. Takata, T. Ohgaki, M. Kobayashi, T. Kobayashi, K. Uesugi, K. Makii and Y. Aruga (2006b). "3-D image-based mechanical simulation of aluminium foams: Effects of internal microstructure." *Adv Eng Mater* **8**: 459–467.

Trtik, P., J. Dual, D. Keunecke, D. Mannes, P. Niemz, P. Stahli, A. Kaestner, A. Groso and M. Stampanoni (2007). "3D imaging of microstructure of spruce wood." *J Struct Biol* **159**: 46–55.

Tu, S., F. Hu, W. Cai, L. Xiao, L. Zhang, H. Zheng, Q. Jiang and L. Chen (2017). "Visualizing polymeric bioresorbable scaffolds with three-dimensional image reconstruction using contrast-enhanced micro-computed tomography." *Int J Cardiovasc Imaging* **33**: 731–737.

Ün, K., G. Bevill and T. M. Keaveny (2006). "The effects of side-artifacts on the elastic modulus of trabecular bone." *J Biomech* **39**: 1955–1963.

Vanis, S., O. Rheinbach, A. Klawonn, O. Prymak and M. Epple (2006). "Numerical computation of the porosity of bone substitution materials from synchrotron microcomputer tomographic data." *Mat-wiss Werkstofftech* **37**: 469–473.

Verhulp, E., B. van Rietbergen and R. Huiskes (2004). "A three-dimensional digital image correlation technique for strain measurements in microstructures." *J Biomech* **37**: 1313–1320.

Verhulp, E., B. van Rietbergen and R. Huiskes (2006). "Comparison of micro-level and continuum-level voxel models of the proximal femur." *J Biomech* **39**: 2951–2957.

Vetter, L. D., V. Cnudde, B. Masschaele, P. J. S. Jacobs and J. V. Acker (2006). "Detection and distribution analysis of organosilicon compounds in wood by means of SEM-EDX and microCT." *Mater Char* **56**: 39–48.

Viot, P., D. Bernard and E. Plougonven (2007). "Polymeric foam deformation under dynamic loading by the use of the microtomographic technique." *J Mater Sci* **42**: 7202–7213.

Waarsing, J. H., J. S. Day, J. C. v. d. Linden, A. G. Ederveen, C. Spanjers, N. D. Clerck, A. Sasov, J. A. N. Verhaar and H. Weinans (2004). "Detecting and tracking local changes in the tibiae of individual rats: A novel method to analyse longitudinal in vivo microCT data." *Bone* **34**: 163–169.

Wang, M., X. F. Hu and X. P. Wu (2006). "Internal microstructure evolution of aluminum foams under compression." *Mater Res Bull* **41**: 1949–1958.

Wang, H., Q. Li, Q. Wang, H. Zhang, W. Shi, H. Gan, H. Song and Z. Wang (2017). "Enhanced repair of segmental bone defects in rabbit radius by porous tantalum scaffolds modified with the RGD peptide." *J Mater Sci Mater Med* **28**: 50.

Wang, B., B. Pan and G. Lubineau (2018). "Morphological evolution and internal strain mapping of pomelo peel using X-ray computed tomography and digital volume correlation." *Mater Design* **137**: 305–315.

Wang, Y. H., E. Rajalakshmanan, C. K. Wang, C. H. Chen, Y. C. Fu, T. L. Tsai, J. K. Chang and M. L. Ho (2017). "PLGA-linked alendronate enhances bone repair in diaphysis defect model." *J Tissue Eng Regen Med* **11**: 2603–2612.

Washburn, N. R., M. Weir, P. Anderson and K. Potter (2004). "Bone formation in polymeric scaffolds evaluated by proton magnetic resonance microscopy and x-ray microtomography." *J Biomed Mater Res Pt A* **69**: 738–747.

Weiss, P., L. Obadia, D. Magne, X. Bourges, C. Rau, T. Weitkamp, I. Khairoun, J. M. Bouler, D. Chappard, O. Gauthier and G. Daculsi (2003). "Synchrotron X-ray microtomography (on a micron scale) provides three-dimensional imaging representation of bone ingrowth in calcium phosphate biomaterials." *Biomaterials* **24**: 4591–4601.

Wilkes, T. E. (2008). Metal/ceramic composites via infiltration of an interconnceted wood-derived ceramic. Ph.D. thesis, Northwestern University.

Willems, N. M. B. K., L. Mulder, G. E. J. Langenbach, T. Grünheid, A. Zentner and T. M. G. J. van Eijden (2007). "Age-related changes in microarchitecture and mineralization of cancellous bone in the porcine mandibular condyle." *J Struct Biol* **158**: 421–427.

Witte, F., J. Fischer, J. Nellesen and F. Beckmann (2006a). "Microtomography of magnesium implants in bone and their degradation." *Developments in X-Ray Tomography V.* U. Bonse. Bellingham (WA), SPIE. **SPIE Proc Vol 6318**: 631806-1–631806-9.

Witte, F., J. Fischer, J. Nellesen, H. A. Crostack, V. Kaese, A. Pisch, F. Beckmann and H. Windhagen (2006b). "In vitro and in vivo corrosion measurements of magnesium alloys." *Biomaterials* **27**: 1013–1018.

Wu, J. P., A. R. Boccaccini and P. D. Lee (2007). "Thermal and mechanical properties of a foamed glass-ceramic material produced from silicate wastes." *Glass Technol A* **48**: 133–141.

Wu, J. P., A. R. Boccaccini, P. D. Lee, M. J. Kershaw and R. D. Rawlings (2006). "Glass ceramic foams from coal ash and waste glass: production and characterization." *Adv Appl Ceram* **105**: 32–39.
Yeh, O. C. and T. M. Keaveny (2001). "Relative roles of microdamage and microfracture in the mechanical behavior of trabecular bone "*J Orthop Res* **19**: 1001–1007.
Yoon, B. H., J. G. Kim, Y. K. Lee, Y. C. Ha, K. H. Koo and Kim, J. H. (2017). "Femoral head trabecular micro-architecture in patients with osteoporotic hip fractures: impact of bisphosphonate treatment." *Bone* **105**: 148–153.
Youssef, S., E. Maire and R. Gaertner (2005). "Finite element modeling of the actual structure of cellular materials determined by x-ray tomography." *Acta Mater* **53**: 719–730.
Yu, W., H. Zhao, Z. Ding, Z. Zhang, B. Sun, J. Shen, S. Chen, B. Zhang, K. Yang, M. Liu, D. Chen and Y. He (2017). "In vitro and in vivo evaluation of MgF(2) coated AZ31 magnesium alloy porous scaffolds for bone regeneration." *Colloids Surf B Biointerf* **149**: 330–340.
Zebaze, R. M. D., A. Jones, M. Knackstedt, G. Maalouf and E. Seeman (2007). "Construction of the femoral neck during growth determines its strength in old age." *J Bone Miner Res* **22**: 1055–1061.
Zhang, Y. T., J. Niu, Z. Wang, S. Liu, J. Wu and B. Yu (2017). "Repair of osteochondral defects in a rabbit model using bilayer poly(lactide-co-glycolide) scaffolds loaded with autologous platelet-rich plasma." *Med Sci Monit* **23**: 5189–5201.

9

Networks

The subject of this chapter is microCT studies of networks. The first section concerns engineered network solids, materials that consist of many fibers consolidated into paper or fiberboard or other materials. Networks of pores and fluid flow and localization therein is the subject of the second section. Coverage flows into microCT studies of animal circulatory systems is the subject in the third section. The chapter concludes with a review of respiratory system studies.

9.1 Engineered Network Solids

Fibrous network solids differ somewhat from most of the cellular materials discussed in Chapter 8. They are collections of long, thin (nominally 1D) objects. While trabecular bone often consists of struts, much of this bone also has plate-like character and so differs from the network solids described in this section.

Cellulosic fibrous networks are encountered in papers and in engineered, low-density, wood-based fiberboards. Synchrotron microCT of low-, medium-, and high-density fiberboards was used to determine fiber network characteristics and to generate a realistic, ABAQUS™-based model for calculating thermal conductivity, a property important in construction materials; a design of experiments analysis of the model parameters revealed that fiber density was responsible for 60% of local conductivity of the network and that fiber orientation and tortuosity represented 25% of the influence of network conductivity (Faessel, Delisée et al. 2005).

Walther et al. examined various medium density fiberboards with synchrotron microCT and provided a particularly clear account of the challenging image analysis needed to differentiate the fiber volume from the surrounding air (Walther, Terzic et al. 2006). With a voxel size of ~2.3 μm, the lignocellulose fiber walls and the hollow lumens were clearly resolved; both needed to be included in the fiber volume but real and apparent (noise-related) breaks in the walls were a complication that was overcome. The maximum fiber thickness was ~14 μm (6 voxels), so the analysis (quantification of fiber diameters and related quantities) should be regarded as quite robust. Quantities such as volume fraction (fiber walls, lumen volume, and interfiber air) and total surface area per unit volume (fiber-lumen and fiber-outer air) were quantified as a function of specimen density. Individual fibers and fiber fragments and their orientations were quantified, and fiber bundles (groups of fibers each with contact areas >10^4 μm^2) were mapped. Figure 9.1 shows 3D renderings of a subvolume of a medium density fiberboard; it compares a segmented image of all of the fibers, a selection of individual fibers identified from the entire volume, and a set of very small fiber fragments. Lux et al. provide analysis similar to that of Walther et al. (but with different image analysis tools) of synchrotron microCT data for a

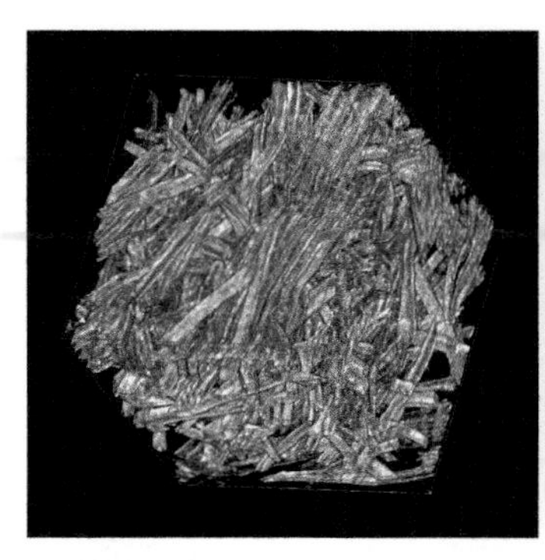

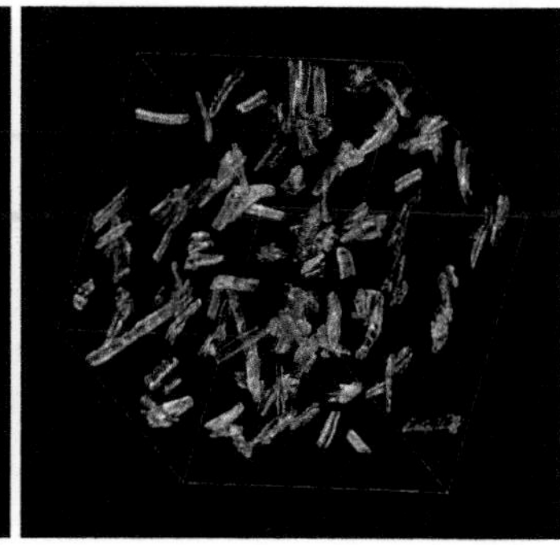

FIGURE 9.1
3D renderings of a $(256)^3$ subvolume of a medium density fiberboard. (left) The complete volume. (middle) A selection of individually labeled fibers. (right) Individually labeled fiber fragments with volume less than 10^3 voxels. The edges of the display cube equal 588 µm (Walther, Terzic et al. 2006).

very low-density fiberboard and found the resulting density agreed with that measured macroscopically (Lux, Delisée et al. 2006).

Synchrotron microCT studies of paper have appeared in which paper's extreme sensitivity to changes in humidity was overcome: shrinkage of even 1% produces displacements of a few micrometers, leading to sample motion artifacts (Antoine, Nygard et al. 2002, Rolland Du Roscoat, Bloch et al. 2005). Yang and Lindquist applied automatic 3D image analysis to synchrotron microCT data from a fibrous polymer mat (Yang and Lindquist 2000). Eberhardt and Clarke advanced another algorithm for analyzing fiber characteristics in cloth (Eberhardt and Clarke 2002).

Porosity and specific surface area of four types of industrial paper materials were quantified (Rolland Du Roscoat, Decain et al. 2007). Two papers contained fillers and two did not. In all four papers, synchrotron microCT revealed two boundary layers (surfaces, totaling about 50% of the volume) with strong gradients in porosity, and a central bulk layer with nearly constant porosity. The microstructure of the four papers was transversely isotropic, and the anisotropy of the filler-containing papers was less pronounced than in the other two materials. Representative elemental volumes were also determined for the material.

Images of a textile sample (denim) were shown in Fig. 6.6. In four different 3D studies of textiles, lab microCT was used to quantify yarn diameter and spacing and the variability in these quantities (Desplentere, Lomov et al. 2005). Incorporating this data into stochastic models of mechanical properties produced good agreement with experiment not only for the value of the in-plane Young's modulus but also for the scatter in results. MicroCT analysis of the structure of polymer fabric used in paper drying has been incorporated into lattice Boltzmann simulations of fluid flow and Monte Carlo simulations of vapor diffusion in the structure (Ramaswamy, Gupta et al. 2004), but the report is unclear (at least to the author) as to whether 2D (possibly very inaccurate) or 3D assessments of pore tortuosity were used in the models. Contaminant classification in cotton fabrics is another area where microCT was applied (Pavani, Dogan et al. 2004).

The relatively good environmental stability of metals, ceramics, and glasses makes fibrous networks of these materials attractive for heat transfer, filtration, and catalyst support applications. Fibrous mats for battery applications, for example, have been studied with microCT (Kok, Jervis et al. 2018, Rawal, Rao et al. 2018). Lab microCT and skeletonization analysis were used to study two bonded stainless steel fiber assemblies (Tan, Elliott et al. 2006). The distribution of fiber segment lengths between the two specimens differed, as did the distribution of fiber orientations (shown on stereographic projections), and this data could certainly be imported into numerical models of stiffness (or other quantities) as in Clyne et al. (2005). One example is the microCT determination of short fiber angular and spatial distribution that was used as input for a multiscale model of composite properties (Pyrz and Schjødt-Thomsen 2006), and a second is the relationship of fiber orientation to transverse stiffness in mineral wool (Chapelle, Lyckegaard et al. 2018). Determination of fiber orientation and fiber length distributions are also important to the behavior of short-fiber reinforced foams, and lab microCT of a phenolic foam incorporating short glass fibers revealed fiber preferred orientation attributed to shear generated during foaming (Shen, Nutt et al. 2004). Fracture mechanisms in sparse networks of fibers are the subject of another microCT study (Krasnoshlyk, du Roscoat et al. 2018). Microstructure of acicular mullite was studied by synchrotron microCT and related to the ceramic's mechanical properties (Hsiung, Pyzik et al. 2012).

Other structures that one does not normally regard as networks can be treated as such and studied by microCT. Even though they are not engineered structures, plant roots can be analyzed with approaches similar to those described above, for example, root networks in two alder specimens (Kaestner, Schneebeli et al. 2006) or in maize (Pan, Ma et al. 2018). Muscle enthuses are a second example (Sartori, Köhring et al. 2018), and networks of second phase in alloys are a third (Zhao, Du et al. 2018).

9.2 Networks of Pores

Channel structures are simply structures complementary to those of cellular solids: once again, the volume of interest occupies the smaller fraction of the total, but here the empty space (or other phase filling the channels) is of interest. Many analyses used for cellular solids can be and are used for channel structures with the solid and empty space subvolumes inverted. Although fluid transport and partition within porous systems could be covered equally appropriately in the chapter on structure evolution (Chapter 10), porosity characterization studies are included in this section.

Plants transport significant amounts of water against the force of gravity, and their systems of pores have been studied with microCT. One study examined whether conifer microarchitecture reflects strategies for avoiding water loss (Sevanto, Ryan et al. 2018). Microarchitecture has been compared to experimental measurements of water transport (Nardini, Savi et al. 2017). MicroCT has shown that storage compartments rarely refill in woody plants (Knipfer, Cuneo et al. 2017). Biofilm development also depends on the 3D pore structure (Carrel, Morales et al. 2018), and structural barriers to diffusion have been observed (Keren-Paz, Brumfeld et al. 2018). Microbial-enhanced oil recovery mechanisms of bioclogging and interfacial tension reduction have been evaluated in the context of pore morphological effects on residual oil blob mobilization (Armstrong, Wildenschild et al. 2015). Retention of soil carbon was studied as a function of pore characteristics (Kravchenko, Negassa et al. 2015).

Pore structures are very important in other bioactive, biological, or biomaterial-derived structures. Pore structure in pharmaceutical tablets was reviewed recently (Markl, Strobel et al. 2018), and the pore structure in several biochars was investigated quantitatively (Hyväluoma, Kulju et al. 2017). The structure of acorn ant nests has been studied (Varoudis, Swenson et al. 2018) as have the network of pores in polysaccharides aerogels (Ghafar,

Parikka et al. 2017) and network of channels in self-healing polymer samples (Postiglione, Alberini et al. 2017).

Understanding the role of porosity, in particular, the pore connectivity, on material transport is crucial in fields as diverse as oil production, filter operation, and composite material densification. *A priori* microstructural knowledge, for example, could aid in the design of a drilling fluid that would cause minimal formation damage from particle invasion (Jasti, Jesion et al. 1993). Precipitation of mineral phases within pore networks can change their flow characteristics (Cil, Xie et al. 2017). A review of cementitious materials concentrated on pore structures (da Silva 2018).

Low-resolution as well as high-resolution tomography systems have been applied to porous materials; the lower resolution, wider field-of-view systems can obtain representative data on larger spatial scales, something that can be very important in highly heterogeneous geological systems. The effect of resolution and thresholding method were examined recently in studies of predicting rocks' permeabilities (Latief, Fauzi et al. 2017). Fluid content and drainage were studied in a wide range of porous specimen sizes using an industrial CT system (76 mm diameter with $(368\ \mu m)^3$ voxels), a medical CT scanner (76 mm diameter with $(150\ \mu m)^2 \times 2$ mm voxels), and synchrotron microCT (27 mm diameter with $(76\ \mu m)^3$ voxels down to 1.5 mm diameter with $(6.7\ \mu m)^3$ voxels) (Wildenschild, Hopmans et al. 2002). Note that similar comparisons were performed for trabecular bone (Peyrin, Salome et al. 1998) and for bone-containing Ti implants (Bernhardt, Scharnweber et al. 2004), and this subject was addressed in more detail in Chapter 5. The industrial and medical systems detected heterogeneous drainage patterns but not the pore spaces. In coarse sand, synchrotron microCT partitioned the specimen volume into KI-doped water, air, and solid and reliably detected different interfaces; changes in fluid distribution after drainage were also imaged clearly (Wildenschild, Hopmans et al. 2002). Contrast sensitivity for water in pores (undoped and doped with tracers such as KI) was established in other works (Altman, Peplinski et al. 2005), and this subject was discussed in Chapter 5. Tortuosity calculations of percolating pore networks have been based on microCT (Shanti, Chan et al. 2014). Including partial voxels in the analysis of interconnected porosity can improve predictions of flow rates (Cid, Carrasco-Núñez et al. 2017).

Disordered packing of idealized particles (e.g., spheres, equilateral cylinders) was examined with tomographic techniques (Aste, Saadatfar et al. 2004, Zhang, Thompson et al. 2006, Vasic, Grobéty et al. 2007). Packing in various multiphase systems was also studied, and properties of pore networks (relatively large open volumes connected by much narrower throats) received attention (Willson, Stacey et al. 2004, Al-Raoush and Willson 2005, Videla, Lin et al. 2006, 2007). Interfacial areas in fluid containing porous materials and water saturation were studied (Culligan, Wildenschild et al. 2004, Brusseau, Peng et al. 2006). The influence of flow rate and porous media microstructure on macroscopic relationships such as capillary pressure vs saturation were examined effectively with synchrotron microCT (Wildenschild, Culligan et al. 2004), and the 3D rendering in Fig. 9.2 shows pendular rings of water in the network, rings that provide a significant contribution to saturation and specific interfacial area on secondary drainage. Detailed examination of aqueous and nonaqueous fluid interactions on quantities like interfacial area suggests that this quantity may be a primary determinant of nonaqueous liquid phase (e.g., oil) removal efficiency (Culligan, Wildenschild et al. 2006). Using one idealized packing system (alumina cylinders, 1 mm diameter, 3–4 mm length), pore-scale modeling and microCT visualization were used to study the spatiotemporal evolution of liquid phase clusters; behavior in untreated (hydrophilic) and silanized (neutral) packings was compared (Kohout, Grof et al. 2006). Observations of the evolution of water distribution with

FIGURE 9.2
3D rendering of packed glass beads showing the KI-containing water as the white phase and orthogonal sectioning planes with empty space in black and glass beads in dark gray. The specimen diameter is 7.0 mm, and the imaging voxel size was 17 µm. (Culligan, Wildenschild et al. 2004) © 2004 American Geophysical Union.

drying were compared with numerical simulations derived from the initial microstructure, and this sort of analysis typifies the direction many microCT investigations are taking. Numerical modeling of equilibrium distributions of water within partially saturated rock has been based on microCT-derived structures and appears to be a very valuable approach (Berkowitz and Hansen 2001).

Sheppard et al. characterized a number of materials with porosities between 18.5% and 54% using microCT and either compared these determinations to results from other techniques or used the data as input to fluid transport or other models (Sheppard, Arns et al. 2006). In addition to quantification of grain size, these authors looked at grain coordination number in different types of specimens, pore connectivity, and local flow paths. One specific approach was to view the porosity as a system of throats and pores, using the microCT data to define the characteristics of the network (throat diameter, throat length, mean throat orientation, etc.); Fig. 9.3 is a 3D rendering of the throats and pores in several very different specimens and shows them to scale and in the proper orientation. Similar analyses were performed by Knackstedt et al. on eight different industrial foams (microcellular polyurethane formed by four different processes), including thermal conductivity, permeability, and elastic properties (Young's modulus) (Knackstedt, Arns et al. 2005). The same group describes a vertically integrated center for the analysis of transport (and mechanical) properties in a wide class of solids (Sakellariou, Senden et al. 2004).

A Fontainebleau sandstone (porosity fraction = 0.197 and experimentally measured permeability K_{exp} = 1,860 mD, where 1 Darcy = 10^{-12} m^2 and the permeability relates the volumetric flow rate divided by the cross-sectional area to the pressure gradient divided by the Newtonian viscosity) was studied (Jasti, Jesion et al. 1993), and the calculated permeability for the imaged sample agreed with that measured directly in larger samples. A second study of Fontainebleau sandstone (Auzerais, Dunsmuir et al. 1996) examined 3.5 mm diameter samples, about seven grains on a side, which was close to the minimum size to estimate permeability with accuracy, and calculated permeabilities from the

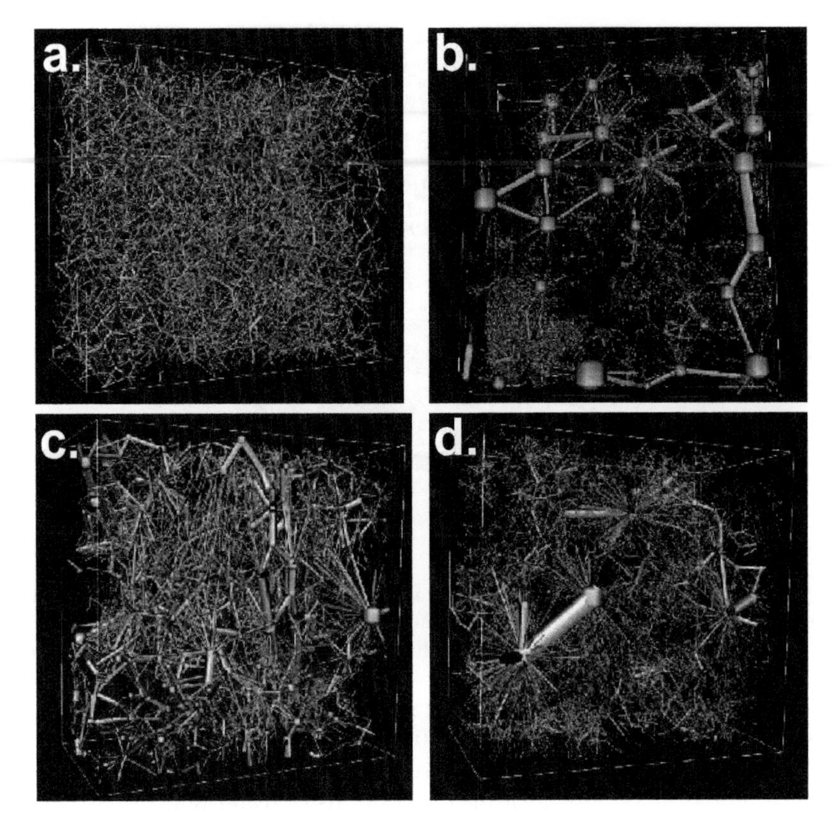

FIGURE 9.3
(a–d) 3D renderings of the throats and pores in subsets of four very different specimens. The variation in structure across the specimens is dramatic. Note that the size of the pores and throats reflects their actual size in the segmented tomographic data set (Sheppard, Arns et al. 2006).

microstructure of 1,000–1,300 mD, in good agreement with experimental values determined with several other techniques. Since Fontainebleau sandstones are reported to be remarkably homogeneous for geological materials, the agreement between these two studies was not surprising. Modeling of fluid permeability in both studies showed large variability in fluid flow depending on which subset of the reconstructed volumes was used to define the 3D channel structure. Because several different numerical methods were used, it appears unlikely that the variability was due to anything other than changes in channel architecture. This demonstrates on the one hand how powerful microCT imaging can be in determining *local* transport conditions and on the other how much care must be taken to ensure representative volumes are imaged and to avoid overinterpretation of limited data. Other microCT studies focused on the role of various local effects on fluid flow through geological materials (Herring, Andersson et al. 2016, Alhammadi, AlRatrout et al. 2017, Singh, Menke et al. 2017a, Singh, Scholl et al. 2017b).

MicroCT results from three Berea sandstones, a Texas creme chalk, and a glass bead sphere pack were reported (Jasti, Jesion et al. 1993); the air permeabilities of the geological samples ranged over nearly three orders of magnitude (from 1,000 to 16 mD), and the results confirmed that pore structures could be followed in materials representative of those found in petroleum reservoirs. When the glass bead pack was infiltrated with a mixture of oil and water, the three fluid phases were resolved. The ability to resolve oil from water or other phases is important in understanding how, during secondary recovery, some oil being swept by the driving fluid becomes immobilized by capillary forces and forms isolated, unextractable blobs.

In natural systems such as sandstone, the extent to which pores are connected dictates not only reservoir characteristics (see Bernard et al. (2000) and the material in Chapter 7) but also the rate of contaminant transport. The distribution of pore diameters and local volume fraction of pores for a Botucatu sandstone were studied with microCT (Appoloni, Fernandes et al. 2007). Many different analysis approaches were outlined for different classes of porous media in a description of a "toolbox" (Levitz 2007). Cluster labeling analysis of synchrotron microCT data of a 14% porous sandstone suggested that isolated pores comprise only 11% of the total porosity, and unresolved connections between pores contributed significantly to fluid transport (Nakashima, Nakano et al. 2004). Multiscale modeling (lattice Boltzmann method simulating single- and multicomponent fluid flow) of real microCT-derived 3D structures of sandstone revealed good agreement with permeability measurements on similar rocks (Martys, Hagedorn et al. 1999, Martys and Hagedorn 2002, Verma, Salem et al. 2007); 2D sampling of 3D volumes of small fragments of sandstone, however, did not yield the porosity and permeability of the sample but rather relative trends requiring calibration by independent data (Kameda, Dvorkin et al. 2006). MicroCT-derived oil and water distributions and pore structures (pore volume, pore surface area, throat surface area principle direction diameters for pores and throats) were used with lattice Boltzmann computations to consider the effect of water-based gels on the relative permeabilities of water and oil (Prodanović, Lindquist et al. 2007).

The presence of organic, water-immiscible liquids, often as blobs, is a key complicating factor in the remediation of hazardous waste sites. The pore-scale morphology of such liquids residing in porous media (sand) was investigated with synchrotron microCT, and the distribution of blob sizes and morphologies were quantified (Schnaar and Brusseau 2005). A follow-on study tracked blob dissolution after each of three column rinsing cycles; changes in blob number, volume, and surface area determined with microCT correlated well with effluent concentrations and validated a first-order mass transfer expression for effluent concentration (Schnaar and Brusseau 2006). Note that these studies used a contrast agent to enhance blob visibility, and a more recent report from these investigators has appeared (Brusseau, Peng et al. 2007). Digital laminography with an x-ray tube source was used to map residual oil and water saturation in porous media (Palchikov, Schemelinin et al. 2006). X-ray imaging in a lab microCT system visualized the dynamic adsorption of organic vapor and water vapor on activated carbon (Lodewyckx, Blacher et al. 2006), but the report left unclear whether microCT or simple radiography was the modality used. Other related studies include (Xu, Long et al. 2018, Zhang, Ramakrishnan et al. 2018).

Natural gas hydrate is a crystalline solid composed of water and natural gas that is stable at high pressure and low temperature conditions, and it occurs as part of a highly dynamic fluid-sediment system. Interest in natural gas hydrate stems from its potential impact on energy resources, its significance as a drilling hazard, etc. The porosity of methane hydrate host sediments was studied with microCT as a first step in determining whether the hydrate is free floating in the sediment, whether it cements and stiffen the bulk sediments, or whether it contacts but does not cement the sediment grains (Jones, Feng et al. 2007a).

Transport during filtration has aspects related to flushing of porous materials, and one microCT study determined filter structure and related it to droplet formation for different fluids (Vladisavljević, Kobayashi et al. 2007). Other filtration studies focused on efficient elimination of water from the remaining solid. Flocculation is one method of extracting suspended solids, and characteristics of the resulting aggregates affect downstream solids recovery. MicroCT was combined with numerical modeling to predict permeability in a model flocculated system (Selomulya, Jia et al. 2005, Selomulya, Tran et al. 2006). Drying of mechanically dewatered wastewater sludges and of filtered particulates in mineral processing (filter cake) are two examples where microCT has suggested how to improve energy efficiency of these processes. MicroCT identified microstructures resulting from drying sludges and their correlation with drying rates and energy consumption (Leonard, Blacher et al. 2002, 2003, 2004). Lin and Miller used numerical modeling to examine flow through well-defined simulated filter cake, and they also simulated flow based on structures directly derived from microCT (Lin and Miller 2000a, 2000b, 2004).

Polymer foams can be used to absorb oil from spills, and the ingestion of fluids into cellular materials such as polyurethane foams therefore has considerable practical importance but has received relatively little attention. Oil uptake (two densities of oil) of polyurethane foams (three densities) was examined for two temperatures (simulating winter and summer seawater temperatures) using lab microCT (Duong and Burford 2006). Weight uptake as a function of time was the main measure of polyurethane foam performance, and a few reconstructions and 3D renderings of the as-received foam were provided. Unfortunately, the foam structure was not analyzed, structure was not correlated with performance, and reconstructions of the foam with absorbed oil were not presented.

Colloid transport in porous media is often treated as a filtration problem, and Li and coworkers observed (with lab microCT) the deposition of 36-μm gold-coated microspheres in two porous media (glass beads and quartz sand, both ~780-μm mean particle size) (Li, Lin et al. 2006). In the absence of an energy barrier (native particle surfaces), the logarithm of the deposited microsphere concentration decreased linearly with increasing transport distance, a result consistent with filtration theory. In the presence of an energy barrier (treatment of the columns with polyoxyethylene laurel ether), colloid deposition was strongly influenced by local geometry (particularly grain–grain contacts) and did not vary monotonically with transport distance.

As was mentioned in Chapter 6, texture analysis has been applied to 3D microCT data sets of five porous specimens (mineral carbon forms from different geographical locations with similar topological structure that differed mainly in textural quality) (Jones, Reztsov et al. 2007b). This approach appears quite promising.

9.3 Circulatory System

Although most of the microCT studies of the circulatory system have been on mammals, there has been some work done on arthropods. Corrosion casts of the circulatory system in the shrimp-like mysidaceans were studied with microCT, and the results were used for phylogenic analyses of a complex mix of characters (Wirkner and Richter 2007). Corrosion casting and microCT were also applied to study the hemolymph vascular system in the major lineages of scorpions (Wirkner and Prendini 2007).

Blood vessel structure in organs is an important topic in medicine and biomedical engineering, and microCT-based analysis of these systems of channels is an efficient quantification method. The characterization of blood vessel infiltration into 3D-printed scaffolds (Fig. 8.6a) was as simple as answering the question of whether a blood vessel had grown into every macropore within the scaffold. Ritman and coworkers described blood vessel quantification for different organ systems including Haversian systems in bone, coronary

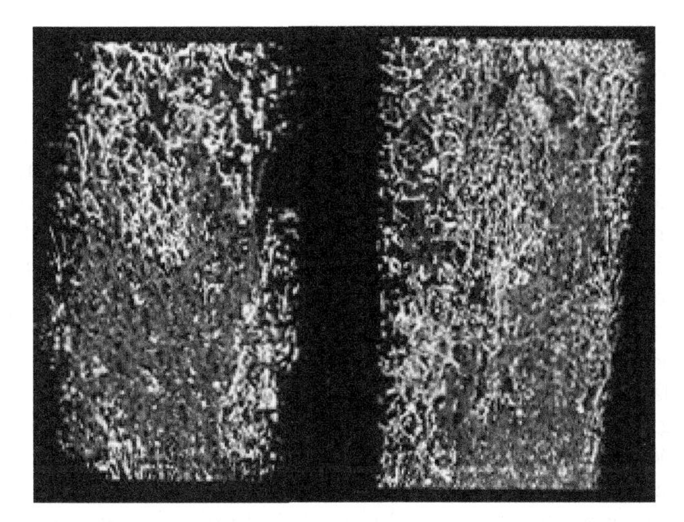

FIGURE 9.4
Haversian canals in a 3D rendering showing only the low-absorption voxels in 1 mm × 1 mm × 3 mm volume. The low absorption voxels that are interconnected are shown in the same color (Ritman, Jorgensen et al. 1997).

arteries, hind limb vascular trees, kidney glomerular microvasculature, and biliary trees (Ritman, Jorgensen et al. 1997, Jorgensen, Demirkaya et al. 1998, Kwon, Sangiori et al. 1998, Wilson, Herrmann et al. 2002, Fortepiani, Ruiz et al. 2003, Masyuk, Huang et al. 2004). The hundreds of branches present in a typical organ are usually filled with a contrast agent, but Haversian canals can be imaged directly. Figure 9.4 shows two views through a volume of bone; each interconnected Haversian canal system is a different color in this inverse 3D rendering, that is, only the low-absorption voxels are shown (Ritman, Jorgensen et al. 1997). Phase imaging can also be used to image blood vessels in soft tissue without contrast agents (Momose, Takeda et al. 2000, Hwu, Tsai et al. 2004).

Regardless of the contrast mechanism, automated analysis routines are required if an adequate number of replicates are to be analyzed, and this is particularly important in biological studies where interindividual variability is very large. In common with analyses of cellular structures, the first step in the analysis is extraction of the vessels from the image and suppression of noise. It is important to note the tree-like nature of the vessel system and branches because this differs from the situation pertaining in cellular materials and conditions analysis algorithms. After extraction of the vessel tree, quantitative data can be computed, and numerical analysis of the tree characteristics can be performed (e.g., partition into mother/child/sibling relationships for the different branches that focuses analysis of functionality and dimensional changes onto equivalent portions of the network).

Wan et al. studied the coronary arterial tree of a rat and focused quantification on arterial lumen cross-sectional area, interbranch segment length, branch surface area at equivalent generation, and interbranch and intrabranch levels (Wan, Ritman et al. 2002). In coronary circulation, arterial blood inflow increases during diastole and venous outflow increases during systole; direct comparison of microvessels in the myocardium for diastolic-arrested and systolic-arrested hearts allowed investigators to identify the characteristics of the capacitance vessels (Toyota, Fujimoto et al. 2001). In vivo microCT with retrospective gating quantified ejection fraction and cardiac output for free-breathing mice (Drangova, Ford et al. 2007).

Anomalies in patterning of coronary arteries are associated with congenital heart disease and can have a profound impact on the outcome of surgical palliation. Patterning of coronary arteries (coronary artery insertion and branching patterns of the proximal coronary stems) was compared in mice (wild-type vs a knockout strain) in order to develop a patterning defect model (Clauss, Walker et al. 2006). Right ventricular dysfunction in patients with congenital heart disease may be due different structural response of microvessels to increased pressure load in the right ventricle, and, using microCT, Ohuchi and collaborators found a significant difference in volume of porcine myocardium perfused per vessel cross-sectional area between right and left ventricle walls at 5 months of age but not at 1 month or neonatally (Ohuchi, Beighley et al. 2007).

Myocardial microvasculature is difficult to image in microCT, and visualizing vessels at the capillary level (~5-µm diameters) led to the examination of novel contrast agents, including the clinically-relevant lyophilic salts $CaSO_4$, $SrSO_4$, and $BaSO_4$ (Müller, Fischer et al. 2006). Incomplete filling remains a problem and the question of whether or not it has occurred should be considered carefully in every analysis. Early morphological changes in coronary arteries of hypercholesterolemic pigs were followed using OsO_4, a novel microCT contrast agent; specifically, coronary artery wall structure, early lesion formation, and changes in microvascular permeability (Zhu, Bentley et al. 2007a). Microvascular permeability accompanying lipopolysaccharide-induced sepsis was investigated using 70-nm particles as contrast agents in rats, and the endothelial defect size was estimated to be larger than 70 nm and smaller than 1 µm (Langheinrich and Ritman 2006). In another ex vivo microCT study, Zhu et al. found simvastin prevented microvascular remodeling and hypertrophy in swine renovascular hypertension (Zhu, Daghini et al. 2007b). Myocardial fatty acid metabolism was measured using a contrast agent and fluorescence microCT in hamsters, and the observed significant decrease in contrast agent uptake in the cardiomyopathic heart compared to the normal heart was not due to cardiomyopathy-related fibrosis but rather to abnormal fatty acid metabolism (Lwin, Takeda et al. 2007). Aortic deposits of mineral were imaged directly with microCT (Coscas, Bensussan et al. 2017).

Pulmonary arterial wall distensibility, decreases of which are important in various lung pathologies, was examined with a liquid contrast agent injected into pulmonary arteries and imaged with lab microCT at different arterial pressures spanning the physiological range (Johnson, Karau et al. 1999a, 2000); main arterial trunk diameter vs distance from the inlet was determined for each of four pressures, and a linear relationship between diameter and pressure was found for a single point on one vessel. This same group used the self-similarity of the arterial trees to improve analysis efficiency (Johnson, Karau et al. 1999b) and employed these tools to compare vascular remodeling for hypoxia-treated rats with rats living under normal oxygen partial pressures (Molthen, Karau et al. 2004). Effects of drug treatment (nintedanib) on microvessel development in a lung fibrosis model were recently investigated (Ackermann, Kim et al. 2017).

Bentley et al. reviewed the use of microCT to study alterations in renal microvasculature caused by development of cirrhosis in a rat model (Bentley, Ortiz et al. 2002); they found changes in microvascular volume fractions in different portions of the kidney that may contribute to changes in salt and water retention that accompany cirrhosis. Sled and coworkers applied semi-automatic analysis to microvasculature in mouse kidneys (simple threshold, distance transformation, special routines for local contact between vessels) (Sled, Marxen et al. 2004); Toyota et al. studied heterogeneity of glomerular volume distribution in a rat model of early diabetic nephropathy and found statistically significant (greater) coefficient of variation within individuals of the model compared to controls

(Toyota, Ogasawara et al. 2004). Recent microCT studies of renal vessels include the following (Lee, Ngo et al. 2017, Lin, Hwu et al. 2017, Markovič, Peltan et al. 2017).

Montet and coworkers investigated iodated liposomes as a contrast agent for microvessels in the murine liver, compared controls with liver-micrometastatase containing animals, and found they could detect 250-µm liver tumors after injection of iodated liposomes at 2 gI/kg body weight (Montet, Pastor et al. 2007). Op Den Buijs et al. hypothesized that hepatic vasculature could more closely approach optimal branching geometry than the vasculature of the lung or myocardium because the liver has fewer anatomical constraints in branching (Op Den Buijs, Bajzer et al. 2006). Using microCT to study rat livers, the investigators found the contrary to be true: the liver showed variation in branching morphology (e.g., branching angles oscillating between that predicted by the optimality principle of minimum power loss and volume and that of minimum shear stress and surface) that was similar to that of other organs. Recent work revisiting segmentation of microCT data on the hepatic blood vessel system has appeared (Peeters, Debbaut et al. 2017).

Quantitative microCT analysis of collateral vessel development after ischemic injury in a mouse model revealed the vascular volume was reconstituted as a series of highly-connected, small-diameter, closely-spaced and isotropically oriented vessels as soon as 3 days after surgical ligation of the femoral artery (Duvall, Taylor et al. 2004). Simple thresholding and vessel diameter measurement based on distance transformation were used.

Corrosion casting of the vasculature system in the brain (perfusion of a polymer followed by maceration of the soft tissue and decalcification of the bone) and a two-resolution approach (lab microCT, 16-µm voxels, to identify volumes of interest for subsequent local tomography reconstruction with synchrotron microCT, 1.4-µm voxels) have been used to compare an Alzheimer's disease model mouse with the wild type using metrics described above (Heinzer, Krucker et al. 2004). Figure 9.5 reproduces a comparison of matching SEM and microCT views of the network of vessels, which highlights several very minor artifacts in the rendering of the microCT data (Heinzer, Krucker et al. 2006). Imaging of primate brains allowed analysis of vascular homogeneity and anisotropy (Kennel, Fonta et al. 2017). MicroCT has also provided important information on fluid flow related to cerebral aneurisms (Levitt, Barbour et al. 2017).

The mammalian placenta is an interesting complex of two interacting but separate sets of blood vessels. Several recent microCT studies have examined these structures (Junaid, Bradley et al. 2017, Pratt, Hutchinson et al. 2017, Rennie, Cahill et al. 2017).

The nutrient canal system (blood vessels) within cortical bone has received extensive attention, but the focus of microCT studies has been on the effect on mechanical properties and on bone remodeling. Figure 9.4, described earlier in this section, shows interconnected Haversian systems in cortical bone, and these systems can be quite complex (Ritman, Jorgensen et al. 1997, Cooper, Turinsky et al. 2007). Further discussion of nutrient canal networks in bone is postponed to Chapter 10.

9.4 Respiratory System

Corrosion casts of a canine lung and a mouse lung were imaged with conventional CT and microCT, respectively, and, due to effects like bubbles in the polymer, a small amount of manual segmentation was required (Chaturvedi and Lee 2005). Analysis of the airway

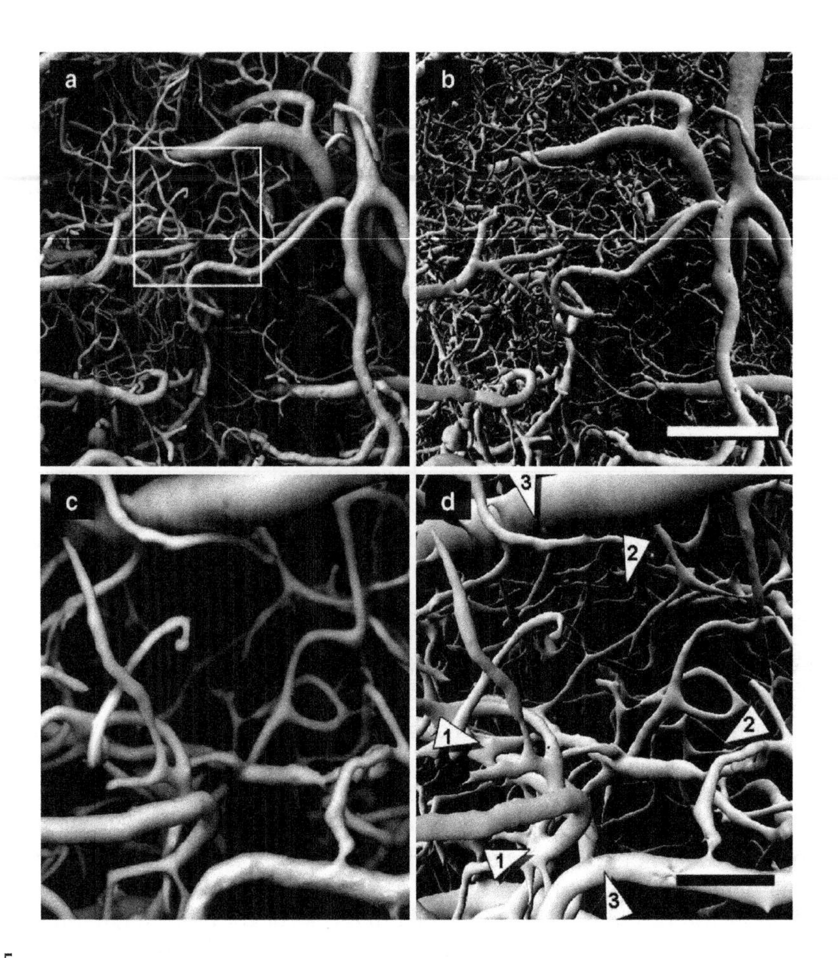

FIGURE 9.5
Validation of microCT rendering of murine brain vascular corrosion castings by comparison to SEM images. (a) SEM image of vasculature at the cortical surface. (b) Surface-rendered microCT data of the same region and with the same viewing perspective as in (a). The scale bar equals 200 μm. (c) SEM detail of the subregion outlined in (a). (d) MicroCT rendering of the same subregion as in (c). At this magnification, three types of minor imaging artifacts become visible (arrows): vessels in close proximity fuse ("1"), some capillaries are too thin or even disappear ("2") and indentations or contusions appear in the walls of the larger vessels ("3") (Heinzer, Krucker et al. 2006).

tree structure in terms of generations, etc., was similar to that reported by others (Wan, Ritman et al. 2002); interestingly, the approach was accurate and efficient for up to six generations for the canine cast and ten generations for the murine cast, presumably because of instrumentation differences. Respiratory systems of insects were observed directly in synchrotron microCT (Ha, Ryu et al. 2017, Wasserthal, Cloetens et al. 2018).

Resolution-related errors in high-resolution (clinical) CT imaging of inflation-fixed porcine lung cubes were evaluated using microCT, and systematic, size-dependent underestimation of lumen area and overestimation of wall area were found for the clinical system (Dame Carroll, Chandra et al. 2006). Respiratory gated microCT of anesthetized free-breathing mice was used to quantify lung tidal volume and functional residual capacity (Ford, Martin et al. 2007). Several versions of respiratory gating were compared with computer-controlled intermittent iso-pressure breath hold technique in in vivo microCT mouse studies, and Namati et al. reported that the last technique yielded the most reproducible lung volume

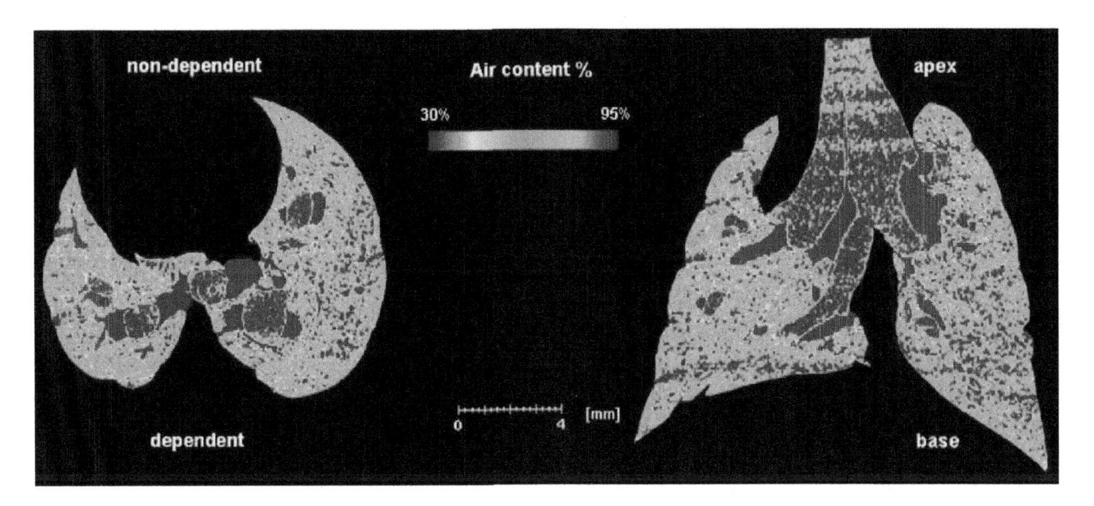

FIGURE 9.6
Axial (left) and coronal (right) 2D sections of a typical mouse lung showing regional air content differences represented by the colors shown in the color bar (Namati, Chon et al. 2006).

and air content measurements (Namati, Chon et al. 2006). Figure 9.6 from Namati et al. shows axial and coronal sections showing air content in a typical mouse lung.

Respiratory patterns have been compared between free-breathing and mechanically ventilated rats (Ford, McCaig et al. 2017). In vivo quantification of regional lung gas volumes in rabbits was reported using synchrotron microCT, mechanical ventilation with Xe-O gas, and K-edge subtraction imaging (Bayat, Le Duc et al. 2001, Monfrais, Bayat et al. 2005). Bayat et al. used Xe-enhanced synchrotron microCT imaging to examine in vivo response to histamine administration in healthy anesthetized and mechanically ventilated rabbits; proximal airway cross-sectional area decreased by 57% by 20 min and recovered gradually but incompletely within 60 min, whereas ventilated alveolar area decreased by 55% immediately after histamine inhalation and recovered rapidly thereafter (Bayat, Porra et al. 2006). Hyperresponsiveness in allergically inflamed mouse lungs was the subject of another in vivo microCT study (Lundblad, Thompson-Figueroa et al. 2007). MicroCT compared in vivo pulmonary compliance of healthy mice with a common pulmonary fibrosis model, damage induced by bleomycin exposure (Cavanaugh, Travis et al. 2006, Shofer, Badea et al. 2007). In vivo imaging with respiratory gating of laboratory rodents was also used to examine lung damage from tumors and from chemotherapy (Cody, Cavanaugh et al. 2004). Preterminal bronchioles in centrilobular and panlobular emphysema were compared in another study (Tanabe, Vasilescu et al. 2017). Lung exposure to metal nanoparticles was examined in one study (Chaurand, Liu et al. 2018), and maternal exposure to fine particulate air pollution–induced lung injury in rat offspring was assessed in another (Tang, Huang et al. 2018).

9.5 Networks of Nerves

Imaging of neural tissue (the brain, spinal cord, networks of nerves, sensory organs) presents a set of challenges different from imaging the respiratory or circulatory systems. In general, phase contrast or staining is required to bring out contrast from neural tissue, and this tissue is much more susceptible to fixation-induced distortions. The 3D strain fields resulting from fixation have been quantified (Schulz, Crooijmans et al. 2011). Neuron networks have been segmented in the mouse brain (Mizutani, Saiga et al. 2016, Fonseca, Araujo et al. 2018). In samples of the human brain, a phase contrast imaging study automatically identified 5,000 Purkinje cells and determined these cells' local surface density averaged 165 cells/mm^2; the 3D data also allowed segmentation of subcellular structures, including dendritic tree and Purkinje cell nucleoli without staining (Hieber, Bikis et al. 2016). Neurite geometry of patients diagnosed with schizophrenia was compared with that from human controls (Mizutani, Saiga et al. 2019).

Some imaging studies have appeared on neural tissue outside the brain. One example is the investigation of vessels and nerve fiber in the rat spinal column (Hu, Li et al. 2017). Nerve tissue in cochleae has been the subject of a series of studies (Moreno, Rajguru et al. 2011, Richter, Young et al. 2018, Tan, Jahan et al. 2018).

References

Ackermann, M., Y. O. Kim, W. L. Wagner, D. Schuppan, C. D. Valenzuela, S. J. Mentzer, S. Kreuz, D. Stiller, L. Wollin and M. A. Konerding (2017). "Effects of nintedanib on the microvascular architecture in a lung fibrosis model." *Angiogenesis* **20**: 359–372.

Alhammadi, A. M., A. AlRatrout, K. Singh, B. Bijeljic and M. J. Blunt (2017). "In situ characterization of mixed-wettability in a reservoir rock at subsurface conditions." *Sci Rep* **7**: 10753.

Al-Raoush, R. I. and C. S. Willson (2005). "Extraction of physically realistic pore network properties from three-dimensional synchrotron x-ray microtomography images of unconsolidated porous media systems." *J Hydrol* **300**: 44–64.

Altman, S. J., W. J. Peplinski and M. L. Rivers (2005). "Evaluation of synchrotron x-ray computerized microtomography for the visualization of transport processes in low porosity materials." *J Contam Hydrol* **78**: 167–183.

Antoine, C., P. Nygard, O. W. Gregersen, R. Holmstad, T. Weitkamp and C. Rau (2002). "3D images of paper obtained by phase-contrast x-ray microtomography: Image quality and binarisation." *Nucl Instrum Meth A* **490**: 392–402.

Appoloni, C. R., C. P. Fernandes and C. R. O. Rodrigues (2007). "X-ray microtomography study of a sandstone reservoir rock." *Nucl Instrum Meth A* **580**: 629–632.

Armstrong, R. T., D. Wildenschild and B. K. Bay (2015). "The effect of pore morphology on microbial enhanced oil recovery." *J Petrol Sci Eng* **130**: 16–25.

Aste, T., M. Saadatfar, A. Sakellariou and T. Senden (2004). "Investigating the geometrical structure of disordered sphere packings." *Physica A* **339**: 16–23.

Auzerais, F. M., J. Dunsmuir, B. B. Ferreol, N. Martys, J. Olson, T. S. Ramakrishnan, D. H. Rothman and L. M. Schwartz (1996). "Transport in sandstones: A study based on three dimensional microtomography." *Geophys Res Lett* **23**: 705–708.

Bayat, S., G. Le Duc, L. Porra, G. Berruyer, C. Nemoz, S. Monfraix, S. Fiedler, W. Thomlinson, P. Suortti, C. G. Standertskjold-Nordenstam and A. R. A. Sovijarvi (2001). "Quantitative functional lung imaging with synchrotron radiation using inhaled xenon as contrast agent." *Phys Med Biol* **46**: 3287–3299.

Bayat, S., L. Porra, H. Suhonen, C. Nemoz, P. Suortti and A. R. A. Sovijärvi (2006). "Differences in the time course of proximal and distal airway response to inhaled histamine studied by synchrotron radiation CT." *J Appl Physiol* **100**: 1964–1973.

Bentley, M. D., M. C. Ortiz, E. L. Ritman and J. C. Romero (2002). "The use of microcomputed tomography to study microvasculature in small rodents." *Am J Physiol Reg Integ Comp Physiol* **282**: R1267–R1279.

Berkowitz, B. and D. P. Hansen (2001). "A numerical study of the distribution of water in partially saturated porous rock." *Transport Porous Media* **45**: 303–319.

Bernard, D., G. L. Vignoles and J. M. Heintz (2000). Modelling porous materials evolution. *X-Ray Tomography in Materials Science*. J. Baruchel, J. Y. Buffière, E. Maire, P. Merle and G. Peix. Paris, Hermes Science: 177–192.

Bernhardt, R., D. Scharnweber, B. Müller, P. Thurnier, H. Schliephake, P. Wyss, F. Beckmann, J. Goebbels and H. Worch (2004). "Comparison of microfocus and synchrotron x-ray tomography for the analysis of osteointegration around Ti6AlV4 implants." *Euro Cells Mater* **7**: 42–51.

Brusseau, M. L., S. Peng, G. Schnaar and M. S. Costanza-Robinson (2006). "Relationships among air-water interfacial area, capillary pressure, and water saturation for a sandy porous medium." *Water Resour Res* **42**: W03501.

Brusseau, M. L., S. Peng, G. Schnaar and A. Murao (2007). "Measuring air-water interfacial areas with X-ray microtomography and interfacial partitioning tracer tests." *Env Sci Technol* **41**: 1956–1961.

Carrel, M., V. L. Morales, M. A. Beltran, N. Derlon, R. Kaufmann, E. Morgenroth and M. Holzner (2018). "Biofilms in 3D porous media: delineating the influence of the pore network geometry, flow and mass transfer on biofilm development." *Water Res* **134**: 280–291.

Cavanaugh, D., E. L. Travis, R. E. Price, G. Gladish, R. A. White, M. Wang and D. D. Cody (2006). "Quantification of bleomycin-induced murine lung damage in vivo with micro-computed tomography." *Acad Radiol* **13**: 1505–1512.

Chapelle, L., A. Lyckegaard, Y. Kusano, C. Gundlach, M. Rosendahl Foldschack, D. Lybye and P. Brøndsted (2018). "Determination of the fibre orientation distribution of a mineral wool network and prediction of its transverse stiffness using X-ray tomography" *J Mater Sci* **53**: 6390–6402.

Chaturvedi, A. and A. Lee (2005). "Three-dimensional segmentation and skeletonization to build an airway tree data structure for small animals." *Phys Med Biol* **50**: 1405–1419.

Chaurand, P., W. Liu, D. Borschneck, C. Levard, M. Auffan, E. Paul, B. Collin, I. Kieffer, S. Lanone, J. Rose and J. Perrin (2018). "Multi-scale X-ray computed tomography to detect and localize metal-based nanomaterials in lung tissues of in vivo exposed mice." *Sci Rep* **8**: 4408.

Cid, H. E., G. Carrasco-Núñez and V. C. Manea (2017). "Improved method for effective rock microporosity estimation using X-ray microtomography." *Micron* **97**: 11–21.

Cil, M. B., M. Xie, A. I. Packman and G. Buscarnera (2017). "Solute mixing regulates heterogeneity of mineral precipitation in porous media." *Geophys Res Lett* **44**: 6658–6666.

Clauss, S. B., D. L. Walker, M. L. Kirby, D. Schimel and C. W. Lo (2006). "Patterning of coronary arteries in wildtype and connexin43 knockout mice." *Dev Dynamics* **235**: 2786–2794.

Clyne, T. W., A. E. Markaki and J. C. Tan (2005). "Mechanical and magnetic properties of metal fibre networks with and without a polymeric matrix." *Compos Sci Technol* **65**: 2492–2499.

Cody, D., D. Cavanaugh, R. E. Price, B. Rivera, G. Gladish and E. Travis (2004). Lung imaging of laboratory rodents in vivo. *Developments in X-Ray Tomography IV*. U. Bonse. Bellingham (WA), SPIE. **SPIE Proc Vol 5535**: 43–52.

Cooper, D., A. Turinsky, C. Sensen and B. Hallgrimsson (2007). "Effect of voxel size on 3D microCT analysis of cortical bone porosity." *Calcif Tiss Int* **80**: 211–219.

Coscas, R., M. Bensussan, M. P. Jacob, L. Louedec, Z. Massy, J. Sadoine, M. Daudon, C. Chaussain, D. Bazin and J. B. Michel (2017). "Free DNA precipitates calcium phosphate apatite crystals in the arterial wall in vivo." *Atherosclerosis* **259**: 60–67.

Culligan, K. A., D. Wildenschild, B. S. B. Christensen, W. G. Gray and M. L. Rivers (2006). "Pore-scale characteristics of multiphase flow in porous media: A comparison of air–water and oil–water experiments." *Adv Water Res* **29**: 227–238.

Culligan, K. A., D. Wildenschild, B. S. B. Christensen, W. G. Gray, M. L. Rivers and A. F. B. Tompson (2004). "Interfacial area measurements for unsaturated flow through a porous medium." *Water Resour Res* **40**: W12413/12411–W12413/12412.

Dame Carroll, J. R., A. Chandra, A. S. Jones, N. Berend, J. S. Magnussen and G. G. King (2006). "Airway dimensions measured from micro-computed tomography and high-resolution computed tomography." *Eur Respir J* **28**: 712–720.

Desplentere, F., S. V. Lomov, D. L. Woerdeman, I. Verpoest, M. Wevers and A. Bogdanovich (2005). "microCT characterization of variability in 3D textile architecture." *Compos Sci Technol* **65**: 1920–1930.

Drangova, M., N. L. Ford, S. A. Detombe, A. R. Wheatley and D. W. Holdsworth (2007). "Fast retrospectively gated quantitative four-dimensional (4D) cardiac micro computed tomography imaging of free-breathing mice." *Invest Radiol* **42**: 85–94.

Duong, H. T. T. and R. P. Burford (2006). "Effect of foam density, oil viscosity and temperature on oil sorption behavior of polyurethane." *J Appl Polymer Sci* **99**: 360–367.

Duvall, C. L., W. R. Taylor, D. Weiss and R. E. Guldberg (2004). "Quantitative microcomputed tomography analysis of collateral vessel development after ischemic injury." *Am J Physiol Heart Circ Physiol* **287**: H302–H310.

Eberhardt, C. N. and A. R. Clarke (2002). "Automated reconstruction of curvilinear fibres from 3D datasets acquired by x-ray microtomography." *J Microsc* **206**: 41–53.

Faessel, M., C. Delisée, F. Bos and P. Castéra (2005). "3D modeling of random cellulosic fibrous networks based on x-ray tomography and image analysis." *Compos Sci Technol* **65**: 1931–1940.

Fonseca, M. C., B. H. S. Araujo, C. S. B. Dias, N. L. Archilha, D. P. A. Neto, E. Cavalheiro, H. J. Westfahl, A. J. R. da Silva and K. G. Franchini (2018). "High-resolution synchrotron-based X-ray microtomography as a tool to unveil the three-dimensional neuronal architecture of the brain." *Sci Rep* **8**: 12074.

Ford, N. L., E. L. Martin, J. F. Lewis, R. A. W. Veldhuizen, M. Drangova and D. W. Holdsworth (2007). "In vivo characterization of lung morphology and function in anesthetized free-breathing mice using micro-computed tomography." *J Appl Physiol* **102**: 2046–2055.

Ford, N. L., L. McCaig, A. Jeklin, J. F. Lewis, R. A. Veldhuizen, D. W. Holdsworth and M. Drangova (2017). "A respiratory-gated microCT comparison of respiratory patterns in free-breathing and mechanically ventilated rats." *Physiol Rep* **5**: 13074.

Fortepiani, L. A., M. C. O. Ruiz, F. Passardi, M. D. Bentley, J. Garcia-Estan, E. L. Ritman and J. C. Romero (2003). "Effect of losartan on renal microvasculature during chronic inhibition of nitric oxide visualized by microCT." *Am J Physiol Renal Physiol* **285**: F852–F860.

Ghafar, A., K. Parikka, D. Haberthür, M. Tenkanen, K. S. Mikkonen and J. P. Suuronen (2017). "Synchrotron reveals the fine three-dimensional porosity of composite polysaccharide aerogels." *Mater (Basel)* **10**: 871.

Ha, Y. R., J. Ryu, E. Yeom and S. J. Lee (2017). "Comparison of the tracheal systems of *Anopheles sinensis* and *Aedes togoi* larvae using synchrotron X-ray microscopic computed tomography (respiratory system of mosquito larvae using SR-μCT)." *Microsc Res Tech* **80**: 985–993.

Heinzer, S., T. Krucker, M. Stampanoni, R. Abela, E. P. Meyer, A. Schuler, P. Schneider and R. Müller (2004). Hierarchical bioimaging and quantification of vasculature in disease models using corrosion casts and microcomputed tomography. *Developments in X-Ray Tomography IV.* U. Bonse. Bellingham (WA), SPIE. **SPIE Proc Vol 5535**: 65–76.

Heinzer, S., T. Krucker, M. Stampanoni, R. Abela, E. P. Meyer, A. Schuler, P. Schneider and R. Müller (2006). "Hierarchical microimaging for multiscale analysis of large vascular networks." *Neuroimage* **32**: 626–636.

Herring, A. L., L. Andersson and D. Wildenschild (2016). "Enhancing residual trapping of supercritical CO2 via cyclic injections." *Geophys Res Lett* **43**: 9677–9685.

Hieber, S. E., C. Bikis, A. Khimchenko, G. Schweighauser, J. Hench, N. Chicherova, G. Schulz and B. Müller (2016). "Tomographic brain imaging with nucleolar detail and automatic cell counting." *Sci Rep* **6**: 32156.

Hsiung, C.-H. H., A. J. Pyzik, F. De Carlo, X. Xiao, S. R. Stock and K. T. Faber (2012). "Microstructure and mechanical properties of acicular mullite." *J Euro Cer Soc* **33**: 503–513.

Hu, J., P. Li, X. Yin, T. Wu, Y. Cao, Z. Yang, L. Jiang, S. Hu and H. Lu (2017). "Nondestructive imaging of the internal microstructure of vessels and nerve fibers in rat spinal cord using phase-contrast synchrotron radiation microtomography." *J Synchrotron Radiat* **24**: 482–489.

Hwu, Y., W. L. Tsai, J. H. Je, S. K. Seol, B. Kim, A. Groso, G. Margaritondo, K. H. Lee and J. K. Seong (2004). "Synchrotron microangiography with no contrast agent." *Phys Med Biol* **49**: 501–508.

Hyväluoma, J., S. Kulju, M. Hannula, H. Wikberg, A. Källi and K. Rasa (2017). "Quantitative characterization of pore structure of several biochars with 3D imaging." *Environ Sci Pollut Res Int* (2017 Mar 24. doi: 10.1007/s11356-017-8823-x. [Epub ahead of print] PubMed PMID: 28342082) **25**: 25648–25658.

Jasti, J. K., G. J. Jesion and L. Feldkamp (1993). "Microscopic imaging of porous media with x-ray computer tomography." *SPE Formation Eval* **8**: 189–193.

Johnson, R. H., K. L. Karau, R. C. Molthen and C. A. Dawson (1999a). Exploiting self-similarity of arterial trees to reduce the complexity of image analysis. *Medical Imaging 2000: Physiology and Function from Multidimensional Images*. C. T. Chen and A. V. Clough. Bellingham (WA), SPIE. **SPIE Proc Vol 3660**: 351–361.

Johnson, R. H., K. L. Karau, R. C. Molthen and C. A. Dawson (1999b). Quantification of pulmonary arterial wall distensibility using parameters extracted from volumetric microCT images. *Developments in X-Ray Tomography II*. U. Bonse. Bellingham (WA), SPIE. **SPIE Proc Vol 3772**: 15–23.

Johnson, R. H., K. L. Karau, R. C. Molthen, S. T. Haworth and C. A. Dawson (2000). MicroCT image-derived metrics quantify arterial wall distensibility reduction in a rat model of pulmonary hypertension. *Medical Imaging 2000: Physiology and Function from Multidimensional Images*. C. T. Chen and A. V. Clough. Bellingham (WA), SPIE. **SPIE Proc Vol 3978**: 320–330.

Jones, K. W., H. Feng, S. Tomov, W. J. Winters, M. Prodanović and D. Mahajan (2007a). "Characterization of methane hydrate host sediments using synchrotron-computed microtomography (CMT)." *J Petrol Sci Eng* **56**: 136–145.

Jones, A. S., A. Reztsov and C. E. Loo (2007b). "Application of invariant grey scale features for analysis of porous minerals." *Micron* **38**: 40–48.

Jorgensen, S. M., O. Demirkaya and E. L. Ritman (1998). "Three-dimensional imaging of vasculature and parenchyma in intact rodent organs with x-ray microCT." *Am J Physiol 275 (Heart Circ Physiol 44)* **275**: H1103–H1114.

Junaid, T. O., R. S. Bradley, R. M. Lewis, J. D. Aplin and E. D. Johnstone (2017). "Whole organ vascular casting and microCT examination of the human placental vascular tree reveals novel alterations associated with pregnancy disease." *Sci Rep* **7**: 4144.

Kaestner, A., M. Schneebeli and F. Graf (2006). "Visualizing three-dimensional root networks using computed tomography." *Geoderma* **136**: 459–469.

Kameda, A., J. Dvorkin, Y. Keehm, A. Nur and W. Bosl (2006). "Permeability-porosity transforms from small sandstone fragments." *Geophysics* **71**: N11–N19.

Kennel, P., C. Fonta, R. Guibert and F. Plouraboué (2017). "Analysis of vascular homogeneity and anisotropy on high-resolution primate brain imaging." *Hum Brain Mapp* **38**: 5756–5777.

Keren-Paz, A., V. Brumfeld, Y. Oppenheimer-Shaanan and I. Kolodkin-Gal (2018). "MicroCT X-ray imaging exposes structured diffusion barriers within biofilms." *NPJ Biofilms Microbiomes* **4**: 8.

Knackstedt, M., C. Arns, M. Saadatfar, T. Senden, A. Sakellariou, A. Sheppard, R. Sok, W. Schrof and H. Steininger (2005). "Virtual materials design: Properties of cellular solids derived from 3D tomographic images." *Adv Eng Mater* **7**: 238–243.

Knipfer, T., I. F. Cuneo, J. M. Earles, C. Reyes, C. R. Brodersen and A. J. McElrone (2017). "Storage compartments for capillary water rarely refill in an intact woody plant." *Plant Physiol* **175**: 1649–1660.

Kohout, M., Z. Grof and F. Stepanek (2006). "Pore-scale modelling and tomographic visualisation of drying in granular media." *J Colloid Interface Sci* **299**: 342–351.

Kok, M. D. R., R. Jervis, D. Brett, P. R. Shearing and J. T. Gostick (2018). "Insights into the effect of structural heterogeneity in carbonized electrospun fibrous mats for flow battery electrodes by X-ray tomography." *Small* **14**: 1703616.

Krasnoshlyk, V., S. R. du Roscoat, P. J. J. Dumont, P. Isaksson, E. Ando and A. Bonnin (2018). "Three-dimensional visualization and quantification of the fracture mechanisms in sparse fibre networks using multiscale X-ray microtomography." *Proc Roy Soc A Math Phys Eng Sci* **474**: 20180175.

Kravchenko, A. N., W. C. Negassa, A. K. Gruber and M. L. Rivers (2015). "Protection of soil carbon within macro-aggregates depends on intra-aggregate pore characteristics." *Sci Rep* **5**: 16261.

Kwon, H. M., G. Sangiori, E. L. Ritman, A. Lerman, C. McKenna, R. Virmani, W. D. Edwards, D. R. Holmes and R. S. Schwartz (1998). "Adventitial vasa vasorum in balloon-injured coronary arteries – visualization and quantitation by microscopic three-dimensional computed tomography technique." *J Am Coll Cardiol* **32**: 2072–2079.

Langheinrich, A. C. and E. L. Ritman (2006). "Quantitative imaging of microvascular permeability in a rat model of lipopolysaccharide-induced sepsis: Evaluation using cryostatic microcomputed tomography." *Invest Radiol* **41**: 645–650.

Latief, F. D., U. Fauzi, Z. Irayani and G. Dougherty (2017). "The effect of X-ray micro computed tomography image resolution on flow properties of porous rocks." *J Microsc* **266**: 69–88.

Lee, C. J., J. P. Ngo, S. Kar, B. S. Gardiner, R. G. Evans and D. W. Smith (2017). "A pseudo-three-dimensional model for quantification of oxygen diffusion from preglomerular arteries to renal tissue and renal venous blood." *Am J Physiol Renal Physiol* **313**: F237–F253.

Leonard, A., S. Blacher, P. Marchot and M. Crine (2002). "Use of x-ray microtomography to follow the convective heat drying of wastewater sludges." *Drying Technol* **20**: 1053–1069.

Leonard, A., S. Blacher, P. Marchot, J. Pirard and M. Crine (2003). "Image analysis of x-ray tomograms of soft materials during convective drying." *J Microsc* **212**: 197–204.

Leonard, A., S. Blacher, P. Marchot, J. Pirard and M. Crine (2004). "Measurement of shrinkage and cracks associated to convective drying of soft materials by x-ray microtomography." *Drying Technol* **22**: 1695–1708.

Levitt, M. R., M. C. Barbour, S. Rolland du Roscoat, C. Geindreau, V. K. Chivukula, P. M. McGah, J. D. Nerva, R. P. Morton, L. J. Kim and A. Aliseda (2017). "Computational fluid dynamics of cerebral aneurysm coiling using high-resolution and high-energy synchrotron X-ray microtomography: Comparison with the homogeneous porous medium approach." *J Neurointerv Surg* **9**: 777–782.

Levitz, P. (2007). "Toolbox for 3D imaging and modeling of porous media: Relationship with transport properties." *Cement Concr Res* **37**: 351–359.

Li, X., C. L. Lin, I. D. Miller and W. P. Johnson (2006). "Pore-scale observation of microsphere deposition at grain-to-grain contacts over assemblage-scale porous media domains using x-ray microtomography." *Env Sci Technol* **40**: 3762–3768.

Lin, Y. C., Y. Hwu, G. S. Huang, M. Hsiao, T. T. Lee, S. M. Yang, T. K. Lee, N. Y. Chen, S. S. Yang, A. Chen and S. M. Ka (2017). "Differential synchrotron X-ray imaging markers based on the renal microvasculature for tubulointerstitial lesions and glomerulopathy." *Sci Rep* **7**: 3488.

Lin, C. L. and J. D. Miller (2000a). "Network analysis of filter cake pore structure by high resolution x-ray microtomography." *Chem Eng J* **77**: 79–86.

Lin, C. L. and J. D. Miller (2000b). "Pore structure and network analysis of filter cake." *Chem Eng J* **80**: 221–231.

Lin, C. L. and J. D. Miller (2004). "Pore structure analysis of particle beds for fluid transport simulation during filtration." *Int J Miner Process* **73**: 281–294.

Lodewyckx, P., S. Blacher and A. Leonard (2006). "Use of x-ray microtomography to visualise dynamic adsorption of organic vapour and water vapour on activated carbon." *Adsorption* **12**: 19–26.

Lundblad, L. K. A., J. Thompson-Figueroa, G. B. Allen, L. Rinaldi, R. J. Norton, C. G. Irvin and J. H. T. Bates (2007). "Airway hyperresponsiveness in allergically inflamed mice: The role of airway closure." *Am J Respir Crit Care Med* **175**: 768–774.

Lux, J., C. Delisée and X. Thibault (2006). "3D characterization of wood based fibrous materials: An application." *Image Anal Stereol* **25**: 25–35.

Lwin, T. T., T. Takeda, J. Win, N. Sunaguchi, T. Murakami, S. Mouri, S. Nasukawa, Q. Huo, T. Yuasa, K. Hyodo and T. Akatsuka (2007). "Preliminary quantitative analysis of myocardial fatty acid metabolism from fluorescent x-ray computed tomography imaging." *J Synchrotron Rad* **14**: 158–162.

Markl, D., A. Strobel, R. Schlossnikl, J. Bøtker, P. Bawuah, C. Ridgway, J. Rantanen, T. Rades, P. Gane, K. E. Peiponen and J. A. Zeitler (2018). "Characterisation of pore structures of pharmaceutical tablets: A review." *Int J Pharm* **538**: 188–214.

Markovič, R., J. Peltan, M. Gosak, D. Horvat, B. Žalik, B. Seguy, R. Chauvel, G. Malandain, T. Couffinhal, C. Duplàa, M. Marhl and E. Roux (2017). "Planar cell polarity genes frizzled4 and frizzled6 exert patterning influence on arterial vessel morphogenesis." *PLoS One* **12**: 0171033.
Martys, N. S. and J. G. Hagedorn (2002). "Multiscale modeling of fluid transport in heterogeneous materials using discrete Boltzmann methods." *Mater Struct* **35**: 650–658.
Martys, N. S., J. G. Hagedorn, D. Goujon and J. E. Devaney (1999). Large scale simulations of single and multi-component flow in porous media. *Developments in X-Ray Tomography II*. U. Bonse. Bellingham (WA), SPIE. **SPIE Proc Vol 3772**: 205–213.
Masyuk, T. V., B. G. Huang, A. I. Masyuk, E. L. Ritman, V. E. Torres, X. Wang, P. C. Harris and N. F. LaRusso (2004). "Biliary dysgenesis in the PCK rat, an orthologous model of autosomal recessive polycystic kidney disease." *Am J Pathol* **165**: 1719–1730.
Mizutani, R., R. Saiga, M. Ohtsuka, H. Miura, M. Hoshino, A. Takeuchi and K. Uesugi (2016). "Three-dimensional X-ray visualization of axonal tracts in mouse brain hemisphere." *Sci Rep* **6**: 35061.
Mizutani, R., R. Saiga, A. Takeuchi, K. Uesugi, Y. Terada, Y. Suzuki, V. De Andrade, F. De Carlo, S. Takekoshi, C. Inomoto, N. Nakamura, I. Kushima, S. Iritani, N. Ozaki, S. Ide, K. Ikeda, K. Oshima, M. Itokawa and M. Arai (2019). "Three-dimensional alteration of neurites in schizophrenia." *Translational psychiatry* **9**: 85.
Molthen, R. C., K. L. Karau and C. A. Dawson (2004). "Quantitative models of the rat pulmonary arterial tree morphometry applied to hypoxia-induced arterial remodeling." *J Appl Physiol* **97**: 2372–2384.
Momose, A., T. Takeda and Y. Itai (2000). "Blood vessels: depiction at phase contrast x-ray imaging without contrast agents in the mouse and rat – feasibility study." *Radiology* **217**: 593–596.
Monfrais, S., S. Bayat, L. Porra, G. Berruyer, C. Nemoz, W. Tomlinson, P. Suortti and A. R. A. Sovijärvi (2005). "Quantitative measurement of regional lung gas volume by synchrotron radiation computed tomography." *Phys Med Biol* **50**: 1–11.
Montet, X., C. M. Pastor, J. P. Vallée, C. D. Becker, A. Geissbuhler, D. R. Morel and P. Meda (2007). "Improved visualization of vessels and hepatic tumors by micro-computed tomography (CT) using iodinated liposomes." *Invest Radiol* **42**: 652–658.
Moreno, L. E., S. M. Rajguru, A. I. Matic, N. Yerram, A. M. Robinson, M. Hwang, S. Stock and C.-P. Richter (2011). "Infrared neural stimulation: beam path in the guinea pig cochlea." *Hear Res* **282**: 289–302.
Müller, B., J. Fischer, U. Dietz, P. J. Thurner and F. Beckmann (2006). "Blood vessel staining in the myocardium for 3D visualization down to the smallest capillaries." *Nucl Instrum Meth B* **246**: 254–261.
Nakashima, Y., T. Nakano, K. Nakamura, K. Uesugi, A. Tsuchiyama and S. Ikeda (2004). "Three-dimensional diffusion of non-sorbing species in porous sandstone: computer simulation based on x-ray microtomography using synchrotron radiation." *J Contam Hydrol* **74**: 253–264.
Namati, E., D. Chon, J. Thiesse, E. A. Hoffman, J. de Ryk, A. Ross and G. McLennan (2006). "In vivo microCT lung imaging via a computer-controlled intermittent iso-pressure breath hold (IIBH) technique." *Phys Med Biol* **51**: 6061–6075.
Nardini, A., T. Savi, A. Losso, G. Petit, S. Pacilè, G. Tromba, S. Mayr, P. Trifilò, M. A. Lo Gullo and S. Salleo (2017). "X-ray microtomography observations of xylem embolism in stems of *Laurus nobilis* are consistent with hydraulic measurements of percentage loss of conductance." *New Phytol* **213**: 1068–1075.
Ohuchi, H., P. E. Beighley, Y. Dong, M. Zamir and E. L. Ritman (2007). "Microvascular development in porcine right and left ventricular walls." *Ped Res* **61**: 676–680.
Op Den Buijs, J., Ž. Bajzer and E. L. Ritman (2006). "Branching morphology of the rat hepatic portal vein tree: A microCT study." *Annal Biomed Eng* **34**: 1420–1428.
Palchikov, E. I., Y. A. Schemelinin, A. G. Skripkin, D. Y. Mekhontsev and N. A. Kondratiev (2006). "Mapping of residual oil and water saturation in porous media by means of digital x-ray laminography." *Part Part Syst Charact* **23**: 254–259.
Pan, X., L. Ma, Y. Zhang, J. Wang, J. Du and X. Guo (2018). "Three-dimensional reconstruction of maize roots and quantitative analysis of metaxylem vessels based on X-ray micro-computed tomography" *Canadian J Plant Sci* **98**: 457–466.

Pavani, S. K., M. S. Dogan, H. Sari-Sarraf and E. F. Hequet (2004). Segmentation and classification of four common cotton contaminants in x-ray microtomographic images. *Machine Vision Applications in Industrial Inspection XII*. J. R. Price and F. Meriaudeau. Bellingham (WA), SPIE. **SPIE Proc Vol 5303**: 1–13.

Peeters, G., C. Debbaut, W. Laleman, D. Monbaliu, I. Vander Elst, J. R. Detrez, T. Vandecasteele, T. De Schryver, L. Van Hoorebeke, K. Favere, J. Verbeke, P. Segers, P. Cornillie and W. H. De Vos (2017). "A multilevel framework to reconstruct anatomical 3D models of the hepatic vasculature in rat livers." *J Anat* **230**: 471–483.

Peyrin, F., M. Salome, P. Cloetens, A. M. Laval-Jeantet, E. Ritman and P. Rüegsegger (1998). "MicroCT examinations of trabecular bone samples at different resolutions: 14, 7 and 2 micron level." *Technol Health Care* **6**: 391–401.

Postiglione, G., M. Alberini, S. Leigh, M. Levi and S. Turri (2017). "Effect of 3D-printed microvascular network design on the self-healing behavior of cross-linked polymers." *ACS Appl Mater Interfaces* **9**: 14371–14378.

Pratt, R., J. C. Hutchinson, A. Melbourne, M. A. Zuluaga, A. Virasami, T. Vercauteren, S. Ourselin, N. J. Sebire, O. J. Arthurs and A. L. David (2017). "Imaging the human placental microcirculation with micro-focus computed tomography: Optimisation of tissue preparation and image acquisition." *Placenta* **60**: 36–39.

Prodanović, M., W. B. Lindquist and R. S. Seright (2007). "3D image-based characterization of fluid displacement in a Berea core." *Adv Water Res* **30**: 214–226.

Pyrz, R. and J. Schjødt-Thomsen (2006). "Bridging the length-scale gap—short fibre composite material as an example." *J Mater Sci* **41**: 6737–6750.

Ramaswamy, S., M. Gupta, A. Goel, U. Aaltosalmi, M. Kataja, A. Koponen and B. V. Ramarao (2004). "The 3D structure of fabric and its relationship to liquid and vapor transport." *Colloids Surf A* **241**: 323–333.

Rawal, A., P. V. K. Rao and V. Kumar (2018). "Deconstructing three-dimensional (3D) structure of absorptive glass mat (AGM) separator to tailor pore dimensions and amplify electrolyte uptake." *J Power Sources* **384**: 417–425.

Rennie, M. Y., L. S. Cahill, S. L. Adamson and J. G. Sled (2017). "Arterio-venous fetoplacental vascular geometry and hemodynamics in the mouse placenta." *Placenta* **58**: 46–51.

Richter, C. P., H. Young, S. V. Richter, V. Smith-Bronstein, S. R. Stock, X. Xiao, C. Soriano and D. S. Whitlon (2018). "Fluvastatin protects cochleae from damage by high-level noise." *Sci Rep* **8**: 3033.

Ritman, E. L., S. M. Jorgensen, P. E. Lund, P. J. Thomas, J. H. Dunsmuir, J. C. Romero, R. T. Turner and M. E. Bolander (1997). Synchrotron-based microCT of in situ biological basic functional units and their integration. *Developments in X-Ray Tomography*. U. Bonse. Bellingham (WA), SPIE. **3149**: 13–24.

Rolland Du Roscoat, S., J. F. Bloch and X. Thibault (2005). "Synchrotron radiation microtomography applied to investigation of paper." *J Phys D* **38**: A78–A84.

Rolland Du Roscoat, S., M. Decain, X. Thibault, C. Geindreau and J. F. Bloch (2007). "Estimation of microstructural properties from synchrotron x-ray microtomography and determination of the REV in paper materials." *Acta Mater* **55**: 2841–2850.

Sakellariou, A., T. J. Senden, T. J. Sawkins, M. A. Knackstedt, M. L. Turner, A. C. Jones, M. Saadatfar, R. J. Roberts, A. Limaye, C. H. Arns, A. P. Sheppard and R. M. Sok (2004). An x-ray tomography facility for quantitative prediction of mechanical and transport properties in geological, biological and synthetic systems. *Developments in X-Ray Tomography IV*. U. Bonse. Bellingham (WA), SPIE. **SPIE Proc Vol 5535**: 473–484.

Sartori, J., S. Köhring, H. Witte, M. S. Fischer and M. Löffler (2018). "Three-dimensional imaging of the fibrous microstructure of Achilles tendon entheses in *Mus musculus*." *J Anat* **233**: 370–380.

Schnaar, G. and M. L. Brusseau (2005). "Pore-scale characterization of organic immiscible-liquid morphology in natural porous media using synchrotron x-ray microtomography." *Env Sci Technol* **39**: 8403–8410.

Schnaar, G. and M. L. Brusseau (2006). "Characterizing pore-scale dissolution of organic immiscible liquid in natural porous media using synchrotron x-ray microtomography." *Env Sci Technol* **40**: 6622–6629.

Schulz, G., H. J. Crooijmans, M. Germann, K. Scheffler, M. Müller-Gerbl and B. Müller (2011). "Three-dimensional strain fields in human brain resulting from formalin fixation." *J Neurosci Meth* **202**: 17–27.

Selomulya, C., X. Jia and R. A. Williams (2005). "Direct prediction of structure and permeability of flocculated structures and sediments using 3D tomographic imaging." *Chem Eng Res Design* **83**: 844–852.

Selomulya, C., T. M. Tran, X. Jia and R. A. Williams (2006). "An integrated methodology to evaluate permeability from measured microstructures." *AIChE J* **52**: 3394–3400.

Sevanto, S., M. Ryan, L. T. Dickman, D. Derome, A. Patera, T. Defraeye, R. E. Pangle, P. J. Hudson and W. T. Pockman (2018). "Is desiccation tolerance and avoidance reflected in xylem and phloem anatomy of two coexisting arid-zone coniferous trees?" *Plant Cell Environment* **41**: 1551–1564.

Shanti, N. O., V. W. L. Chan, S. R. Stock, F. DeCarlo, K. Thornton and K. T. Faber (2014). "X-ray micro-computed tomography and tortuosity calculations of percolating pore networks." *Acta Mater* **71**: 126–135.

Shen, H., S. Nutt and D. Hull (2004). "Direct observation and measurement of fiber architecture in short fiber-polymer composite foam through microCT imaging." *Compos Sci Technol* **64**: 2113–2120.

Sheppard, A. P., C. H. Arns, A. Sakellariou, T. J. Senden, R. M. Sok, H. Averdunk, M. Saadatfar, A. Limaye and M. A. Knackstedt (2006). Quantitative properties of complex porous materials calculated from x-ray μCT images. *Developments in X-Ray Tomography V*. U. Bonse. Bellingham (WA), SPIE. **SPIE Proc Vol 6318**: 631811-1–631811-5.

Shofer, S., C. Badea, S. Auerbach, D. A. Schwartz and G. A. Johnson (2007). "A micro-computed tomography-based method for the measurement of pulmonary compliance in healthy and bleomycin-exposed mice." *Exp Lung Res* **33**: 169–183.

da Silva, Í. B. (2018). "X-ray computed microtomography technique applied for cementitious materials: A review." *Micron* **107**: 1–8.

Singh, K., H. Menke, M. Andrew, Q. Lin, C. Rau, M. J. Blunt and B. Bijeljic (2017a). "Dynamics of snap-off and pore-filling events during two-phase fluid flow in permeable media." *Sci Rep* **7**: 5192.

Singh, K., H. Scholl, M. Brinkmann, M. Di Michiel, M. Scheel, S. Herminghaus and R. Seemann (2017b). "The role of local instabilities in fluid invasion into permeable media." *Sci Rep* **7**: 444.

Sled, J. G., M. Marxen and R. M. Henkelman (2004). Analysis of microvasculature in whole kidney specimens using microCT. *Developments in X-Ray Tomography IV*. U. Bonse. Bellingham (WA), SPIE. **SPIE Proc Vol 5535**: 53–64.

Tan, J. C., J. A. Elliott and T. W. Clyne (2006). "Analysis of tomography images of bonded fibre networks to measure distributions of fiber segment length and fiber orientation." *Adv Eng Mater* **8**: 495–500.

Tan, X., I. Jahan, Y. Xu, S. R. Stock, C. C. Kwan, C. Soriano, X. Xiao, B. Fritzsch, J. García-Añoveros and C.-P. Richter (2018). "Auditory neural activity in congenital deaf mice induced by infrared neural stimulation." *Sci Rep* **8**: 388.

Tanabe, N., D. M. Vasilescu, J. E. McDonough, D. Kinose, M. Suzuki, J. D. Cooper, P. D. Paré and J. C. Hogg (2017). "Micro-computed tomography comparison of preterminal bronchioles in centrilobular and panlobular emphysema." *Am J Respir Crit Care Med* **195**: 630–638.

Tang, W., S. Huang, L. Du, W. Sun, Z. Yu, Y. Zhou, J. Chen, X. Li, X. Li, B. Yu and D. Chen (2018). "Expression of HMGB1 in maternal exposure to fine particulate air pollution induces lung injury in rat offspring assessed with microCT." *Chem Biol Interact* **280**: 64–69.

Toyota, E., K. Fujimoto, Y. Ogasawara, T. Kajita, F. Shigeto, T. Matsumoto, M. Goto and F. Kajiya (2001). "Dynamic changes in three-dimensional architecture and vascular volume of transmural coronary microvasculature between diastolic- and systolic-arrested rat hearts." *Circulation* **105**: 621–626.

Toyota, E., Y. Ogasawara, K. Fujimoto, T. Kajita, F. Shigeto, T. Asano, N. Watanabe and F. Kajiya (2004). "Global heterogeneity of glomerular volume distribution in early diabetic nephropathy." *Kidney Int* **66**: 855–861.

Varoudis, T., A. G. Swenson, S. D. Kirkton and J. S. Waters (2018). "Exploring nest structures of acorn dwelling ants with X-ray microtomography and surface-based three-dimensional visibility graph analysis." *Philos Trans R Soc Lond B Biol Sci* **373**: 20170237.

Vasic, S., B. Grobéty, J. Kuebler, T. Graule and L. Baumgartner (2007). "X-ray computed micro tomography as complementary method for the characterization of activated porous ceramic preforms." *J Mater Res* **22**: 1414–1424.

Verma, N., K. Salem and D. Mewes (2007). "Simulation of micro- and macro-transport in a packed bed of porous adsorbents by lattice Boltzmann methods." *Chem Eng Sci* **62**: 3685–3698.

Videla, A., C. L. Lin and J. D. Miller (2006). "Watershed functions applied to a 3D image segmentation problem for the analysis of packed particle beds." *Part Part Syst Charact* **23**: 237–245.

Videla, A. R., C. L. Lin and J. D. Miller (2007). "3D characterization of individual multiphase particles in packed particle beds by x-ray microtomography (XMT)." *Int J Miner Process* **84**: 321–326.

Vladisavljević, G. T., I. Kobayashi, M. Nakajima, R. A. Williams, M. Shimizu and T. Nakashima (2007). "Shirasu porous glass membrane emulsification: Characterisation of membrane structure by high-resolution X-ray microtomography and microscopic observation of droplet formation in real time." *J Membrane Sci* **302**: 243–253.

Walther, T., K. Terzic, T. Donath, H. Meine, F. Beckmann and H. Thoemen (2006). Microstructural analysis of lignocellulosic fiber networks. *Developments in X-Ray Tomography V.* U. Bonse. Bellingham (WA), SPIE. **SPIE Proc Vol 6318**: 631812-1–631812-10.

Wan, S. Y., E. L. Ritman and W. E. Higgins (2002). "Multi-generational analysis and visualization of the vascular tree in 3D microCT images." *Computers Biol Med* **32**: 55–71.

Wasserthal, L. T., P. Cloetens, R. H. Fink and L. K. Wasserthal (2018). "X-ray computed tomography study of the flight-adapted tracheal system in the blowfly *Calliphora vicina*, analysing the ventilation mechanism and flow-directing valves." *J Exp Biol* **221**: 176024.

Wildenschild, D., K. A. Culligan and B. S. B. Christensen (2004). Application of x-ray microtomography to environmental fluid flow problems. *Developments in X-Ray Tomography IV.* U. Bonse. Bellingham (WA), SPIE. **SPIE Proc Vol 5535**: 432–441.

Wildenschild, D., J. W. Hopmans, C. M. P. Vaz, M. L. Rivers, D. Rickard and B. S. B. Christensen (2002). "Using x-ray tomography in hydrology: systems, resolutions and limitations." *J Hydrol* **267**: 285–297.

Willson, C. S., R. W. Stacey, K. Ham and K. E. Thompson (2004). Investigating the correlation between residual nonwetting phase liquids and pore-scale geometry and topology using synchrotron x-ray tomography. *Developments in X-Ray Tomography IV.* U. Bonse. Bellingham (WA), SPIE. **SPIE Proc Vol 5535**: 101–111.

Wilson, S. H., J. Herrmann, L. O. Lerman, D. R. Homes Jr, C. Napoli, E. L. Ritman and A. Lerman (2002). "Simvastatin preserves the structure of coronary adventitial vasa vasorum in experimental hypercholesterolemia independent of lipid lowering." *Circulation* **105**: 415–418.

Wirkner, C. S. and L. Prendini (2007). "Comparative morphology of the hemolymph vascular system in scorpions-A survey using corrosion casting, MicroCT, and 3D-reconstruction." *J Morphol* **268**: 401–413.

Wirkner, C. S. and S. Richter (2007). "The circulatory system in Mysidacea - implications for the phylogenetic position of Lophogastrida and Mysida (Malacostraca, Crustacea)." *J Morphol* **268**: 311–328.

Xu, Q., W. Long, H. Jiang, B. Maa, C. Zan, D. Mac and L. Shi (2018). "Quantification of the microstructure, effective hydraulic radius and effective transport properties changed by the coke deposition during the crude oil in-situ combustion." *Chem Eng J* **331**: 856–869.

Yang, H. and B. W. Lindquist (2000). Three-dimensional image analysis of fibrous materials. *Applications of Digital Image Processing XXIII.* A. G. Tesher. Bellingham (WA), SPIE. **SPIE Proc Vol 4115**: 275–282.

Zhang, H., T. S. Ramakrishnan, A. Nikolov and D. Wasan (2018). "Enhanced oil displacement by nanofluid's structural disjoining pressure in model fractured porous media." *J Colloid Interf Sci* **511**: 48–56.

Zhang, W., K. E. Thompson, A. H. Reed and L. Beenken (2006). "Relationship between packing structure and porosity in fixed beds of equilateral cylindrical particles." *Chem Eng Sci* **61**: 8060–8074.

Zhao, Y., W. Du, B. Koe, T. Connolley, S. Irvine, P. K. Allan, C. M. Schlepütz, W. Zhang, F. Wang, D. G. Eskin and J. Mi (2018). "3D characterisation of the Fe-rich intermetallic phases in recycled Al alloys by synchrotron X-ray microtomography and skeletonisation." *Scripta Mater* **146**: 321–326.

Zhu, X. Y., M. D. Bentley, A. R. Chade, E. L. Ritman, A. Lerman and L. O. Lerman (2007a). "Early changes in coronary artery wall structure detected by microcomputed tomography in experimental hypercholesterolemia." *Am J Physiol Heart Circ Physiol* **293**: H1997–H2003.

Zhu, X. Y., E. Daghini, A. R. Chade, C. Napoli, E. L. Ritman, A. Lerman and L. O. Lerman (2007b). "Simvastatin prevents coronary microvascular remodeling in renovascular hypertensive pigs." *J Am Soc Nephrol* **18**: 1209–1217.

10

Evolution of Structures

This chapter covers microCT studies in the broad area of evolution of microstructure. Being able to follow the 3D structure of a specimen as it changes is an enormous benefit of microCT. To be sure, great value also exists for single observations of specimens taken from within a sequence of processing steps or of time points, especially for a) the class of specimens whose preparation for observation by other means is impractical (too laborious or fraught with artifacts) or b) samples requiring volumes to be interrogated in order to avoid sampling bias from unanticipated microstructural anisotropy (e.g., Fig. 6.1). The greatest impact occurs, however, when the sample itself can act as its own control during longitudinal studies.

Since the first longitudinal microCT observations of a specimen in 1989 (crack opening at different loads) with data collection times of several hours per loading increment, in situ microCT has evolved to the point where a set of projections can be collected in 1 s or less. Now the tendency is to describe these studies as 4D microCT, that is, the three physical dimensions plus time as the fourth. Many of the studies conducted in the first half of the 2000s, and covered in the first edition of this book, remain relevant and are included in this edition, even if data acquisition was not as rapid as in later studies.

Evolving microstructure studies covered in other chapters include changes in cellular materials (Chapter 8), changes in networks (Chapter 9), and changes in deformed or cracked solid materials (Chapter 11). Some papers that might have been covered in Chapters 8, 9, and 11 have found their way into this chapter. The examples of the current chapter are grouped into four sections: food and pharmaceuticals (Section 10.1), materials processing (Section 10.2), environmental interactions (Section 10.3), and hard and soft tissue adaptation (Section 10.4). In Section 10.2, microCT studies of solidification, vapor phase processing, plastic forming, and particle packing and sintering are reviewed. Section 10.3 covers geological applications, construction materials, degradation of biological structures, and corrosion of metals. Section 10.4 organizes the material into one subsection on mineralized tissue and implants, bone healing, bone mineral levels, and remodeling and a second subsection on degradation of tissue, primarily soft tissue, including studies on arthritis, on pathological calcification characterization, and tumor generation and growth.

10.1 Food and Pharmaceuticals

MicroCT has been used to study bubbles in noodle dough (Guillermic, Koksel et al. 2018) and the dynamics of gas cell coalescence in bread (Mis, Nawrocka et al. 2018). Food suspensions have also been investigated (Islam, Wysokinski et al. 2018). Banana drying (Léonard, Blacher et al. 2008) and complex changes in microstructure during drying of other plant-based food materials (Rahman, Joardder et al. 2018) have been examined. Drug migration during granule drying has been investigated quantitatively (Kataria, Oka et al. 2018). The physical properties of coffee beans have been followed during roasting (Bstos-Vanegas, Correa et al. 2018).

10.2 Materials Processing

Difficulty in exactly reproducing conditions and a plethora of adjustable variables bedevil studies aimed at understanding materials processing. For metals, for example, not only processing temperature and time at temperature but also the rate of cooling and rate of heating figure into the microstructure produced. MicroCT is an extremely valuable tool, therefore, in this area of engineering/science, especially as longitudinal comparative data can improve predictive numerical models and, if accurate, enable virtual interrogation of many combinations of variables, thereby focusing attention on the most promising avenues.

10.2.1 Solidification

Solidification is a profitable processing application for microCT. One application that can be studied readily is the inhomogeneous distribution of particles in a discontinuously reinforced composite. Clustering of reinforcement particles can be deleterious from a fracture resistance perspective or can be used to great effect to provide a component with a hard, wear-resistant (albeit low-toughness) outer surface and a tough internal volume.

There are a number of situations like restorative dentistry or orthopedics where implants or other structures must be fabricated for an exact match to the remaining hard tissue. MicroCT can be used to provide the numerical coordinates for computer-aided machining. If the structure is to be cast, then microCT can also provide the geometrical input for numerical models predicting how to avoid porosity and validate the success of each specific casting (Atwood, Lee et al. 2007).

In an in situ composite of Al and TiB_2, Watson and coworkers used a novel sampling procedure to withdraw material from the melt and observed boride particle clustering with synchrotron microCT as a function of melt hold time (Watson, Forster et al. 2005). Small amounts of melt were drawn off at times up to 2.6×10^5 s and quickly solidified. MicroCT revealed the maximum cluster size decreased from an initial value of 50 to 10 μm at the end of the experiment. Even though the sampling technique might have biased the clustering results somewhat, use of the same method throughout means that the changes observed almost surely reflected changes occurring in the melt vessel. In an Al/SiC_p functionally graded composite fabricated by centrifugal casting, Velhinho et al. observed slight gradients in particle volume fraction away from the SiC-rich surface (Velhinho, Sequeira et al. 2003). Because their mass attenuation coefficients are very similar, SiC and Al are difficult to distinguish based on absorption alone, and phase-enhanced interface contrast in synchrotron microCT can help segmentation considerably.

Solidification with segregation of atoms with different absorptivities is another area where microCT has been applied. This segregation, and the accompanying range of solidification temperatures, can lead to undesirable excess porosity or cracks (hot-tearing) appearing at the end of solidification. Ludwig et al. followed in situ solidification of Al-4 wt% Cu in 3D with ultrafast synchrotron microCT (Ludwig, Di Michiel et al. 2005).

A complete 512 (perpendicular to the rotation axis) × 256 scan (500 projections over 180°) was recorded *every 10 seconds* using polychromatic wiggler radiation; this was a case where contrast sensitivity was sacrificed for temporal resolution. A cooling rate of 0.1°C/s was used for these in situ experiments; despite the reconstructions encompassing structures averaged over a ~1°C temperature range, the slices were quite clear, showed growth and linkage of the solid phase particles, and showed increasing Cu content in the liquid phase. Evolution of the experimentally determined solid volume fraction with temperature was compared to different solidification models; shrinkage vs solid fraction was linear; S_V evolved as expected and mean wt.% Cu in the solid and liquid phases (determined from analysis of the linear attenuation coefficients) agreed reasonably well with that predicted by the liquidus and solidus temperatures of the equilibrium phase diagram (Ludwig, Di Michiel et al. 2005). In an Al-15.8 wt.% Cu alloy, the investigators were also able to show that rapid quenching produces a higher volume fraction of solid phase than was present at the starting temperature (Baruchel, Buffiere et al. 2006); therefore, models based on quenching data may contain a bias that must be corrected. An earlier report of microCT materials quenched from the solid state appeared elsewhere (Verrier, Braccini et al. 2000).

The grain size distribution and number of faces per grain were measured in synchrotron microCT-reconstructed volumes of solidified Al–Sn with 1, 2, or 3 at.% Sn (Krill III, Dobrich et al. 2001, Dobrich, Rau et al. 2004); Sn is immiscible in Al and segregates to the Al–Al grain boundaries. Some areas of the grain boundaries appeared to be free of Sn, a possible effect of the sensitivity limit, and a special algorithm was derived to fill in the missing boundaries. The authors concluded that size distribution agreed with and the number of faces per grain differed from metallography data in the literature (Dobrich, Rau et al. 2004). In situ microCT of coarsening of a model binary semi-solid alloy (Al-32 wt.% Ge) revealed the particle network followed theoretical predictions (Zabler, Rueda et al. 2007).

Increasingly, high temporal and spatial resolutions are being obtained in in situ studies of solidification. Twin mediated growth (interface dynamics and faceting) of Si particles within an Al–Si–Cu melt have been followed with 30 s temporal resolution (Shahani, Gulsoy et al. 2016) as has Ostwald ripening in the same system (Shahani, Xiao et al. 2016). The morphology of growing dendrites is another study (Gibbs, Mohan et al. 2015).

Retained porosity in cast Al–Si was studied as a function of H_2/Ar gas ratio introduced during stirring of the melt (Buffière, Savelli et al. 2000). Two populations of voids were observed. The smaller voids were associated with microshrinkage when the metal solidified, and the population characteristics (equivalent size vs sphericity) did not vary with gas composition nor did the volume fraction of microshrinkage pores. The larger voids were from artificial incorporation of gas bubbles, and the volume fraction of gas pores increased exponentially with H_2 content.

Evolution of micropores during homogenization of direct chill cast Al (2, 4, and 6 wt.% Mg) was studied with microCT (Chaijaruwanich, Lee et al. 2007); the specimens were machined from 2.5-mm diameter rods of as-cast and heat-treated ingots (0, 1, 10, and 100 h at 530°C). These investigators found that the tortuosity of the pore networks was very complex and that there was no increase in maximum pore length. Intrapore Ostwald ripening of the networks produced coarsening of both the asperities and interconnects. Hot tear damage was studied in four specimens of a direct chill cast commercial Al–Mg alloy (AA5182) tested to four tensile loads at 528°C (fraction solid ~0.98); growth of preexisting voids and formation of damage-related voids were observed and interpreted quantitatively (Phillion, Cockcroft et al. 2006). Solidification cracking has also been studied in welds (Aucott, Huang et al. 2017). Void growth and coalescence were studied in an Mg alloy (Kondori, Morgeneyer et al. 2018).

Porosity formation in epoxy-based composites used in filling cavities is one concern in restorative dentistry, and a second is marginal debonding of the light-cured material from dentin cavity walls. MicroCT was used to study volume fraction of micropores and interfacial gap in a model system (8-mm diameter human dentin disks in which a 2-mm diameter, 1-mm deep cavity was filled with two types of composites) (Kakaboura, Rahiotis et al. 2007).

Highly nonequilibrium solidification occurs in spray deposition of thermal barrier coatings, and synchrotron microCT is invaluable for determining pore shapes and for quantifying pore volume fraction in the coating as a function of distance from the substrate (Kulkarni, Sampath et al. 2000, Kulkarni, Gutleber et al. 2004a, Kulkarni, Herman et al. 2004b, Kulkarni, Goland et al. 2005). Different processes produced different pore shapes that can be directly visualized and compared with the results of techniques such as small-angle neutron scattering (SANS) and nanoindentation for elastic moduli determination (Kulkarni, Sampath et al. 2000, Kulkarni, Goland et al. 2005). Gradients in porosity were also correlated with indentation-derived moduli and SAXS (Kulkarni, Gutleber et al. 2004a, Kulkarni, Herman et al. 2004b).

10.2.2 Vapor Phase Processing

Similar pore topology exists in ceramic matrix composites partially densified by chemical vapor infiltration (CVI) (Butts 1993, Kinney, Breunig et al. 1993, Lee 1993, Kinney, Henry et al. 1994, Kinney and Haupt 1997, Lee, Stock et al. 1998). The goal with ceramic composites is to maximize the density produced by processing so that the toughening effect of the reinforcement is fully realized; an evolving structure must be understood in this case. Production of complex, near net shape parts results for materials that are very difficult to machine; cost savings for SiC/SiC composites can be enormous. The preforms for these composites are often constructed by laying layers of cloth to form the required thickness or by wrapping multiple layers of cloth around a mandrel. In CVI, the preform is infiltrated by a vapor precursor that decomposes and deposits the desired phase within the composite. Another approach, reaction sintering, fills the preform with a very fine powder and forms the matrix by reacting the precursor with a gas. While reaction-sintered samples were studied with microCT for different processing conditions (Stock, Guvenilir et al. 1989), the principle work has been in CVI of SiC/SiC composites based on cloths of Nicalon amorphous SiC fibers.

Use of a portable reaction chamber allowed Nicalon/SiC composites to be transported to a synchrotron radiation source between increments of CVI and to be imaged multiple times with microCT (Butts 1993, Kinney, Breunig et al. 1993, Lee 1993, Lee, Stock et al. 1998). The graphite chambers were cylindrical with a 10-mm outside diameter, and the Nicalon cloth preforms, each 6.2-mm in diameter and 6.0-mm high, consisted of 20 layers stacked within a chamber. Each cloth was woven of bundles or tows of ~500, 10–20 μm diameter Nicalon fibers, and there were approximately 6.3 tows per cm. Two architectures were examined: 0°/90° lay-ups where the tows in adjacent cloth layers are parallel, and 0°/45° where the tows in adjacent layers are rotated 45°. The preforms were infiltrated with methyltrichlorosilane at 975°C, and the flow was adjusted so that after 3, 6, and 9 h densification would be 33%, 67%, and 100% complete, respectively.

Samples were imaged prior to and after each increment of CVI, 400 contiguous slices were recorded, and the reconstructions consisted of isotropic 15.8-μm voxels. Data could not be collected at higher resolution because of the need to keep the entire reaction chamber in the field of view, that is, as a direct consequence of the number of elements in the CCD detector. If the experiment were repeated now, one expects that local tomography would be performed including only SiC-containing volume; voxel sizes might be ~3 μm.

In the 0°/90° preform, the structure of the channels between cloths depended on the relative displacement of the holes in the cloths bounding the channel (Butts 1993, Kinney, Breunig et al. 1993, Lee 1993, Lee, Stock et al. 1998), and this relationship is illustrated in Fig. 10.1. Offsets 45° to the tow axes resulted in very closed channels (Fig. 10.1, top), hole displacements along the tow directions produced an array of parallel, 1D open

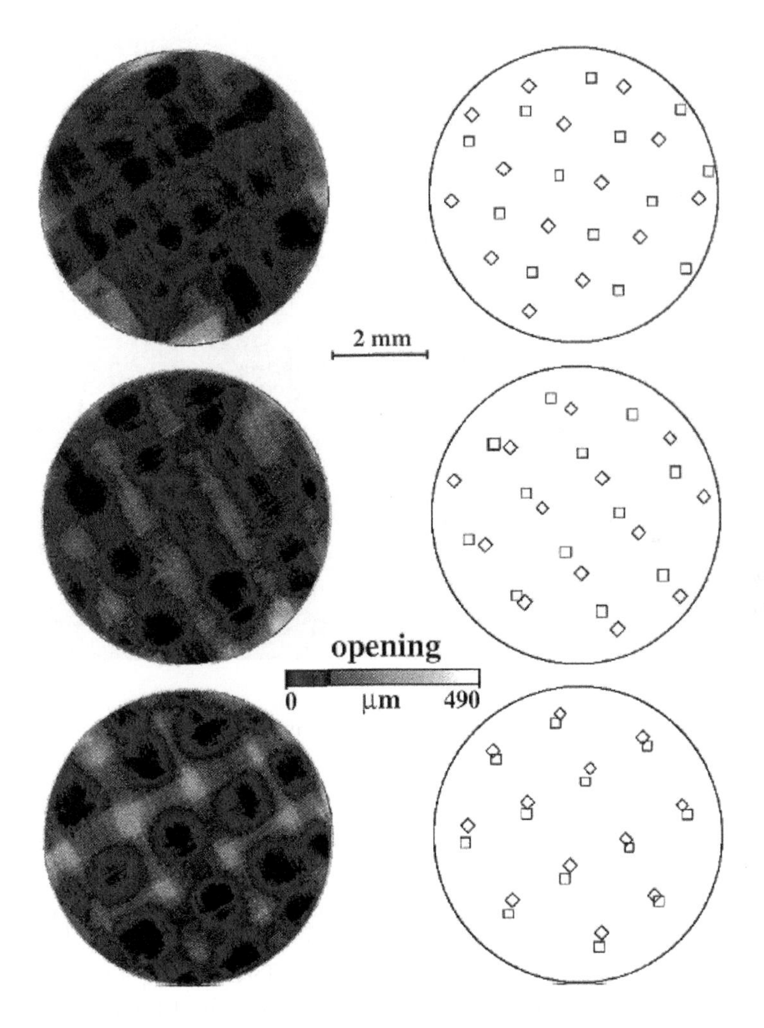

FIGURE 10.1
Maps of three patterns of channel width (left column) as a function of relative positions of the holes (squares and diamonds) in plane weave SiC cloths on either side of each channel (right column). The total channel width is projected onto a plane. Black pixels show position with no opening, with red, blue, green, and white showing increasing opening. (top) Poorly defined, relatively closed channel due to hole displacements 45° to the tow (fiber bundle) direction. (middle) Hole displacements along the tow direction and the resulting set of 1D channels. (bottom) Closely aligned holes and the accompanying 2D network of opening. (Lee 1993, Lee, Stock et al. 1998) © Materials Research Society.

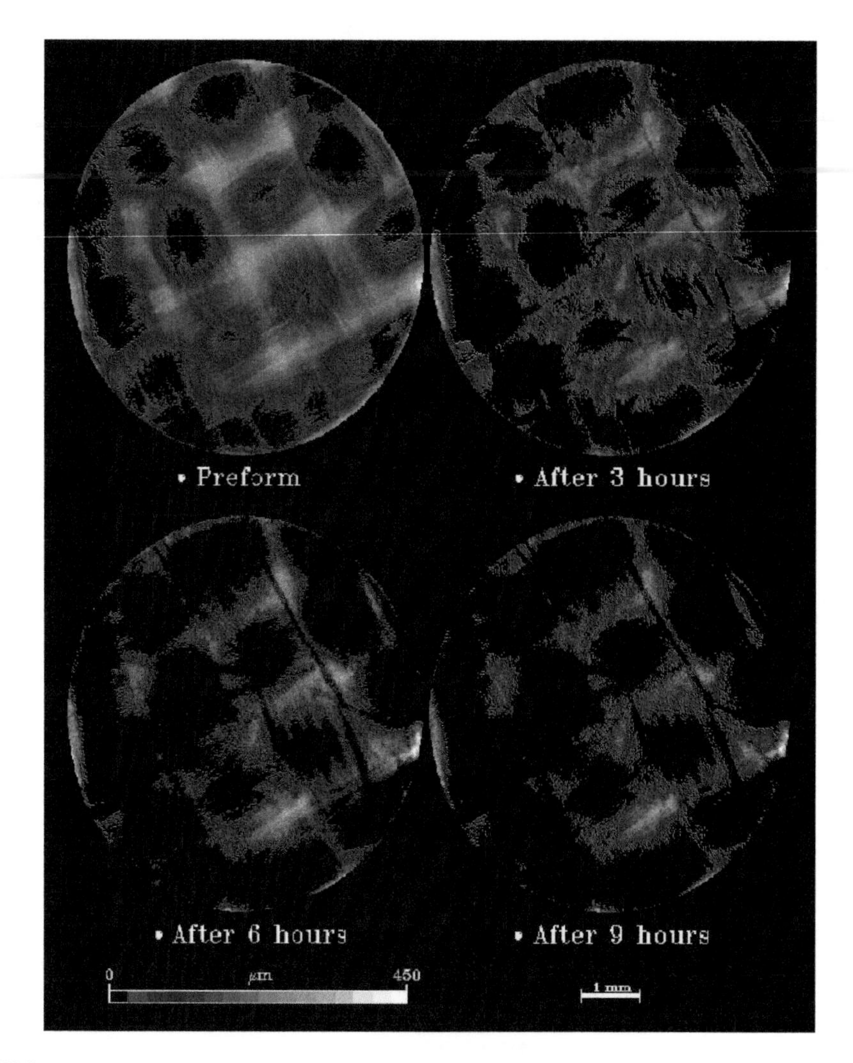

FIGURE 10.2
Variation of channel width in the SiC cloth composite described in Fig. 10.1 as a function of SiC CVI infiltration time. The holes on either side of the channel are well aligned (2D array of pipes), the colors have the same meaning as in Fig. 10.1, and the infiltration times are given below each map (Lee 1993).

pipes (Fig. 10.1, middle) and when the holes were aligned, a 2D network of openings was observed in the channels (Fig. 10.1, bottom). Fig. 10.2 shows how the 2D pipe network closes after infiltration. Circle cuts, numerical sections along the surface of a cylinder unwrapped to provide a flat representation, are used to show where SiC was deposited within or on the SiC tows after three increments of infiltration: difference maps of the same cylinder through the composite are shown in Fig. 10.3. The gas inlet is at the bottom of each image of Fig. 10.3, and this is where the greatest change appears (the color bar shows increasing linear attenuation coefficient from left to right).

A consequence of the voxel size was that individual micropores within the tows could not be resolved, even in the preforms. Nonetheless, accurate volume fractions of Nicalon, deposited SiC, and micropores were obtained by taking averages over several subvolumes

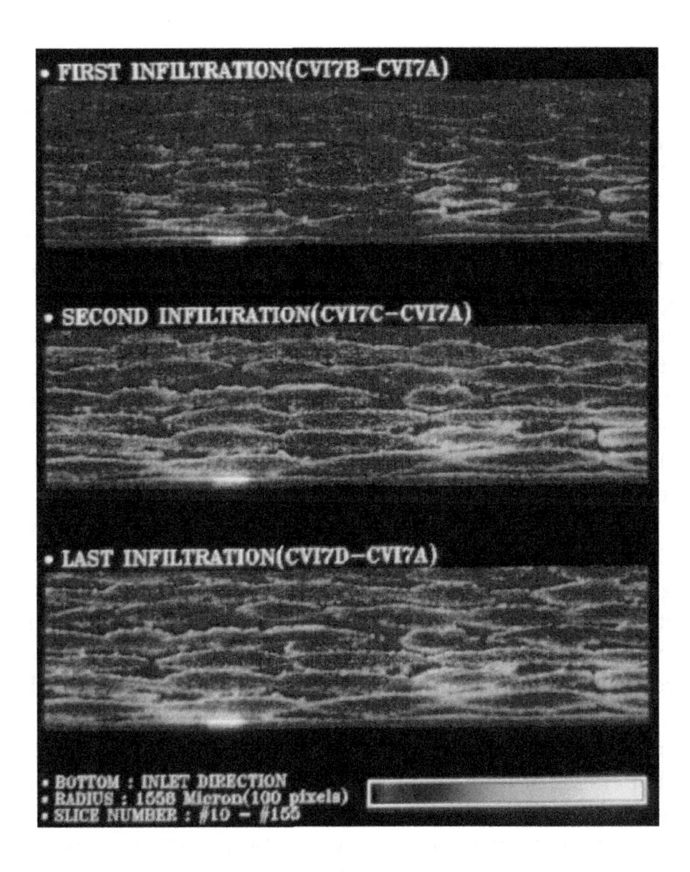

FIGURE 10.3
Circle cuts through a SiC/SiC cloth preform composite showing where SiC was deposited. (top) Difference image showing SiC deposited in the first 3 h of CVI. (middle) Difference image showing the total SiC deposited after 6 h. (bottom) Difference image showing the total SiC deposited after 9 h. Red, blue, and white show increasing change from the preform. The open areas between tows, where there are no fibers onto which SiC can be deposited, remain red throughout. The gas enters the reaction chamber from the bottom, and the white shells seen on many tows show that the interior microporosity between individual ~15 μm diameter fibers remains incompletely densified (Lee 1993).

within tows for preform and for 3 and 6 h infiltration (Kinney, Breunig et al. 1993, Lee 1993). Comparison of a microCT section through a densified Nicalon/SiC composite with an optical micrograph of nearly the same section demonstrated the accuracy of the reconstruction (Kinney and Nichols 1992). Measurement of surface area per unit volume as a function of fractional density (Kinney, Breunig et al. 1993, Lee 1993) agreed well with the uniform deposition model used to describe the CVI process (Starr 1992); the slight differences could be attributed to the fact that the surface area per unit volume of macroporosity (channels and holes through cloth layers) could be measured and the measurements could not include microporosity.

Bernard and coworkers reported some results of isobaric CVI of a C/C composite (Bernard, Vignoles et al. 2000). Kang et al. showed some synchrotron microCT images of cracks in C fibers grown from the vapor phase (Kang, Johnson et al. 2004). Other composites processed via a vapor phase route are mentioned in passing elsewhere in this volume.

10.2.3 Plastic Forming

Superplastic forming (very high strains in certain alloys without rupture of the starting material) is finding application in aerospace and automotive fields. Superplastic deformation is generally limited by strain-induced cavitation leading to fracture. Using synchrotron microCT, Martin et al. found that the number of cavities per unit volume vs strain ($\sim 1 < \varepsilon < \sim 1.7$) in AA5083 followed model predictions and observed developing cavity linkage (Martin, Blandin et al. 2000a, Martin, Josserond et al. 2000b). Pore evolution from rolling of an Al-6Mg alloy was studied with lab microCT, and the authors demonstrated that the highly tortuous pores would be difficult to detect in polished sections (Youssef, Chaijaruwanich et al. 2006). The tortuous pores spheroidized during homogenization, and accelerated centerline intrapore coarsening observed during initial, low-reduction-ratio rolling passes was attributed (through finite element modeling) to local tensile conditions, a counterintuitive but not unreasonable result (Youssef, Chaijaruwanich et al. 2006).

Warm drawing of a 6-mm diameter PE rod down to ~5-mm diameter was studied with SAXS microCT (Schroer, Kuhlmann et al. 2006, Stribeck, Camarilla et al. 2006). Circular zones of differing lamellar sizes (longitudinal and lateral) were clear in the reconstructed slice presented (Stribeck, Camarilla et al. 2006), and undoubtedly much more can be done with this approach, especially in specimens that would appear featureless in absorption tomographs. It would be interesting to compare phase tomography to the SAXS-derived reconstruction.

Synchrotron microCT of an Al–Mg industrial alloy (AA5182) tracked the size distribution and spatial dispersion of intermetallic particles (iron rich, Mg_2Si) and of voids from as-cast + homogenized to hot rolled to cold rolled to tensile tested state (Maire, Grenier et al. 2006). In addition to 3D views of a small portion of the structure (four thresholds with Al rendered transparent, voids as black and the intermetallics in two different grays), numerical values for mean, minimum, and maximum equivalent radii of the intermetallic particles showed the progression of fragmentation expected for the different processing steps. Similar data for volume fraction, number density, and mean equivalent radii were measured for pores. One limitation in the study with 0.7-μm voxels was that very small particles (those with smallest dimension <1 μm) could not be included in the analysis; the authors expected that this would be improved with imaging with 0.3-μm voxels (Maire, Grenier et al. 2006).

Movement of marker particles within metals or in unconsolidated powders can be used very effectively to map displacement fields in response to deformation (Nielsen, Poulsen et al. 2003, Nielsen, Beckmann et al. 2004, McDonald, Schneider et al. 2006, Zettler, Donath et al. 2006). Specimens of Al – 1 vol.% Ti (particle diameters between 1 and 10 μm) were imaged with synchrotron microCT at the 2-μm level, and displacement gradients for deformations up to 9.5% compression were quantified (Nielsen, Poulsen et al. 2003, Nielsen, Beckmann et al. 2004). Bulk material flow has also been studied in friction-stirred welds in Al using the particle technique and synchrotron microCT (Zettler, Donath et al. 2006). Closed die compaction and sintering of powders have long been used to fabricate metal and ceramic components, but constitutive models for loose powder behavior under intense shear deformation need to be developed. McDonald et al. used lab microCT and image correlation to follow particle displacements when a cylindrical punch was pushed into a somewhat wider diameter cylindrical die (McDonald, Schneider et al. 2006); they reported uncertainties in strains of ~0.05% and correlated particle displacement vectors with dilational strains calculated from the particle displacements (Fig. 10.4). Generally speaking,

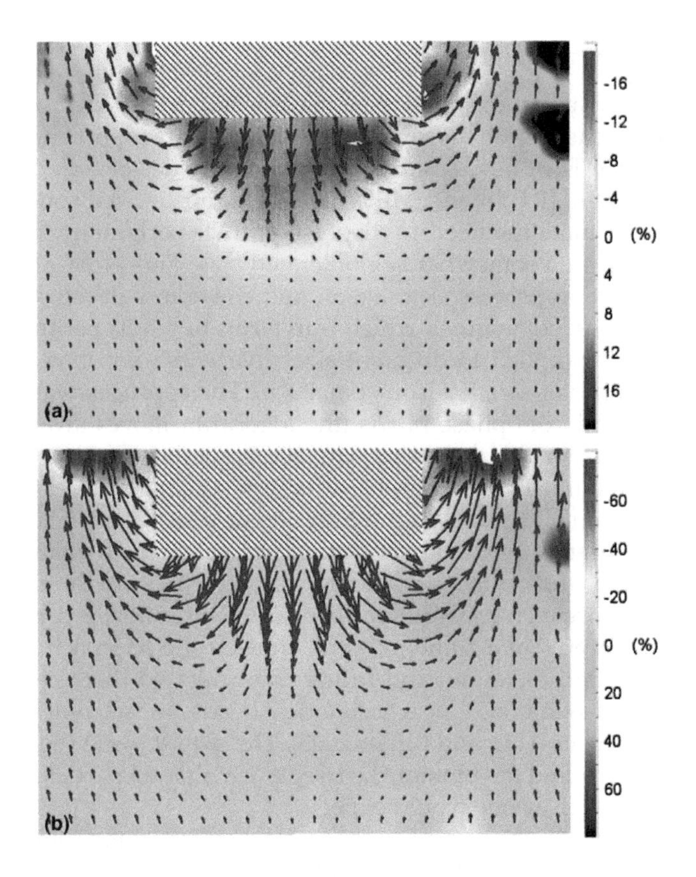

FIGURE 10.4
Powder displacement during compression. The arrows show particle displacement vectors around the downward moving punch calculated from image correlation over: (a) four 0.5-mm displacement steps (2.0 mm total) and (b) eight steps (4 mm total). The background color scale represents the dilational (volumetic) strain calculated from the particle displacements, effectively showing the change in density across the diametral section. The powder immediately under the punch is compacted (negative strain), but dilation (positive strain) is observed as the loosely packed powder is sheared, particularly as it flows around the corners of the punch and upwards against the sides of the container. Reprinted from McDonald et al. (2006), Copyright (2006), with permission from Elsevier.

these studies found that high-contrast particles (e.g., Ti, Sn, or W in Al) need to have diameters of several voxels for reliable detection and automated tracking of displacement.

Pressure gradients driving resin flow during liquid composite molding can be very low when large composite parts are manufactured, and capillary pressure can become the dominant force for tow infiltration, producing meso-scale voids. Mesostructure in glass fiber composites was characterized (Schell, Renggli et al. 2006), and experimental validation of a numerical method for predicting meso-scale voids in the glass-fiber, epoxy resin composite was provided by microCT, and, due to the voxel size of 6 μm, microvoids were ignored (Schell, Deleglise et al. 2007). Aspects of these two papers are reminiscent of the SiC/SiC CVI processing study described in Section 10.2.2.

Laser micromachining of hardened Portland cement paste was also studied with very high-resolution synchrotron microCT (Trtik and Hauri 2007). The 0.35-μm voxel data revealed that the hydrated cement was ablated, but large unhydrated particles were relatively unaffected.

10.2.4 Particle Packing and Sintering

Particle shapes, size distributions, and packing are important in processing and also in fluid transport; discussion of those studies related to transport in the open phases appeared in Chapter 9. The 3D size and shape characteristics of collections of particles were studied with microCT (Lin and Miller 2005, Thompson, Willson et al. 2006); note that microCT can eliminate the need to disperse particles originally in a dense assembly or packing or even in a somewhat agglomerated state and avoid artifacts inevitable in any physical separation process. Thompson and coworkers developed their analysis algorithm in a dimensionless manner, which is in terms of voxels per unit particle diameter, so that it could be applied to different sized particles' distributions studied with different voxel sizes (Thompson, Willson et al. 2006). The approach was based on simultaneous solid-phase and void-phase burns (the algorithm moves away from the interface and assigns a value to each voxel that equals the number voxels it is away from the interface). The local maxima of burn number was used as a particle center, and all voxels previously identified as solid were assigned to one or another particle with additional steps required to accurately partition contacting grains. Particle volume, surface area, orientation, aspect ratio, and contact statistics flowed directly from the assigned particles; computation performance metrics were supplied; and the results for computer-generated structures and for microCT of a standard sand showed the approach works quite well (Thompson, Willson et al. 2006). Lin and Miller characterized three collections of well-defined particles with lab microCT: nearly spherical beads, an isotropic quartz sand, and quite jagged rock fragments (Lin and Miller 2005). The data showed the expected surface area vs volume behavior, and the plot for the irregularly shaped particles was offset from that of the beads and sand, as one would expect.

MicroCT data sets have been used as the basis for simulating random packing of polydisperse particles (Al-Raoush and Alsaleh 2007). Fu and coworkers examined in situ compaction of powder with lab microCT (Fu, Milroy et al. 2005, Fu, Elliott et al. 2006); Richard et al. observed granular packing resulting from vibration with synchrotron microCT (Richard, Philippe et al. 2003), and Seidler et al. characterized the distribution of granules packed under the influence of gravity (Seidler, Martinez et al. 2000). MicroCT-derived particle characteristics from a well-characterized ore were also compared with 2D measures (Gay and Morrison 2006). Agglomeration of particles and the breakage of agglomerates during compression or extrusion have begun to be studied with microCT combined with numerical modeling (Golchert, Moreno et al. 2004a, 2004b, McGuire, Blackburn et al. 2007). Agglomerates, disruptions, and flashes were observed in micro powder injection molding of a silica model system (Heldele, Rath et al. 2006). In another study, microCT evaluated different injection parameters and the resulting defects in the green body of an Al alloy, and used these results to predict the proper parameters for steel (Yang, Zhang et al. 2015).

Sintering/cementation of powders has also been studied by microCT (Bernard, Vignoles et al. 2000, Lame, Bellet et al. 2003, 2004, Bernard, Gendron et al. 2005, Fu, Milroy et al. 2005, Pfister, Walz et al. 2005, Vagnon, Lame et al. 2006, Vaucher, Unifantowicz et al. 2007); research areas include analysis of rapid prototyped material and evolution of materials with nanoparticulate precursors. Bernard and coworkers employed local tomography to quantify porosity elimination and neck evolution in a glass powder and in a lithium borate powder over 1.6×10^4 s at 700°C and 720°C, respectively; growth of necks were particularly well illustrated by matched pairs of renderings, one showing the solid phase and the second the complementary void space (Bernard 2005, Bernard, Gendron et al. 2005). Topological evolution during sintering was described by Euler characteristics as a function of relative density (Okuma, Kadowaki et al. 2017). Sintering of copper and steel (Distaloy AE) powders was also studied with synchrotron microCT, and changes in pore geometries in copper and elimination of very thin interparticle voids but not large pores in Distaloy were observed with increasing sintering time (Lame, Bellet et al. 2003, 2004). Final dimensional changes are strongly anisotropic in Distaloy AE Densmix™ (axial swelling during delubrication and axial shrinkage during sintering), and repeated in situ synchrotron microCT observations during the different sintering stages (initial structure, lubricant burn-off, sintering, cooled structure) were used to investigate these changes (Vagnon, Lame et al. 2006). Analysis of the orientation distribution of three populations of pores (highly elongated, near circular, and intermediate geometry) was one probe; the second was an image correlation-based, local strain mapping comparison for the different directions in the compact.

Evaluation of suitable porosity for sintered porous β-tricalcium phosphate as a bone substitute was the subject of another study (Park, Bae et al. 2012). Porosity vs compressive and corrosion behaviors were examined in a bone scaffold formed by a mixture of Zn and NaCl powders; the latter was dissolved subsequent to compaction (Hou, Jia et al. 2018).

10.3 Environmental Interactions

Whether the materials are natural or engineered or whether hard or soft tissue is affected, the interaction of these substances with their environment is an area of research where microCT offers important insights that are difficult or impossible to obtain by other methods. This is true in particular because these interactions often are very local or are in inaccessible locations, or at least begin that way. After discussing geological applications, construction materials are covered. The third subsection concerns degradation of biological structures, and the fourth the corrosion of metals.

10.3.1 Geological Applications

Consideration of the transition of snow to firn to ice (porosities of >95%, ~40%, <10%, respectively) follows naturally from the discussion of sintering of engineering materials. The particles (snow) are at a relatively high homologous temperature (i.e., a large fraction of melting temperature) and are frequently under pressure from the weight of subsequent snowfalls: thus, the process and the structures are very similar to those covered above. Properties of snow/firn/ice are important in understanding avalanches. If porosity is closed in specimens cored from ice, an air archive exists (atmospheric information) that can be compared to the ice archive (climatic information). Synchrotron microCT of a snow sample collected at the failure site of a slab avalanche, for example, showed a cohesive layer above a weak layer (Coléou and Barnola 2001), a situation encountered in skier-triggered avalanches. The metamorphosis of snow to ice depends on local 3D curvature and solid–vapor surface area; under isothermal conditions, minimization of local curvature (minimization of surface energy) is thought to govern the densification of the structure, that is, a structure with many sharp, flat grains transforms to a much more rounded structure. Results of synchrotron microCT of snow at different points during transformation have been compared with numerical models evolving from real (microCT-derived)

3D microstructures, and the agreement between actual and modeled structures was quite encouraging (Flin, Brzoska et al. 2003, 2005). One should note that considerable technical challenges were overcome in the preparation of unaltered snow specimens for microCT examination, but these will not be discussed here (Coléou and Barnola 2001). More recently, crystal growth in snow was followed by 4D microCT and compared to models (Krol and Loewe 2018).

Understanding reactive percolation (CO_2) through porous limestone is important if, as some have proposed, rises in atmospheric CO_2 are to be combated by CO_2 sequestration in subsurface reservoirs. Synchrotron microCT followed changes in limestone porosity during flow of CO_2-charged water over 3×10^4 s and related changing microstructure to increasing permeability (Bernard 2005). Other studies directly compared successive structures in the same solid–fluid system over 8×10^4 s (Noiriel, Gouze et al. 2004, Noiriel, Bernard et al. 2005). As limestone is an important construction material in historic structures, its attack by CO_2 and by industrial organic pollutants and repair of such damage is an important area of research (Brunetti, Bidali et al. 2004). Vapor transport through two types of bricks was examined with synchrotron microCT, and the relationship between permeability, diffusivity, and pore size agreed with analytical expressions (Bentz, Quenard et al. 2000). Water transport in the proton exchange membrane fuel cell was also tracked with microCT (Sinha, Halleck et al. 2006).

Several studies of soils have appeared (Macedo, Vaz et al. 1999, Hansel, Force et al. 2002, Altman, Rivers et al. 2005, Feeney, Crawford et al. 2006, Hettiarachchi, Lombi et al. 2006, Nunan, Ritz et al. 2006). Methane bubble growth and migration in aquatic soils have been studied (Liu, De Kock et al. 2018)+. Metal contaminant absorption characterization is a current area of activity (Altman, Rivers et al. 2005). In Altman et al. (2005), iron content was mapped, relative amounts of iron oxyhydroxides and iron-bearing clays were determined semiquantitatively, and the relationship of Cs adsorption with iron-bearing materials was imaged using absorption edge difference techniques. MicroCT showed that application of granule and liquid phosphorus–based fertilizer produced different effects on local density: over time, the former produced increased density within ~1 mm but the latter did not (Hettiarachchi, Lombi et al. 2006). Properties of harbor sediment were also studied (Jones, Feng et al. 2006).

Soil microbes and plant roots microengineer their habitats by changing porosity and pore cluster characteristics (Rabbi, Tighe et al. 2018). Synchrotron microCT revealed soil modified by root action over 30 days had significantly greater pore volume and pore connectivity than in soil without roots (Feeney, Crawford et al. 2006, Nunan, Ritz et al. 2006). Changing pore structure due to organic material decomposition was studied in several soils (De Gryze, Jassogne et al. 2006). Direct microCT observation over hours of the movement of the root-feeding clover weevil showed constant average speed toward root nodules, guided by a chemical signal released by the nodules, and indicated that the larvae respond to the gradient direction of the chemical signal but not its magnitude; these data were used to adjust numerical models of larva survival, a major negative effect on sustaining white clover in a mixed sward (Zhang, Johnson et al. 2006). These and other results demonstrated that the soil ecosystem exhibits self-organization in relatively short periods of time. Some aquatic plant species sequester metal(loid) species on root surfaces, and synchrotron microCT and other techniques showed As associated with regions of enhanced Fe content (Hansel, Force et al. 2002). Progressive leaf dehydration is another area of interest (Scoffoni, Albuquerque et al. 2017).

The location of the metal species of interest within or at the surface of particles governs the extent to which that metal may be recovered by leaching operations. In order to determine how well lab microCT might function as an assaying tool (i.e., predicting leaching performance for comparison with actual efficiency), Miller et al. quantified the volume fraction of copper-bearing minerals in contact with the surfaces of ore particles (Miller, Lin et al. 2003). Standards of natural particles of copper-bearing mineral phases (known to exist in the ore of interest) were used to set thresholds for segmentation and allowed 3D assessment of mineral exposure vs particle size to be determined for the ore of interest. MicroCT-predicted recovery generally underestimated actual recoveries from column leaching tests, an unsurprising result given the finite resolution and sensitivity of the microCT system and the additional exposure that partial leaching might produce (Miller, Lin et al. 2003). Silver is associated with pores in certain ores, and Chen et al. used lab microCT to determine that porosity was insufficient to allow silver to be leached from the interior of a coarsely ground ore sample (Chen, Sasov et al. 2006).

A considerable number of 4D studies of geological phenomena have appeared recently. Generation of porosity during olivine carbonation via dissolution channels and expansion cracks is one example (Xing, Zhu et al. 2018). Pore-scale imaging of the reaction of dolomite reaction with supercritical CO_2 acidified brine is a second example (Al-Khulaifi, Lin et al. 2018). Metamorphic dehydration was followed in a heating experiment of gypsum (Bedford, Fusseis et al. 2017). Particle paths through porous media were observed by in situ time-resolved microCT (Blankenburg, Rack et al. 2017). Bubble nucleation and growth were followed in andesitic magmas (Plese, Higgins et al. 2018) as was crystallization in basaltic magma (Polacci, Arzilli et al. 2018).

10.3.2 Construction Materials

Construction materials frequently are exposed to aggressive environments over a period of years or even centuries, and, as much of the resulting damage is subsurface, a number of reports of microCT applied to degradation of building materials have appeared (Cnudde and Jacobs 2004). One of the most important construction materials is concrete, and Portland cement used in this composite has been studied from several perspectives. A collection of useful baseline cement paste (synchrotron microCT) images were published (Bentz, Mizell et al. 2002), and features such as unhydrated cement particles, regions of hydrated cement, and ettringite needles were identified. Based on this data, cement particle shapes were analyzed and were expected to improve computational models of cement hydration (Garboczi and Bullard 2004) and to serve as a resource for those needing microstructures for modeling.

The curing of cement is perhaps more of a processing topic than an environmental interaction subject, but it seems more appropriate to deviate in this one particular from the overall plan and to mention a microCT study of the time course of the anhydrous cement particle–water interaction in conjunction with other studies of Portland cement. Gallucci and coworkers found there was a real conflict between resolving the narrow diameter of the capillary pores controlling the process (i.e., the required small specimen diameter) and between obtaining representative volumes (Gallucci, Scrivener et al. 2007). This tension occurs again and again in microCT and nanoCT studies, and, for specimens with appreciable densities (i.e., those other than foams), progress in this area remains a challenge.

The pore structure of cement-based materials seriously affects resistance to environmental attack as well as mechanical properties. Often pozzolans (fine natural volcanic ash, fly ash from power generation, diatomaceous earth, etc.) are added to cement to produce a fine-structured composite. A study of the pore structure of Portland cement composites with pozzolan (neat cement vs 25 wt.% fly ash vs 10 wt.% metakaolin) found mean pore size and maximum pore diameters decreased for the composites compared to the simple cement (Rattanasak and Kendall 2005). Chloride permeability is one cement durability issue, and microCT-measured pore structure of a reference concrete and of fly ash, silica fume, and slag-modified concretes were compared with results of a rapid chloride permeability test (Lu, Landis et al. 2006). With 4-μm voxels, little pore connectivity over distances of 100–200 μm was documented for any of the four conditions; with 1-μm voxels the reference concrete showed deep pore penetration while the modified concretes showed clear gaps in connected pore spaces. The somewhat limited data showed chloride permeability correlated clearly with disconnected pore distance (Lu, Landis et al. 2006).

Leaching of cement in mortar (sand plus Portland cement) was studied by repeated synchrotron microCT of the same specimen over 2.2×10^5 s (the 4-day scheduling window at ESRF dictated the length of the experiment) (Burlion, Bernard et al. 2006). Because of the time constraint, calcium efflux was increased by a factor of 300× compared to deionized water by using an ammonium nitrate solution (which produces the same mineral end products as water alone). Variation of linear attenuation coefficient μ in the cement phase was followed as a function of depth from the specimen surface for four exposure times. Within a zone near the surface, μ decreased rapidly during the first 24 h (8.6×10^4 s), and the authors inferred that this was due to decalcification of Portlandite crystals (calcium hydroxide, CH) with little C–S–H (calcium silicate hydrate) involvement. Between 24 and 48 h of leaching (1.5×10^5 s), μ decreased more slowly, indicating that CH was completely removed from this portion of the specimen and that the less soluble (than CH) C–S–H was being removed, a process continuing to the end of the test. The variation of the thickness of the leached region agreed with diffusion-controlled kinetics (i.e., square root of time dependence).

Naik and coworkers studied sulfate attack of Portland cement paste using lab microCT and repeated observations on each specimen (Stock, Naik et al. 2002, Naik 2003, Jupe, Stock et al. 2004, Naik, Kurtis et al. 2004, Naik, Jupe et al. 2006). Two cement types, different water-to-cement ratios, two different cation sulfates, and the presence/absence of aggregates were examined. Figure 10.5 shows how cracks develop in the same slice over 52 weeks of sulfate exposure (pores within the cement paste are used as fiducials). Figure 10.6 shows 3D renderings of sulfate-induced cracking (that develops into spalling) in one specimen over 32 weeks of exposure. As the interpretation of the results depended to a significant extent on use of an additional x-ray modality (position-resolved energy-dispersive x-ray diffraction), further discussion is postponed until the section on multimode studies. A more recent 4D study followed CO_2 attack of cement (Chavez Panduro, Torsæter et al. 2017).

Synchrotron microCT has proven very valuable in studying wood degradation (Illman and Dowd 1999). Fungi enzymes and metabolites degrade the structural integrity of tracheids (thick-walled, tubular structures with hollow centers containing air and/or water), and significant strength can be lost early in the decay process. Over the 96 h following fungal inoculation, tracheid pore volume increased somewhat as did pore interconnectivity (Illman and Dowd 1999). Siloxanes/silanes mixtures are often applied as wood preservatives, and microCT was used to study penetration of the preservative into two types of wood using brominated silane as a contrast agent (Vetter, Cnudde et al. 2006). Boundaries between treated and untreated wood were clear, and one expects that with further work improved wood preservation protocols will result.

10.3.3 Degradation of Biological Structures

This fairly long subsection covers several topics from dissolution of drug tablets to clinical and basic science aspects of tooth decay to degradation of implant materials. Coverage of

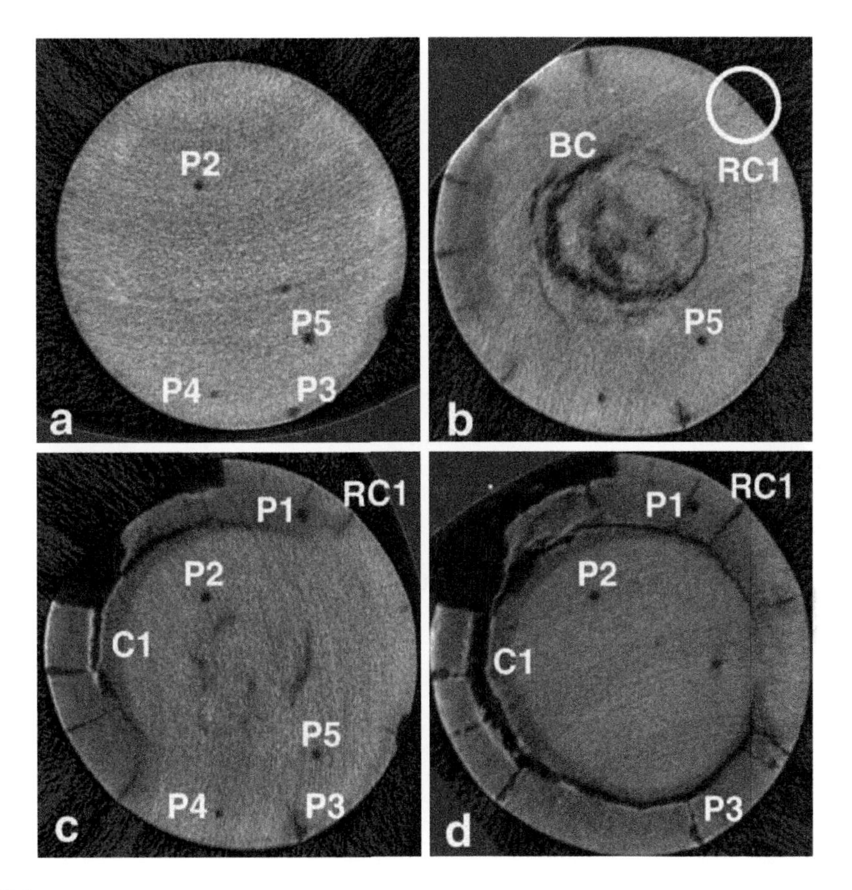

FIGURE 10.5
Matching lab microCT slices of a cement paste sample produced with a water-to-cement ratio of 0.485 and exposed to 10,000 ppm of sulfate ions in a sodium sulfate solution for (a) 21 weeks, (b) 36 weeks, (c) 42 weeks, and (d) 52 weeks. Crack C1, radial crack RC1, cracks within the body of the specimen BC, and pores P1–P5 are labeled. The horizontal field of view is (a) 15.3 mm, (b) 15.1 mm, (c) 16.1 mm, and (d) 15.2 mm, and reconstruction was with 37-μm isotropic voxels. The lighter the pixel, the more absorbing is the voxel. Reprinted from Naik et al. (2006), Copyright (2006), with permission from Elsevier.

topics such as formation of mineral at undesirable locations in soft tissue, growth of soft or hard tissue tumors, destruction of bone (development of high levels of porosity in cortical bone or erosion of bone surfaces), healing of damaged bone, and bone remodeling are covered in Section 10.4 below. Some aspects of bone loss and tumor formation were covered in previously (Chapters 8 and 9, respectively).

Dissolution of drug tablets fits into this subsection in a very loose sense, but the microCT studies of this process deserve brief mention before treating the main subject. MicroCT and modeling were applied to granules and tablets (Ansari and Stepanek 2006, Jia and Williams 2007): as discussed above, this can be a very powerful combination. MicroCT was employed to study the surface membrane responsible for slow release rate characteristics in a tablet (Chauve, Ravenelle et al. 2007).

Teeth consist of an outer covering of enamel and an internal hard tissue dentin, and several microstructural features have been studied with microCT. Before reviewing microCT studies of changing tooth structure, it is sensible to consider microCT data on baseline microstructure and mineral levels.

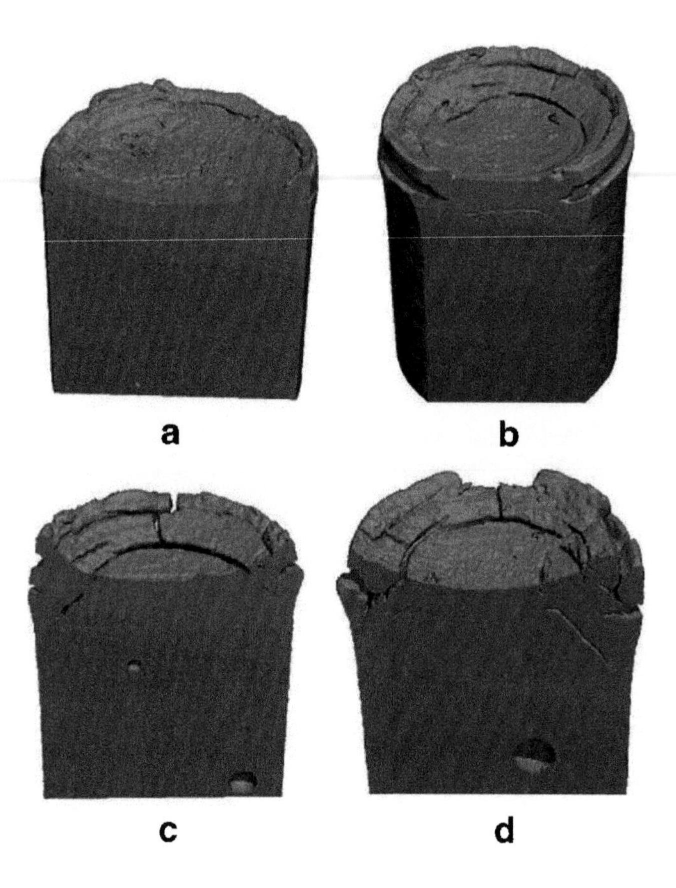

FIGURE 10.6
3D renderings of a single cement paste sample with water-to-cement ratio of 0.435 and exposed to the same solution as the specimens in Fig. 10.5. (a) Nine weeks exposure showing onset of cracking, (b) 16 weeks exposure showing widening of preexisting cracks and formation of new cracks, (c) 22 weeks exposure showing continued cracking, and (d) 32 weeks exposure showing spalling has begun. Reprinted from Naik et al. (2006), Copyright (2006), with permission from Elsevier.

The dentinoenamel junction (DEJ) divides enamel and dentin, has a graded interface structure, and, quite remarkably, couples these elastically very dissimilar materials under the application of complex loading patterns. Tafforeau and coworkers used synchrotron microCT and a multiresolution approach to study details of rhinoceros enamel structure (rods, incremental lines, mineral levels, etc.) and the structure of the DEJ (Tafforeau, Bentaleb et al. 2007). Similar features were examined in hominid enamel with phase contrast microCT (Tafforeau and Smith 2008). Stock et al. studied the structure of bovine DEJ and surrounding tissue with the goal of understanding how the microstructure of the graded interface might contribute to its functionality (Stock, Vieira et al. 2008). In dentin, tubules form during growth and run from the DEJ to pulp chamber; a collar of hypermineralized dentin surrounds the fluid or soft tissue filled tubule and is the subject of considerable interest. Stock and coworkers were able to image the tubules, but higher resolution microCT is needed to observe their structure, such as the ptychography study cited above (Zanette, Enders et al. 2015). Phase microCT revealed tubules in dentin, features whose diameter are only slightly larger than the voxel size in the reconstructions (Zabler, Riesemeier et al. 2006); submicrometer-sized tubule features were subsequently investigated by careful consideration of phase contributions to image formation (Zabler, Cloetens et al. 2007).

By measuring the linear attenuation coefficients of enamel and dentin, Elliott et al. appear to have been the first to quantify differences in mineral content (Elliott, Bromage et al. 1989). Wong et al. demonstrated that a molar and developing incisor could be distinguished quantitatively within a mouse mandible (Wong, Elliott et al. 1991). Mineral content of enamel and dentin has been compared in human molars and enamel pearls (Anderson, Elliott et al. 1996). Gradients in mineral concentration in enamel but not in dentin were found from apex to incisal end of a lower rat incisor (Wong, Elliott et al. 1995a, 1995b). Elliott et al. and Kinney et al. studied demineralization at a carious lesion and subsequent remineralization at the same site (Elliott, Anderson et al. 1992, Kinney, Marshall Jr. et al. 1994). Deciduous enamel defects in low-birthweight children were easily distinguished from normal regions in the same teeth (Fearne, Elliott et al. 1994). Transparent dentin forms gradually with aging and can have different restoration-related fracture properties; measurement of mineral concentration from microCT, crystallite size from small-angle scattering, and fracture toughness were combined in one study of this human tissue (Kinney, Nalla et al. 2005). Mercer and Anderson reported effects of applying CO_2 laser pulses to human enamel. The ability of microCT to quantify microstructure repeatedly can be a key to following processes in biological structures whose microstructure varies greatly between individuals (Mercer and Anderson 1995).

Wong and coworkers used microCT to address an interesting conundrum in clinical dentistry (Wong, Willmott et al. 2006). On the one hand, carious lesions in dentin have long been regarded as conical in shape with the base at the DEJ. On the other hand, experience leads UK clinicians to teach students to remove softened carious dentin with a round bur or a curved excavator instead of a conical bur which would more closely resemble the purported shape of the lesion. MicroCT of ten carious primary molars revealed that the lesions were bowl shaped, explaining the clinical tool preference; quite effective use was made of color 3D renderings showing sound dentin and enamel as nearly transparent grays and mineral levels of the lesions as different colors in the 3D interior of the tooth.

Environmental attack of apatite in tooth enamel has been studied with lab microCT and scanning microradiography for a number of years by Elliott and coworkers (Elliott, Bollet-Quivogne et al. 2005). Over 70 days, these investigators periodically examined (with microCT) packed powders of carbonated apatite in an acidic buffer; these data were supplemented by infrared spectroscopy and Rietveld analysis of x-ray diffraction patterns of the dissected internal surface. This same group also carefully considered microCT-derived mineral levels in sound and carious enamel, and these papers will repay careful study (Elliott, Wong et al. 1998, Dowker, Elliott et al. 2004). A de/remineralization model using small coupons of bovine tooth (enamel plus dentin) was studied with synchrotron and lab microCT; analysis concentrated on quantifying the gradients of mineral content (Delbem, Vieira et al. 2006, Vieira, Delbem et al. 2006) and on one type of treatment (Vieira, Danelon et al. 2017). A 4D model system for investigating tooth demineralization has also been described (Davis, Mills et al. 2018), and a rat model of caries has been developed (Free, DeRocher et al. 2017).

The papers cited at the end of the preceding paragraph illustrate two approaches to use of values of the linear attenuation coefficients in quantitative analyses. Elliott and coworkers took the fundamental approach of relating the linear attenuation coefficients back to mineral standards and expressing measured quantities in terms of absolute amounts of mineral present per unit volume of tissue (Elliott, Wong et al. 1998, Dowker, Elliott et al. 2004). Similar approaches were utilized in studies of bone (Nuzzo, Lafage-Proust et al. 2002,

Borah, Ritman et al. 2005). Although this approach allows for direct comparison between studies conducted under different conditions or with different techniques, extreme care must be taken to avoid systematic errors that might bias comparisons. The quantification of de/remineralization of enamel employed an operational approach, assuming that linear attenuation coefficient values away from the surface were identical from specimen to specimen, and scaling all values to this presumed reference (Delbem, Vieira et al. 2006, Vieira, Delbem et al. 2006). This might be thought to be a poor assumption because enamel mineral levels can differ by several percent or more between the tooth surface and volumes near the DEJ, but the profiles were normalized to values of the linear attenuation coefficient at essentially the same depth from the tooth surface, thereby rendering this consideration moot. Very little of either sort of analysis has appeared to date, although, to be fair, some investigators explicitly verified that experimental values of linear attenuation corresponded to values expected for the material being studied (Kinney, Stock et al. 1990, Stock, Barss et al. 2002).

The dentin–resin interface in tooth restorations is a zone of potential structural and/or chemical weakness; microhardness and microCT were combined to study this zone (Marshall, Balooch et al. 1998). Another study of a dentin adhesive interface paired static loading (via synchrotron microCT) with FEM (Mollica, Santis et al. 2004); fatiguing in a solution of silver ions (and imaging at an energy just above the silver absorption edge) allowed interface leakage to be studied. Efficacy of endodontic seals was studied with synchrotron microCT (Contardo, Luca et al. 2005). Fissures in enamel and changes in mineralization were the subject of another report (Dowker, Elliott et al. 2006).

Balloon dilation followed by stent implantation is a frequently applied percutaneous coronary intervention, but closure within the stented region occurs in a significant number of cases. Stent-related vessel closure may be triggered by introduction of foreign material or the presence of mechanical stresses from the rigid stent within a structure, the artery wall, designed to flex during pulsatile flow. Self-destructing stents that degrade before they are covered by the vascular cell walls are a possible solution, and microCT was used to follow Mg stent degradation over 56 days in minipigs (Loos, Rohde et al. 2007). Note that the biologically necessary ion Mg^{2+} is well tolerated and features antithrombic activity, which is helpful in vessel applications. The in vivo degradation of Mg implants in bone were also studied (Witte, Fischer et al. 2006a). In vivo Mg corrosion rates were about four orders of magnitude lower than those in ASTM in vitro corrosion tests, and relative rates in two Mg implant alloys were opposite in in vivo vs in vitro tests (Witte, Fischer et al. 2006b).

10.3.4 Corrosion of Metals

Materials transport is also important in solids that do not have well-defined porosity, and microCT was very effective for nondestructively quantifying the spatial distribution of corrosion products within printed wiring boards (Dollar 1992, Stock, Dollar et al. 1994). In the presence of high-voltage gradients, copper can diffuse and form conductive anodic filaments (CAF) and short out circuit elements, and microCT quantified the amount of copper within CAF. Results of microCT agreed with those of serial sectioning and with the amount of copper removed from the copper anode (Stock, Dollar et al. 1994). It is important to emphasize that the copper–halide deposits comprising the CAF are typically 20–40 μm below the surface of the printed wiring board, that the translucent boards allow only indistinct visualization of the CAF with optical microscopy, and that an x-ray method is required for nondestructive characterization. Because the contrast from the printed wiring board is so variable and the contrast of the CAF along any particular viewing direction is

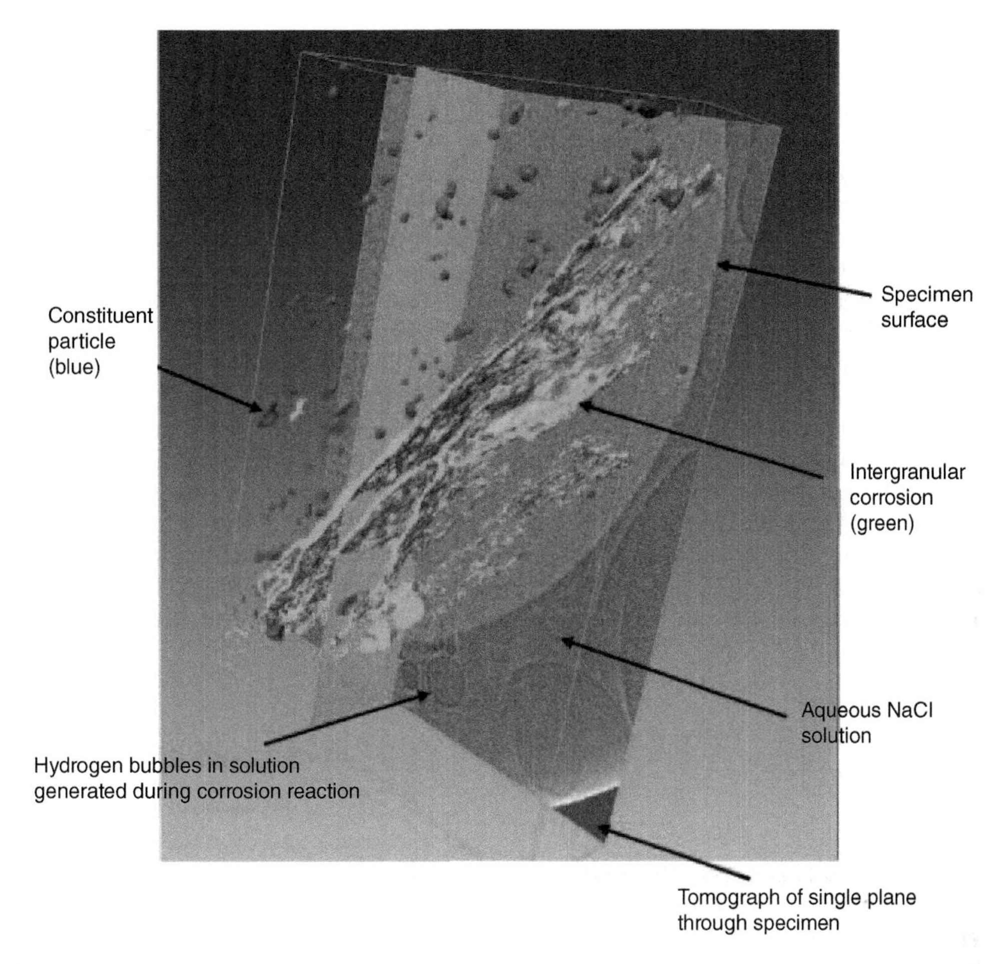

FIGURE 10.7
3D rendering of the heat-affected zone near a weld in AA2024 after 40 h exposure to 0.6 M NaCl. About 200 slices are used, showing the sample transparent except for intergranular corrosion (green) and intermetallic particles (blue). The 0.5-mm diameter specimen was imaged in situ with 20.5-keV synchrotron x-radiation and reconstructed with 0.7-µm isotropic voxels. Reprinted from Connolly et al. (2006), Copyright (2006), with permission of Maney Publishing (http://www.ingentaconnect.com/content/maney/mst).

so weak, radiography cannot provide images better than those from optical microscopy. This work demonstrated how subsurface CAF formation kinetics could be deduced nondestructively with microCT.

Repeated observations of the same specimen were also performed on stainless steel specimens undergoing localized corrosion (Connolly, Homer et al. 2006) and intergranular stress corrosion cracking (Babout, Marrow et al. 2006, Marrow, Babout et al. 2006). Synchrotron microCT observed localized corrosion morphology within Al specimens exposed *in situ* to a chloride environment (Fig. 10.7), and lab microCT investigated the morphology and quantified the transition from localized corrosion to stress corrosion cracking in steel specimens exposed *ex situ* to a simulated corrosive condensate environment (Connolly, Homer et al. 2006). A 302 stainless steel wire was heat treated to produce a stress-free, fully-sensitized microstructure (i.e., one with grain boundary chromium carbides) and examined with synchrotron microCT after three increments of stress in an acidic environment (Babout, Marrow et al. 2006). Analysis centered on identifying bridging ligaments formed during the first increment of crack propagation and on following the progressive failure of the ligaments using a combination of 2D sections and 3D renderings (Fig. 10.8), and these authors noted the presence of unresolved cracks that could still be detected through their phase contrast (Babout, Marrow et al. 2006). Three-dimensional finite element models were devised to investigate the development of crack bridging and its effects on crack propagation and crack coalescence in intergranular stress corrosion cracking (Marrow, Babout et al. 2006).

Synchrotron microCT was used to study the relationship between the spatial distribution of Y in Mg alloy WE43 and the morphology of corrosion attack (Davenport, Padovani et al. 2007). Imaging above and below the Y K-edge (17 keV) increased sensitivity to this element. Repeated observations showed that Y-rich regions slowed local propagation of corrosion. Homogenization of the distribution of Y decreased the corrosion rate, but both general attack and pitting were still observed.

Liquid metal embrittlement was studied through use of the model system of liquid Ga applied to polycrystalline Al (Ludwig, Bouchet et al. 2000). As discussed elsewhere, Ga penetrated many but not all of the grain boundaries (Ludwig, Buffière et al. 2003). Most studies of Ga on Al grain boundaries, however, focused on its use as a decoration so that grain boundary geometry could be correlated with fatigue crack path; see the deformation subsection above.

10.4 Bone and Soft Tissue Adaptation

This section has two parts. The first is on mineralized tissue, the second on soft tissue. Increasingly, in vivo microCT is used for longitudinal studies, some of which were described in the chapter on cellular materials, and various authors have offered guidance on how best to perform these studies (Schambach, Bag et al. 2010, Osborne, Kuntner et al. 2017). Minimizing dose, and thus accurate dosimetry, is a consideration in these types of imaging experiments (Meganck and Liu 2017).

10.4.1 Mineralized Tissue: Implants, Healing, Mineral Levels, and Remodeling

Bone and implants have recently been reviewed (Neldam and Pinholt 2014). NanoCT of bone ultrastructure has also been summarized (Langer and Peyrin 2016). Another review considered longitudinal observation of resorption and formation (Christen and Müller 2017).

Bone formation around implants and the resulting structural integrity (or lack thereof) is a biomedical engineering topic of increasing importance as the number of hip (and knee) replacements increases. Examples of microCT studies include the 3D analysis of bone formation around Ti implants (Bernhardt, Scharnweber et al. 2006), study of bone around dental implants (Cattaneo, Dalstra et al. 2004), and investigation of human tooth–alveolar bone complex (Dalstra, Cattaneo et al. 2006). One complication in many such imaging studies is that the high absorptivity of many implants such as stainless steel or Ti washes out contrast in bone. In the author's experience, this can cripple analysis if contrast is confined to linear, 256-level grayscale; such an 8-bit approach seems to be favored in many different analyses, so this is not an academic concern. Use of nonlinear contrast scales or high-end clipping can be effective; contrast enhancements for bone structures adjacent to

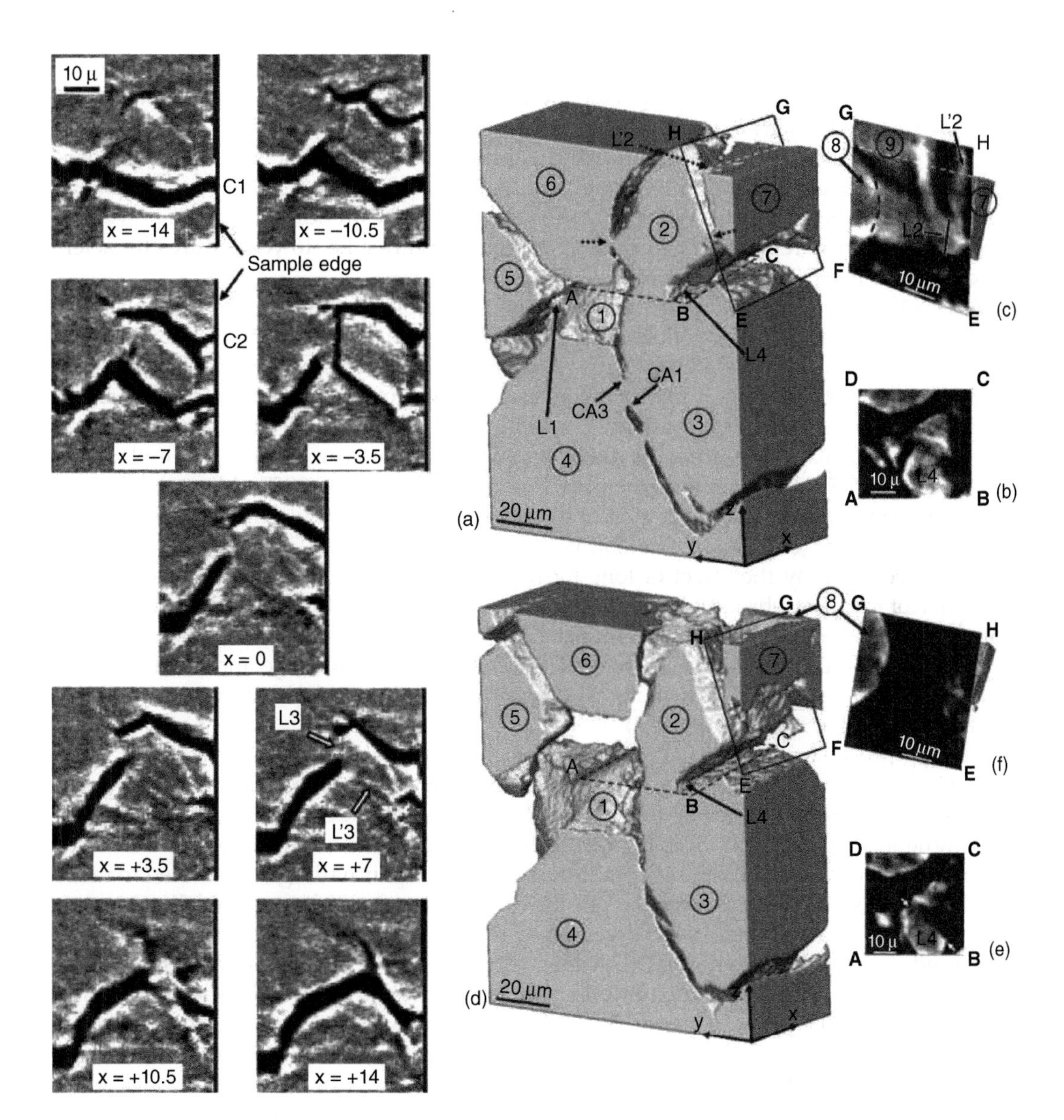

FIGURE 10.8
Intergranular stress corrosion cracks in sensitized 302 stainless steel immersed in 0.15 mol $K_2S_4O_6$ acidifed with dilute H_2SO_4 to pH 2. Imaging was with 30-keV synchrotron x-radiation and with the detector 40 mm from the specimen, and the reconstruction was with 0.7-μm isotropic voxels. Labels have the following meanings: CA1, CA3 indicate positions of crack arrest; L1–L4 uncracked ligaments; C1, C2 crack segments. The nine rectangular images on the left side of the figure are parallel numerical sections (cuts) through the stack of slices; the vertical direction is the direction along which the stress was applied. These data were recorded after the initial increment of cracking under applied stress of 100 MPa and with an open electrical circuit. The relative separation between the cuts (in μm) is given by the value of x. The white blurred edging along the cracks is from phase contrast. The renderings in the right-hand portion of the figure show the same volume of the specimen after the second increment of crack growth, which was done with applied stress of 60 MPa ((a), top) and after the third increment under 40 MPa applied stress ((d), bottom). The circled numbers identify different grains, and the rectangular images at the far right at oblique cuts through the volume indicated on the renderings (i.e., oblique cut ABCD in (b) shows ligament L4 in (a)). Reprinted from Babout et al. (2006), Copyright (2006), with permission of Maney Publishing (http://www.ingentaconnect.com/content/maney/mst).

implants are described elsewhere (Tesei, Casseler et al. 2005, DaPonte, Clark et al. 2006). In synchrotron microCT, phase contrast at edges of different low-absorption tissue types may also aid segmentation in the presence of a very absorbing material.

Wear particles can produce inflammation at bone–implant surfaces and bone resorption or pain. Ren et al. found inflammation but not osteoclastic bone resorption in a study incorporating microCT (Ren, Wu et al. 2006). Cracks and the amount of wear in polyethylene from total disc replacements excised from patients after 1.8–16 years were characterized with microCT (Kurtz, van Ooij et al. 2007).

Weiss et al. performed human trials of an injectable bone substance (biphasic calcium phosphate in a water soluble cellulose polymer) used to fill tooth sockets after extractions; after 3 years microCT revealed bone growth between the artificial mineral particles (Weiss, Layrolle et al. 2007). Seashell material converted to hydroxyapatite was placed in femoral defects in rats, and in vivo microCT scanning (27-μm voxels) after implantation and after 6 weeks revealed the implants did not move, and bone grew toward the implant but untreated control defects remained empty (Vecchio, Zhang et al. 2007). MicroCT showed thread integrity was maintained and bone contact established for titanium-coated PMMA screws placed into goat tibiae (Shalabi, Wolke et al. 2007). MicroCT was one of the techniques used to study the effect of long-term static load on degradation and mechanical integrity of Mg alloys-based biodegradable metals (Koo, Jang et al. 2017). Treatment with bisphosphonates has been linked with osteonecrosis of the jaw bone and has been studied with microCT (Vilarinho, Ferrare et al. 2017). Damage to cartilage and its underlying subchondral bone and subsequent repair in a minipig model is another topic that has been addressed (Pfeifer, Fisher et al. 2017).

Healing processes in bone have also been the subject of multiple investigations. A rat ulnar loading model was studied with pQCT, with FEM, and with attached strain gages (to assess load sharing between ulna and radius), and greatest bone formation in response to fatigue loading was found in regions of high compressive strain (Kotha, Hsieh et al. 2004). Callus formation following tibial osteotomy and repair nailing was studied in mice; application of small (but not large) amplitude cyclic loads after a short delay following surgery improved fracture healing as assessed by callus strength and by callus organization observed qualitatively by microCT (Gardner, van der Meulen et al. 2006). Fracture healing in a rat model of bone union/nonunion was quantified using microCT (Schmidhammer, Zandieh et al. 2006). Although bone healing stimulation was reported via administration of adenoviral vectors encoding bone morphogeneic protein-2 (BMP-2) in small animal models (Betz, Betz et al. 2007), Egermann et al. used microCT of standardized iliac crest defects and tibial osteotomies to find the opposite to be true in sheep (Egermann, Lill et al. 2006). The role of osteopontin (OPN) on bone healing in a mouse femoral fracture model, specifically angiogenesis and fracture callus formation, was examined via microCT in wild-type and $OPN^{-/-}$ mice (Duvall, Taylor et al. 2007). MicroCT data suggested that controlled release of a thrombin peptide may enhance healing of critical and noncritical (sized) defects in rabbits (Sheller, Crowther et al. 2004) and that BMP-silk composite matrices may be useful in healing critical-sized femoral defects in nude rats (Kirker-Head, Karageorgiou et al. 2006). MicroCT was also used to study an extreme case of healing, the regeneration of limbs in adult newts (Stock, Blackburn et al. 2003, Stock, Ignatiev et al. 2004).

It is sometimes medically necessary to fuse the bones of the spine, and microCT was used to evaluate the effectiveness of a growth factor in a rat spinal fusion model (Lu, Zhang et al. 2007). In a rat model, the effect of estrogen loss (Ghodasra, Nickoli et al. 2016) and vancomycin (Mendoza, Sonn et al. 2016) on bone healing in posteriolateral spine fusion mediated by human recombinant BMP-2 have been examined using microCT.

Nanoengineered materials for spine fusion have been investigated (Lee, Fyrner et al. 2017) as have 3D-printed hydroxyapatite-polymer scaffolds (Jakus, Rutz et al. 2016); see Fig. 8.6. In osteoporosis, vertebral bodies can collapse, and one treatment option is to use an artificial spacer: for this purpose, a composite interbody spacer was examined by microCT in an ovine model of interbody lumbar fusion (McGilvray, Waldorff et al. 2017).

The material bone is a discontinuously reinforced composite of apatite (mineral) nanocrystallites dispersed in a matrix primarily of collagen, and it is important to realize that mineral content varies considerably from older, mature areas (highest mineral content) to newly remodeled osteons (lower mineral content). The sensitivity limits of microCT make it ill-suited for quantifying small differences in composition, but a number of studies have shown that the mineralization levels in bone can be mapped. A recent quantitative study used bones and hydroxyapatite standards and compared backscattered electron (BSE) images with microCT to conclude that mean mineralization and most parameters used to characterize bone mineral density distributions can be assessed with microCT in rat-sized cortical bone samples, but caution should be used when reporting the heterogeneity (Mashiatulla, Ross et al. 2017). The cortices of rabbit tibiae have been imaged and little, if any, change in mineral level was observed across the bone (Engelke, Graeff et al. 1993, Wong, Elliott et al. 1995c). Spaceflight has a considerable effect on bone, and microCT was used to map the resulting "striking regional differences" in the distribution of mineral content as a function of distance from the end of femora of growing rats (Mechanic, Arnaud et al. 1990). Lactation also has an enormous effect on bone mineral, and a recent review examined whether the mineral is ever fully replaced (Kovacs 2017).

Synchrotron microCT of low mineralized vs high mineralized volumes in trabecular as a function of bisphosphonate treatment was the subject of one study (Borah, Ritman et al. 2005). Microradiography of thin sections of bone has long been used to show remodeling, and determination of the degree of bone mineralization via this method was compared to that with synchrotron microCT (Nuzzo, Lafage-Proust et al. 2002). Synchrotron microCT was also used to study the degree of mineralization in human normal vertebra, osteoblastic metastases and sites with degenerative osteosclerosis (Sone, Tamada et al. 2004), and in different mouse genetic strains (Martín-Badosa, Amblard et al. 2003, Bayat, Apostol et al. 2005).

Cortical bone from human femora was the subject of a number of early microCT studies, and Haversian and Volkmann canals (for blood supply and innervation axial and transverse to long bone axes, respectively) were resolved. The Haversian/Volkman canals serve as sites for bone remodeling, that is, the replacement of old, damaged bone with new bone. The process involves the concerted action of osteoclasts gouging out old bone followed by osteoblasts depositing bone matrix and mineral. The cylinders of remodeled bone are termed osteons and for some time after they are formed contain lower mineral density than mature bone. The decreased mineral levels in newly formed osteons have long been imaged with microradiographs of thin sections of bone, and synchrotron microCT was used more recently to good effect to reveal the qualitative spatial distribution of mineral density. There are a number of studies of mineral density differences in addition to those cited in the previous paragraph (Ritman, Bolander et al. 1998, Nuzzo, Meneghini et al. 2003, Bossy, Talmant et al. 2004, Bousson, Peyrin et al. 2004, Scherf, Beckmann et al. 2004). Dyck et al. discussed calibration of mineral levels in lab microCT for various calcified tissues, but with the instrumentation available at the time it was difficult to see remodeled osteons in lab microCT (Dyck, Postnov et al. 2002). With the current generation of commercial microCT systems, one expects to detect newly remodeled osteons within the older, fully mineralized bone. Standard measurements can be used to remove systematic biases in values of linear attenuation coefficients resulting from beam hardening (Prevrhal 2004, 2006); although this is important for correctly interpreting mean mineral level, it does not help with detection of differences between recently formed and mature osteons.

Reports were published on determination of Ca/P ratios for bone from synchrotron microCT scans: linear attenuation coefficients from bone were compared to those of two phantoms of calcium phosphate salts with different Ca/P ratios (Speller, Pani et al. 2005, Tzaphlidou, Speller et al. 2006). Although differences in bone mineral density certainly appear to be supported by this group's data, interpretation of differences in terms of Ca/P ratios appears problematic at best. In bone, the relative amounts of collagen and mineral are expected to vary, the density of osteocyte lacunae can alter the apparent linear attenuation coefficients, and other variations will be present (see the discussion of Elliott, Wong et al. 1998, Kozul, Davis et al. 1999, Dowker, Elliott et al. 2004). The variation of ostecyte number density in different human bones has been examined (Andronowski, Mundorff et al. 2017), and a second study found a bimodal distribution of osteocyte lacuna sizes in the human femoral cortex (Hannah, Thomas et al. 2010).

Thin bands of high absorptivity were found in synchrotron microCT data of bone. In animals dosed with $SrCl_2$, these were interpreted as zones of bone with high Sr replacement of Ca in the mineral (Ritman, Bolander et al. 1998). Similar bands of high absorptivity were observed in specimens of human femoral neck bone in extraosteonal areas near to and parallel to the external surface of the cortex; these features in aged bone were interpreted as regions of high mineral content (Cooper, Matyas et al. 2004). The author observed similar high absorptivity bands within cortical bone of animals including newts and mice, bands that examination of adjacent slices could not be due to out-of-plane geometrical features producing unexpected phase contrast. The magnitude of the difference is quite striking: linear attenuation coefficients in the mouse cortical bone were 40% or more greater within the very small volume of hypermineralized tissue than in the "normal" bone, the difference being over twice the standard deviation seen for areas of "normal" bone (Stock, Xiao et al. 2006). It is difficult to imagine that these features, observed by several investigators working at different synchrotron radiation facilities, are an artifact of some sort, but their origin remains obscure.

The current model of bone remodeling holds that Haversian systems are most active where bone undergoes the most extreme loading, that is, where it accumulates the greatest amount of damage, such as microcracking (see Schneider et al. (2006) for examples synchrotron microCT imaging of microcracking associated with bone porosity including osteocyte lacunae). Several studies of channels in cortical bone focused on establishing the equivalence of microCT-based measures of porosity and osteon dimensions with long-standing methods such as microradiography or histology (Wacheter, Augat et al. 2001, Cooper, Matyas et al. 2004). In a study of pore networks in human cortical bone, resolution was shown to have a significant biasing effect on various measurements when it was incommensurate with the structural scale of the pores (Cooper, Turinsky et al. 2007). Pore content and mineral levels determined from microCT correlated with axial ultrasound velocity; the data showed the structure in the outer 1 mm of cortex (about one-half of the cortex thickness) affected velocity and suggest clinical usefulness of this noninvasive monitoring method (Bossy, Talmant et al. 2004).

Clustering of osteons might result from remodeling due to concentrated damage, particularly in bone of aged individuals, and the presence of such a significant stress concentrator could lead to unexpected fracture (Jordan, Loveridge et al. 2000). Intracortical porosity measured in compact tension samples of human bone was quantified and imported into a finite element model of aging bone; keeping nonporosity factors constant, the model showed that a 4% increase in porosity produced a 6% decrease in initiation toughness and a 62% decrease in propagation toughness, while incorporation of other age-related changes into the model affected primarily the initiation toughness (Ural and Vashishth 2007). The hypothesis that the density and range (length) of basic multicellular unit related resorption spaces in human cortical bone vary with age and sex was examined with lab microCT, and the investigators found that the range did not vary and that the age-dependent apparent decrease in resorption space density was due to cortical rarefaction leading to difficulty in detecting resorption spaces with microCT rather than to a decrease in remodeling (Cooper, Thomas et al. 2006). Cooper et al. also showed a rendering of cutting cones (Cooper, Thomas et al. 2006). Nutrient canal network rarefaction (decreased canal cross-sectional area, sectional canal number density, etc.) in disuse-mediated vs control rat tibial diaphyses was confirmed using synchrotron microCT (Matsumoto, Yoshino et al. 2006); positional variation and changes with animal growth were documented in subsequent work (Matsumoto, Yoshino et al. 2007). Renders and coworkers determined the 3D distribution of the cortical canal network in the human mandibular condyle and related this to the directions of principal stresses and strains (Renders, Mulder et al. 2007).

Bone is a hierarchically structured tissue, and, to this point, discussion has focused on microstructural and higher levels. NanoCT, with absorption and with phase contrast, and techniques such as ptychography have advanced to the point where results are becoming available at the ultrastructural level, that is, with voxels substantially smaller than 100 nm. Results include visualization of osteocyte–canaliculi networks, evidence of mineralized collagen fibril orientation, and data on mineral density variation in the vicinity of osteocyte lacunae; see Langer and Peyrin (2016) for a review. While tissue alteration due to x-ray dose is always a consideration when studying biological tissues, the concerns about tissue damage affecting the observations are even greater with nanoCT.

10.4.2 Soft Tissue and Soft Tissue Interfaces

Degradation of biological structures can occur by the unwanted addition of mineral to soft tissue structures or erosion of bone at soft tissue interfaces. Cancers of various types affect bone and soft tissue sites; these microCT studies, especially those with a longitudinal component, have been very informative and are covered after discussion of errant calcification and bone erosion.

Lack of small-diameter synthetic vascular grafts limits the possibility of life- or limb-saving surgery for many individuals, and a mouse model for evaluation of these grafts was developed and assessed with microCT (Lopez-Soler, Brennan et al. 2007). Bioprosthetic valves calcify, and assessing the tendency of new valve designs (including materials) to calcify in vitro is an important first step toward clinical use. MicroCT was used to assess porcine heart valve calcification in a more physically realistic accelerated testing model involving pH control of the mineralizing solution and millions cycles of valve flexure (Krings, Kanellopoulou et al. 2006). Calcification in excised human valves has been studied by microCT (Rajamannan, Nealis et al. 2005, Rajamannan, Subramanium et al. 2005) (see Fig. 6.16), and in vivo microCT of vascular calcification in a rat model of chronic renal failure produced interesting results (Persy, Postnov et al. 2006). Valve calcification in a large number of ex vivo specimens was investigated using microCT in order to quantify the amount of calcification in aortic stenosis and differences between bicuspid and tricuspid aortic valves (Mazur, Wypasek et al. 2017). Microcalcifications in the cap of coronary plaques from autopsy specimens were observed with microCT, and numerical modeling suggested that they provide enough stress concentration to rupture the plaque in an unexpected location (Vengrenyuk, Carlier et al. 2006).

Rheumatoid arthritis involves bone destruction, whereas bone is formed in inappropriate locations in osteoarthritis. Figure 4.6 compares matching tube-based and synchrotron microCT slices of the dysfunctional mineralized tissue formed in a mouse model of osteoarthritis. Appleton and coworkers studied a rat model of preclinical osteoarthritis using in vivo microCT at several time points up to 20 weeks and found forced mobilization after surgery (knee joint destabilization by anterior cruciate ligament transection and partial medial meniscectomy) accelerated pathogenesis (Appleton, McErlain et al. 2007). The dose dependence of a therapeutic agent for rheumatoid arthritis was established in a rat model using microCT to measure changes in total bone volume (Schopf, Savinainen et al. 2006). Bone erosion in a preclinical model of rheumatoid arthritis was determined via local roughness quantification for the rat talus bone (Silva, Ruan et al. 2006). The approach of Silva and coworkers, outlined in Chapter 6, appeared to be much more sensitive to the early stages of damage than measurements of quantities such as the total bone volume and seems to be quite robust; the method also allowed mapping on the surface of 3D renderings of the locations of greatest roughness. MicroCT observed that whole-body vibration of mice induced progressive degeneration of intervertebral discs that was associated with increased expression of Il-1β and multiple matrix-degrading enzymes (McCann, Veras et al. 2017).

There have been quite a number of microCT studies of tumor formation in animal models. In a rat model of bone cancer pain, osteolytic bone loss in tibiae from the rat mammary tumor cells was examined (Roudier, Bain et al. 2006; see also He et al. (2017); osteoblastic as well as osteolytic lesions can form (Hu, Gerseny et al. 2011). In a murine model of cancer-induced bone pain, behavioral manifestations of pain were found to emerge in parallel with progression of bone destruction revealed by microCT, and the effectiveness of pain reduction drugs was assessed (El Mouedden and Meert 2005). The effect of a noncalcemic vitamin D analog on prostate cancer metastasis was examined in mice with microCT (Peleg, Khan et al. 2005). Squamous cell carcinoma and its spread to murine mandibles was the subject of another study (Henson, Li et al. 2007). In a melanoma metastasis mouse model, PET and longitudinal microCT imaging were used to study development of lung, mandibular, long bone, and subcutaneous metastases (Winkelmann, Figueroa et al. 2006). Longitudinal microCT-derived biomarkers for lung metastasis detection in a syngeneic mouse model added value to bioluminescence imaging (Marien, Hillen et al. 2017). Development of metastases in the murine liver was studied longitudinally with microCT using a hepatobiliary-specific contrast agent (Ohta, Lai et al. 2006).

In vivo microCT imaging is especially challenging for lung and heart studies, and many of these papers were reviewed in Chapter 9. Two studies that relate to damaged structures are mentioned here. Cody et al. used in vivo microCT to observe lung damage in a mouse model of pulmonary fibrosis; breath-hold imaging and respiratory gating were two methods used to eliminate motion artifacts (Cody, Cavanaugh et al. 2004, Cody, Nelson et al. 2005). Model murine myocardial infarct size was determined in another study using a gating approach, and observations 5 or 35 days after the damage was introduced (Nahrendorf, Badea et al. 2007). The cardiac studies of mice were particularly challenging because of the small heart size and its rapid motion.

Staining is often used to increase contrast in soft tissue samples for histology or for microCT imaging. Typically, staining times are determined through lengthy trial-and-error processes. A recent dual-energy microCT study quantified the uptake of staining agents in animal organs (murine heart and lung) in 3D (Martins de Souza E Silva, Utsch et al. 2017).

References

Al-Khulaifi, Y., Q. Lin, M. J. Blunt and B. Bijeljic (2018). "Reservoir-condition pore-scale imaging of dolomite reaction with supercritical CO_2 acidified brine: Effect of pore-structure on reaction rate using velocity distribution analysis." *Int J Greenhouse Gas Control* **68**: 99–111.

Al-Raoush, R. and M. Alsaleh (2007). "Simulation of random packing of polydisperse particles." *Powder Technol* **176**: 47–55.

Altman, S. J., M. L. Rivers, M. D. Reno, R. T. Cygan and A. A. Mclain (2005). "Characterization of adsorption sites on aggregate soil samples using synchrotron x-ray computerized microtomography." *Env Sci Technol* **39**: 2679–2685.

Anderson, P., J. C. Elliott, U. Bose and S. J. Jones (1996). "A comparison of the mineral content of enamel and dentine in human premolars and enamel pearls measured by x-ray microtomography." *Archs Oral Biol* **41**: 281–290.

Andronowski, J. M., A. Z. Mundorff, I. V. Pratt, J. M. Davoren and D. M. Cooper (2017). "Evaluating differential nuclear DNA yield rates and osteocyte numbers among human bone tissue types: A synchrotron radiation microCT approach." *Forensic Sci Int Genet* **28**: 211–218.

Ansari, M. A. and F. Stepanek (2006). "Design of granule structure: Computational methods and experimental realization." *AIChE J* **52**: 3762–3774.

Appleton, C. T. G., D. D. McErlain, V. Pitelka, N. Schwartz, S. M. Bernier, J. L. Henry, D. W. Holdsworth and F. Beier (2007). "Forced mobilization accelerates pathogenesis: Characterization of a preclinical surgical model of osteoarthritis." *Arth Res Therap* **9**: R13.

Atwood, R. C., P. D. Lee, R. V. Curtis and D. M. Maijer (2007). "Modeling the investment casting of a titanium crown." *Dent Mater* **23**: 60–70.

Aucott, L., D. Huang, H. B. Dong, S. W. Wen, J. A. Marsden, A. Rack and A. C. Cocks (2017). "Initiation and growth kinetics of solidification cracking during welding of steel." *Sci Rep* **7**: 40255.

Babout, L., T. J. Marrow, D. Engelberg and P. J. Withers (2006). "X-ray microtomographic observation of intergranular stress corrosion cracking in sensitised austenitic stainless steel." *Mater Sci Technol* **22**: 1068–1075.

Baruchel, J., J. Y. Buffiere, P. Cloetens, M. D. Michiel, E. Ferrie, W. Ludwig, E. Maire and L. Salvo (2006). "Advances in synchrotron radiation microtomography." *Scripta Mater* **55**: 41–46.

Bayat, S., L. Apostol, E. Boller, T. Borchard and F. Peyrin (2005). "In vivo imaging of bone microarchitecture in mice with 3D synchrotron radiation microtomography." *Nucl Instrum Meth A* **548**: 247–252.

Bedford, J., F. Fusseis, H. Leclère, J. Wheeler and D. Faulkner (2017). "A 4D view on the evolution of metamorphic dehydration reactions." *Sci Rep* **7**: 6881.

Bentz, D. P., S. Mizell, S. Satterfield, J. Devaney, W. George, P. Ketcham, J. Graham, J. Porterfield, D. Quenard, F. Vallee, H. Sallee, E. Boller and J. Baruchel (2002). "The visible cement data set." *J Res NIST* **107**: 137–148. See also www.visiblecement.nist.gov.

Bentz, D. P., D. A. Quenard, H. M. Kunzel, J. Baruchel, F. Peyrin, N. S. Martys and E. J. Garboczi (2000). "Microstructure and transport properties of porous building materials. II: Three-dimensional x-ray tomographic studies." *Mater Struct* **33**: 147–153.

Bernard, D. (2005). "3D quantification of pore scale geometrical changes using synchrotron computed microtomography." *Oil Gas Sci Technol* **60**: 747–762.

Bernard, D., D. Gendron, J. M. Heintz, S. Bordere and J. Etourneau (2005). "First direct 3D visualisation of microstructural evolutions during sintering through x-ray computed microtomography." *Acta Mater* **53**: 121–128.

Bernard, D., G. L. Vignoles and J. M. Heintz (2000). Modelling porous materials evolution. *X-Ray Tomography in Materials Science*. J. Baruchel, J. Y. Buffière, E. Maire, P. Merle and G. Peix. Paris, Hermes Science: 177–192.

Bernhardt, R., D. Scharnweber, B. Müller, F. Beckmann, J. Goebbels, J. Jansen, H. Schliephake and H. Worch (2006). 3D analysis of bone formation around Ti implants using microcomputed tomography (μCT). *Developments in X-Ray Tomography V.* U. Bonse. Bellingham (WA), SPIE. **SPIE Proc Vol 6318**: 631807-1–631807-10.

Betz, V. M., O. B. Betz, V. Glatt, L. C. Gerstenfeld, T. A. Einhorn, M. L. Bouxsein, M. S. Vrahas and C. H. Evans (2007). "Healing of segmental bone defects by direct percutaneous gene delivery: Effect of vector dose." *Hun Gene Therap* **18**: 907–915.

Blankenburg, C., A. Rack, C. Daul and J. Ohser (2017). "Torsion estimation of particle paths through porous media observed by in-situ time-resolved microtomography." *J Microsc* **266**: 141–152.

Borah, B., E. L. Ritman, T. E. Dufresne, S. M. Jorgensen, S. Liu, J. Sacha, R. J. Phipps and R. T. Turner (2005). "The effect of risedronate on bone mineralization as measured by microcomputed tomography with synchrotron radiation: correlation to histomorphometric indices of turnover." *Bone* **37**: 1–9.

Bossy, E., M. Talmant, F. Peyrin, L. Akrout, P. Cloetens and P. Laugier (2004). "An in vitro study of the ultrasonic axial transmission technique at the radius: 1 MHz velocity measurement are sensitive to both mineralization and intracortical porosity." *J Bone Miner Res* **19**: 1548–1556.

Bousson, V., F. Peyrin, C. Bergot, M. Hausard, A. Sautet and J. D. Laredo (2004). "Cortical bone in the human femoral neck: Three-dimnsional appearance and porosity using synchrotron radiation." *J Bone Miner Res* **19**: 794–801.

Brunetti, A., S. Bidali, A. Mariani and R. Cesareo (2004). X-ray tomography for the visualization of monomer and polymer filling inside wood and stone. *Developments in X-Ray Tomography IV.* U. Bonse. Bellingham (WA), SPIE. **SPIE Proc Vol 5535**: 191–200.

Bstos-Vanegas, J. D., P. C. Correa, M. A. Martins, F. M. Baptestini, R. C. Camposa, G. H. H. de Oliveira and E. H. M. Nunes (2018). "Developing predictive models for determining physical properties of coffee beans during the roasting process." *Ind Crops Products* **112**: 839–845.

Buffière, J. Y., S. Savelli and E. Maire (2000). Characterisation of MMC_P and cast aluminum alloys. *X-Ray Tomography in Materials Science.* J. Baruchel, J. Y. Buffière, E. Maire, P. Merle and G. Peix. Paris, Hermes Science: 103–114.

Burlion, N., D. Bernard and D. Chen (2006). "X-ray microtomography: Application to microstructure analysis of a cementitious material during leaching process." *Cement Concr Res* **36**: 346–357.

Butts, M. D. (1993). Nondestructive examination of nicalon fiber composite preforms using x-ray tomographic microscopy. MS thesis, Georgia Institute of Technology.

Cattaneo, P. M., M. Dalstra, F. Beckmann, T. Donath and B. Melsen (2004). Comparison of conventional and synchrotron-radiation-based microtomography of bone around dental implants. *Developments in X-Ray Tomography IV.* U. Bonse. Bellingham (WA), SPIE. **SPIE Proc Vol 5535**: 757–764.

Chaijaruwanich, A., P. D. Lee, R. J. Dashwood, Y. M. Youssef and H. Nagaumi (2007). "Evolution of pore morphology and distribution during the homogenization of direct chill cast Al–Mg alloys." *Acta Mater* **55**: 285–293.

Chauve, G., F. Ravenelle and R. H. Marchessault (2007). "Comparative imaging of a slow-release starch excipient tablet: Evidence of membrane formation." *Carbohyd Poly* **70**: 61–67.

Chavez Panduro, E. A., M. Torsæter, K. Gawel, R. Bjørge, A. Gibaud, Y. Yang, S. Bruns, Y. Zheng, H. O. Sørensen and D. W. Breiby (2017). "In-situ X-ray tomography study of cement exposed to CO(2) saturated brine." *Environ Sci Technol* **51**: 9344–9351.

Chen, T. T., A. Sasov, J. F. Dutrizac, P. Kondos and G. Poirier (2006). "The x-ray tomography of a siliceous silver ore." *J Metals* **58**: 41–44.

Christen, P. and R. Müller (2017). "In vivo visualisation and quantification of bone resorption and bone formation from time-lapse imaging." *Curr Osteoporos Rep* **15**: 311–317.

Cnudde, V. and P. J. S. Jacobs (2004). "Monitoring of weathering and conservation of building materials through non-destructive x-ray computed microtomography." *Env Geol* **46**: 477–485.

Cody, D., D. Cavanaugh, R. E. Price, B. Rivera, G. Gladish and E. Travis (2004). Lung imaging of laboratory rodents in vivo. *Developments in X-Ray Tomography IV.* U. Bonse. Bellingham (WA), SPIE. **SPIE Proc Vol 5535**: 43–52.

Cody, D. D., C. L. Nelson, W. M. Bradley, M. Wislez, D. Juroske, R. E. Price, X. Zhou, B. N. Bekele and J. M. Kurie (2005). "Murine lung tumor measurement using respiratory-gated microComputed tomography." *Invest Radiol* **40**: 263–269.

Coléou, C. and J. M. Barnola (2001). "3-D snow and ice images by x-ray microtomography." *ESRF Newsletter*: 24–26.

Connolly, B. J., D. A. Homer, S. J. Fox, A. J. Davenport, C. Padovani, S. Zhou, A. Turnbull, M. Preuss, N. P. Stevens, T. J. Marrow, J. Y. Buffiere, E. Boller, A. Groso and M. Stampanoni (2006). "X-ray microtomography studies of localised corrosion and transitions to stress corrosion cracking." *Mater Sci Technol* **22**: 1076–1085.

Contardo, L., M. D. Luca, M. Biasotto, R. Longo, A. Olivo, S. Pani and R Di Lenarda (2005). "Evaluation of the endodontic apical seal after post insertion by synchrotron radiation tomography." *Nucl Instrum Meth A* **548**: 253–256.

Cooper, D. M. L., J. R. Matyas, M. A. Katzenberg and B. Hallgrimsson (2004). "Comparison of micro-computed tomographic and microradiaographic measurements of cortical bone porosity." *Calcif Tiss Int* **74**: 437–447.

Cooper, D. M. L., C. D. L. Thomas, J. G. Clement and B. Hallgrimsson (2006). "Three-dimensional microComputed tomography imaging of basic multicellular unit-related resoprtion spaces in human cortical bone." *Anat Rec* **288A**: 806–816.

Cooper, D., A. Turinsky, C. Sensen and B. Hallgrimsson (2007). "Effect of voxel size on 3D microCT analysis of cortical bone porosity." *Calcif Tiss Int* **80**: 211–219.

Dalstra, M., P. M. Cattaneo, F. Beckmann, M. T. Sakima, G. Lemor, M. G. Laursen and B. Melsen (2006). Microtomography of the human tooth-alveolar bone complex. *Developments in X-Ray Tomography V.* U. Bonse. Bellingham (WA), SPIE. **SPIE Proc Vol 6318**: 631804-1–631804-9.

DaPonte, J. S., M. Clark, P. Nelson, T. Sadowski and E. Wood (2006). Quantitative confirmation of visual improvements to microCT bone density images. *Visual Information Processing XV.* Z. Rahman, S. E. Reichenbach and M. A. Neifeld. Bellingham (WA), SPIE. **SPIE Proc Vol 6246**: 62460D-1–62460D-9.

Davenport, A. J., C. Padovani, B. J. Connolly, N. P. C. Stevens, T. A. W. Beale, A. Groso and M. Stampanoni (2007). "Synchrotron X-ray microtomography study of the role of Y in corrosion of magnesium alloy WE43." *Electrochem Sol State Lett* **10**: C5–C8.

Davis, G. R., D. Mills and P. Anderson (2018). "Real-time observations of tooth demineralization in 3 dimensions using X-ray microtomography." *J Dent* **69**: 88–92.

De Gryze, S., L. Jassogne, J. Six, H. Bossuyt, M. Wevers and R. Merckx (2006). "Pore structure changes during decomposition of fresh residue: X-ray tomography analyses." *Geoderma* **134**: 82–96.

Delbem, A. C. B., A. E. M. Vieira, K. T. Sassaki, M. L. Cannon, S. R. Stock, X. Xiao and F. De Carlo (2006). Quantitative analysis of mineral content in enamel using synchrotron microtomography and microhardness analysis. *Developments in X-Ray Tomography V.* U. Bonse. Bellingham (WA), SPIE. **SPIE Proc Vol 6318**: 631824-1–631824-5.

Dobrich, K. M., C. Rau and C. E. Krill III (2004). "Quantitative characterization of the three-dimensional microstructure of polycrystalline Al-Sn using x-ray microtomography." *Metall Mater Trans A* **35**: 1953–1961.

Dollar, L. L. (1992). Evaluation of nondestructive x-ray techniques for electronic packaging materials. MS thesis, Georgia Institute of Technology.

Dowker, S. E., J. C. Elliott, G. R. Davis, R. M. Wilson and P. Cloetens (2004). "Synchrotron x-ray microtomographic investigation of mineral concentrations at micrometer scale in sound and carious enamel." *Caries Res* **38**: 514–522.

Dowker, S. E., J. C. Elliott, G. R. Davis, R. M. Wilson and P. Cloetens (2006). "Three-dimensional study of human dental fissure enamel by synchrotron x-ray microtomography." *Eur J Oral Sci* **114**: Suppl 1, 353–359; discussion 375–376, 382–383.

Duvall, C. L., W. R. Taylor, D. Weiss, A. M. Wojtowicz and R. E. Guldberg (2007). "Impaired angiogenesis, early callus formation, and late stage remodeling in fracture healing of osteopontin-deficient mice." *J Bone Miner Res* **22**: 286–297.

Dyck, D., A. Postnov, S. Saveliev, A. Saso and N. M. D. Clerck (2002). Definition of local density in biological calcified tissues using x-ray microtomography. *Medical Imaging 2002: Visualization, Image-Guided Procedures and Display.* S. K. Mun. Bellingham (WA), SPIE. **SPIE Proc Vol 4681**: 749–755.

Egermann, M., C. A. Lill, K. Griesbeck, C. H. Evans, P. D. Robbins, E. Schneider and A. W. Baltzer (2006). "Effect of BMP-2 gene transfer on bone healing in sheep." *Gene Therap* **13**: 1290–1299.

El Mouedden, M. and T. D. Meert (2005). "Evaluation of pain-related behavior, bone destruction and effectiveness of fentanyl, sufentanil and morphine in a muring model of cancer pain." *Pharmacol Biochem Behav* **82**: 109–119.

Elliott, J. C., P. Anderson, G. R. Davis, F. S. L. Wong, X. J. Gao, S. D. Dover and A. Boyde (1992). X-ray microtomographic studies of bone and teeth. *X-Ray Microscopy III*. A. Michette, G. R. Morrison and C. J. Buckley. New York, Springer: 461–464.

Elliott, J. C., F. R. G. Bollet-Quivogne, P. Anderson, S. E. P. Dowker, R. M. Wilson and G. R. Davis (2005). "Acidic demineralization of apatites studied by scanning x-ray microradiography and microtomography." *Mineralogical Mag* **69**: 643–652.

Elliott, J. C., T. G. Bromage, P. Anderson, G. Davis and S. D. Dover (1989). Application of microtomography to the study of dental hard tissues. *Enamel V*. R. W. Feanhead. Tsurumi, Florence Publishers: 429–433.

Elliott, J. C., F. S. L. Wong, P. Anderson, G. R. Davis and S. E. P. Dowker (1998). "Determination of mineral concentration in dental enamel from x-ray attenuation measurements." *Conn Tiss Res* **38**: 61–72.

Engelke, K., W. Graeff, L. Meiss, M. Hahn and G. Delling (1993). "High spatial resolution imaging of bone mineral using computed microtomography: Comparison with microradiography and undecalcified histologic sections." *Invest Radiol* **28**: 341–349.

Fearne, J. M., J. C. Elliott, F. S. Wong, G. R. Davis, A. Boyde and S. J. Jones (1994). "Deciduous enamel defects in low-birth-weight children: Correlated X-ray microtomographic and backscattered electron imaging study of hypoplasia and hypomineralization." *Anat Embryol* **189**: 375–381.

Feeney, D. S., J. W. Crawford, T. Daniell, P. D. Hallett, N. Nunan, K. Ritz, M. Rivers and I. M. Young (2006). "Three-dimensional microorganization of the soil-root-microbe system." *Microb Ecol* **52**: 151–158.

Flin, F., J. B. Brzoska, D. Coeurjolly, R. A. Pieritz, B. Lesaffre, C. Coléou, P. Lamboley, O. Teytaud, G. L. Vignoles and J. F. Delesse (2005). "Adaptive estimation of normals and surface area for discrete 3-D objects: Application to snow binary data from x-ray tomography." *IEEE Trans Image Process* **14**: 585–596.

Flin, F., J. B. Brzoska, B. Lesaffre, C. Coléou and R. A. Pieritz (2003). "Full three-dimensional modeling of curvature-dependent snow metamorphism: First results and comparison with experimental tomographic data." *J Phys D* **36**: A49–A54.

Free, R. D., K. DeRocher, S. R. Stock, D. Keane, A. K. Scott, W. H. Bowen and D. Joester (2017). "Characterization of enamel caries lesions in rat molars using synchrotron X-ray microtomography." *J Synchrotron Radiat* **24**: 1056–1064.

Fu, X., J. A. Elliott, A. C. Bentham, B. C. Hancock and R. E. Cameron (2006). "Application of x-ray microtomography and image processing to the investigation of a compacted granular system." *Part Part Syst Character* **23**: 229–236.

Fu, X., G. E. Milroy, M. Dutt, A. C. Bentham, B. C. Hancock and J. A. Elliott (2005). Quantitative analysis of packed and compacted granular system by x-ray microtomography. *Medical Imaging 2005 - Image Processing*. J. M. Fitzpatrick and J. M. Reinhardt. Bellingham (WA), SPIE. **SPIE Proc Vol 5747, n III**: 1955–1964.

Gallucci, E., K. Scrivener, A. Groso, M. Stampanoni and G. Margaritondo (2007). "3D experimental investigation of the microstructure of cement pastes using synchrotron X-ray microtomography (μCT)." *Cement Concr Res* **37**: 360–368.

Garboczi, E. J. and J. W. Bullard (2004). "Shape analysis of a reference cement." *Cement Concr Res* **34**: 1933–1937.

Gardner, M. J., M. C. H. van der Meulen, D. Demetrakopoulos, T. M. Wright, E. R. Myers and M. P. Bostrom (2006). "In vivo cyclic axial compression affects bone healing in the mouse tibia." *J Orthop Res* **24**: 1679–1686.

Gay, S. L. and R. D. Morrison (2006). "Using two dimensional sectional distributions to infer three dimensional volumetric distributions – validation using tomography." *Part Part Syst Charact* **23**: 246–253.

Ghodasra, J. H., M. S. Nickoli, S. Z. Hashmi, J. T. Nelson, M. C. Mendoza, S. S. Bellary, K. A. Sonn, A. Ashtekar, C. Park, C. Yun, A. Kannan, S. Stock, E. L. Hsu and W. K. Hsu (2016). "Ovariectomy-induced osteoporosis does not impact fusion rates in an rhBMP-2-dependent rat posterolateral arthrodesis model." *Global Spine J* **6**: 60–68.

Gibbs, J. W., K. A. Mohan, E. B. Gulsoy, A. J. Shahani, X. Xiao, C. A. Bouman, M. D. Graef and P.W. Voorhees (2015). "The three-dimensional morphology of growing dendrites." *Sci Rep* **5**: 11824.

Golchert, D., R. Moreno, M. Ghadiri and J. Litster (2004a). "Effect of granule morphology on breakage behaviour during compression." *Powder Technol* **143–144**: 84–96.

Golchert, D. J., R. Moreno, M. Ghadiri, J. Litster and R. Williams (2004b). "Application of x-ray microtomography to numerical simulations of agglomerate breakage by distinct element method." *Adv Powder Technol* **15**: 447–457.

Guillermic, R.-M., F. Koksel, X. Sun, D. W. Hatcher, M. T. Nickerson, G. S. Belev, M. A. Webb, J. H. Page and M. G. Scanlon (2018). "Bubbles in noodle dough: Characterization by X-ray microtomography." *Food Res Int* **105**: 548–555.

Hannah, K. M., C. D. Thomas, J. G. Clement, F. De Carlo and A. G. Peele (2010). "Bimodal distribution of osteocyte lacunar size in the human femoral cortex as revealed by microCT." *Bone* **47**: 866–871.

Hansel, C. M., M. J. L. Force, S. Fendorf and S. Sutton (2002). "Spatial and temporal association of As and Fe species on aquatic plant roots." *Env Sci Technol* **36**: 1988–1994.

He, F., A. E. Chiou, H. C. Loh, M. Lynch, B. R. Seo, Y. H. Song, M. J. Lee, R. Hoerth, E. L. Bortel, B. M. Willie, G. N. Duda, L. A. Estroff, A. Masic, W. Wagermaier, P. Fratzl and C. Fischbach (2017). "Multiscale characterization of the mineral phase at skeletal sites of breast cancer metastasis." *Proc Natl Acad Sci U S A* **114**: 10542–10547.

Heldele, R., S. Rath, L. Merz, R. Butzbach, M. Hagelstein and J. Haußelt (2006). "X-ray tomography of powder injection moulded micro parts using synchrotron radiation." *Nucl Instrum Meth B* **246**: 211–216.

Henson, B., F. Li, D. D. Coatney, T. E. Carey, R. S. Mitra, K. L. Kirkwood and N. J. D'Silva (2007). "An orthotopic floor-of-mouth model for locoregional growth and spread of human squamous cell carcinoma." *J Oral Pathol Med* **36**: 363–370.

Hettiarachchi, G. M., E. Lombi, M. J. McLaughlin, D. Chittleborough and P. Self (2006). "Density changes around phosphorus granules and fluid bands in a calcareous soil "*Soil Sci Soc Am J* **70**: 960–966.

Hou, Y., G. Jia, R. Yue, C. Chen, J. Pei, H. Zhang, H. Huang, M. Xiong and G. Yuan (2018). "Synthesis of biodegradable Zn-based scaffolds using NaCl templates: relationship between porosity, compressive properties and degradation behavior." *Mater Char* **137**: 162–169.

Hu, Z., H. Gerseny, Z. Zhang, Y.-J. Chen, A. Berg, Z. Zhang, S. Stock and P. Seth (2011). "Oncolytic adenovirus expressing soluble TGFβ receptor II-Fc-mediated inhibition of established bone metastases: A safe and effective systemic therapeutic approach for breast cancer." *Mol Ther* **19**: 1609–1618.

Illman, B. and B. Dowd (1999). High resolution microtomography for density and spatial information about wood structures. *Developments in X-Ray Tomography II*. U. Bonse. Bellingham (WA), SPIE. **SPIE Proc Vol 3772**: 198–204.

Islam, S. F., T. W. Wysokinski, G. Belev, R. V. Sundara, S. Whitehouse, S. Palzer, M. J. Hounslow and A. D. Salman (2018). "Food suspensions study with SR microtomography." *Chem Eng Sci* **175**: 208–219.

Jakus, A. E., A. L. Rutz, S. W. Jordan, A. Kannan, S. M. Mitchell, C. Yun, K. D. Koube, S. C. Yoo, H. E. Whiteley, C.-P. Richter, R. D. Galiano, W. K. Hsu, S. R. Stock, E. L. Hsu and R. N. Shah (2016). "Hyperelastic 'bone': A highly versatile, growth factor-free, osteoregenerative, scalable, and surgically friendly biomaterial." *Sci Transl Med* **358**: 358ra127.

Jia, X. and R. A. Williams (2007). "A hybrid mesoscale modelling approach to dissolution of granules and tablets." *Chem Eng Res Design* **85**(7A): 1027–1038.

Jones, K. W., H. Feng, E. A. Stern, U. Neuhäusler, N. Marinkovic and Z. Song (2006). "Properties of New York/New Jersey harbor sediments." *Acta Phys Polonica A* **109**: 279–286.

Jordan, G. R., N. Loveridge, K. L. Bell, K. Power, N. Rushton and J. Reeve (2000). "Spatial clustering of remodeling osteons in the femoral neck cortex: A cause of weakness in hip fracture?" *Bone* **26**: 305–313.

Jupe, A. C., S. R. Stock, P. L. Lee, N. N. Naik, K. E. Kurtis and A. P. Wilkinson (2004). "Phase composition depth profiles using spatially resolved energy dispersive x-ray diffraction." *J Appl Cryst* **37**: 967–976.

Kakaboura, A., C. Rahiotis, D. Watts, N. Silikas and G. Eliades (2007). "3D-marginal adaptation versus setting shrinkage in light-cured microhybrid resin composites." *Dent Mater* **23**: 272–278.

Kang, Z., R. Johnson, J. Mi, S. Bondi, M. Jiang, W. J. Lackey, S. Stock and K. More (2004). "Microstructure of carbon fibers prepared laser CVD." *Carbon* **42**: 2721–2727.

Kataria, A., S. Oka, D. Smrcka, Z. Grof, F. Sťepánek and R. Ramachandran (2018). "A quantitative analysis of drug migration during granule drying." *Chem Eng Res Design* **136**: 199–206.

Kinney, J. H., T. M. Breunig, T. L. Starr, D. Haupt, M. C. Nichols, S. R. Stock, M. D. Butts and R. A. Saroyan (1993). "X-ray tomographic study of chemical vapor infiltration processing of ceramic composites." *Science* **260**: 789–792.

Kinney, J. H. and D. L. Haupt (1997). "Evidence of critical scaling behavior during vapor phase synthesis of continuous filament composites." *J Mater Res* **12**: 610–612.

Kinney, J. H., C. Henry, D. L. Haupt and T. L. Starr (1994). "The topology of percolating porosity in woven fiber ceramic matrix composites." *Appl Compos Mater* **1**: 325–331.

Kinney, J. H., G. W. Marshall Jr. and S. J. Marshall (1994). "Three-dimensional mapping of mineral densities in carious dentin: theory and method." *Scanning Microsc* **8**: 197–205.

Kinney, J. H., R. K. Nalla, J. A. Pople, T. M. Breunig and R. O. Ritchie (2005). "Age-related transparent root dentin: Mineral concentration, crystallite size and mechanical properties." *Biomater* **26**: 3363–3376.

Kinney, J. H. and M. C. Nichols (1992). "X-ray tomographic microscopy (XTM) using synchrotron radiation." *Annu Rev Mater Sci* **22**: 121–152.

Kinney, J. H., S. R. Stock, M. C. Nichols, U. Bonse, T. M. Breunig, R. A. Saroyan, R. Nusshardt, Q. C. Johnson, F. Busch and S. D. Antolovich (1990). "Nondestructive investigation of damage in composites using x-ray tomographic microscopy." *J Mater Res* **5**: 1123–1129.

Kirker-Head, C., V. Karageorgiou, S. Hofmann, R. Fajardo, O. Betz, H. P. Merkle, M. Hilbe, B. von Rechenberg, J. McCool, L. Abrahamsen, A. Nazarian, E. Cory, M. Curtis, D. Kaplan and L. Meinel (2006). "BMP-silk composite matrices heal critically sized femoral defects." *Bone* **41**: 247–255.

Kondori, B., T. F. Morgeneyer, L. Helfen and A. A. Benzerga (2018). "Void growth and coalescence in a magnesium alloy studied by synchrotron radiation laminography." *Acta Mater* **155**: 80–94.

Koo, Y., Y. Jang and Y. Yun (2017). "A study of long-term static load on degradation and mechanical integrity of Mg alloys-based biodegradable metals." *Mater Sci Eng B Solid State Mater Adv Technol* **219**: 45–54.

Kotha, S. P., Y. F. Hsieh, R. M. Strigel, R. Müller and M. J. Silva (2004). "Experimental and finite element analysis of the rat ulnar loading model – correlations between strain and bone formation following fatigue loading." *J Biomech* **37**: 541–548.

Kovacs, C. S. (2017). "The skeleton is a storehouse of mineral that is plundered during lactation and (fully?) replenished afterwards." *J Bone Miner Res* **32**: 676–680.

Kozul, N., G. R. Davis, P. Anderson and J. C. Elliott (1999). "Elemental quantification using multiple-energy x-ray absorptiometry." *Meas Sci Technol* **10**: 252–259.

Krill III, C. E., K. Dobrich, D. Michels, A. Michels, C. Rau, T. Weitkamp, A. Snigirev and R. Birringer (2001). Tomographic characterization of grain-size correlations in polycrystalline Al-Sn. *Developments in X-Ray Tomography III*. U. Bonse. Bellingham (WA), SPIE. **SPIE Proc Vol 4503**: 205–212.

Krings, M., D. Kanellopoulou, D. Mavrilas and B. Glasmacher (2006). "In vitro pH-controlled calcification of biological heart valve prostheses." *Mat-wiss. Werkstofftech* **37**: 432–435.

Krol, Q. and H. Loewe (2018). "Upscaling ice crystal growth dynamics in snow: Rigorous modeling and comparison to 4D X-ray tomography data." *Acta Mater* **151**: 478–487.

Kulkarni, A., A. Goland, H. Herman, A. J. Allen, J. Ilavsky, G. G. Long and F De Carlo (2005). "Advanced microstructural characterization of plasma-sprayed zirconia coatings over extended length scales." *J Therm Spray Technol* **14**: 239–250.

Kulkarni, A., J. Gutleber, S. Sampath, A. Goland, W. Lindquist, H. Herman, A. Allen and B Dowd (2004a). "Studies of the microstructure and properties of dense ceramic coatings produced by high-velocity oxygen-fuel combustion spraying." *Mater Sci Eng A* **369**: 124–137.

Kulkarni, A., H. Herman, F. D. Carlo and R. Subramanian (2004b). "Microstructural characterization of electron beam physical vapor deposition thermal barrier coatings through high-resolution computed microtomography." *Metall Mater Trans* **35A**: 1945–1952.

Kulkarni, A., S. Sampath, A. Goland, H. Herman and B Dowd (2000). "Computed microtomography studies to characterize microstructure-property correlations in thermal sprayed alumina deposits." *Scripta Mater* **43**: 471–476.

Kurtz, S. M., A. van Ooij, R. Ross, J. de Waal Malefijt, J. Peloza, L. Ciccarelli and M. L. Villarraga (2007). "Polyethylene wear and rim fracture in total disc arthroplasty." *Spine J* **7**: 12–21.

Lame, O., D. Bellet, M. D. Michiel and D. Bouvard (2003). "In situ microtomography investigation of metal powder compacts during sintering." *Nucl Instrum Meth B* **200**: 287–294.

Lame, O., D. Bellet, M. D. Michiel and D. Bouvard (2004). "Bulk observation of metal powder sintering by x-ray synchrotron microtomography." *Acta Mater* **52**: 977–984.

Langer, M. and F. Peyrin (2016). "3D X-ray ultra-microscopy of bone tissue." *Osteopor Int* **27**: 441–455.

Lee, S. B. (1993). Nondestructive examination of chemical vapor infiltration of 0°/90° SiC/Nicalon composites. Ph.D. thesis, Georgia Institute of Technology.

Lee, S., T. Fyrner, F. Chen, Z. Álvarez, E. Sleep, D. Chun, J. Weiner, R. Cook, R. Freshman, M. Schallmo, K. Katchko, A. Schneider, J. Smith, C. Yun, G. Singh, S. Hashmi, M. McClendon, Z. Yu, S. Stock, W. Hsu, E. Hsu and S. Stupp (2017). "Sulfated glycopeptide nanostructures for multipotent protein activation." *Nature Nanotechnol* **12**: 821–829.

Lee, S. B., S. R. Stock, M. D. Butts, T. L. Starr, T. M. Breunig and J. H. Kinney (1998). "Pore geometry in woven fiber structures: 0°/90° plain-weave cloth lay-up preform." *J Mater Res* **13**: 1209–1217.

Léonard, A., S. Blacher, C. Nimmol and S. Devahastin (2008). "Effect of far-infrared radiation assisted drying on microstructure of banana slices: An illustrative use of X-ray microtomography in microstructural evaluation of a food product." *J Food Eng* **85**: 154–162.

Lin, C. L. and J. Miller (2005). "3D characterization and analysis of particle shape using x-ray microtomography (XMT)." *Powder Technol* **154**: 61–69.

Liu, L., T. De Kock, J. Wilkinson, V. Cnudde, S. Xiao, C. Buchmann, D. Uteau, S. Peth and A. Lorke (2018). "Methane bubble growth and migration in aquatic sediments observed by X-ray microCT." *Env Sci Technol* **52**: 2007–2015.

Loos, A., R. Rohde, A. Haverich and S. Barlach (2007). "In vitro and in vivo biocompatibility testing of absorbable metal stents." *Macromol Symp* **253**: 103–108.

Lopez-Soler, R. I., M. P. Brennan, A. Goyal, Y. Wang, P. Fong, G. Tellides, A. Sinusas, A. Dardik and C. Breuer (2007). "Development of a mouse model for evaluation of small diameter vascular grafts." *J Surg Res* **139**: 1–6.

Lu, S., E. N. Landis and D. T. Keane (2006). "X-ray microtomographic studies of pore structure and permeability in Portland cement concrete." *Mater Struct* **39**: 611–620.

Lu, S. S., X. Zhang, C. Soo, T. Hsu, A. Napoli, T. Aghaloo, B. M. Wu, P. Tsou, K. Ting and J. C. Wang (2007). "The osteoinductive properties of Nell-1 in a rat spinal fusion model." *Spine J* **7**: 50–60.

Ludwig, W., S. Bouchet, D. Bellet and J. Y. Buffière (2000). 3D observation of grain boundary penetration in Al alloys. *X-Ray Tomography in Materials Science*. J. Baruchel, J. Y. Buffière, E. Marie, P. Merle and G. Peix. Paris, Hermes Science: 155–164.

Ludwig, W., J. Y. Buffière, S. Savelli and P. Cloetens (2003). "Study of the interaction of a short fatigue crack with grain boundaries in a cast Al alloy using x-ray microtomography." *Acta Mater* **51**: 585–598.

Ludwig, O., M. Di Michiel, L. Salvo, M. Suery and P. Falus (2005). "In-situ three-dimensional microstructural investigation of solidification of an Al-Cu alloy by ultrafast x-ray microtomography." *Metall Mater Trans A* **36**: 1515–1523.

Macedo, A., C. M. P. Vaz, J. M. Naime, P. E. Cruvinel and S. Crestana (1999). "X-ray microtomography to characterize the physical properties of soil and particulate systems." *Powder Technol* **101**: 178–182.

Maire, E., J. C. Grenier, D. Daniel, A. Baldacci, H. Klöcker and A. Bigot (2006). "Quantitative 3D characterization of intermetallic phases in an Al-Mg industrial alloy by x-ray microtomography." *Scripta Mater* **55**: 123–126.

Marien, E., A. Hillen, F. Vanderhoydonc, J. V. Swinnen and G. Vande Velde (2017). "Longitudinal microcomputed tomography-derived biomarkers for lung metastasis detection in a syngeneic mouse model: added value to bioluminescence imaging." *Lab Invest* **97**: 24–33.

Marrow, T. J., L. Babout, A. P. Jivkov, P. Wood, D. Engelberg, N. Stevens, P. J. Withers and R. C. Newman (2006). "Three dimensional observations and modeling of intergranular stress corrosion cracking in austenitic stainless steel." *J Nucl Mater* **352**: 62–74.

Marshall, S. J., M. Balooch, T. Breunig, J. H. Kinney, A. P. Tomsia, N. Inai, L. G. Watanabe, I. C. Wu-Magidi and G. W. J. Mashall (1998). "Human dentin and the dentin-resin adheisve interface." *Acta Mater* **46**: 2529–2539.

Martin, C. F., J. L. Blandin, L. Salvo, C. Josserond and P. Cloetens (2000a). Study of damage during superplastic deformation. *X-Ray Tomography in Materials Science*. J. Baruchel, J. Y. Buffière, E. Marie, P. Merle and G. Peix. Paris, Hermes Science: 193–204.

Martin, C. F., C. Josserond, L. Salvo, J. J. Blandin, P. Cloetens and E. Boller (2000b). "Characterization by x-ray micro-tomography of cavity coalescence during superplastic deformation." *Scripta Mater* **42**: 375–381.

Martín-Badosa, E., D. Amblard, S. Nuzzo, A. Elmoutaouakkilo, L. Vico and F. Peyrin (2003). "Excised bone structures in mice: imaging at three-dimensional synchrotron microCT." *Radiology* **229**: 921–928.

Martins de Souza E Silva, J., J. Utsch, M. A. Kimm, S. Allner, M. F. Epple, K. Achterhold and F. Pfeiffer (2017). "Dual-energy microCT for quantifying the time-course and staining characteristics of ex-vivo animal organs treated with iodine- and gadolinium-based contrast agents." *Sci Rep* **7**: 17387.

Mashiatulla, M., R. D. Ross and D. R. Sumner (2017). "Validation of cortical bone mineral density distribution using micro-computed tomography." *Bone* **99**: 53–61.

Matsumoto, T., M. Yoshino, T. Asano, K. Uesugi, M. Todoh and M. Tanaka (2006). "Monochromatic synchrotron radiation μCT reveals disuse-mediated canal network rarefaction in cortical bone of growing rat tibiae." *J Appl Physiol* **100**: 274–280.

Matsumoto, T., M. Yoshino, K. Uesugi and M. Tanaka (2007). "Biphasic change and disuse-mediated regression of canal network structure in cortical bone of growing rats." *Bone* **41**: 239–246.

Mazur, P., E. Wypasek, B. Gawęda, D. Sobczyk, P. Kapusta, J. Natorska, K. P. Malinowski, J. Tarasiuk, M. Bochenek, S. Wroński, K. Chmielewska, B. Kapelak and A. Undas (2017). "Stenotic bicuspid and tricuspid aortic valves - micro-computed tomography and biological indices of calcification." *Circ J* **81**: 1043–1050.

McCann, M. R., M. A. Veras, C. Yeung, G. Lalli, P. Patel, K. M. Leitch, D. W. Holdsworth, S. J. Dixon and C. A. Séguin (2017). "Whole-body vibration of mice induces progressive degeneration of intervertebral discs associated with increased expression of Il-1β and multiple matrix degrading enzymes." *Osteoarthr Cartilage* **25**: 779–789.

McDonald, S. A., L. C. R. Schneider, A. C. F. Cocks and P. J. Withers (2006). "Particle movement during the deep penetration of a granular material studied by x-ray microtomography." *Scripta Mater* **54**: 191–196.

McGilvray, K. C., E. I. Waldorff, J. Easley, H. B. Seim, N. Zhang, R. J. Linovitz, J. T. Ryaby and C. M. Puttlitz (2017). "Evaluation of a polyetheretherketone (PEEK) titanium composite interbody spacer in an ovine lumbar interbody fusion model: Biomechanical, microcomputed tomographic, and histologic analyses." *Spine J* **17**: 1907–1916.

McGuire, P. A., S. Blackburn and E. M. Holt (2007). "An X-ray micro-computed tomography study of agglomerate breakdown during the extrusion of ceramic pastes." *Chem Eng Sci* **62**: 6451–6456.

Mechanic, G. L., S. B. Arnaud, A. Boyde, T. G. Bromage, P. Buckendahl, J. C. Elliott, E. P. Katz and G. N. Durnova (1990). "Regional distribution of mineral and matrix in the femurs of rats flown on Cosmos 1887 biosatellite." *FASEB J* **4**: 34–40.

Meganck, J. A. and B. Liu (2017). "Dosimetry in micro-computed tomography: A review of the measurement methods, impacts, and characterization of the quantum GX imaging system." *Mol Imaging Biol* **19**: 499–511.

Mendoza, M. C., K. Sonn, A. S. Kannan, S. S. Bellary, S. M. Mitchell, G. Singh, C. Park, A. Ghosh, J. Yun, S. R. Stock, E. L. Hsu and W. K. Hsu (2016). "The effect of vancomycin powder on bone healing in a rat spinal rhBMP-2 model." *J Neurosurg Spine* **25**: 147–153.

Mercer, C. E. and P. Anderson (1995). "X-ray microtomographic quantification of the effects on enamel following CO_2 laser application." *J Dent Res* **74**: 849.

Miller, J. D., C. L. Lin, C. Garcia and H. Arias (2003). "Ultimate recovery in heap leaching operations as established from mineral exposure analysis by x-ray microtomography." *Int J Miner Process* **72**: 331–340.

Mis, A., A. Nawrocka, K. Lamorski and D. Dziki (2018). "Dynamics of gas cell coalescence during baking expansion of leavened dough." *Food Res Int* **103**: 30–39.

Mollica, F., R. D. Santis, L. Ambrosio, L. Nicolais, D. Prisco and S. Rengo (2004). "Mechanical and leakage behaviour of the dentin - adhesive interface." *J Mater Sci Mater Med* **15**: 485–492.

Nahrendorf, M., C. Badea, L. W. Hedlund, J. L. Figueiredo, D. E. Sosnovik, G. A. Johnson and R. Weissleder (2007). "High-resolution imaging of murine myocardial infarction with delayed-enhancement cine microCT." *Am J Physiol Heart Circ Physiol* **292**: H3172–H3178.

Naik, N. (2003). Sulfate attack on portland cement-based materials: mechanisms of damage and long term performance. Ph.D. thesis, Georgia Institute of Technology.

Naik, N. N., A. C. Jupe, S. R. Stock, A. P. Wilkinson, P. L. Lee and K. E. Kurtis (2006). "Sulfate attack monitored by microCT and EDXRD: Influence of cement type, water-to-cement ratio, and aggregate." *Cement Concr Res* **36**: 144–159.

Naik, N. N., K. E. Kurtis, A. P. Wilkinson, A. C. Jupe and S. R. Stock (2004). Sulfate deterioration of cement-based materials examined by x-ray microtomography. *Developments in X-Ray Tomography IV.* U. Bonse. Bellingham (WA), SPIE. **SPIE Proc Vol 5535**: 442–452.

Neldam, C. A. and E. M. Pinholt (2014). "Synchrotron μCT imaging of bone, titanium implants and bone substitutes - a systematic review of the literature." *J Craniomaxillofac Surg* **42**: 801–805.

Nielsen, S. F., F. Beckmann, R. B. Godiksen, K. Haldrup, H. F. Poulsen and J. A. Wert (2004). Measurement of the components of plastic displacement gradients in three dimensions. *Developments in X-Ray Tomography IV.* U. Bonse. Bellingham (WA), SPIE. SPIE Proc **Vol 5535**: 485–492.

Nielsen, S. F., H. F. Poulsen, F. Beckmann, C. Thorning and J. Wert (2003). "Measurements of plastic displacement gradient components in three dimensions using marker particles and synchrotron x-ray absorption microtomography." *Acta Mater* **51**: 2407–2415.

Noiriel, C., D. Bernard, P. Gouze and X. Thibault (2005). "Hydraulic properties and microgeometry evolution accompanying limestone dissolution by acidic water." *Oil Gas Sci Technol* **60**: 177–192.

Noiriel, C., P. Gouze and D. Bernard (2004). "Investigation of porosity and permeability effects from microstructure changes during limestone dissolution." *Geophys Res Lett* **31**: L24603/1–L24603/4.

Nunan, N., K. Ritz, M. Rivers, D. S. Feeney and I. M. Young (2006). "Investigating microbial micro-habitat structure using x-ray computed tomography." *Geoderma* **133**: 398–407.

Nuzzo, S., M. H. Lafage-Proust, E. Martin-Badosa, G. Boivin, T. Thomas, C. Alexandre and F. Peyrin (2002). "Synchrotron radiation microtomography allows the analysis of three-dimensional microarchtiecture and degree of mineralization of human iliac crest biopsy specimens: Effects of etidronate treatment." *J Bone Miner Res* **17**: 1372–1382.

Nuzzo, S., C. Meneghini, P. Braillon, R. Bouvier, S. Mobilio and F. Peyrin (2003). "Microarchitectural and physical changes during fetal growth in human vertebral bone." *J Bone Miner Res* **18**: 760–768.

Ohta, S., E. W. Lai, J. C. Morris, D. A. Bakan, B. Klaunberg, S. Cleary, J. F. Powers, A. S. Tischler, M. Abu-Asab, D. Schimel and K. Pacak (2006). "MicroCT for high-resolution imaging of ectopic pheochromocytoma tumors in the liver of nude mice." *Int J Cancer* **119**: 2236–2241.

Okuma, G., D. Kadowaki, T. Hondo, S. Tanaka and F. Wakai (2017). "Interface topology for distinguishing stages of sintering." *Sci Rep* **7**: 11106.

Osborne, D. R., C. Kuntner, S. Berr and D. Stout (2017). "Guidance for efficient small animal imaging quality control." *Mol Imaging Biol* **19**: 485–498.

Park, J.-H., J.-Y. Bae, J. Shim and I. Jeon (2012). "Evaluation of suitable porosity for sintered porous β-tricalcium phosphate as a bone substitute." *Mater Char* **71**: 103–111.

Peleg, S., F. Khan, N. M. Navone, D. D. Cody, E. M. Johnson, C. S. Van Pelt and G. H. Posner (2005). "Inhibition of porstate cancer-mediated osteoblastic bone lesions by the low-calcemic analog 1alpha-hydroxymethyl-16-ene-26,27-bishomo-25-hydroxy vitamin D_3." *J Steroid Biochem Mole Biol* **97**: 203–211.

Persy, V., A. Postnov, E. Neven, G. Dams, M. De Broe, P. D'Haese and N. De Clerck (2006). "High-resolution X-ray microtomography is a sensitive method to detect vascular calcification in living rats with chronic renal failure." *Arterio Thomb Vasc Biol* **26**: 2110–2116.

Pfeifer, C. G., M. B. Fisher, V. Saxena, M. Kim, E. A. Henning, D. A. Steinberg, G. R. Dodge and R. L. Mauck (2017). "Age-dependent subchondral bone remodeling and cartilage repair in a minipig defect model." *Tissue Eng Part C Methods* **23**: 745–753.

Pfister, A., U. Walz, A. Laib and R. Mulhaupt (2005). "Polymer ionomers for rapid prototyping and rapid manufacturing by means of 3D printing." *Macromol Mater Eng* **290**: 99–113.

Phillion, A. B., S. L. Cockcroft and P. D. Lee (2006). "X-ray micro-tomographic observations of hot tear damage in an Al-Mg commercial alloy." *Scripta Mater* **55**: 489–492.

Plese, P., M. D. Higgins, L. Mancini, G. Lanzafame, F. Brun, J. L. Fife, J. Casselman and D. R. Bake (2018). "Dynamic observations of vesiculation reveal the role of silicate crystals in bubble nucleation and growth in andesitic magmas." *Lithos* **296**: 532–546.

Polacci, M., F. Arzilli, G. La Spina, N. Le Gall, B. Cai, M. E. Hartley, D. Di Genova, N. T. Vo, S. Nonni, R. C. Atwood, E. W. Llewellin, P. D. Lee and M. R. Burton (2018). "Crystallisation in basaltic magmas revealed via in situ 4D synchrotron X-ray microtomography." *Sci Rep* **8**: 8377.

Prevrhal, S. (2004). Beam hardening correction and quantitative microCT. *Developments in X-Ray Tomography IV.* U. Bonse. Bellingham (WA), SPIE. **SPIE Proc Vol 5535**: 152–161.

Prevrhal, S. (2006). Simulation of trabecular mineralization measurements in microCT. *Developments in X-Ray Tomography V.* U. Bonse. Bellingham (WA), SPIE. **SPIE Proc Vol 6318**: 631808-1–631808-10.

Rabbi, S. M. F., M. K. Tighe, R. J. Flavel, B. N. Kaiser, C. N. Guppy, X. Zhang and I. M. Young (2018). "Plant roots redesign the rhizosphere to alter the three-dimensional physical architecture and water dynamics." *New Phytol* **219**: 542–550.

Rahman, M. M., M. U. H. Joardder and A. Karim (2018). "Non-destructive investigation of cellular level moisture distribution and morphological changes during drying of a plant-based food material." *Biosys Eng* **169**: 126–138.

Rajamannan, N. M., T. B. Nealis, M. Subramaniam, S. R. Stock, K. I. Ignatiev, T. J. Sebo, J. W. Fredericksen, S. W. Carmichael, T. K. Rosengart, T. C. Orszulak, W. D. Edwards, R. O. Bonow and T. C. Spelsberg (2005). "Calcified rheumatic valve neoangiogenesis is associated with VEGF expression and osteoblast-like bone formation." *Circulation* **111**: 3296–3301.

Rajamannan, N. M., M. Subramanium, S. Stock, F. Caira and T. C. Spelsberg (2005). "Atorvastatin inhibits hypercholesterolemia-induced calcification in the aortic valves via the Lrp5 receptor pathway." *Circulation* **112** (suppl I): I-229–I-234.

Rattanasak, U. and K. Kendall (2005). "Pore structure of cement/pozzolan composites by x-ray microtomography." *Cement Concr Res* **35**: 637–640.

Ren, W., B. Wu, X. Peng, J. Hua, H. N. Hao and P. H. Wooley (2006). "Implant wear induces inflammation, but not osteoclastic bone resorption, in RANK-/- mice." *J Orthop Res* **24**: 1575–1586.

Renders, G. A. P., L. Mulder, L. J. van Ruijven and T. M. G. J. van Eijden (2007). "Porosity of human mandibular condylar bone." *J Anat* **210**: 239–248.

Richard, P., P. Philippe, F. Barbe, S. Bourles, X. Thibault and D. Bideau (2003). "Analysis by x-ray microtomography of a granular packing undergoing compaction." *Phys Rev E* **68**: 020301/1–020301/4.

Ritman, E. L., M. E. Bolander, L. A. Fitzpatrick and R. T. Turner (1998). "MicroCT imaging of structure-to-function relationship of bone microstructure and associated vascular involvement." *Technol Health Care* **6**: 403–412.

Roudier, M. P., S. D. Bain and W. C. Dougall (2006). "Effects of the RANKL inhibitor, osteoprotegerin, on the pain and histopathology of bone cancer in rats." *Clin Exp Metast* **23**: 167–175.

Schambach, S. J., S. Bag, L. Schilling, C. Groden and M. A. Brockmann (2010). "Application of microCT in small animal imaging." *Methods* **50**: 2–13.

Schell, J. S. U., M. Deleglise, C. Binetruy, P. Krawczak and P. Ermanni (2007). "Numerical prediction and experimental characterisation of meso-scale-voids in liquid composite moulding." *Compos Pt A* **38**: 2460–2470.

Schell, J. S. U., M. Renggli, G. H. van Lenthe, R. Müller and P. Ermanni (2006). "Micro-computed tomography determination of glass fibre reinforced polymer meso-structure." *Compos Sci Technol* **66**: 2016–2022.

Scherf, H., F. Beckmann, J. Fischer and F. Witte (2004). Internal channel structures in trabecular bone. *Developments in X-Ray Tomography IV.* U. Bonse. Bellingham (WA), SPIE. **SPIE Proc Vol 5535**: 792–798.

Schmidhammer, R., S. Zandieh, R. Mittermayr, L. E. Pelinka, M. Leixnering, R. Hopf, A. Kroepfl and H. Redl (2006). "Assessment of bone union/nonunion in an experimental model using micro-computed technology." *J Trauma Inj Infec Crit Care* **61**: 199–205.

Schneider, P., R. Voide, M. Stuaber, M. Stampanoni, L. R. Donahue, P. Wyss, U. Sennhauser and R. Müller (2006). Assessment of murine bone ultrastructure using synchrotron light: Towards nanocomputed tomography. *Developments in X-Ray Tomography V.* U. Bonse. Bellingham (WA), SPIE. **SPIE Proc Vol 6318**: 63180C-1–63180C-9.

Schopf, L., A. Savinainen, K. Anderson, J. Kujawa, M. DuPont, M. Silva, E. Siebert, S. Chandra, J. Morgan, P. Gangurde, D. Wen, J. Lane, Y. Xu, M. Hepperle, G. Harriman, T. Ocain and B. Jaffee (2006). "IKKbeta inhibition protects against bone and cartilage destruction in a rat model of rheumatoid arthritis." *Arth Rheum* **54**: 3163–3173.

Schroer, C. G., M. Kuhlmann, T. F. Gunzler, B. Benner, O. Kurapova, J. Patormmel, B. Lengeler, S. V. Roth, R. Gehrke, A. Snigirev, I. Snigireva, N. Stribeck, A. Almendarez-Camarillo and F. Beckmann (2006). Full-field and scanning microtomography based on parabolic refractive x-ray lenses. *Developments in X-Ray Tomography V.* U. Bonse. Bellingham (WA), SPIE. **SPIE Proc Vol 6318**: 63181H-1–63181H-9.

Scoffoni, C., C. Albuquerque, C. R. Brodersen, S. V. Townes, G. P. John, M. K. Bartlett, T. N. Buckley, A. J. McElrone and L. Sack (2017). "Outside-xylem vulnerability, not xylem embolism, controls leaf hydraulic decline during dehydration." *Plant Physiol* **173**: 1197–1210.

Seidler, G. T., G. Martinez, L. H. Seeley, K. H. Kim, E. A. Behne, S. Zaranek, B. D. Chapman, S. M. Heald and D. L. Brewe (2000). "Granule-by-granule reconstruction of a sandpile from x-ray microtomography data." *Phys Rev E* **62**: 8175–8181.

Shahani, A. J., E. B. Gulsoy, S. O. Poulsen, X. Xiao and P. W. Voorhees (2016). "Twin-mediated crystal growth: An enigma resolved." *Sci Rep* **6**: 28651.

Shahani, A. J., X. Xiao, K. Skinner, M. Peters and P. W. Voorhees (2016). "Ostwald ripening of faceted Si particles in an Al-Si-Cu melt." *Mater Sci Eng A* **673**: 307–320.

Shalabi, M. M., J. G. C. Wolke, V. M. J. I. Cuijpers and J. A. Jansen (2007). "Evaluation of bone response to titanium-coated polymethylmethacrylate resin (PMMA) implants by x-ray tomography." *J Mater Sci Mater Med* **18**: 2033–2039.

Sheller, M. R., R. S. Crowther, J. H. Kinney, J. Yang, S. Di Jorio, T. Breunig, D. H. Carney and J. T. Ryaby (2004). "Repair of rabbit segmental defects with thrombin peptide, TP508." *J Orthop Res* **22**: 1094–1099.

Silva, M. D., J. Ruan, E. Siebert, A. Savinainen, B. Jaffee, L. Schopf and S. Chandra (2006). "Application of surface roughness analysis on micro–computed tomographic images of bone erosion: Examples using a rodent model of rheumatoid arthritis." *Mole Imaging* **5**: 475–484.

Sinha, P. K., P. Halleck and C. Y. Wang (2006). "Quantification of liquid water saturation in a PEM fuel cell diffusion medium using x-ray microtomography." *Electrochem Sol State Lett* **9**: A344–A348.
Sone, T., T. Tamada, Y. Jo, H. Miyoshi and M. Fukunaga (2004). "Analysis of three-dimensional microarchitecture and degree of mineralization in bone metastases from prostate cancer using synchrotron microcomputed tomography." *Bone* **35**: 432–438.
Speller, R., S. Pani, M. Tzaphlidou and J. Horrocks (2005). "MicroCT analysis of calcium/phosphorus ratio maps at different bone sites." *Nucl Instrum Meth A* **548**: 269–273.
Starr, T. L. (1992). Advances in modeling of the chemical infiltration process. *Chemical Vapor Deposition of Refractory Metals and Ceramics II*. T. M. Besman, B. M. Gallois and J. Warren, Materials Research Society **250**: 207–214.
Stock, S. R., J. Barss, T. Dahl, A. Veis and J. D. Almer (2002). "X-ray absorption microtomography (microCT) and small beam diffraction mapping of sea urchin teeth." *J Struct Biol* **139**: 1–12.
Stock, S. R., D. Blackburn, M. Gradassi and H. G. Simon (2003). "Bone formation during forelimb regeneration: a microtomography (microCT) analysis." *Dev Dyn* **226**: 410–417.
Stock, S. R., L. L. Dollar, G. B. Freeman, W. J. Ready, L. J. Turbini, J. C. Elliott, P. Anderson and G. R. Davis (1994). Characterization of conductive anodic filament (CAF) by x-ray microtomography and by seial sectioning. *Electronic Packaging Materials Science VII*. P. Børgesen, K. F. Jansen and R. A. Pollak. Pittsburgh, Mater Res Soc. **323**: 65–69.
Stock, S. R., A. Guvenilir, T. L. Starr, J. C. Elliott, P. Anderson, S. D. Dover and D. K. Bowen (1989). "Microtomography of silicon nitride/silicon carbide composites." *Ceram Trans* **5**: 161–170.
Stock, S. R., K. I. Ignatiev, H. G. Simon and F. D. Carlo (2004). Newt limb regeneration studies with synchrotron microCT. *Developments in X-Ray Tomography IV*. U. Bonse. Bellingham (WA), SPIE. **SPIE Proc Vol 5535**: 748–756.
Stock, S. R., N. N. Naik, A. P. Wilkinson and K. E. Kurtis (2002). "X-ray microtomography (microCT) of the progression of sulfate attack of cement paste." *Cement Concr Res* **32**: 1673–1675.
Stock, S. R., A. E. M. Vieira, A. C. B. Delbem, M. L. Cannon, X. Xiao and F. De Carlo (2008). "Synchrotron microComputed Tomography of the bovine dentinoenamel junction." *J Struct Biol* **161** 162–171.
Stock, S. R., X. Xiao and F. D. Carlo (2006). Unpublished data 2-BM, APS.
Stribeck, N., A. A. Camarilla, U. Nochel, C. Schroer, M. Kuhlmann, S. V. Roth, R. Gehrke and R. K. Bayer (2006). "Volume-resolved nanostructure survey of a polymer part by means of SAXS microtomography." *Macromol Chem Phys* **207**: 1139–1149.
Tafforeau, P., I. Bentaleb, J. J. Jaeger and C. Martin (2007). "Nature of laminations and mineralizationin rhinoceros enamel using histology and x-ray synchrotron microtomography: Potential implications for palaeoenvironmental isotopic studies." *Palaeogeo Palaeoclim Palaeoecol* **246**: 206–227.
Tafforeau, P. and T. M. Smith (2008). "Nondestructive imaging of hominoid dental microstructure using phase contrast x-ray synchrotron microtomography." *J Hum Evol* **54**: 272–278.
Tesei, L., F. Casseler, D. Dreossi, L. Mancini, G. Tromba and F. Zanini (2005). "Contrast-enhanced x-ray microtomography of the bone structure adjacent to oral implants." *Nucl Instrum Meth A* **548**: 257–263.
Thompson, K. E., C. S. Willson and W Zhang, AH Reed, L Beenken, (2006). "Quantitative computer reconstruction of particulate materials from microtomography images." *Powder Technol* **163**: 169–182.
Trtik, P. and C. P. Hauri (2007). "Micromachining of hardened Portland cement pastes using femtosecond laser pulses." *Mater Struct* **40**: 641–650.
Tzaphlidou, M., R. Speller, G. Royle and J. Griffiths (2006). "Preliminary estimates of the calcium/phosphorus ratio at different cortical bone sites using synchrotron microCT." *Phys Med Biol* **51**: 1849–1855.
Ural, A. and D. Vashishth (2007). "Effects of intracortical porosity on fracture toughness in aging human bone: a μCT-based cohesive finite element study." *J Biomech Eng* **129**: 625–631.
Vagnon, A., O. Lame, D. Bouvard, M. D. Michiel, D. Bellet and G. Kapelski (2006). "Deformation of steel powder compacts during sintering: Correlation between macroscopic measurement and in situ microtomography analysis." *Acta Mater* **54**: 513–522.

Vaucher, S., P. Unifantowicz, C. Ricard, L. Dubois, M. Kuball, J. M. Catala-Civera, D. Bernard, M. Stampanoni and R. Nicula (2007). "On-line tools for microscopic and macroscopic monitoring of microwave processing." *Phys B* **398**: 191–195.

Vecchio, K. S., X. Zhang, J. B. Massie, M. Wang and C. W. Kim (2007). "Conversion of bulk seashells to biocompatible hydroxyapatite for bone implants." *Acta Biomater* **3**: 910–918.

Velhinho, A., P. D. Sequeira, R. Martins, G. Vignoles, F. B. Fernandes, J. D. Botas and L. A. Rocha (2003). "X-ray tomographic imaging of Al/SiCp functionally graded composites fabricated by centrifugal casting." *Nucl Instrum Meth B* **200**: 295–302.

Vengrenyuk, Y., S. Carlier, S. Xanthos, L. Cardoso, P. Ganatos, R. Virmani, S. Einav, L. Gilchrist and S. Weinbaum (2006). "A hypothesis for vulnerable plaque rupture due to stress-induced debonding around cellular microcalcifications in thin fibrious caps." *PNAS* **103**: 14678–14683.

Verrier, S., M. Braccini, C. Josserond, L. Salvo, M. Suèry, W. Ludwig, P. Cloetens and J. Baruchel (2000). Study of materials in the semi-solid state. *X-Ray Tomography in Materials Science*. J. Baruchel, J. Y. Buffière, E. Maire, P. Merle and G. Peix. Paris, Hermes Science: 77–88.

Vetter, L. D., V. Cnudde, B. Masschaele, P. J. S. Jacobs and J. V. Acker (2006). "Detection and distribution analysis of organosilicon compounds in wood by means of SEM-EDX and microCT." *Mater Char* **56**: 39–48.

Vieira, A. E. M., M. Danelon, D. M. D. Camara, E. R. Rosselli, S. R. Stock, M. L. Cannon, X. Xiao, F. De Carlo and A. C. B. Delbem (2017). "In vitro effect of amorphous calcium phosphate paste applied for extended periods of time on enamel remineralization." *J Appl Oral Sci* **25**: 596–603.

Vieira, A. E. M., A. C. B. Delbem, K. T. Sassaki, M. L. Cannon and S. R. Stock (2006). Quantitative analysis of mineral content in enamel using laboratory microtomography and microhardness analysis. *Developments in X-Ray Tomography V*. U. Bonse. Bellingham (WA), SPIE. **SPIE Proc Vol 6318**: 631823-1–631823-5.

Vilarinho, J. L. P., N. Ferrare, A. M. R. Moreira, H. F. Moura, A. C. Acevedo, S. B. Chaves, N. S. Melo, A. F. Leite, S. B. Macedo, M. P. de Souza, A. T. B. Guimarães and P. T. Figueiredo (2017). "Early bony changes associated with bisphosphonate-related osteonecrosis of the jaws in rats: A longitudinal in vivo study." *Arch Oral Biol* **82**: 79–85.

Wacheter, N. J., P. Augat, G. D. Krischak, M. Mentzel, L. Kinzl and L. Claes (2001). "Prediction of cortical bone porosity in vitro by microcomputed tomography." *Calcif Tiss Int* **68**: 38–42.

Watson, I. G., M. F. Forster, P. D. Lee, R. J. Dashwood, R. W. Hamilton and A. Chirrazi (2005). "Investigation of the clustering behaviour of titanium diboride particles in aluminium." *Compos Pt A* **36**: 1177–1187.

Weiss, P., P. Layrolle, L. P. Clergeau, B. Enckel, P. Pilet, Y. Amouriq, G. Daculsi and B. Giumelli (2007). "The safety and efficacy of an injectable bone substitute in dental sockets demonstrated in a human clinical trial." *Biomaterials* **28**: 3295–3305.

Winkelmann, C. T., S. D. Figueroa, T. L. Rold, W. A. Volkert and T. J. Hoffman (2006). "Microimaging characterization of a B16-F10 melanoma metastasis mouse model." *Mole Imaging* **5**: 105–114.

Witte, F., J. Fischer, J. Nellesen and F. Beckmann (2006a). Microtomography of magnesium implants in bone and their degradation. *Developments in X-Ray Tomography V*. U. Bonse. Bellingham (WA), SPIE. **SPIE Proc Vol 6318**: 631806-1–631806-9.

Witte, F., J. Fischer, J. Nellesen, H. A. Crostack, V. Kaese, A. Pisch, F. Beckmann and H. Windhagen (2006b). "In vitro and in vivo corrosion measurements of magnesium alloys." *Biomaterials* **27**: 1013–1018.

Wong, F. S. L., J. C. Elliott, P. Anderson and G. R. Davis (1991). "X-ray microtomographic study of the mineral content and structure of a mouse mandible." *J Dent Res* **70**: 691.

Wong, F. S. L., J. C. Elliott, P. Anderson and G. R. Davis (1995a). "Mineral concentration in rat femoral diaphyses measured by x-ray microtomography." *Calcif Tissue Int* **56**: 62–70.

Wong, F. S. L., J. C. Elliott, P. Anderson and G. R. Davis (1995b). "Three dimensional mineral distribution in the dentine of a rat incisor measured by x-ray microtomography." *J Dent Res* **74**: 849.

Wong, F. S. L., J. C. Elliott, P. Anderson and G. R. Davis (1995c). "Three dimensional mineral distribution in the enamel of a rat incisor measured by x-ray microtomography." *Bone* **16**: 690.

Wong, F. S. L., N. S. Willmott and G. R. Davis (2006). "Dentinal carious lesion in three dimensions." *Int J Paed Dent* **16**: 419–423.
Xing, T., W. Zhu, F. Fusseis and H. Lisabeth (2018). "Generating porosity during olivine carbonation via dissolution channels and expansion cracks." *Solid Earth* **9**: 879–896.
Yang, S., R. Zhang and X. Qu (2015). "Optimization and evaluation of metal injection molding by using X-ray tomography." *Mater Char* **104**: 107–115.
Youssef, Y. M., A. Chaijaruwanich, R. W. Hamilton, H. Nagaumi, R. J. Dashwood and P. D. Lee (2006). "X-ray microtomographic characterisation of pore evolution during homogenisation and rolling of Al-6Mg." *Mater Sci Technol* **22**: 1087–1093.
Zabler, S., P. Cloetens and P. Zaslansky (2007). "Fresnel-propagated submicrometer x-ray imaging of water-immersed tooth dentin." *Opt Lett* **32**: 2987–2989.
Zabler, S., H. Riesemeier, P. Fratzl and P. Zaslansky (2006). "Fresnel-propagated imaging for the study of human tooth dentin by partially coherent x-ray tomography." *Optics Express* **14**: 8584–8597.
Zabler, S., A. Rueda, A. Rack, H. Riesemeier, P. Zaslansky, I. Manke, F. Garcia-Moreno and J. Banhart (2007). "Coarsening of grain-refined semi-solid Al–Ge32 alloy: X-ray microtomography and in situ radiography." *Acta Mater* **55**: 5045–5055.
Zanette, I., B. Enders, M. Dierolf, P. Thibault, R. Gradl, A. Diaz, M. Guizar-Sicairos, A. Menzel, F. Pfeiffer and P. Zaslansky (2015). "Ptychographic X-ray nanotomography quantifies mineral distributions in human dentine." *Sci Rep* **5**: 9210.
Zettler, R., T. Donath, J. F. d. Santos, F. Beckman and D. Lohwasser (2006). "Validation of marker material flow in 4mm thick friction stir welded Al 2024-T351 through computer microtomography and dedicated metallographic techniques." *Adv Eng Mater* **8**: 487–490.
Zhang, X., S. N. Johnson, P. J. Gregory, J. W. Crawford, I. M. Young, P. J. Murray and S. C. Jarvis (2006). "Modelling the movement and survival of the root-feeding clover weevil, Sitona lepidus, in the root-zone of white clover." *Eco Model* **190**: 133–146.

11

Mechanically Induced Damage, Deformation, and Cracking

This chapter covers mechanically induced deformation and damage (pore formation, cracking) as well as quantification of crack characteristics. Deformation of cellular materials (metal and ceramic foams, trabecular bone, etc.) and environmentally assisted cracking are not subjects here, as they were covered earlier (Chapters 7 and 10, respectively), nor is deformation-based materials processing (Chapter 10). Some material on measurement of the 3D spatial distribution of crack openings was presented in Chapter 6 and on detection limits of crack openings in Chapter 5.

Investigators commonly use microCT to observe a given sample four or more times during its evolution, and this thread runs through the present chapter. Accounts illustrating the scope microCT of deformation open this chapter. Section 11.2 examines crack characterization with microCT. Section 11.3 covers microCT studies of deformation of composite materials, including cortical bone and tooth, and natural composites.

It is pertinent to mention several reviews before diving into the details. Summaries of digital volume correlation applied to strain measurement in bone (Roberts, Perilli et al. 2014), used with other materials (Fedele, Ciani et al. 2014) and employed for 3D strain mapping (Toda, Maire et al. 2011), have appeared. Results on 3D cavitation in high-strength steels (Gupta, Toda et al. 2015) and the extensive literature on synchrotron microCT imaging of failure in structural materials (Wu, Xiao et al. 2017) have been covered.

11.1 Deformation Studies

Functionality of nonmineralized tissues often depends on calibrated response to applied loads, and microCT can be the basis of numerical models. Modeling of the response of the tympanic membrane to pressure is one example (Rohani, Ghomashchi et al. 2017). MicroCT has shown the hinged structure of the dragonfly wing offering important functional advantages (Rajabi, Ghoroubi et al. 2017). MicroCT data were the basis for computational aerodynamic analysis of a biorealistic fruit fly wing (Brandt, Doig et al. 2015). Small diameter natural fibers (<150 μm) and their deformation-related damage mechanisms have also been studied with microCT (Beaugrand, Guessasma et al. 2017), as has fracture of arterial tissue (Helfenstein-Didier, Taïnoff et al. 2018).

Strain localization, the concentration of deformation into narrow bands of intense shear, occurs not only in cellular solids, as noted in Chapter 8, but also in most geomaterials. Viggiani et al. observed a stiff soil at several strains with synchrotron microCT; a single shear band was formed at an axial strain of 2.7% (Viggiani, Lenoir et al. 2004). Development of shear bands in argillaceous rock during compressive triaxial deformation was followed with microCT using a volumetric digital correlation method (Lenoir, Bornert et al. 2007).

In situ loading and microCT of concrete have shown that the fracture processes must be treated not as a 2D crack but as a system of smaller 3D cracks (Landis and Keane 1999). The nonrecoverable work of loading in compression of mortar, calculated using a linear elastic fracture mechanics approach, was determined by measuring the incremental changes in crack surface area revealed by microCT as load was increased (Landis and Keane 1999, Landis, Nagy et al. 1999, Landis and Nagy 2000); the incremental fracture energy rose with increasing damage, indicating secondary toughening mechanisms such as friction made up a greater fraction of the measured energy (Landis, Nagy et al. 2003, Landis, Zhang et al. 2007). Breakage of reinforcing particles in recycled construction materials is an important aspect of their performance, and this has been studied with synchrotron microCT (Afshar, Disfani et al. 2018).

Additive manufacturing is being used increasingly to produce a range of components and structures, and microCT is an important part of the characterization of these materials. Synchrotron microCT examined the role of heat treatment and build orientation in the microstructure-sensitive deformation characteristics of IN718 produced by additive manufacturing via selective laser melting (Sangid, Book et al. 2018). The elastic modulus/impedance/porosity and microstructural features of 3D-printed hydrogel cardiac implants were investigated via synchrotron phase-contrast microCT (Izadifar, Babyn et al. 2017).

Several microCT studies are focused on the role of porosity in failure of metallic samples. Model copper specimens were manufactured with a regular array of interior pores, and their coalescence in the final stages preceding fracture was studied with synchrotron microCT (Weck, Wilkinson et al. 2006). MicroCT revealed considerable void growth but not nucleation of new voids (i.e., no strain localization between the preexisting manufactured voids). Evolution of pore size distribution during creep was studied by microCT and x-ray diffraction in a three-phase copper alloy (Pyzalla, Camin et al. 2005); details of this study are discussed later in the section on multimode studies. High-velocity impacts generate refraction waves within the target, that superimpose and generate zones of high density of pores; synchrotron microCT was applied to small specimens cut from impact tested Ta disks (Bontaz-Carion and Pellegrini 2006). These data showed that pore volume distribution obeyed a power law, at least for the larger pores, and that results from 2D methods, even with correction, did not extrapolate to the actual 3D distribution. Lab microCT of specimens from five positions in a high-pressure die-casting of Mg alloy AM60B revealed considerable variability in the amount and distribution of microporosity; location of the fracture plane, the fracture strain and the fracture strain agreed with predictions of a critical strain model based on the initial pore distribution (Weiler, Wood et al. 2005). Void growth near a notch was followed at several loads with synchrotron microCT (~0.5 μm isotropic voxels) of a cast A356 Al specimen (Al grains surrounded by a Si-rich eutectic phase), and this study noted constraint effects on the specimen faces (compared to the middle of the specimen) as well as considerable void-Si particle spatial correlation (Qian, Toda et al. 2005).

Fatigued commercial bone cement (PMMA beads in a $BaSO_4$-filled PMMA matrix) was studied with synchrotron microCT; macroscopic failure was found to be linked to the presence of large voids, crack deflection was observed at matrix beads, and crack arrest was found within beads (Sinnett-Jones, Browne et al. 2005). In an Al engineering alloy, the distribution of pore sizes determined from individual lab microCT slices and from the 3D stack of slices was compared: the 2D measurement showed a significantly lower mean pore diameter, even after Saltykov-type correction for sampling biases (Underwood 1968), than the 3D measurement (Li, Lee et al. 2006). The interrupted fatigue test specimens showed the crack tended to deviate toward pores that were near the transverse plane (normal to the load axis) that the crack was following across the specimen (Li, Lee et al. 2006).

Lab microCT with rather large voxels (0.08–0.12 mm) was used to investigate the mechanisms responsible for rising crack growth resistance with increased crack length (R curve behavior) in a fairly large specimen of polygranular graphite (Hodgkins, Marrow et al. 2006). After crack extension, the crack was held open by inserting a wedge. Significant crack face contact was observed behind the crack tip, and the authors reported a zone of discrete, low attenuation features around the crack tip and in the crack wake, features that were interpreted as microcracks hypothesized to be responsible for the R-curve behavior. It would be interesting to confirm this interpretation by performing either phase-enhanced microradiography plus 3D stereometry (Ignatiev 2004, Ignatiev, Lee et al. 2005) or performing synchrotron microCT of small sections cut from one of the samples.

11.2 Cracks and Failure – Monolithic Materials

Cracks in metals or ceramics are very high-contrast features and are relatively easy to detect with microCT. Given the importance of cracking in engineered structures, it is not surprising that microCT imaging of small cracked specimens under load was performed and published very early on (Breunig, Stock et al. 1992b). Reviews have covered microCT of cracks previously (Stock 1999, 2008, Maire and Withers 2014), and a recent review focused on synchrotron microCT of cracks in light metals (Luo, Wu et al. 2018).

During fatigue crack propagation, application of a cyclically varying load drives crack extension, and many variables can affect crack growth rates. Processes such as fatigue crack closure greatly alter crack propagation rates, and understanding these has been the focus of considerable research. X-ray microCT with in situ loading of samples was used to study how, at what stress/stress intensities, and where crack faces come into contact or separate during a fatigue cycle. After a brief description of the crack closure phenomenon, results obtained on an Al-Li alloy are discussed. Subsequent to this discussion, other results on cracks are reviewed.

Fatigue crack closure describes situations where the crack faces come into contact prematurely during unloading of a sample (i.e., before the minimum stress of a fatigue cycle is reached) or where the crack faces remain in contact much longer than expected during loading. Without crack closure, one would expect the crack faces either to contact when the cycles' minimum stress σ_{min} is reached or not to touch at all; without this effect, contact should occur over the entire crack face simultaneously. A variety of mechanisms of crack closure have been proposed, including oxide- or [other] particle-induced closure (Christensen 1963), plasticity-induced closure (Elber 1971), and roughness-induced closure (Suresh and Ritchie 1982). When portions of the crack faces touch at stress $\sigma > \sigma_{min}$, further displacements of the crack faces in the vicinity of the crack tip are resisted by this local contact, so that the driving "force" for crack extension decreases from the nominal value of the stress intensity range $\Delta K = K_{max} - K_{min}$ to $\Delta K_{eff} < \Delta K$. The fact that crack closure occurs and its importance is widely accepted, even though details of what constitutes the closure load P_{Cl} used to calculate K_{eff} remain less than crystal clear. Normally, load–deflection curves of samples exhibiting crack closure show two stages, a lower slope at higher stresses and a higher slope at lower stresses; P_{Cl} is taken to be the point where the tangents to the upper and lower portions of the curves intersect.

A series of papers and theses employed microCT during in situ loading of specimens of AA2090 T8E41 to study changing pattern of crack opening as a function of applied load, that is, at different points during a fatigue cycle. This alloy is particularly interesting to study because fatigue crack growth rates are much lower in the L-T orientation than in other Al alloys used in aerospace applications. Roughness-induced crack closure is very pronounced in this alloy and persists to much higher stress intensity ranges than in other alloys.

Notched tensile samples with 2.9 and ~2.0 mm gage and notch tip diameters, respectively, were examined at four or five loads spanning the unloading portion of the fatigue cycle (Breunig, Stock et al. 1992b, Guvenilir 1995, Stock, Guvenilir et al. 1995, Guvenilir, Breunig et al. 1997, 1999, Guvenilir and Stock 1998). One set of observations was done with 22 keV synchrotron x-radiation, and the second with x-rays from an Ag sealed source tube operated at 40 kV; both were reconstructed with ~6 μm isotropic voxels. The fatigue cracks were rough in some places and quite planar in other parts, as expected for this alloy. Crack openings were quantified as a function of position for each load (Fig. 6.13 shows the openings measured for one sample projected onto a plane perpendicular to the load axis), and crack contact, even at the maximum stress, was observed behind the crack tip, particularly at some, but not all, positions where the crack was at its most nonplanar (compare the left and right sides of the valley labeled "d,s" in Fig. 6.14). The more planar sections of the cracks zipped shut well before the microscopic closure load was reached (e.g., at "c" in Fig. 6.13), whereas portions of the cracks remained open in the nonplanar sections at stresses below the microscopic closure load. In other words, mixed-mode contact was important, and this is best seen in three-dimensional meshes showing crack position on which the openings (indicated by different colors) are superimposed (Fig. 6.14). Further, the fraction of voxels (of the original crack) open was observed to remain nearly constant upon reducing the load from just above to just below the closure load; this was direct evidence that the mixed-mode surface began to carry significant load at the point where the load–displacement curve starts to deflect (Guvenilir, Breunig et al. 1997), that is, where the samples started to stiffen during unloading. These observations could not have been made without microCT.

MicroCT quantification of fatigue crack opening as a function of position has also been performed in the more conventional compact tension sample geometry in AA2090 T8E41 (Guvenilir, Stock et al. 1994, Guvenilir 1995, Morano 1998, Morano, Stock et al. 2000, Ignatiev, Davis et al. 2006). While these samples were much smaller than normal (2–2.5 mm thick and 25.4 mm from notch tip to back face), they were scaled according to ASTM specifications, and, because the entire width must be kept within the beam, this limited the voxel size to between 20 and 59 μm, depending on the data acquisition parameters. In this series of papers, fatigue cracks grown under different ratios of minimum to maximum applied stress were compared; patterns and amounts of crack opening were compared for other specimens before and after increments of extension.

The relationship of different spatial scales of crystallographic texture in AA2090 T8E41 to crack path has been examined in some detail. These studies, combining microCT and microbeam x-ray diffraction mapping, are described in Chapter 12.

A necessary component of the in situ microCT of crack closure is the use of a loading apparatus. The standard load frame design of two posts on either side of the specimen blocks projections over a significant fraction of the 180° required for exact reconstruction. The simplest alternative is to use a x-ray transparent standoff to hold the two grips apart; thin-walled polycarbonate (Breunig, Stock et al. 1993a) or aluminum (Morano 1998, Morano, Stock et al. 1999, 2000) tubes can provide enough rigidity and allow the sample to be viewed around 360°. The force on the sample was, in the data presented above, applied pneumatically and can also be applied by a screw mechanism much in the manner of commercial mechanical testing apparatus. Simple, small, and portable loading apparatus do not need to be designed for growing fatigue cracks, and the studies cited earlier, where the cracked specimen is dismounted from a conventional servohydraulic apparatus and remounted in the loading apparatus, demonstrated the efficacy and economy of this approach. Another approach employs support posts and a pair of synchronized rotators on either side of the sample; the rotators allow samples views over 360° to be obtained without moving the posts into the beam (Breunig, Nichols et al. 1994, Hirano, Usami et al. 1995). The apparatus of Breunig et al. had the capability of testing samples to 15.6 kN.

One of the constraints of the AA2090 compact tension specimen studies was the limited crack opening sensitivity in absorption microCT dictated by the large voxel size. Ignatiev and coworkers indicated that phase-contrast stereometry (tracking features' relative displacements vs specimen rotation in multiple radiographs and computing 3D positions from this data; see Section 3.8) allowed one to measure (albeit laboriously) small changes in opening for special positions on the crack faces (Ignatiev 2004, Ignatiev, Lee et al. 2005, Ignatiev, Davis et al. 2006). Others addressed this sensitivity issue by cutting a small volume of material from around the crack tip and imaging this with the highest available spatial resolution (see the following paragraphs).

Several papers reported synchrotron microCT of small section of Al specimens cut to contain the tip of the fatigue crack (Ludwig and Bellet 2000, Ludwig, Bouchet et al. 2000, Ludwig, Buffière et al. 2003, Toda, Sinclair et al. 2003, 2004, Ohgaki, Toda et al. 2005, Buffiere, Ferrie et al. 2006, Khor, Buffière et al. 2006, Zhang, Toda et al. 2007). The resulting small voxel size and the strong phase contrast in these reconstructions increased crack visibility enormously compared to reconstructions with pure absorption contrast and in intact specimens. One is never quite sure how much the change of constraint (from removed material) affects the observations, so some caution should be exercised in interpreting these results. Buffiere et al. summarized their experience in visualizing cracks in these types of specimens, as well as the drawbacks and advantages of microCT in the presence of significant phase contrast, and a few comments on their studies and others' follow (Buffiere, Ferrie et al. 2006).

Toda et al. found that the large transients in contrast from the Fresnel fringes parallel to fatigue crack surfaces dictated that robust crack opening measurements required use of features somewhat displaced from the crack plane (Toda, Sinclair et al. 2003, Toda, Sinclair et al. 2004). Therefore, near-crack-tip opening displacements in in situ loaded AA2024-T351 were measured using small microvoids (a small distance away from the crack faces) as fiducials, in much the same way that Breunig and coworkers worked with C cores in SiC fibers on either side of a crack in an Al/SiC monofilament composite (Breunig, Elliott et al. 1992a, Breunig, Stock et al. 1993b). The high-resolution closure observations of Toda et al. were in agreement with earlier work (Guvenilir, Breunig et al. 1997, 1999, Guvenilir and Stock 1998, Morano, Stock et al. 1999, 2000, Ignatiev, Davis et al. 2006), namely, the loss of surface contact occurred gradually up to the maximum load of the fatigue cycle, mixed-mode surface contact was very important, and near-tip contact was suggested as producing crack growth resistance.

Decoration of grain boundaries in Al with Ga liquid (the melting point of gallium is 30°C) allowed grain boundary positions to be correlated with the 3D crack geometry and changes in crack path, without sectioning the specimen (Ludwig and Bellet 2000, Ludwig, Bouchet et al. 2000, Babout, Ludwig et al. 2003, Ludwig, Buffière et al. 2003,

Ferrie, Buffiere et al. 2005, 2006, Ohgaki, Toda et al. 2005, Khor, Buffière et al. 2006). In this approach, the authors first performed the in situ loading experiments on the material cut from larger specimens; subsequently, Ga was applied. Paths of short cracks are well known to be dominated by crystallography of the few grains cut by the crack, and electron back scattering diffraction (EBSD) was used to provide the crystallographic information needed to understand which changes in grain orientation produced large deflections in the crack path (Ludwig, Buffière et al. 2003). Analysis of crystallographic character of several branches of a short crack in a cast AS7G03 Al-Si alloy specimen containing artificial pores illustrated the power of this approach (Ludwig, Buffière et al. 2003). In the same material, eleven observations for different crack extensions were made of a short fatigue crack that had nucleated in a narrow ligament between a pore and the specimen surface (Ferrie, Buffiere et al. 2005), and the evolution of the crack front shape was interpreted with respect to the surrounding grain microstructure and pore positions. Figure 11.1 shows a 3D view of the tip of a fatigue crack in an AA2024-T351 specimen; the solid material is rendered transparent and only the Ga-labeled grain boundaries and crack are shown. Large portions of the crack (cut from the larger specimen to include only the near-tip region of the long fatigue crack) were within five degrees of {100} or {111}, with steeply inclined sections following {111} (Khor, Buffière et al. 2006); these crack paths are those observed in AA2090 T8E41 (Yoder, Pao et al. 1989).

Ferrie et al. (2006) studied fatigue propagation in an ultrafine-grained, powder metallurgy alloy (AA5091 material system with mechanical properties equivalent to the T1 condition). This alloy was selected because fatigue cracks follow highly planar paths, and an in situ fatiguing apparatus was mounted directly on the synchrotron microCT rotation stage. Radiography was used to monitor crack initiation, and, after each of nine increments of crack extension, data for reconstruction were collected with the specimen under maximum applied load. The crack grew more elliptical with increasing number of cycles with the major axis perpendicular to the specimen surface, and the authors attributed this to differences in closure stresses along the crack front. Local stress intensity range ΔK was calculated via FEM for the portion of the crack growing parallel to the surface and for the portion growing perpendicular to the surface. Plots of crack growth rate da/dN vs ΔK showed comparable power law exponents, but the surface curve was displaced to lower ΔK relative to the bulk curve, and the latter followed the experimental long crack growth curve. The authors concluded that, for the specimen geometry studied, a single Paris equation can predict the observed crack growth anisotropy provided variation in closure stress along the crack front was assumed. Experiments of the sort described in this paragraph are enormously difficult to conduct and are not to be undertaken lightly.

Zhang and coworkers characterized the interaction of a fatigue crack with pores resulting from flow forming of AA A356 T6 (Zhang, Toda et al. 2007). Marrow and coworkers performed synchrotron microCT on a very small ductile cast iron specimen and focused on characterizing changing crack front geometry as the short cracks interacted with pores and graphite nodules (Marrow, Buffiere et al. 2004). A gage diameter of ~0.35 mm was required to give adequate transmission through the iron specimen at the highest photon energy that was practical for use with the high-resolution x-ray detector (30 keV). The difficulty of working with such fragile specimens is undoubtedly the reason that relatively few experiments are performed with steel, copper, or still more attenuating metals or composites. Even Ti poses challenges for microCT imaging.

Cracking and failure in nonmetallic samples also can be studied profitably by microCT. In one in situ study of a geological specimen, damage evolution leading to fracture was quantified (Renard, Weiss et al. 2018). Effects of heating rate and confinement on

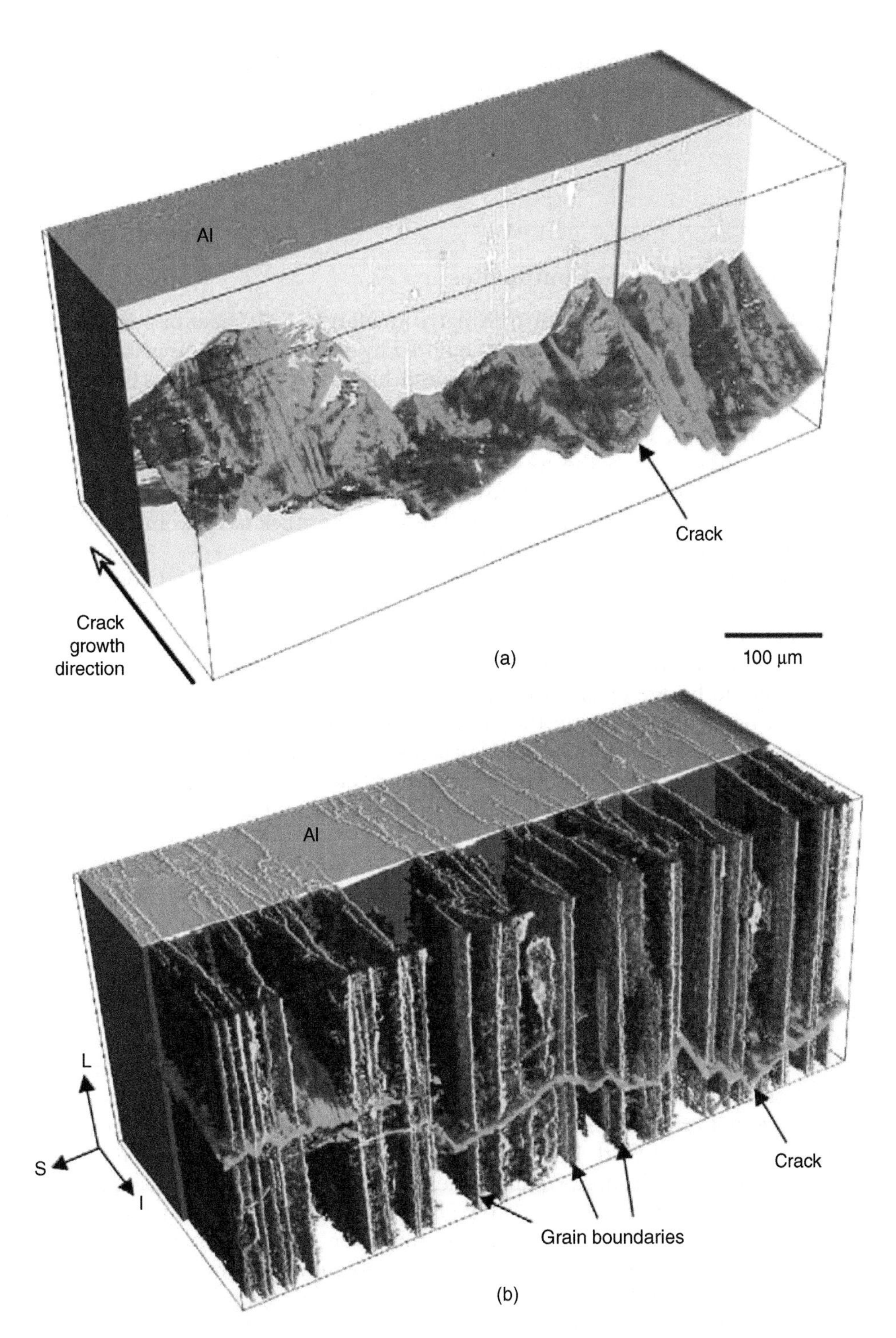

FIGURE 11.1
Three-dimensional representations of a fatigue crack in AA2024-T351. (a) Crack volume (green) and (b) grain boundaries decorated with Ga (gold). Reprinted from Khor et al. (2006), copyright (2006), with permission from Elsevier.

fracture development were examined in in situ microCT of an organic rich shale (Panahi, Kobchenko et al. 2018). Impact loading and fragmentation of SiC ceramics were studied with microCT-based image analysis (Forquin and Ando 2017).

11.3 Cracks and Failure – Composites

Metal matrix composites employ a ductile matrix (metal) and stiff fiber or particulate reinforcements; the resulting properties can be tailored by design of the composite system (e.g., by aligning fibers along the direction of desired high tensile stiffness). The reinforcing phase is generally a ceramic such as Si C or Al_2O_3, materials that have high elastic moduli but very low toughness. The metal matrix surrounds and protects the reinforcements and carries only a small fraction of the applied stress. In the case of fiber reinforcement, cracks generated by low-amplitude fatigue loading can propagate in the metal and bypass the fibers; the intact fibers bridging the crack will restrain the crack faces from opening as far as they otherwise would and thereby lower the stress intensity.

In ceramic matrix composites, the reinforcements provide increased toughness and damage tolerance and are normally in the form of fibers. One strategy is to use the fibers to absorb energy by causing a network of microcracks to form, energy that might otherwise contribute to catastrophic propagation of a single, main crack. A second strategy is to use the fibers to bridge cracks. Ceramic matrix composites have reinforcements and matrices with negligible ductility; therefore, the faces of microcracks and large cracks tend to resume contact once applied loads decrease to zero. Even with load applied, crack openings typically remain below a few micrometers; being able to detect such small features has been the main impediment to applying microCT to quantification of damage in ceramic matrix composites.

This section follows the rather traditional scheme of separating the material into particle- and fiber-reinforced subsections. It may be possible to organize the material another way, but the kinds of questions asked (and the analysis approaches adopted) strongly favor the approach used here.

11.3.1 Particle-Reinforced Composites

In an early microCT study, the pore structure in degassed and nondegassed reaction-bonded silicon nitride/silicon carbide was studied with 10-μm voxels in a 1 mm × 1 mm cross-section, and it is not surprising that the 15-μm diameter Nicalon fibers could not be resolved (Breunig, Elliott et al. 1990). Other early studies include those of metal matrix composites reinforced with short, small diameter fibers (Bonse, Nusshardt et al. 1991) and with particulates (Mummery, Derby et al. 1995, Peix, Cloetens et al. 1997); the latter study followed damage accumulation. Research continues on these types of composites, for example, a study of damage evolution in an Al alloy containing aluminum oxide particles (Nellesen, Laquai et al. 2018).

A pencil beam microtomography system was used to investigate several aspects of ceramic particulate-reinforced aluminum composites (Mummery, Derby et al. 1995). Within a given cross-section of the powder processed composite, the content of 12-μm TiB_2 particles was found to deviate substantially from the nominal 20 vol.% of reinforcement: contents as low as 10 vol.% were reported. The local void volume fraction within the necked regions of a set of composite samples with 5, 10, and 20 vol.% 30-μm SiC particles was also measured as a function of true strain; these types of measurements focused attention on the portion of the sample where damage was concentrated; they offered a very sensitive probe of damage and allowed the same volume to be interrogated in three dimensions multiple times during its evolution. Unfortunately, the sections were not completely contiguous, a limitation imposed by the quite low data acquisition rates with pencil beam systems.

The characteristics of a model discontinuously reinforced composite system (0, 5, and 10 vol.% Ni particles blended with AA2124 powder and hot extruded) were studied with lab microCT (Watson, Lee et al. 2006). Particle clustering was quantified by 2D (SEM) and 3D (microCT) methodologies, and good agreement was found. Subvolumes of the microCT-reconstructed volume were meshed and incorporated into FE simulations of the three materials, and the actual and simulated stress–strain curves showed quite good agreement (Watson, Lee et al. 2006).

Buffière et al. used synchrotron microCT to study a more challenging composite system than Ni-Al: 10 vol.% SiC particles in a matrix of AA6061-T4 (Buffiere, Maire et al. 1999). As the mass attenuation coefficients of SiC and Al differ by less than 3% at 23 keV, phase-enhanced imaging (detector-specimen separation of 830 mm instead of a few tens of millimeters) was used to provide contrast between particle and matrix and to enhance crack visibility. The same volume was compared at five strains (initial microstructure, at the yield point of the stress–strain curve and at three strains up to 13%), the fraction of broken particles was greater in the bulk than near the surface and FE calculations (normal stress, total stored elastic energy in particles) were supplied to explain the observations (Buffiere, Maire et al. 1999). The spatiotemporal distribution of fractured SiC particles was mapped in a subsequent study (Buffière, Savelli et al. 2000). Interest in microCT-based FEM analyses of particulate-reinforced composites continues (Crostock, Nellesen et al. 2006).

The association between deformation-induced porosity in several Al matrices and particulate reinforcements (ZrO_2) was investigated with lab microCT using a dual-energy reconstruction technique (Justice, Derby et al. 2000, 2003). This was an example where an energy-sensitive x-ray detector (and a translate-rotate or pinhole data collection scheme) was required. Variance analysis was used to show that little if any clustering of the particles was present and to determine that voids were also not clustered. A direct relationship between volume fraction of particles and void volume fraction was demonstrated (Justice, Derby et al. 2000, 2003). Synchrotron microCT and in situ loading was used to study damage in aluminum-zirconia composites (Babout, Ludwig et al. 2003, Babout, Maire et al. 2004).

Before turning the discussion to the most widespread discontinuously-reinforced composite, the cAp-collagen system in tooth dentin and bone, it is interesting to examine a microCT study of a model for an energetic material (read explosive), a nonstructural material whose high strain rate deformation properties determine its functionality. McDonald and coworkers examined the shock response of a model energetic material consisting of a polymer matrix loaded with two sizes of glass spheres, 300 and 30 μm diameters (McDonald, Millett et al. 2007). They found very few of the spheres were damaged in processing prior to testing. The microCT-derived microstructure of a representative volume was incorporated into a numerical simulation, and the investigators found that the shock front was not planar (on the order of 100 μm) and that flow was inhomogeneous at this length scale.

Most microCT deformation studies of mineralized tissue, for example, bone and dentin, both nanocrystalline particle-reinforced composites, concern trabecular bone; these were covered in Chapter 8. There have been a number of microCT studies of the deformation, cracking, and fracture of cortical bone and dentin; these have been particularly important as they have the potential to directly test the extent to which microcracking and/or crack bridging toughen these materials. A rather unusual study examined microCT for quantitative toolmark analysis of sharp force trauma to bone (Norman, Watson et al. 2018).

Animal limbs have skeletal structures that possess some flexibility; this allows the structures to return energy when applied loads are removed (springback), a useful attribute when the animal's limbs are moving rapidly such as in running or in flight. Insect studies were mentioned earlier in this chapter. The reinforcements in bird wings were examined and used as the basis for numerical modeling (Novitskaya, Ruestes et al. 2017). Strain measurement in the mouse forearm using subject-specific finite element models, strain gaging, and digital image correlation were compared in another study (Begonia, Dallas et al. 2017). As one would expect from a structure evolved to return energy, elastic modulus was found to vary along the length of the bovine femur, but microCT results did not correlate with mechanical measurements (Nobakhti, Katsamenis et al. 2017). This suggests that effects other than those measured by microCT are acting to produce macroscopic stiffness.

Nanoindentation is increasingly popular for applications such as quantification of the anisotropy of elastic moduli in bone, and such moduli were shown to agree well with moduli from load–deflection curves (Hengsberger, Enstroem et al. 2003). Some assumptions are required in the analysis, and Hengsberger et al. used synchrotron microCT to provide some of this information (specimen mineral levels, porosity, and cross-sectional dimensions).

One bovine cortical bone study used lab microCT to examine short-rod chevron-notched tension specimens for fracture toughness determination (Santis, Anderson et al. 2000). The V-shaped notch allowed steady-state crack propagation in a sample diameter rather smaller than a standard compact tension specimen, an important advantage given limited dimensions available even in the long bones of large animals. In principle, fracture toughness for this specimen geometry does not require measurement of the crack length, but practically realizable geometry did not meet the assumption for the calculation and compliance tests and crack length measurements (via microCT) were used for more robust determination of the plain strain-stress intensity factor (Santis, Anderson et al. 2000).

MicroCT was combined with in situ scattering and digital image correlation in notched cortical bone (Gustafsson, Mathavan et al. 2018).

Cortical bone exhibits good toughness, and two views of the source of toughness are bridging by uncracked ligaments in the crack wake and microcracking ahead of the crack tip. Both absorb energy that would otherwise be used to extend the crack. Establishing the relative importance of these mechanisms would suggest treatment strategies for osteoporosis prevention. Nalla and coworkers examined mechanistic aspects of crack growth resistance in human cortical bone by determining crack growth-resistance curves (R-curves) and using synchrotron microCT to image the 3D crack structure (Nalla, Kruzic et al. 2004, 2005). Fracture toughness rose linearly with crack length, but there were clear differences in behavior between bone from young mature adults (age < 41 years, described as young bone below) and that from aged individuals (age > 85 years, described as aged bone in what follows). Figure 11.2 shows differences in crack paths revealed by synchrotron microCT for the young bone compared to the aged bone (Nalla, Kruzic et al. 2004). Ex vivo crack initiation toughness decreased 40% from young to aged bone, and crack growth toughness present in the young bone was essentially eliminated over this period. Quantification of the amount of crack bridging vs crack extension (practical only with microCT) revealed considerable initial bridging for both young and aged bone (Fig. 11.3);

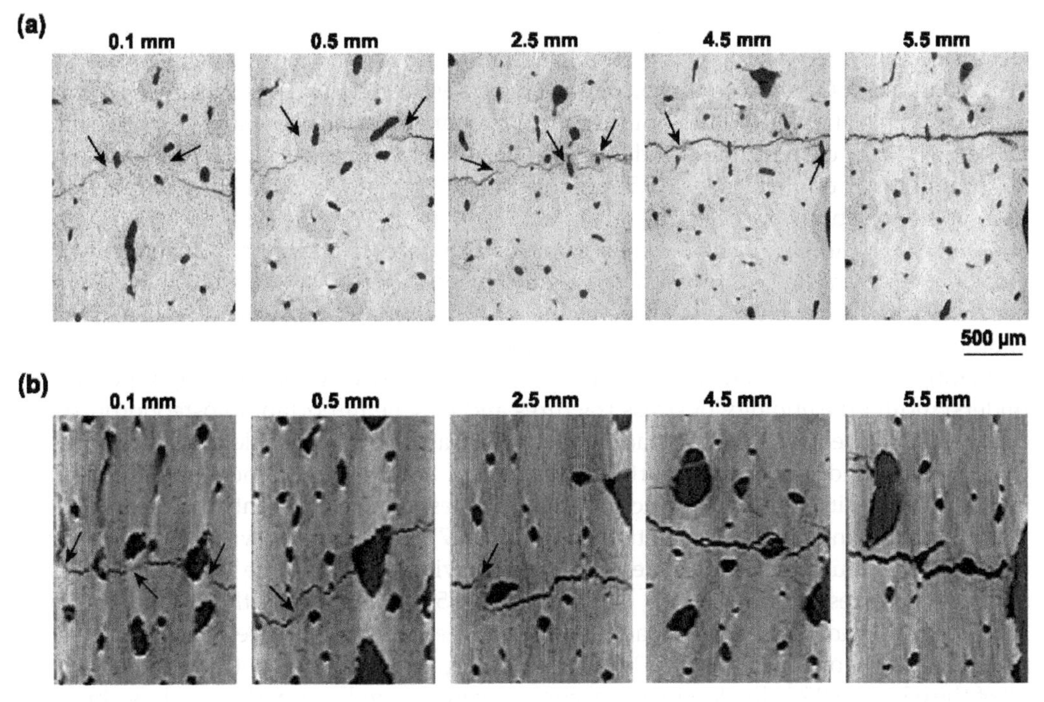

FIGURE 11.2
Synchrotron microCT sections through human cortical bone showing typical cracks in (a) young (34 years) and (b) aged (85 years) groups. The numbers above each column of images give the distance from the crack tip, and the black arrows indicate uncracked ligaments bridging the crack. The darker the shade of orange, the lower the x-ray absorption and mineral content (Nalla, Kruzic et al. 2004).

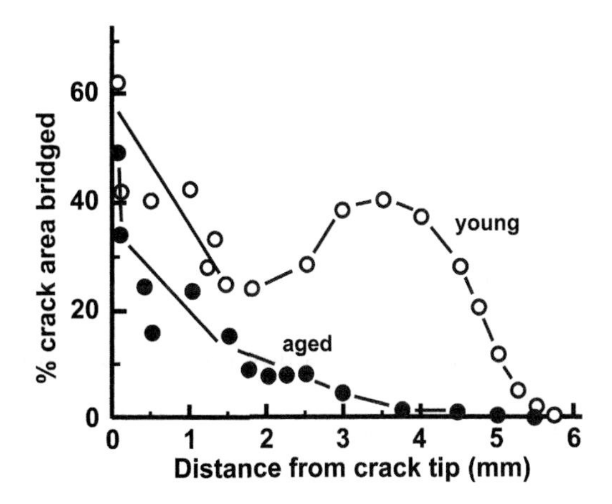

FIGURE 11.3
The fraction of crack-bridging zones in human cortical bone measured from microCT images as a function of distance from the crack tip. The plots for young and aged bone demonstrate the decline in size and percentage area of the cracked bridged zones in the older bone, and particularly the much lower bridging in the aged bone beyond ~2 mm from the crack tip. Plotted from data in Nalla et al. (2004) and Ritchie et al. (2006) with the trends lines shown here differing from what is shown in the original presentation.

after some crack extension, bridging for young bone remained comparable to the initial levels but for the aged bone was virtually absent (Nalla, Kruzic et al. 2004, Ritchie, Nalla et al. 2006). MicroCT showed that bridges were present throughout the specimen thickness (demonstrating that SEM data for bridging at specimen surfaces is valid) and that cracks tended to follow cement lines bordering osteons (Nalla, Kinney et al. 2003). The bridging zone length was on the order of 5.5 mm long for this human cortical bone. While toughness values for bone and dentin, related collagen-apatite composites, were comparable and were thought to reflect the nanoscale structure, differences in time-dependent crack blunting between the two mineralized tissues were thought to reflect the very different micrometer-level structures. Fracture, aging, and disease in bone were also discussed in another review by the same group (Ager, Balooch et al. 2006).

Quantifying damage (cracking) produced by in vivo loading of bones is important in understanding the etiology of stress fractures and may be important in osteoporotic fractures as well. The in vivo ulnar loading model for rats allows controlled levels of damage to be applied, and this is an important advantage if the response of bone to damage is to be followed longitudinally. The micro-damage process zone was recently studied in transverse cortical bone fracture (Willett, Josey et al. 2017). Lab microCT was used to quantify deformation-induced cortical bone cracking following ulnar fatigue with displacement amplitudes corresponding to 30%, 45%, 65%, and 85% of fracture (Uthgenannt and Silva 2007). Control ulnae and bones loaded to 30% of the fracture displacement did not show cracking, nearly one-half of the 45% displacement group exhibited detectable cracks, and all of the 65% and 85% displacement specimens were cracked – in particular, large branching cracks in the medial side of the 85% bones. Further, statistically significant differences in crack length density were found for the four displacement groups. In another in vivo ulnar loading study employing microCT, woven bone repair of fatigue damage was found to restore whole bone strength and to enhance resistance to further fatigue damage (Silva and Touhey 2007).

Diffraction-enhanced imaging (microradiography) was used along with bone's diffraction peak widths in an attempt to identify damage in cortical bone (Connor, Sayers et al. 2005). Neither method revealed damage: even with the ~tenfold increase in sensitivity to small cracks compared to absorption-based imaging; this is not surprising because microcracks are very tiny features and the effect of overlapping depths will obscure even larger features. This conclusion should not be taken to demonstrate that x-ray diffraction cannot reveal useful information for damaged bone; as examples in Chapter 12 on multimode studies demonstrate, this is not the case.

11.3.2 Fiber-Reinforced Composites

Images of a SiC monofilament – Si_3N_4 (91 wt.% Si_3N_4, 6 wt.% Y_2O_3, 3 wt.% Al_2O_3) composite were produced in an early microCT study (Hirano, Usami et al. 1989); in 111-μm thick slices, the 140-μm diameter fibers and their 30-μm diameter cores were quite visible, and the radial variation of linear attenuation coefficient at 24 keV for the SiC fibers agreed with others' 21 keV measurements of similar fibers (Kinney, Stock et al. 1990). Thermomechanical fatigue of ceramic and metal matrix composites were investigated by others (Baaklini, Bhatt et al. 1995).

Indentation damage was studied in carbon-fiber-reinforced plastic composites, but this study was limited to qualitative comparisons with results of ultrasonic characterization (Symons 2000). A more complete focus on detection limits for different types of damage in fiber-reinforced polymer matrix composites was provided by a second study that examined the same cracks before and after (high x-ray absorption) dye penetrants were added (Schilling, Karedla et al. 2005); in this lab microCT study, the authors reported crack detection limits without penetrant that were similar to those determined by Breunig et al. (1992a, 1993a), but cracks open 0.5–1 μm in ~20-μm voxels (opening <5% of the voxel size) could be detected when penetrant was added. Other polymeric composite systems containing cracks and studied with microCT were the elastomeric material of auto tires (Dunsmuir, Dias et al. 1999) and aged dental composites (Drummond, De Carlo et al. 2006). An optical fiber sensor was processed into a carbon-fiber thermoplastic composite, and microCT and the fiber sensor monitored strain during fatigue testing (De Baere, Voet et al. 2007). A short-fiber-reinforced polymeric matrix composite was recently studied with synchrotron microCT (Rolland, Saintier et al. 2018).

Fracture of unidirectional fiber composites is generally thought to occur when a cluster of broken fibers reaches a critical number N*, and microCT is an ideal tool for assessing whether the critical cluster concept is valid and, if so, what N* might be for a given composite system. Synchrotron microCT of uniaxially aligned quartz fiber, epoxy matrix composites investigated this concept with in situ loading, and simple stochastic failure models were reported to underpredict N* by a factor of 3–5 (Aroush, Maire et al. 2006).

Early reports of microCT of metal matrix composites focused on Al and Ti matrices and on SiC monofilament reinforcements (Armistead 1988, Hirano, Usami et al. 1989, Breunig, Elliott et al. 1990, 1992a, Elliott, Anderson et al. 1990, Kinney, Stock et al. 1990, London, Yancey et al. 1990, Stock 1990, Breunig, Stock et al. 1991,1993b, Kinney, Saroyan et al. 1991, Breunig 1992, Stock, Breunig et al. 1992, Baaklini, Bhatt et al. 1995, Hirano, Usami et al. 1995, Peix, Cloetens et al. 1997), not only because these composites offer important performance gains over other composites but also because the size of the fibers allows unambiguous identification of damage modes at individual fibers and because these results can be linked to work on damage mode characterization in optically transparent glass composites employing the same monofilaments.

MicroCT of SiC monofilament-Ti matrix composites presents a somewhat different challenge than imaging small diameter particulate or chopped fiber Al matrix composites, or, for that matter, imaging the same monofilament in an Al matrix. While there is relatively little contrast between Al and SiC (their linear attenuation coefficients are quite similar (Kinney, Stock et al. 1990)), the large absorption of Ti makes it difficult to detect changes within the SiC or C portions of the monofilaments (Stock, Breunig et al. 1992). Matrices of Ti_3Al and Ti-6-4 (6% Al, 4% V) with Textron SCS-6 SiC monofilament reinforcements were imaged using tube-generated (London, Yancey et al. 1990) or synchrotron x-radiation (Stock, Breunig et al. 1992). In these three- and eight-ply, aligned-fiber composites, matrix cracks and broken fibers were clearly seen.

The bulk of the microCT work on monofilament-reinforced composites, however, was on Al matrix materials and included at least one Ph.D. thesis (Breunig 1992). With the exception of one damaged* sample of $[0_2/\pm45]_s$ eight-ply SCS-2 SiC/Al (Stock, Breunig et al. 1992), the samples consisted of aligned monofilaments. In an eight-ply as-processed SCS-8 SiC/6061-0 Al matrix composite (see Fig. 5.10 for a more recently recorded slice of this same material), considerable intra-ply, processing-related porosity was noted in reconstructions (Breunig, Elliott et al. 1990, Kinney, Stock et al. 1990, Breunig, Stock et al. 1991, Breunig 1992). Monochromatic synchrotron x-radiation with energy between 20 and 22 keV was used to collect the data for these reconstructions with 6-μm isotropic voxels.

* The sample (the outer two plies on each side of the sample were parallel to and the four central plies were oriented at ±45° to the load axis) experienced 5.0×10^6 cycles with $R = \sigma_{miu}/\sigma_{max} = 0.3$ and load range of 413 MPa.

The contrast between SiC and Al was quite small at these energies, so that porosity and the monofilaments' carbon cores were the most visible features, and differentiating between fiber cores and porosity required careful scrutiny and a close eye for the weak contrast between SiC and Al. Based on examination of many slices, loading to 828 MPa eliminated much of the porosity. Likewise, measurement of the separation between fiber cores of first- and higher-order neighbors revealed that the fiber centers became more regularly spaced as the load increased. Thus, as monotonic load increased, the porosity disappeared and fiber spacing became more regular (Breunig, Stock et al. 1991, Breunig 1992). It is doubtful whether fiber rearrangement and porosity elimination would have been uncovered in studies of polished sections: too much labor would be required to interrogate enough serial sections.

Reconstructions with isotropic 6- to 7-μm voxels of ~1.5 mm × ~1.5 mm cross-sections cut from coupons of the same composite panel (as described above) revealed increased mechanical stiffness after the first few fatigue cycles corresponded to elimination of the processing-related matrix porosity (initially 2–7 vol.%) and to displacement of the fibers from somewhat irregular arrangement into a more nearly hexagonal array (Breunig 1992, Breunig, Kinney et al. 2006). This study showed the fibers rearranged and the porosity disappeared by the time the load reached 828 MPa. Fracture of the C cores of the SiC fibers appeared to occur before 828 MPa and to nucleate the subsequent SiC fracture, but SiC cracking could not be observed except after fracture (σ = 1,448 MPa), no doubt because the cracks are pulled closed by the relaxation of adjacent fibers at lower stresses. Spiral and planar cracks within SiC fibers were visible with microCT in fractured samples of the Al/SiC metal matrix composite (Breunig 1992, Kinney and Nichols 1992): Fig. 11.4 includes only the low absorption voxels from within the volume containing a single SiC fiber. The carbon core extends vertically to the sample surface (top) where the fiber pulled out of the matrix. The observation of C core fracture in the SiC fibers illustrates one of the advantages of tomography: this observation could not be made with serial sectioning and optical or scanning electron microscopy since any such fractures observed would undoubtedly be attributed to polishing damage.

Three-point bending of a notched eight-ply SiC/Al monofilament composite produced broken fibers, matrix cracking, and fiber-matrix disbonds that were observed with microCT (Breunig, Elliott et al. 1990, Breunig, Stock et al. 1991, Breunig 1992). The notch was 0.6 mm deep and ~0.8 mm wide, and it was on the tensile side of the 1.5 mm × 1.5 mm cross-section of the sample. Outside the notch only a small amount of fiber disbonding was present adjacent to the end of the notch. Most of the damage was confined to within the material ahead of the notch. Large lengths of fibers pulled from the matrix, and the authors noted that the matrix-fiber shear zone extended 200–300 μm beyond both ends of fractured fibers. The fiber fracture surfaces were not planar, and significant plastic deformation was inferred to have occurred after fiber fracture and pullout.

In situ observation of monotonic deformation in an Al/SiC monofilament composite also was reported (Hirano, Usami et al. 1995). In this study, the 140-μm diameter fibers were cut into 1-mm lengths and included in a 1-mm diameter sample at a volume fraction of 0.10. These dimensions are rather unrealistic when compared to typical monofilament-reinforced composites, but it appears that they were convenient for microtomography and, in the case of the monofilament length, avoided the need to apply large loads and avoided unstable fracture once the ultimate tensile stress was exceeded. Observations were reported at zero stress, at a stress just before failure, and after the sample failed, but only 40 slices were recorded at each stress. The slice width and interval between slices were 31 μm and 79 μm, respectively. Despite the difficulty of working with a set of noncontiguous

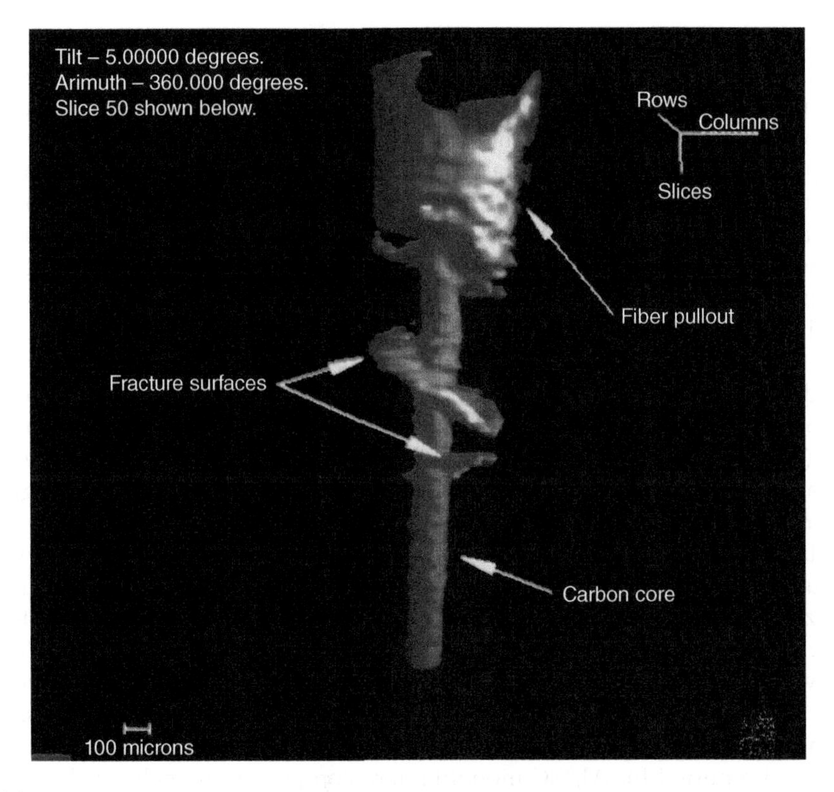

FIGURE 11.4
Three-dimensional rendering showing the surface of low-absorption voxels within a SiC monofilament in a fractured Al–SiC composite specimen. The fracture surface is at the top, and the C fiber core extends down from the fracture surface. The larger diameter cylinder at the top and concentric with the C core shows where the SiC fiber pulled from the Al surface in this half of the specimen. A spiral crack appears midway down the fiber (Breunig 1992, Kinney and Nichols 1992).

slices and the decreased crack sensitivity due to the relatively large slice thickness, the authors reported observing fiber-matrix disbonding, fiber pullout, and matrix cracking only at the maximum stress (and, of course, in the failed condition). From observation of the failed composite, the authors inferred that the matrix crack, imaged at the maximum stress, propagated close to several monofilament segments and led to the sample's failure. Because of the small monofilament lengths, the relatively low volume fraction of reinforcement, and the relatively small sample diameter (compared to the monofilament diameter), it is unclear whether the conclusions represent what occurs during fracture of composites containing the much longer monofilaments typically employed.

The rapid change of attenuation coefficients across most interfaces poses a particular challenge for accurate reconstruction. This is a greater problem for SiC reinforcements in a Ti matrix than in an Al matrix, for example, and an experimenter's choice of x-ray energy needs to consider the desired difference in contrast between reinforcement and matrix as well as optimum sample transmissivity. There is a large area of internal interfaces, and detecting cracks at these interfaces, an important failure mode, presents a very real challenge. Little can be done if the interface cracks are tightly closed, except if in situ loading can be applied to open the cracks or if phase contrast is significant. Even when these cracks are open, it is difficult to detect the change of attenuation from the crack superimposed on the transition in absorption across the interface. Considerable caution must be exercised, therefore, in quantification of interface cracking using x-ray microCT reconstructions.

In a uniaxial monofilament Al/SiC composite, microCT of mechanically induced damage was compared with unloading modulus (Breunig 1992, Breunig, Kinney et al. 2006). Macroscopic measures of damage (changes in unloading compliance and in unrecovered strain) correlated with microCT quantification of microstructural changes (fiber separation, fiber misorientation relative to the load axis, fiber carbon core fracture). More recent generations of SiC monofilaments have improved properties, so these results are not indicative of current performance. Few complete fractures of SiC fibers were observed except after specimen failure; the authors concluded SiC fiber fractures were responsible for decreased compliance but, upon unloading, residual stresses from intact fibers presumably pulled fracture surfaces back together in the damaged fibers. Because these synchrotron microCT data were obtained under imaging conditions where phase contrast was negligible (i.e., during the early 1990s at CHESS and SSRL), crack visibility in the SiC fibers was substantially lower than that in the studies reported in the following paragraph, and it is not surprising that tightly closed cracks might be invisible.

Uniaxially aligned monofilament Ti/SiC specimens were imaged with synchrotron microCT under in situ loads, and fiber fracture geometry and spatial distribution were characterized (Maire and Buffière 2000, Maire, Owen et al. 2001, McDonald, Preuss et al. 2002, 2003, Preuss, Rauchs et al. 2003, Sinclair, Preuss et al. 2004, 2005, Withers and Lewandowski 2006). Single-fiber, single-ply, and multiple-ply specimens were studied; artificially fractured fibers within the one-ply specimen and fiber bridging across a fatigue crack in the multiple-ply specimen were studied. The SiC fiber fractures were similar to what had been reported for Al/SiC monofilament composites, namely, wedge cracks and spiral cracks. Careful consideration of synchrotron phase-enhanced microCT renderings of the fractured fibers identified with wedge cracks (Figs. 5b and 6 of McDonald et al. (2003)) revealed complex contrast between the wedge edges (where contrast was strongest), and the discussion in this paper clearly identified small fragments that gave rise to the complex contrast. The fainter contrast regions suggested that the SiC material between the wedge edges contained additional (albeit more tightly closed) crack segments (small fragments of the locally shattered fiber?). Although this may seem to be a minor point, the fine details of fiber fracture could provide important insight into interface bonding or into stress wave interactions during fiber fracture. The longitudinal sections through the fiber centers also revealed that the fibers curve along the length of the specimen (smaller apparent width of the fiber's core at top and bottom of the sections). As an integral part of the study was microbeam diffraction mapping of strains in the Ti matrix and of the longitudinal fiber strains, more detailed discussion is postponed until Chapter 12.

References

Afshar, T., M. M. Disfani, G. A. Narsilio and A. Arulrajah (2018). "Post-breakage changes in particle properties using synchrotron tomography." *Powder Technol* **325**: 530–544.

Ager, J. W., G. Balooch and R. O. Ritchie (2006). "Fracture, aging, and disease in bone." *J Mater Res* **21**: 1878–1892.

Armistead, R. A. (1988). "CT: quantitative 3-D inspection." *Adv Mater Process Inc Met Prog* (Mar): 41–49.

Aroush, D. R. B., E. Maire, C. Gauthier, S. Youssef, P. Cloetens and H. D. Wagner (2006). "A study of fracture of unidirectional composites using in situ high-resolution synchrotron x-ray microtomography." *Compos Sci Technol* **66**: 1348–1353.

Baaklini, G. Y., R. T. Bhatt, A. J. Eckel, P. Engler, M. G. Castelli and R. W. Rauser (1995). "X-ray microtomography of ceramic and metal matrix composites." *Mater Eval* **53**: 1040–1044.

Babout, L., W. Ludwig, E. Maire and J. Y. Buffière (2003). "Damage assessment in metallic structural materials using high resolution synchrotron x-ray tomography." *Nucl Instrum Meth B* **200**: 303–307.

Babout, L., E. Maire and R. Fougères (2004). "Damage initiation in model metallic materials: X-ray tomography and modelling." *Acta Mater* **52**: 2475–2487.

Beaugrand, J., S. Guessasma and J. E. Maigret (2017). "Damage mechanisms in defected natural fibers." *Sci Rep* **7**: 14041.

Begonia, M., M. Dallas, M. L. Johnson and G. Thiagarajan (2017). "Comparison of strain measurement in the mouse forearm using subject-specific finite element models, strain gaging, and digital image correlation." *Biomech Model Mechanobiol* **16**: 1243–1253.

Bonse, U., R. Nusshardt, F. Busch, R. Pahl, J. H. Kinney, Q. C. Johnson, R. A. Saroyan and M. C. Nichols (1991). "X-ray tomographic microscopy of fibre-reinforced materials." *J Mater Sci* **26**: 4076–4085.

Bontaz-Carion, J. and Y. P. Pellegrini (2006). "X-ray microtomography analysis of dynamic damage in tantalum." *Adv Eng Mater* **8**: 480–486.

Brandt, J., G. Doig and N. Tsafnat (2015). "Computational aerodynamic analysis of a microCT based bio-realistic fruit fly wing." *PLOS ONE* **10**: 124824.

Breunig, T. M. (1992). Nondestructive evaluation of damage in SiC/Al metal/matrix composite using x-ray tomographic microscopy. Ph.D. thesis, Georgia Institute of Technology.

Breunig, T. M., J. C. Elliott, P. Anderson, G. Davis, S. R. Stock, A. Guvenilir and S. D. Dover (1990). Application of x-ray microtomography to the study of SiC/Al metal matrix composite material. *New Materials and their Applications*. D. Holland. London, Institute of Physics. **111**: 53–60.

Breunig, T. M., J. C. Elliott, S. R. Stock, P. Anderson, G. R. Davis and A. Guvenilir (1992a). Quantitative characterization of damage in a composite material using x-ray tomographic microscopy. *X-Ray Microscopy III* A. G. Michette, G. R. Morrison and C. J. Buckley. New York, Springer: 465–468.

Breunig, T. M., J. H. Kinney and S. R. Stock (2006). "MicroCT (microtomography) quantification of microstructure related to macroscopic behavior. Part 2: Damage in SiC-Al monofilament composites tested in monotonic tension and fatigue." *Mater Sci Technol* **22**: 1059–1067.

Breunig, T. M., M. C. Nichols, J. S. Gruver, J. H. Kinney and D. L. Haupt (1994). "Servo-mechanical load frame for in situ, non-invasive, imaging of damage development." *Ceram Eng Sci Proc* **15**: 410–417.

Breunig, T. M., S. R. Stock, S. D. Antolovich, J. H. Kinney, W. N. Massey and M. C. Nichols (1992b). A framework relating macroscopic measures and physical processes of crack closure of Al-Li Alloy 2090. *Fracture Mechanics: Twenty-Second Symposium (Vol. 1)*. H. A. Ernst, A. Saxena and D. L. McDowell. Philadelphia, ASTM. **ASTM STP 1131**: 749–761.

Breunig, T. M., S. R. Stock and R. C. Brown (1993a). "Simple load frame for in situ computed tomography and x-ray tomographic microscopy." *Mater Eval* **51**: 596–600.

Breunig, T. M., S. R. Stock, A. Guvenilir, J. C. Elliott, P. Anderson and G. R. Davis (1993b). "Damage in aligned fibre SiC/Al quantified using a laboratory x-ray tomographic microscope." *Composites* **24**: 209–213.

Breunig, T. M., S. R. Stock, J. H. Kinney, A. Guvenilir and M. C. Nichols (1991). Impact of X-ray tomographic microscopy on deformation studies of a SiC/Al MMC. *Advanced Tomographic Imaging Methods for the Analysis of Materials*. J. L. Ackerman and W. A. Ellingson. Pittsburgh, Mater Res Soc. **217**: 135–141.

Buffiere, J. Y., E. Ferrie, H. Proudhon and W. Ludwig (2006). "Three-dimensional visualization of fatigue cracks in metals using high resolution synchrotron x-ray micro-tomography." *Mater Sci Technol* **22**: 1019–1024.

Buffiere, J. Y., E. Maire, P. Cloetens, G. Lormand and R. Fougeres (1999). "Characterization of internal damage in a MMCp using x-ray synchrotron phase contrast microtomography." *Acta Mater* **47**: 1613–1625.

Buffière, J. Y., S. Savelli and E. Maire (2000). Characterisation of MMCP and cast aluminum alloys. *X-Ray Tomography in Materials Science*. J. Baruchel, J. Y. Buffière, E. Maire, P. Merle and G. Peix. Paris, Hermes Science: 103–114.

Christensen, R. H. (1963). "Fatigue crack growth affected by metal fragments wedged between opening-closing crack surfaces." *Appl Mater Res* **24**(October): 207–210.

Connor, D. M., D. Sayers, D. R. Sumner and Z. Zhong (2005). "Identification of fatigue damage in cortical bone by diffraction enhanced imaging." *Nucl Instrum Meth A* **548**: 234–239.

Crostock, H. A., J. Nellesen, G. Fischer, S. Schmauder, U. Weber and F. Beckmann (2006). Tomographic analysis and FE-simulations of MMC microstructures under load. *Developments in X-Ray Tomography V*. U. Bonse. Bellingham (WA), SPIE. **SPIE Proc Vol 6318**: 63181A-1–63181A-2.

De Baere, I., E. Voet, W. Van Paepegem, J. Vlekken, V. Cnudde, B. Masschaele and J. Degrieck (2007). "Strain monitoring in thermoplastic composites with optical fiber sensors: Embedding process, visualization with micro-tomography, and fatigue results." *J Thermoplast Comp Mater* **20**: 453–472.

Drummond, J. L., F. De Carlo, K. B. Sun, A. Bedran-Russo, P. Koin, M. Kotche and B. Super (2006). Tomography of dental composites. *Developments in X-Ray Tomography V*. U. Bonse. Bellingham (WA), SPIE. **SPIE Proc Vol 6318**: 63182B-1–63182B-8.

Dunsmuir, J. H., A. J. Dias, D. G. Peiffer, R. Kolb and G. Jones (1999). Microtomography of elastomers for tire manufacture. *Developments in X-Ray Tomography II*. U. Bonse. Bellingham (WA), SPIE. **SPIE Proc Vol 3772**: 87–96.

Elber, W. (1971). The significance of fatigue crack closure. *Damage Tolerance in Aircraft Structures*. M. S. Rosenfeld. West Conshohocken (PA), ASTM. **STP 486**: 230–242.

Elliott, J. C., P. Anderson, S. D. Dover, S. R. Stock, T. M. Breunig, A. Guvenilir and S. D. Antolovich (1990). "Application of X-ray microtomography in materials science illustrated by a study of a continuous fiber metal matrix composite." *J X-ray Sci Technol* **2**: 249–258.

Fedele, R., A. Ciani and F. Fiori (2014). "X-ray Microtomography under loading and 3D-volume digital image correlation. A review." *Fund Info* **135**: 171–197.

Ferrie, E., J. Y. Buffiere and W. Ludwig (2005). "3D characterisation of the nucleation of a short fatigue crack at a pore in a cast Al alloy using high resolution synchrotron microtomography." *Int J Fatigue* **27**: 1215–1220.

Ferrie, E., J. Y. Buffiere, W. Ludwig, A. Gravouil and L. Edwards (2006). "Fatigue crack propagation: in situ visualization using x-ray microtomography and 3D simulation using the extended finite element method." *Acta Mater* **54**: 1111–1122.

Forquin, P. and E. Ando (2017). "Application of microtomography and image analysis to the quantification of fragmentation in ceramics after impact loading." *Philos Trans A Math Phys Eng Sci* **375**: 20160166.

Gupta, C., H. Toda, P. Mayr and C. Sommitsch (2015). "3D creep cavitation characteristics and residual life assessment in high temperature steels: a critical review." *Mater Sci Technol* **31**: 603–626.

Gustafsson, A., N. Mathavan, M. J. Turunen, J. Engqvist, H. Khayyeri, S. A. Hall and H. Isaksson (2018). "Linking multiscale deformation to microstructure in cortical bone using in situ loading, digital image correlation and synchrotron X-ray scattering." *Acta Biomater* **69**: 323–331.

Guvenilir, A. (1995). Investigation into asperity induced closure in an Al-Li alloy using X-ray tomography. Ph.D. thesis, Georgia Institute of Technology.

Guvenilir, A., T. M. Breunig, J. H. Kinney and S. R. Stock (1997). "Direct observation of crack opening as a function of applied load in the interior of a notched tensile sample of Al-Li 2090." *Acta Mater* **45**: 1977–1987.

Guvenilir, A., T. M. Breunig, J. H. Kinney and S. R. Stock (1999). "New direct observations of crack closure processes in Al-Li 2090 T8E41." *Phil Trans Roy Soc (Lond)* **357**: 2755–2775.

Guvenilir, A. and S. R. Stock (1998). "High resolution computed tomography and implications for fatigue crack closure modeling." *Fatigue Fract Eng Mater Struct* **21**: 439–450.

Guvenilir, A., S. R. Stock, M. D. Barker and R. A. Betz (1994). *Aluminum Alloys: Their Physical Properties and Mechanical Properties*. T. H. Sanders and E. A. Starke. Atlanta, Georgia Institute of Technology. **II**: 413–419.

Helfenstein-Didier, C., D. Taïnoff, J. Viville, J. Adrien, É. Maire and P. Badel (2018). "Tensile rupture of medial arterial tissue studied by X-ray micro-tomography on stained samples." *J Mech Behav Biomed Mater* **78**: 362–368.

Hengsberger, S., J. Enstroem, F. Peyrin and P. Zysset (2003). "How is the indentation modulus of bone tissue related to its macroscopic elastic response? A validation study." *J Biomech* **36**: 1503–1509.

Hirano, T., K. Usami and K. Sakamoto (1989). "High resolution monochromatic tomography with x-ray sensing pickup tube." *Rev Sci Instrum* **60**: 2482–2485.

Hirano, T., K. Usami, Y. Tanaka and C. Masuda (1995). "In situ x-ray CT under tensile loading using synchrotron radiation." *J Mater Res* **10**: 381–385.

Hodgkins, A., T. J. Marrow, P. Mummery, B. Marsden and A. Fok (2006). "X-ray tomography observation of crack propagation in nuclear graphite." *Mater Sci Technol* **22**: 1045–1051.

Ignatiev, K. I. (2004). Development of x-ray phase contrast and microtomography methods for the 3D study of fatigue cracks. Ph.D. thesis, Georgia Institute of Technology.

Ignatiev, K. I., G. R. Davis, J. C. Elliott and S. R. Stock (2006). "MicroCT (microtomography) quantification of microstructure related to macroscopic behavior. Part 1: Fatigue crack closure measured in situ in AA 2090 compact tension samples." *Mater Sci Technol* **22**: 1025–1037.

Ignatiev, K. I., W. K. Lee, K. Fezzaa and S. R. Stock (2005). "Phase contrast stereometry: Fatigue crack mapping in 3D." *Phil Mag* **83**: 3273–3300.

Izadifar, M., P. Babyn, M. E. Kelly, D. Chapman and X. Chen (2017). "Bioprinting pattern-dependent electrical/mechanical behavior of cardiac alginate implants: Characterization and ex vivo phase-contrast microtomography assessment." *Tissue Eng Part C Methods* **23**: 548–564.

Justice, I., B. Derby, G. Davis, P. Anderson and J. Elliott (2000). Characterisation of void and reinforcement distributions by edge contrast. *X-Ray Tomography in Materials Science*. J. Baruchel, J. Y. Buffière, E. Maire, P. Merle and G. Peix. Paris, Hermes Science: 89–102.

Justice, I., B. Derby, G. Davis, P. Anderson and J. Elliott (2003). "Characterisation of void and reinforcement distributions in a metal matrix composite by x-ray edge-contrast microtomography." *Scripta Mater* **48**: 1259–1264.

Khor, K. H., J. Y. Buffière, W. Ludwig and I. Sinclair (2006). "High resolution x-ray tomography of micromechanisms of fatigue crack closure." *Scripta Mater* **55**: 47–50.

Kinney, J. H. and M. C. Nichols (1992). "X-ray tomographic microscopy (XTM) using synchrotron radiation." *Annu Rev Mater Sci* **22**: 121–152.

Kinney, J. H., R. A. Saroyan, W. N. Massey, M. C. Nichols, U. Bonse and R. Nusshardt (1991). "X-ray tomographic microscopy for nondestructive characterization of composites." *Rev Prog Quant NDE* **10A**: 427–433.

Kinney, J. H., S. R. Stock, M. C. Nichols, U. Bonse, T. M. Breunig, R. A. Saroyan, R. Nusshardt, Q. C. Johnson, F. Busch and S. D. Antolovich (1990). "Nondestructive investigation of damage in composites using x-ray tomographic microscopy." *J Mater Res* **5**: 1123–1129.

Landis, E. and D. T. Keane (1999). X-ray microtomography for fracture studies in cement-based materials. *Developments in X-Ray Tomography II*. U. Bonse. Bellingham (WA), SPIE. **SPIE Proc Vol 3772**: 105–113.

Landis, E. N. and E. N. Nagy (2000). "Three-dimensional work of fracture for mortar in compression." *Eng Fract Mech* **65**: 223–234.

Landis, E. N., E. N. Nagy and D. T. Keane (2003). "Microstructure and fracture in three dimensions." *Eng Fract Mech* **70**: 911–925.

Landis, E. N., E. N. Nagy, D. T. Keane and G. Nagy (1999). "Technique to measure 3D work-of-fracture of concrete in compression." *J Exp Mech* **125**: 599–605.

Landis, E. N., T. Zhang, E. N. Nagy, G. Nagy and W. R. Franklin (2007). "Cracking, damage and fracture in four dimensions." *Mater Struct* **40**: 357–364.

Lenoir, N., M. Bornert, J. Desrues, P. Bésuelle and G. Viggiani (2007). "Volumetric digital image correlation applied to X-ray microtomography images from triaxial compression tests on Argillaceous rock." *Strain* **43**: 193–205.

Li, P., P. D. Lee, T. C. Lindley, D. M. Maijer, G. R. Davis and J. C. Elliott (2006). "X-ray microtomographic characterisation of porosity and its influence on fatigue crack growth." *Adv Eng Mater* **8**: 476–479.

London, B., R. N. Yancey and J. A. Smith (1990). "High-resolution x-ray computed tomography of composite materials." *Mater Eval* **48**: 604–608.

Ludwig, W. and D. Bellet (2000). "Penetration of liquid gallium into the grain boundaries of aluminum: A synchrotron radiation microtomographic investigation." *Mater Sci Eng A* **281**: 198–203.

Ludwig, W., S. Bouchet, D. Bellet and J. Y. Buffière (2000). 3D observation of grain boundary penetration in Al alloys. *X-Ray Tomography in Materials Science*. J. Baruchel, J. Y. Buffière, E. Marie, P. Merle and G. Peix. Paris, Hermes Science: 155–164.

Ludwig, W., J. Y. Buffière, S. Savelli and P. Cloetens (2003). "Study of the interaction of a short fatigue crack with grain boundaries in a cast Al alloy using x-ray microtomography." *Acta Mater* **51**: 585–598.

Luo, Y., S. C. Wu, Y. N. Hu and Y. N. Fu. (2018). "Cracking evolution behaviors of lightweight materials based on in situ synchrotron X-ray tomography: A review." *Front Mech Eng* **13**: 461–481.

Maire, E. and J. Buffière (2000). X-ray tomography of aluminium foams and Ti/SiC composites. *X-Ray Tomography in Materials Science*. J. Baruchel, J. Y. Buffière, E. Maire, P. Merle and G. Peix. Paris, Hermes Science: 115–126.

Maire, E., A. Owen, J. Y. Buffiere and P. J. Withers (2001). "A synchrotron X-ray study of a Ti/SiC_f composite during in situ straining." *Acta mater* **49**: 153–163.

Maire, E. and P. J. Withers (2014). "Quantitative X-ray tomography." *Inter Mater Rev* **59**: 1–43.

Marrow, T. J., J. Y. Buffiere, P. J. Withers, G. Johnson and D. Engelberg (2004). "High resolution x-ray tomography of short fatigue crack nucleation in austempered ductile cast iron." *Int J Fatigue* **26**: 717–725.

McDonald, S. A., J. C. F. Millett, N. K. Bourne, K. Bennett, A. M. Milne and P. J. Withers (2007). "The shock response, simulation and microstructural determination of a model composite material." *J Mater Sci* **42**: 9671–9678.

McDonald, S. A., M. Preuss, E. Maire, J. Y. Buffiere, P. M. Mummery and P. J. Withers (2002). "Synchrotron x-ray study of micromechanics of Ti/SiCf composites with fibres containing defects introduced by laser drilling." *Mater Sci Technol* **18**: 1497–1503.

McDonald, S. A., M. Preuss, E. Marie, J. Y. Buffiere, P. M. Mummery and P. J. Withers (2003). "X-ray tomographic imaging of Ti/SiC composites." *J Microsc* **209**: 102–112.

Morano, R. (1998). Effect of R-ratio on crack closure in Al-Li 2090 T8E41, investigated nondestructively with x-ray microtomography. MS thesis, Georgia Institute of Technology.

Morano, R., S. R. Stock, G. R. Davis and J. C. Elliott (1999). Macrotexture-related fatigue crack closure in Al-Li 2090 studied by x-ray microtomography. *Proceedings of the Twelfth International Conference on Textures of Metals, Vol. 2*. J. A. Szpunar. Ottawa, National Research Council of Canada: 1106–1111.

Morano, R., S. R. Stock, G. R. Davis and J. C. Elliott (2000). X-ray microtomography of fatigue crack closure as a function of applied load in Al-Li 2090 T8E41 samples. *Nondestructive Methods for Materials Characterization*. Warrendale (PA), MRS. **MRS Symp Proc Vol 591**: 31–35.

Mummery, P. M., B. Derby, P. Anderson, G. R. Davis and J. C. Elliott (1995). "X-ray microtomographic studies of metal matrix composites using laboratory x-ray sources." *J Microsc* **177**: 399–406.

Nalla, R. K., J. H. Kinney and R. O. Ritchie (2003). "Mechanistic fracture criteria for the failure of human cortical bone." *Nature Mater* **2**: 164–168.

Nalla, R. K., J. J. Kruzic, J. H. Kinney and R. O. Ritchie (2004). "Effect of aging on the toughness of human cortical bone: evaluation by R-curves." *Bone* **35**: 1240–1246.

Nalla, R. K., J. J. Kruzic, J. H. Kinney and R. O. Ritchie (2005). "Mechanistic aspects of fracture and R-curve behavior in human cortical bone." *Biomaterials* **26**: 217–231.

Nellesen, J., R. Laquai, B. R. Mueller, A. Kupsch, M. P. Hentschel, N. B. Anar, E. Soppa, W. Tillmann and G. Bruno (2018). "In situ analysis of damage evolution in an Al/MMC under tensile load by synchrotron X-ray refraction imaging." *J Mater Sci* **53**: 6021–6032.

Nobakhti, S., O. L. Katsamenis, N. Zaarour, G. Limbert and P. J. Thurner (2017). "Elastic modulus varies along the bovine femur." *J Mech Behav Biomed Mater* **71**: 279–285.

Norman, D. G., D. G. Watson, B. Burnett, P. M. Fenne and M. A. Williams (2018). "The cutting edge - microCT for quantitative toolmark analysis of sharp force trauma to bone." *Forensic Sci Int* **283**: 156–172.

Novitskaya, E., C. J. Ruestes, M. M. Porter, V. A. Lubarda, M. A. Meyers and J. McKittrick (2017). "Reinforcements in avian wing bones: Experiments, analysis, and modeling." *J Mech Behav Biomed Mater* **76**: 85–96.

Ohgaki, T., H. Toda, I. Sinclair, J. Y. Buffiere, W. Ludwig, T. Kobayashi, M. Niinomi and T. Akahori (2005). "Quantitative assessment of liquid Ga penetration into an aluminium alloy by high-resolution x-ray tomography." *Mater Sci Eng A* **406**: 261–267.

Panahi, H., M. Kobchenko, P. Meakin, D. K. Dysthea and F. Renard (2018). "In-situ imaging of fracture development during maturation of an organic rich shale: effects of heating rate and confinement." *Mar Petrol Geol* **95**: 314–327.

Peix, G., P. Cloetens, M. Salome, J. Y. Buffiere, J. Baruchel, F. Peyrin and M. Schlenker (1997). Hard x-ray phase tomographic investigation of materials using Fresnel diffraction of synchrotron radiation. *Developments in X-Ray Tomography*. U. Bonse. Bellingham (WA), SPIE. **3149**: 149–159.

Preuss, M., G. Rauchs, T. J. A. Doel, A. Steuwer, P. Bowen and P. J. Withers (2003). "Measurements of fibre bridging during fatigue crack growth in Ti/SiC fibre metal matrix composites." *Acta Mater* **51**: 1045–1057.

Pyzalla, A., B. Camin, T. Buslaps, M. D. Michiel, H. Kaminiski, A. Kottar, A. Pernack and W. Reimers (2005). "Simultaneous tomography and diffraction analysis of creep damage." *Science* **308**: 92–95.

Qian, L., H. Toda, K. Uesugi, T. Kobayashi, T. Ohgaki and M. Kobayashi (2005). "Application of synchrotron x-ray microtomography to investigate ductile fracture in Al alloys." *Appl Phys Lett* **87**: 241907.

Rajabi, H., N. Ghoroubi, K. Stamm, E. Appel and S. N. Gorb (2017). "Dragonfly wing nodus: a one-way hinge contributing to the asymmetric wing deformation." *Acta Biomater* **60**: 330–338.

Renard, F., J. Weiss, J. Mathiesen, Y. Ben-Zion, N. Kandula and B. Cordonnier (2018). "Critical evolution of damage toward system-size failure in crystalline rock." *J Geophys Res Solid Earth* **123**: 1969–1986.

Ritchie, R. O., R. K. Nalla, J. J. Kruzic, J. W. Ager III, G. Balooch and J. H. Kinney (2006). "Fracture and ageing in bone: toughness and structural characterization." *Strain* **42**: 225–232.

Roberts, B. C., E. Perilli and K. J. Reynolds (2014). "Application of the digital volume correlation technique for the measurement of displacement and strain fields in bone: A literature review." *J Biomech* **47**: 923–934.

Rohani, S. A., S. Ghomashchi, S. K. Agrawal and H. M. Ladak (2017). "Estimation of the Young's modulus of the human pars tensa using in-situ pressurization and inverse finite-element analysis." *Hear Res* **345**: 69–78.

Rolland, H., N. Saintier, I. Raphael, N. Lenoir, A. King and G. Rober (2018). "Fatigue damage mechanisms of short fiber reinforced PA66 as observed by in-situ synchrotron X-ray microtomography." *Composites Pt B Eng* **143**: 217–229.

Sangid, M. D., T. A. Book, D. Naragani, J. Rotella, P. Ravi, A. Finch, P. Kenesei, J.-S. Park, H. Sharma, J. Almer and X. Xiao (2018). "Role of heat treatment and build orientation in the microstructure sensitive deformation characteristics of IN718 produced via SLM additive manufacturing." *Additive Manufacturing* **22**: 479–496.

Santis, R. D., P. Anderson, K. E. Tanner, L. Ambrosio, L. Nicolais, W. Bonfield and G. R. Davis (2000). "Bone fracture analysis on the short rod chevron-notch specimens using the x-ray computer micro-tomography." *J Mater Sci Mater Med* **11**: 629–636.

Schilling, P. J., B. P. R. Karedla, A. K. Tatiparthi, M. A. Verges and P. D. Herrington (2005). "X-ray computed microtomography of internal damage in fiber reinforced polymer matrix composites." *Compos Sci Technol* **65**: 2071–2078.

Silva, M. J. and D. C. Touhey (2007). "Bone formation after damaging in vivo fatigue loading results in recovery of whole-bone monotonic strength and increased fatigue life." *J Orthop Res* **25**: 252–261.

Sinclair, R., M. Preuss, E. Marie, J. Y. Buffiere, P. Bowen and P. J. Withers (2004). "The effect of fibre fractures in the bridging zone of fatigue cracked Ti-6Al-4V/SiC fibre composites." *Acta Mater* **52**: 1423–1438.

Sinclair, R., M. Preuss and P. J. Withers (2005). "Imaging and strain mapping fibre by fibre in the vicinity of a fatigue crack in a Ti/SiC fibre composite." *Mater Sci Technol* **21**: 27–34.

Sinnett-Jones, P. E., M. Browne, W. Ludwig, J. Y. Buffiere and I. Sinclair (2005). "Microtomography assessment of failure in acrylic bone cement." *Biomater* **26**: 6460–6466.

Stock, S. R. (1990). X-Ray methods for mapping deformation and damage. *Micromechanics: Experimental Techniques*. W. N. J. Sharpe. New York, ASME. **AMD 102**: 147–162.

Stock, S. R. (1999). "Microtomography of materials." *Int Mater Rev* **44**: 141–164.

Stock, S. R. (2008). "Recent advances in x-ray microtomography applied to materials." *Inter Mater Rev* **58**: 129–181.

Stock, S. R., T. M. Breunig, A. Guvenilir, J. H. Kinney and M. C. Nichols (1992). Nondestructive X-ray tomographic microscopy of damage in various continuous-fiber metal matrix composites. *Damage Detection in Composite Materials*. J. E. Masters. West Conshohocken (PA), ASTM. **STP 1128**: 25–34.

Stock, S. R., A. Guvenilir, T. M. Breunig, J. H. Kinney and M. C. Nichols (1995). "Computed tomography. Part III: Volumetric, high-resolution x-ray analysis of fatigue crack closure." *J Metals* Jan, **47**: 19.

Suresh, S. and R. O. Ritchie (1982). "A geometric model for fatigue crack closure induced by fracture surface roughness." *Met Trans* **13A**: 1627–1631.

Symons, D. D. (2000). "Characterization of indentation damage in 0/90 lay-up T300/914 CFRP." *Compos Sci Technol* **60**: 391–401.

Toda, H., E. Maire, Y. Aoki and M. Kobyashi (2011). "Three-dimensional strain mapping using in situ X-ray synchrotron microtomography." *J Strain Anal Eng Design* **46**: 549–561.

Toda, H., I. Sinclair, J. Y. Buffiere, E. Maire, T. Connolley, M. Joyce, K. H. Khor and P. Gregson (2003). "Assessment of the fatigue crack closure phenomenon in damage-tolerant aluminium alloy by in-situ high-resolution synchrotron x-ray microtomography." *Phil Mag* **83**: 2429–2448.

Toda, H., I. Sinclair, J. Y. Buffiere, E. Maire, K. H. Khor, P. Gregson and T. Kobayashi (2004). "A 3D measurement procedure for internal local crack driving forces via synchrotron x-ray microtomography." *Acta Mater* **52**: 1305–1317.

Underwood, E. E. (1968). Particle size distribution. *Quantitative Microscopy*. R. T. DeHoff and F. N. Rhines. New York, McGraw-Hill: 149–200.

Uthgenannt, B. A. and M. J. Silva (2007). "Use of the rat forelimb compression model to create discrete levels of bone damage in vivo." *J Biomech* **40**: 317–324.

Viggiani, G., N. Lenoir, P. Bésuelle, M. D. Michiel, S. Marello, J. Desrues and M. Kretzschmer (2004). "X-ray microtomography for studying localized deformation in fine-grained geomaterials under triaxial compression." *C R Mech* **332**: 819–826.

Watson, I. G., P. D. Lee, R. J. Dashwood and P. Young (2006). "Simulation of the mechanical properties of an aluminum matrix composite using x-ray microtomography." *Metall Mater Trans A* **37**: 551–558.

Weck, A., D. S. Wilkinson, H. Toda and E Maire (2006). "2D and 3D visualization of ductile fracture." *Adv Eng Mater* **8**: 469–472.

Weiler, J. P., J. T. Wood, R. J. Klassen, E. Maire, R. Berkmortel and G. Wang (2005). "Relationship between internal porosity and fracture strength of die-cast magnesium AM60B alloy." *Mater Sci Eng A* **395**: 315–322.

Willett, T., D. Josey, R. X. Z. Lu, G. Minhas and J. Montesano (2017). "The micro-damage process zone during transverse cortical bone fracture: no ears at crack growth initiation." *J Mech Behav Biomed Mater* **74**: 371–382.

Withers, P. J. and J. J. Lewandowski (2006). "Three-dimensional imaging of materials by microtomography." *Mater Sci Technol* **22**: 1009–1010.

Wu, S. C., T. Q. Xiao and P. J. Withers (2017). "The imaging of failure in structural materials by synchrotron radiation X-ray microtomography." *Eng Fract Mech* **182**: 127–156.

Yoder, G. R., P. S. Pao, M. A. Imam and L. A. Cooley (1989). Micromechanisms of fatigue fracture in Al-Li Alloy 2090. *Aluminum-Lithium Alloys, Proceedings of the Fifth Aluminum-Lithium Conference.* J. T.H. Sanders, E. A. Starke, Jr. Birmingham (UK), Materials and Component Engineering Publications Ltd.: 1033–1041.

Zhang, H., H. Toda, H. Hara, M. Kobayashi, T. Kobayashi, D. Sugiyama, N. Kuroda and K. Uesugi (2007). "Three-dimensional visualization of the interaction between fatigue crack and micropores in an aluminum alloy using synchrotron X-ray microtomography." *Met Mater Trans A* **38**: 1774–1785.

12

Multimode Studies and Nonabsorption Modalities

A number of groups have employed x-ray microCT and another x-ray or non-x-ray modality to gain a more complete understanding than either method could have provided separately. The first section of this chapter describes various multimode studies. The second section covers scattering tomography.

In Section 12.1, the first example is microstructural characterization of sea urchin teeth: x-ray microbeam mapping and precision lattice parameter determination were combined with microCT. The second example is sulfate attack of Portland cement paste, where energy-dispersive x-ray diffraction mapping of reaction products was used along with microCT. Three combined diffraction and microCT studies of mechanical responses of specimens provide the next examples: x-ray mesotexture analysis of crack path and microCT quantification of changes in fatigue crack opening, microCT plus x-ray diffraction of creep damage, and diffraction-based strain mapping plus microCT of load redistribution in monofilament metal matrix composites. The composite material bone is covered in Section 12.6: internal stress analysis plus microCT of loaded bone and microCT plus diffraction or strain gage analysis or other modalities. The final section looks at networks analyzed with microCT plus other methods.

12.1 Multimode Studies

12.1.1 Sea Urchin Teeth

Sea urchin teeth were discussed in some detail in Section 7.2.3, and the interested reader is referred to that section and the references therein for a start to reading about these fascinating structures. What is important here is that sea urchin teeth mineralize in two stages. The primary, secondary, and carinar process plates, as well as the needles/prisms are termed the primary mineralized tissue and, in *Lytechinus variegatus*, are nonequilibrium calcite $Ca_{1-x}Mg_xCO_3$ with $x \approx 0.13$, as determined by synchrotron x-ray diffraction (Stock, Barss et al. 2002). X-ray diffraction of a living *L. variegatus* tooth showed that the mineral was crystalline at the earliest stages of growth, and synchrotron microCT showed the expected tooth plate structure (Stock, Veis et al. 2012). At about the point where the keel develops, the primary skeletal elements begin to be linked by the secondary skeletal elements, which are termed columns or disks. The columns are much higher Mg calcite with $x \approx 0.33$ (Stock, Barss et al. 2002). The relationship between first- and second-stage mineral is found across the regular sea urchins (Stock, Ignatiev et al. 2014). These columns, linking adjacent primary structural elements into a rigid structure, were often described as being polycrystalline, but the transmission x-ray diffraction data clearly showed that the high Mg ($x \approx 0.13$) and very high Mg ($x \approx 0.33$) phases had their crystallographic axes identically aligned (Stock, Ignatiev et al. 2014). The view of Stock and coworkers, therefore, is that the tooth is a compositionally modulated crystal much like multiple quantum well structures from molecular beam epitaxy (Stock, Barss et al. 2002), although, of course, of much lower crystal quality than the artificial material. X-ray diffraction mapping (precision lattice parameter and crystallite size/microstrain broadening determinations) provided additional structural information supplementing 3D microCT-derived geometric information. Sea urchin teeth grow continuously, so examining their structure at different places along their length provides a window into growth processes, and this has been exploited in *L. variegatus* with synchrotron microCT within intact teeth and higher resolution SEM imaging and chemical analysis of structures exposed by fracture (Stock, Seto et al. 2018).

12.1.2 Sulfate Ion Attack of Portland Cement

Study of sulfate ion attack of Portland cement via microCT was introduced briefly in Chapter 10, where it was noted that microCT showed the results of the attack but provided little information about the reaction phases producing the damage (i.e., the softening, cracking, loss of adhesion, etc., of the cement). Position-resolved x-ray diffraction with high-energy synchrotron x-radiation is a good method of mapping phase content as a function of depth, and using this method and microCT provided much more information than either technique by itself (Naik 2003, Jupe, Stock et al. 2004, Wilkinson, Jupe et al. 2004, Naik, Jupe et al. 2006). Energy-dispersive x-ray diffraction was used instead of the more normal single wavelength methods because the former allowed more precise definition of the sampling volume combined with simultaneous collection of diffraction patterns from multiple phases within this same volume. The reader is directed elsewhere for more details specific to this application (Jupe, Stock et al. 2004).

Sulfate attack is (simplistically) described in the literature by one of two classes of reaction and associated damage. The first is gypsum formation that is associated with loss of adhesion and strength. The second is ettringite formation, associated with expansion and cracking. After considerable sulfate exposure, energy-dispersive x-ray diffraction identified an ettringite-rich, gypsum-free layer outside of cylindrical cracks paralleling the outer surface of the cylindrical specimens (i.e., C1 in Fig. 10.5). Inside the crack, that is, closer to the cylinder center, a gypsum-containing volume was identified (Naik 2003, Jupe, Stock et al. 2004, Wilkinson, Jupe et al. 2004, Naik, Jupe et al. 2006). While the same identification might have been performed by destructive specimen preparation (with considerably more effort), the results could have been criticized as affected by exposure to the atmosphere, etc.

A similar approach was used to study phase transformation in a liquid Al foam (Jiménez, Paeplow et al. 2018). One difference from the cement pasted corrosion experiments was that foam experiments were conducted very rapidly.

12.1.3 Fatigue Crack Path and Mesotexture

MicroCT of cracked AA2090 specimens revealed complex 3D patterns of crack face contact as a function of applied load (see the deformation subsection). Roughness of the crack faces produced the closure effects and is intrinsically related to the low fatigue crack propagation rate for this material. The underlying question is what drives the crack to assume this highly nonplanar path: fracture mechanics indicates that the energetically favorable path would be more or less directly across the specimen (i.e., a path perpendicular to the applied tensile load). Yoder and coworkers related the average texture or

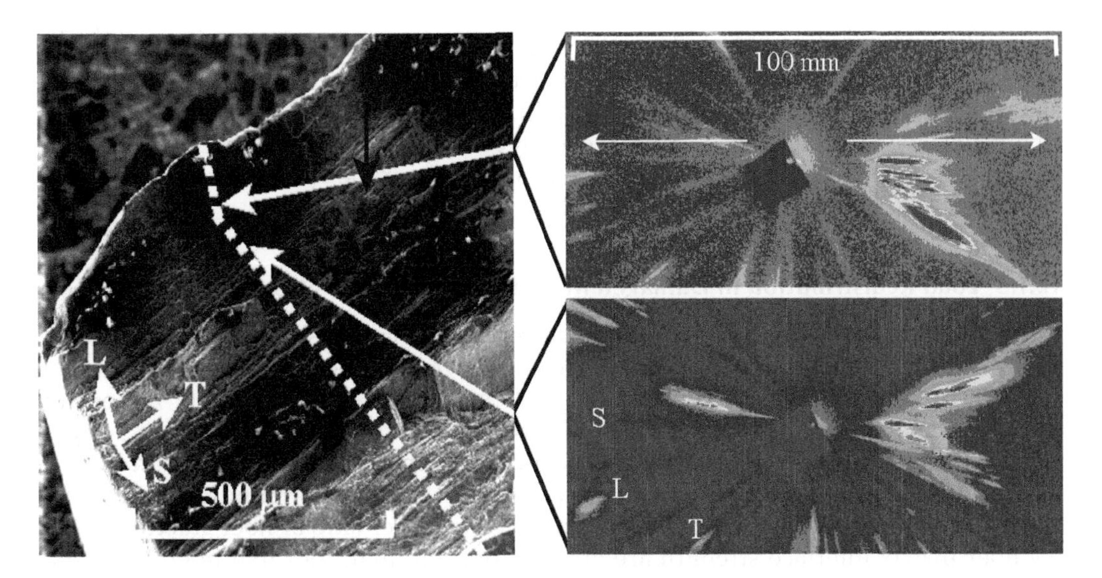

FIGURE 12.1
SEM fractograph (left) of a fatigue crack surface in a small compact tension specimen of AA2090 viewed at a large angle of tilt. A large asperity on the crack face appears at the top. The directions indicate directions in the plate from which the specimen was machined: longitudinal (L), transverse (T), and short transverse (S) directions. The dashed line indicates the positions at which a series of (synchrotron x-ray) microbeam Laue patterns were recorded. The central portions of two transmission Laue patterns, from the positions indicated by the arrows, are shown at the right. The streaks are 111 reflections, and increasing x-ray diffracted intensity is indicated by the following colors: purple (lowest), dark blue, light blue, green, yellow, orange, and red (highest). The diamond shape near the center of the image is the beam stop. The orientation of the streaks changes abruptly from within the asperity to outside the asperity. See Haase et al. (1998).

macrotexture to the faces of asperities (large peaks) on the fracture surface (Yoder, Pao et al. 1989): while this data explained the geometry of asperities and why they, on the average, formed, these observations did not identify the cause of the transition between an asperity and a relatively planar section of the crack. Microbeam Laue pattern mapping revealed the scale of crystallographic texture between that of individual grains (microtexture) and the average specimen texture (Haase, Guvenilir et al. 1998, 1999). This particular type of mesotexture consisted of groups of 5–10 adjacent pancake-shaped grains with nearly identical orientations, that is, these adjacent grains comprised near single-crystal volumes. Asperities formed when the fatigue crack passed through the border between near single-crystal volumes with different orientations (Fig. 12.1). Further, a large fraction of the volume of the plate centers of AA2090 T8E41 consisted of near single-crystal domains, and this differentiated AA2090 from other Al-Li alloys with similar macrotextures (Ignatiev, Rek et al. 2000) and produced decreased fatigue crack growth rates compared to the other alloys.

12.1.4 Creep and Corrosion Damage

Pyzalla et al. studied creep of a three-phase copper alloy (Pb particles in a mixture of α- and β- brass) using synchrotron microCT and x-ray diffraction (Pyzalla, Camin et al. 2005). Three detector systems were positioned so that microCT, energy-dispersive x-ray diffraction, and angle-dispersive x-ray diffraction could be performed sequentially without realignment or recalibration. The microCT-determined pore size distribution was reported for 10 time intervals and agreed with an exponential growth dependence. The decrease in diffraction peak FWHM ended when the voids start to reach appreciable size, and changes in peak intensity after this point in the creep test revealed texture formation. Another study threaded together x-ray computed tomography, serial section FIB-SEM tomography, EBSD, and TEM elemental analysis for the same 3D region of a corroded stainless steel (Burnett, McDonald et al. 2014).

12.1.5 Load Redistribution in Damaged Monofilament Composites

Failure of uniaxially aligned, monofilament-reinforced composites depends on many factors. MicroCT allows one to study where and at what applied stresses the reinforcements fail; repeated observations of the same specimen are particularly important because fibers such as the SCS series SiC monofilaments will often fracture several times within a 10-mm gage section; the location of each successive break is an important input for modeling. The increasing strain within the fiber, longitudinal and transverse strains within the matrix, strain relaxation to either side of fiber fractures, and the fiber matrix interface strength are other important quantities that microCT alone cannot define. As demonstrated by a series of reports on Ti/SiC_f composites, combining high-energy x-ray microbeam diffraction mapping with microCT was a powerful approach to measuring these quantities (McDonald, Preuss et al. 2002, 2003, Preuss, Withers et al. 2002, Sinclair, Preuss et al. 2004, 2005, Aroush, Maire et al. 2006, Withers, Bennett et al. 2006).

Preuss et al. studied deformation of a single SiC fiber in a Ti-6Al-4V matrix: synchrotron microCT revealed the position and morphology of SiC fiber fractures and, as a function of applied stress, microbeam diffraction quantified the matrix and fiber strains (Preuss, Withers et al. 2002). At each of 19 loading steps, mapping with transmission x-ray diffraction along the length of the fiber (100 diffraction profiles spaced by 50-μm steps at each load) revealed SiC longitudinal strains of at least 1.5% before the fiber cracked at the first point (equivalent to a failure stress of at least 6 GPa for E = 400 GPa). Above the nominal yield stress for the matrix, strains in the matrix became only slightly nonlinear, but the fiber longitudinal strain rose very rapidly. At 790 MPa, the load preceding first fracture, two local strain maxima were observed along the length of the monofilament; the next deformation increment produced a load drop and local strains approaching zero at the positions of the two maxima, positions that corresponded to fiber fracture revealed in microCT. Longitudinal strains in the matrix were relatively uniform at 790 MPa but, at the next deformation state, rose sharply at the positions where the fiber fractured. In other words, localized strain concentration occurred in the matrix in the vicinity of the SiC breaks. Fitting the data to a partial sliding model allowed the authors to estimate a constant interfacial friction shear stress of ~200 MPa that was significantly higher than results from fiber push-out tests. The authors note that blind application of conventional full-fragmentation postmortem analysis of fragment lengths would suggest a significantly higher interfacial strength (~700 MPa) and suggest that fiber strength decreases after the first fast fracture event.

Well-defined defects were introduced into a single-ply Ti/SiC_f composite, and redistribution of loads from damaged fibers to neighboring ones was investigated with microbeam diffraction mapping (McDonald, Preuss et al. 2002). Load redistribution around damage sites increased the load in the nearest neighbor fibers by ~25% and second nearest neighbors by ~10%. The interfacial fractional shear stress was found to be similar to or slightly larger than that cited in the previous paragraph. Reverse sliding was observed during unloading and produced compressive residual stresses near the fiber ends. Wedge crack geometry was frequently observed (McDonald, Preuss et al. 2002, 2003).

In the examples of a single-fiber and a single-ply composite described above, simple phase-enhanced microradiography would have sufficed to correlate fiber fracture and maxima/minima in the fiber and matrix strain profiles. In multiple-ply composites, the overlapping fiber images necessitated use of microCT, and microCT plus microbeam diffraction mapping were applied to multiple-ply Ti/SiC_f to determine the stress partition between fiber bridging a fatigue crack and broken fibers (Sinclair, Preuss et al. 2004, 2005, Withers, Bennett et al. 2006). Initially, strain mapping averaged over the entire thickness of the specimen (Sinclair, Preuss et al. 2004), but subsequent experiments used a narrow receiving slit and 2θ-scanning to limit the gage volume to a single SiC fiber plus the surrounding matrix material (Sinclair, Preuss et al. 2005). Strain distribution in an intact fiber in the crack wake was compared for maximum and minimum applied loads; for example, strain distribution as a function of distance from the crack plane was analyzed using a partial debonding shear lag model (Sinclair, Preuss et al. 2005). Measurements of crack opening displacements showed that the fatigue crack front bowed out between fibers when it emerged from a ply and advanced preferentially toward fibers when the front was between plies (Withers, Bennett et al. 2006). Further the 3D distributions of crack opening were measured for three stress intensities characteristic of a fatigue cycle (K_{max}, K_{min}, and K_{mid}), and very little variation in crack opening was observed parallel to the crack front irrespective of the proximity to bridging fibers (Withers, Bennett et al. 2006).

12.1.6 Bone and Other Mineralized Tissues in Mammals

X-ray scattering measurement of internal strains (and conversion to internal stresses) in loaded bone (or tooth dentin and enamel) is a relatively uninvestigated research area and one where a combination with microCT will provide valuable insight. For bone, the collagen *D*-period (~67 nm) along the fibril axis produces SAXS peaks, and the Angstrom-level periodicities of carbonated apatite (cAp) crystallites produce diffraction peaks in the WAXS regime. While the mineral nanoparticles in long bone have a pronounced crystallographic texture, there are still enough orientations present to produce more-or-less complete cones of diffracted intensity for monochromatic x-rays; Debye cones from different *hkl* exist simultaneously and produce rings of increased intensity on area detectors. Force applied to a specimen distorts the unit cells and alters the Debye cones. Hydrostatic applied stresses (those with equal magnitude in all directions) uniformly alter the diameter of cones, whereas deviatoric stresses (those with directionality) change the shape of diffraction rings. Similarly, SAXS peak positions from (the collagen *D*-period) alter in response to applied stress.

X-ray scattering measures quantities such as d_{hkl} in cAp or the *D*-period in collagen, and changes in these quantities define the internal strain imposed during loading, that is, strain in cAp is $\varepsilon_{cAp} = (d - d_{initial})/d_{initial}$ and in collagen is $\varepsilon_{collagen} = (D - D_{initial})/D_{initial}$. Internal stress is a quantity derived from internal strain, and stress σ_{ij} and strain ε_{kl} are second rank tensors related through the fourth rank elastic constants C_{ijkl}, that is, $\sigma_{ij} = C_{ijkl}\,\varepsilon_{kl}$. Describing the conversion of deviatoric strains to deviatoric stresses is beyond the scope of this chapter, and the reader is directed elsewhere for details as pertains to bone internal stress measurements (Almer and Stock 2005, 2007).

The inverse of the slope of macroscopic strain ε_{macro} (measured by an attached strain gage) as a function of $\sigma_{applied}$ (measured by the load cell of the mechanical testing apparatus) is

Young's modulus for the specimen. Such slopes for the WAXS and SAXS data (ε_{cAp} and $\varepsilon_{collagen}$ vs $\sigma_{applied}$) reflect Young's modulus for the individual constituent phase of bone. While this last extension may not be strictly correct, it does provide a numerical operational probe of how the individual phases differ from pure (inorganic) cAp or collagen. For a section of canine fibula, the resulting moduli (90% confidence limit) were: E_{macro} = 24.7(0.2) GPa, E_{cAp} = 41(1.0) GPa, and $E_{collagen}$ = 18(1.2) GPa (Almer and Stock 2007). The value for E_{macro} was in good agreement with moduli of similar bone types reported in the literature; for cAp was about one-third of that of inorganic apatite; for collagen was at least nine times higher than one would expect (see Almer and Stock (2007) for details). The data demonstrated the extent to which the local environment affects the different phases' responses to applied load.

In the studies described in the preceding pair of paragraphs, microCT was used to measure cross-sectional area and to account for internal porosity. This is a rather trivial use of microCT because of the simple geometry. In specimens containing bone in complex geometries, that is, specimens containing trabecular bone or cortical plus trabecular bone, determination of the spatial distribution of bone segments and of their orientation relative to the load axis will be essential for proper interpretation of the scattering data. Obtaining such 3D maps is impractical except through microCT, especially when one considers that that the different bone segments may suffer significant relative displacements during loading, displacements that may change from load to load and that may not be preserved during post-testing serial sectioning. Recently, microCT was combined with in situ scattering and digital image correlation in notched cortical bone (Gustafsson, Mathavan et al. 2018).

The junction between dentin and enamel in teeth is an extraordinary structure. Both dentin and enamel are based upon the mineral hydroxyapatite, and the former has a composition quite close to bone while the latter is essentially all mineral. The junction, therefore, couples materials with a factor of four difference in Young's modulus, a remarkably evolved structure. In one study of the DEJ functionality, synchrotron microCT complemented in situ x-ray diffraction mapping (under different applied loads) of the distribution of strains across this junction in bovine tooth (Stock, Vieira et al. 2008, Almer and Stock 2010).

X-ray diffraction and synchrotron microCT were combined in a study of changing mineralization during human fetal vertebrae growth (Nuzzo, Meneghini et al. 2003). From microCT, trabecula were much thicker and more widely spaced in the interior of the vertebrae than in the peripheral volume, and bone volume fraction increased linearly over gestational ages of 16 through 24 weeks. X-ray diffraction revealed a linear increase in crystallite size with age and a linear increase in lattice parameter ratio c/a (both over the same gestational range as above). As complete understanding of bone mechanical properties depends not only on properly incorporating microarchitecture but also inclusion of proper materials properties (in bone these include crystallite dimensions and distortions of the apatite unit cell), more studies of this sort need to be completed and the data imported into numerical models of elastic moduli, etc. At the other age extreme, bone degeneration in senile fish was studied with diffraction and microCT (Aguilera, Rocha et al. 2017). Diffraction plus phase microCT have been used to study porous ceramic scaffolds (for implantation in bone) seeded with cells (Cedola, Campi et al. 2014).

The role of insulin-like growth factor-1 (IGF-1) was studied in fetal mouse bone with microCT, with Fourier transform infrared spectroscopy (FTIR) and von Kossa staining, a histology technique designed to reveal bone (Burghardt, Wang et al. 2007). Tibiae and lumbar vertebrae were examined from knock-out (IGF-1 -/-) and wild-type animals at the

18th gestational day. The degree of mineralization (in terms of mg hydroxyapatite per cubic centimeter) and the morphology of the bone were determined with synchrotron microCT. FTIR was used to infer the ratio of mineral to matrix using the ratio of the phosphate peak integral to that of the amide I band, and to measure the carbonate-to-phosphate ratio using a carbonate band, all standard methods in bone research. FTIR data at both bone sites and both animal populations were characteristic of poorly crystalline mineral. Mineral-to-matrix ratio was systematically lower in IGF-1 -/- animals than wild types for both sites. Carbonate-to-phosphate ratio was significantly lower in the IGF-1 -/- spine but not the tibia. Interestingly, von Kossa staining failed to reveal mineral at the spinal ossification center of the IGF-1 -/- animals, even though the synchrotron microCT clearly reveals high attenuation components within this structure, indicating a subtle difference in the mineral's environment that affects the staining process.

The last result cited in the previous paragraph, von Kossa staining failing to reveal mineral shown to be present from two other techniques, suggests that the results of studies that employ only von Kossa staining and that fail to find mineral despite the presence of bone matrix need to be reexamined. Further investigation of the necessary environment for the staining to occur may reveal a new way to probe the mineral's local environment. One study addressing this type of question used dual-energy microCT to quantify the time-course and staining characteristics of ex vivo animal organs treated with iodine- and gadolinium-based contrast agents (Martins de Souza E Silva, Utsch et al. 2017).

MicroCT is a powerful method for mapping 3D differences in mineral content of bone. Results of BSE compared quantitatively with microCT suggest, however, that one must exercise caution in interpreting microCT-observed heterogeneities in mineral content (Mashiatulla, Ross et al. 2017).

In another study, diffraction with small diameter synchrotron x-ray beams was combined with synchrotron microCT to characterize coronary atherosclerosis in vitro (Jin, Ham et al. 2002). It may be that diffraction-based measures of crystallite size, microstrain, or a/c ratio will help define the natural history of such pathological mineralization processes or even the rate of development of these dangerous structures. Mineral within ligaments was located with microCT, and techniques such as x-ray diffraction and fluorescence were used to identify the material present (Orzechowska, Wróbel et al. 2018). Characterization of the mineral phase was done at multiple scales for skeletal sites of bone cancer metastasis (He, Chiou et al. 2017). X-ray diffraction was used with microCT to infer mechanisms of growth and crystal aggregation in kidney stones (Manzoor, Agrawal et al. 2019). Optically functional crystals are an integral part of the eyes of some marine invertebrates, and microCT as well as diffraction and other techniques have been used to study these very interesting organs (Palmer, Hirsch et al. 2018).

The role of sutures between bones was discussed earlier in the context of analysis of effect on jaw structure of the hardness or softness of food. Another study employed microCT, strain gages, and high-speed photography to investigate in vivo the mechanical role of cranial sutures and their morphology during feeding in a fish; the goal was to determine what inferences, if any, could be gathered about skull function in living and fossil fish (Markey, Main et al. 2006). These investigators studied *Polypterus*, a fish that feeds by suction or by biting, and determined that peak suture strains are higher during suction than biting. They also found that interfrontal and frontoparietal sutures, typically loaded in tension, were less interdigitated in cross-section than the interparietal suture which experiences compression.

MicroCT can often provide data to validate the results of other techniques. Ultrasonic analysis, a noninvasive, inexpensive, and nonharmful modality, is being developed as a clinical tool, for example, not only for fetal imaging and for heart morphology and fluid flow imaging but also for cortical and cancellous bone analysis. Interpretation of in vivo ultrasonic velocity measurements in cortical bone depends on microCT calibration, an active area of research. MicroCT data also were used to validate conclusions developed from lower-resolution MRI studies of cancellous bone (Krug, Carballido-Garmio et al. 2007).

In animal models, increasing use is being made of bioluminescence imaging, that is, detection of photons emitted through the skin by luciferase expressing cells in the living animal. Recently, this approach was demonstrated for tumor progression in deep tissues including bone. MicroCT plus bioluminescence were combined in a bone tumor study (Fritz, Louis-Plence et al. 2007), and one expects the number of such studies to grow substantially (Marien, Hillen et al. 2017).

As was mentioned in Chapter 8, scaffold design and evaluation of bone growth into scaffolds is an active area of research. Cancedda et al. reviewed application of microCT and of microdiffraction to this problem; only a very few studies combined the two modalities. Combined SAXS and WAXS were used to map the orientations of the cAp crystallites' c-axes and the collagen fibril axes relative to the hydroxyapatite scaffold material seeded with bone stromal cells (Cedola, Mastrogiacomo et al. 2006, Cancedda, Cedola et al. 2007). This approach extended to 3D diffraction mapping, for example, as described elsewhere (Stock, De Carlo et al. 2008), and combined with microCT should be very powerful.

12.1.7 Networks and Porosity

Fibers form the basis of some network solids, and biodegradable fibers are used in applications as diverse as fishing line and sutures. Tanaka and coworkers applied synchrotron microCT and SAXS to study the internal structure of as formed and isothermally crystallized fibers of poly[(R)-hydroxybutyrate] (Tanaka, Uesugi et al. 2007). Without crystallization, the drawn fibers produced strong SAXS peaks along the meridian (fiber axis), suggesting that there were lamellar crystals of systematic long period along the fiber axis. The meridian reflections could barely be detected in the crystallized fibers, but equatorial streak scattering was much stronger than for the uncrystallized fibers, suggesting voids were an important component of these fibers. MicroCT revealed the inner structure of the 100- to 120-μm diameter fibers with 1-μm spatial resolution. Few micrometer-sized voids were seen in the uncrystallized material. In the crystallized fiber, voids comprised nearly 50% of the cross-sectional area, 3D renderings revealed the voids were highly elongated along the drawing direction, and the voids were principally 1–6 μm in cross-section, with a mean of 2.3 μm and a strong peak in the size distribution of ~1.5 μm.

Blood vessels in the brain comprise a network of great interest in a variety of disciplines. Dorr et al. combined MRI and microCT imaging of the brains of mice to produce a comprehensive vasculature atlas of this organ, emphasizing the location of vessels with respect to neuroanatomical structures and watershed regions associated with specific arteries (Dorr, Sled et al. 2007). Given the much higher spatial resolution of microCT, it was used to image the blood vessels that had previously been filled with an x-ray absorbing polymer (Microfil), whereas MRI mapped the different neuroanatomical structures using another contrast agent.

Pores with additive manufactured components can have an extremely deleterious effect of properties such as strength or service lifetime. Selective laser melting consolidation of powder metallurgy specimens produced voids, which were characterized by high-energy synchrotron microCT, and these voids' effect on deformation was studied with a variety of electron beam techniques, digital image correlation, and in situ high-energy x-ray diffraction (Sangid, Book et al. 2018).

12.2 Reconstruction Other than with Absorption or Phase Contrast

There was some coverage of reconstruction with x-ray signals other than differences in absorption or in phase in Chapter 4. Various examples of other microCT modalities appeared in the different applications chapters, and this scattered coverage might hinder the reader from appreciating the advantages of reconstructions using signals other than the "conventional" ones. There is considerable value, therefore, in grouping together examples of these nonconventional reconstructions, which are, to date, limited in number. Section 12.2.1 presents scattering tomography studies for small grained or for noncrystalline materials. Section 12.2.2 briefly describes 3D diffraction mapping for samples with grains large enough to resolve on detectors. Results from coherent diffraction imaging and ptychography appear in Section 12.2.3 and from fluorescence microCT in Section 12.2.4.

12.2.1 X-Ray Scattering Tomography

As was mentioned in Section 4.12, microCT using scattered x-rays, specifically WAXS and SAXS, provides spatial information much different from that in absorption or phase-contrast microCT. Quite a bit more WAXS tomography results have appeared than studies with SAXS tomography, but examples of both follow. In the case of diffraction (WAXS) microCT, the information that can be obtained includes the spatial distribution of the crystalline phases present (Stock, De Carlo et al. 2008, Stock and Almer 2012), crystallographic texture (Gursoy, Bicer et al. 2015), lattice constants and their variation within the sampled volume (Frølich, Leemreize et al. 2016), macrostrain (e.g., residual stresses), and crystallite size and microstrain (Birkbak, Nielsen et al. 2017). In addition, diffraction tomography can also reconstruct the diffraction pattern from each voxel within the slice (Leemreize, Almer et al. 2013). There has been enough activity that a review has recently appeared (Birkbak, Leemreize et al. 2015).

The approach to mapping crystallographic texture via diffraction microCT is relatively straightforward. In specimens whose texture consists of specific crystallographic directions <*h'k'l'*> taking preferred orientations relative to processing directions or to in vivo loading directions, intensity in *h'k'l'* diffraction rings varies azimuthally. If one reconstructs using a specific sector of a diffraction ring, say a 20° azimuthal range encompassing the maximum diffracted intensity, instead of the average of the entire ring, one can map preferred orientation within the specimen. Typically, for comparison, one also reconstructs with a sector normal to the sector containing the maximum. In bone, the mineral nanoplatelets are strongly oriented with their *c*-axes lying along the direction of maximum in vivo principal stress, and 00.2 is the most strongly affected reflection. In a specimen cut from a porcine spinous process and reconstructed with two orthogonal 00.2 sectors (Fig. 12.2), comparison of the two reconstructions showed much greater concentration of *c*-axis orientations along the cortical bone loading axis (Gursoy, Bicer et al. 2015). One approach to mapping lattice constants and their variation within the sampled volume is to azimuthally

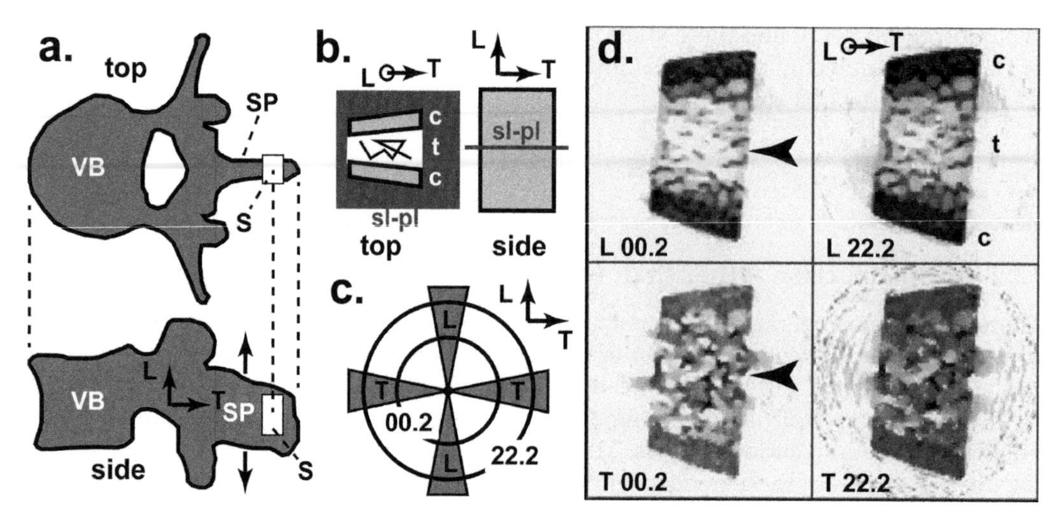

FIGURE 12.2
Diffraction tomography of a specimen cut from a porcine spinous process. (a) Schematic representation of a vertebra (top and side views) showing the area from which the specimen was cut. The arrows on the spinous process (side view) show the direction of principal strain in vivo (axis L); the transverse direction, normal to the loading direction, is the axis T. VB – vertebral body; SP – spinous process; S – sample. (b) Schematic top and side views of the specimen. c – cortical bone; t – trabecular bone; sl-pl – slice plane. (c) Schematic of the diffraction pattern showing 00.2 and 22.2 diffraction rings. The azimuthally integrated ring intensities within the sectors labeled L are used to reconstruct the top pair of slices in (d); the azimuthally integrated ring intensities within the sectors labeled T are used to reconstruct the top pair of slices in (d). (d) Slices reconstructed with hydroxyapatite 00.2 diffracted intensities (left column) and with 22.2 diffracted intensities (right column). The darker the pixel in the reconstructions, the greater the diffracted intensity. The cortices (c) of the process show uniform high intensities indicating greater hydroxyapatite content and little porosity, whereas the more spotty intensities between the two cortices reflect the trabecular structure (t), which is highly fenestrated. The arrowheads show that the trabecular intensities in the T-orientation 00.2 reconstruction are comparable to those of the nearby cortices, whereas those in the L-orientation 00.2 reconstruction are smaller than those of the nearby cortices. This shows that the hydroxyapatite nanocrystal *c*-axes in the cortices are highly aligned along the loading direction L, whereas the trabecular nanocrystal *c*-axes have many different orientations reflecting the alignments of the struts of this structure. See Gursoy et al. (2015) for more details.

average intensity for all diffraction angles 2θ and to reconstruct separate slices for each 2θ. For each x,y within the series of slices, one can recombine the intensities for the different 2θ into a 1D diffraction pattern for that voxel and extract lattice parameter(s) using fitting approaches such as Rietveld refinement. This approach was used to map Mg content in a mineralized bivalve attachment system and lattice parameters in a hydroxyapatite sample containing two powder types (Leemreize, Almer et al. 2013, Frølich, Leemreize et al. 2016). Somewhat later, and surprisingly unaware of the earlier work, others made similar determinations on a fossil specimen (Mürer, Sanchez et al. 2018). An entire tooth has been reconstructed (Egn, Jacques et al. 2013). With a laboratory setup, the tissue-specific microfibril orientation was reconstructed with diffraction tomography for bamboo; other techniques including lab microCT were used to confirm these results (Ahvenainen, Dixon et al. 2017).

Diffraction microCT can also be used to map nanocrystallite size of different phases and the spatial distribution of amorphous phases. Like the hydroxyapatite specimen mentioned in the preceding paragraph, which was designed to test the ability to differentiate materials with slightly different lattice parameters and to determine the statistical limits of the reconstruction method, a sample made from several known nanocrystalline and amorphous phases was used to demonstrate that diffraction microCT could provide accurate and precise quantification (Birkbak, Nielsen et al. 2017).

Different phases have been mapped in Portland cement paste (Artioli, Cerulli et al. 2010, Claret, Grangeon et al. 2018). Batteries and fuel cells have been studied by diffraction tomography (Sottmann, Di Michiel et al. 2017, Li, Heenan et al. 2019). In the area of cultural heritage, diffraction tomography has been used to study a small sample from a van Gogh painting (Vanmeert, Van der Snickt et al. 2015) and one from Rembrandt's *Homer* (Price, Van Loon et al. 2019).

Most scattering tomography studies scan the entire sample volume and use synchrotron radiation. Region-of-interest diffraction tomography, however, has been investigated (Zhu, Katsevich et al. 2019), and a laboratory system for diffraction tomography has been demonstrated (Zhu, Katsevich et al. 2018). Collecting interlaced projections (at relatively large angular separations but over multiple revolutions) for diffraction tomographic reconstruction allows post-collection choices to be made in terms of spatial vs temporal resolution (Vamvakeros, Jacques et al. 2016).

Signals from SAXS and the translate/rotate data collection have been used to study tumors in the murine brain, specifically to reconstruct slices and to determine SAXS patterns for individual voxels within the slices (Jensen, Bech et al. 2011). A slightly different approach was used to obtain 3D orientation information from raster scanning of thin specimens and recording the 2D SAXS patterns (Schaff, Bech et al. 2015, Liebi, Georgiadis et al. 2018).

Typically, x-ray scattering tomography uses monochromatic radiation, and this requires that many different projections be recorded over 180°. If a polychromatic beam, a 1D energy sensitive detector (Rumaiz, Kuczewski et al. 2018), and a slit between sample and detector are used, energy-dispersive diffraction (EDD) tomography can be performed without rotating the specimen and by only translating the specimen (Fig. 4.13). This approach was demonstrated recently, and Fig. 12.3 compares a reconstructed EDD slice of an archeological human second metacarpal bone with an absorption microCT slice at approximately the same position of the same bone (Stock, Okasinski et al. 2017). This method of acquiring

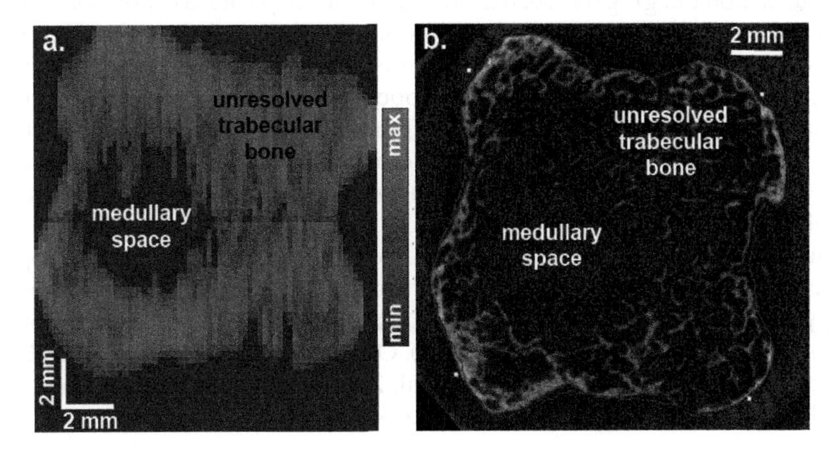

FIGURE 12.3
EDD tomographic slice (a) of an archeological human second metacarpal bone compared to an absorption microCT slice (b). (a) EDD slice using the 21.1+ diffraction peak. The slice is slightly foreshortened vertically. The grayscale bar to the right of the slice shows the linear intensity range. (b) MicroCT slice recorded with 21-μm isotropic voxels and at approximately the same position of the Mc2 bone as the EDD reconstructed slice. The lighter the pixel, the more attenuating the corresponding voxel. See Stock et al. (2017).

depth information without sample rotation is ideally suited for large plate-like samples where projections cannot be recorded over a significant angular range; depth resolution with EDD tomography and the present generation of detectors are poorer than with monochromatic methods.

12.2.2 Diffraction Tomography of Large-Grained Specimens

The scattering tomography approaches described in the previous section applied to crystalline specimens where there are many grains, or to amorphous specimens. Alternative approaches are needed where grains are relatively large, and, instead of a pencil beam, area beams are typically used. The diffraction peaks from individual grains can be resolved on the detector and cover quite a number of detector pixels. As the specimen is rotated, the individual grains appear and disappear as different *hkl* satisfy the Bragg condition within the solid angle sampled by the detector. There are several approaches for determining the volume occupied by the diffracting grain and even the spatial variation of various microstructural quantities, and considerable progress has taken place at synchrotron radiation sources (Poulsen, Nielsen et al. 2001, Poulsen 2004, Schuren, Shade et al. 2015). A Laue diffraction approach was employed with large-grained solid oxide fuel cell sample (Ferreira Sanchez, Villanova et al. 2015).

A laboratory system for diffraction tomography of large-grained specimens has been demonstrated (McDonald, Reischig et al. 2015), and this method was used to study sintering of a Cu powder specimen (McDonald, Holzner et al. 2017). Tomosynthesis data acquisition using an annular beam scanned across the specimen is another way of obtaining 3D maps of crystals (Evans, Rogers et al. 2014). Two different approaches for determining the location of grain boundaries were compared in another study (Renversade, Quey et al. 2016).

12.2.3 Coherent Diffraction Imaging and Ptychography

Coherent diffraction imaging has been used to map the shape of Au (Robinson, Vartanyants et al. 2001) and barium titanate (Harder and Robinson 2013) nanoparticles. This technique also allows a nanoparticle's full strain field to be quantified, for example, as a function of processing paths, time or strain, including topological effects in a battery particle, and strains/rotations around a single dislocation (Pfeifer, Williams et al. 2006, Newton, Leake et al. 2010, Proudhon, Vaxelaire et al. 2010, Hofmann, Abbey et al. 2013, Ulvestad, Clark et al. 2015, Robinson, Clark et al. 2016). Spatial resolutions can approach several nm and strain resolutions 1×10^{-6}.

The related technique, ptychography, has also been used for a variety of applications. X-ray ptychographic microscopy combines the advantages of raster scanning X-ray microscopy with the more recently developed techniques of coherent diffraction imaging, and has been reviewed recently (Pfeiffer 2018). One study reconstructed peritubular dentin surrounding tubules (Zanette, Enders et al. 2015). Another examined the strain field around a dislocation in silicon (Takahashi, Suzuki et al. 2013).

12.2.4 Fluorescence Tomography

The basics of data collection and reconstruction in fluorescence tomography were covered in Section 4.11. It is important to remember that quantitative results depend on the attenuation of the material along the path to the voxel in question as well as along the path from the voxel to the energy-discriminating detector. Signal from bone ($E_{K\alpha}$ = 3.7 keV, $E_{K\beta}$ = 4.0 keV) within a millimeter-sized specimen of soft tissue would be detectable. Signal from voxels of titanium ($E_{K\alpha}$ = 4.5 keV, $E_{K\beta}$ = 4.9 keV) near the center of a millimeter-sized bone specimen would not. When the signal can escape from the sample interior, a joint reconstruction algorithm (fluorescence and absorption simultaneously) seems promising (Di, Chen et al. 2017), and colocalization of fluorescence and absorption microCT provided interesting results (Laforce, Masschaele et al. 2017). Spectroscopy related to absorption edges has been used in chemical tomographic mapping (Owens, Butler et al. 2001, Rau, Somogyi et al. 2001, 2004, Schroer, Kuhlmann et al. 2003b).

Reports combining fluorescence and phase microCT in biomedical applications (organs labeled with iodine-containing contrast agents) have appeared (Takeda, Wu et al. 2004, 2006, Wu, Takeda et al. 2006), as has a map of labeled bacteria in a soil particle (Antipova, Kemner et al. 2018). Light elements have been mapped in a biological specimen (Vincze, Vekemans et al. 2004), and other reports of element distributions in organs include (Takeda, Yu et al. 1999). Elemental mapping within zebrafish embryos has been reported (Bourassa, Gleber et al. 2014, 2016). Elemental maps in slices of specimens of roots have been published: K, Fe, Rb, Cl in mahogany (Schroer, Benner et al. 2001, Simionovici, Chukalina et al. 2001) and K, Fe, Zn in tomato (Schroer, Benner et al. 2003a).

Mapping in small particles has been of interest, both for those of terrestrial origin (fly ash particles (Simionovici, Chukalina et al. 2001), sediment particles (Vincze, Vekemans et al. 2001), diatoms (Simionovici, Golosio et al. 2004)), and extraterrestrial origin (Si, Ca, Fe, and Cr maps in a microfragment of the Tatahouine meteriorite (Simionovici, Chukalina et al. 2001, Lemelle, Simionovici et al. 2003); S, Ca, Cr, Mn, Fe, Ni, Cu, and Zn maps in a cosmic dust particle (Schroer, Gunzler et al. 2004); Fe, Ni maps in micrometeriorites (Chukalina, Simionovici et al. 2003, Ignatyev, Huwig et al. 2006)). Other studies include Fe nanocatalyst spatial distribution (Jones, Feng et al. 2005), trace elements in a SiC shell of a nuclear fuel particle (Naghedolfeizi, Chung et al. 2003), and metal elemental maps within inclusions in diamond and quartz (Vincze, Vekemans et al. 2004).

References

Aguilera, O., I. Rocha, M. S. Lopes, I. Lima, R. T. Lopes, A. S. Machado, R. B. Guimarães, M. A. C. Crapez, M. C. Tenório and A. Nepomuceno (2017). "The bone degenerative processes in senile fishes from Holocene Brazilian shell mounds." *J Fish Dis* **40**: 1869–1881.

Ahvenainen, P., P. G. Dixon, A. Kallonen, H. Suhonen, L. J. Gibson and K. Svedström (2017). "Spatially-localized bench-top X-ray scattering reveals tissue-specific microfibril orientation in Moso bamboo." *Plant Meth* **13**: 5.

Almer, J. D. and S. R. Stock (2005). "Internal strains and stresses measured in cortical bone via high-energy x-ray diffraction." *J Struct Biol* **152**: 14–27.

Almer, J. D. and S. R. Stock (2007). "Micromechanical response of mineral and collagen phases in bone." *J Struct Biol* **157**: 365–370.

Almer, J. D. and S. R. Stock (2010). "Loading-related strain gradients spanning the mature bovine dentinoenamel junction (DEJ): Quantification using high energy x-ray scattering." *J Biomech* **43**: 2294–2300.

Antipova, O., K. Kemner, C. Roehrig, S. Vogt, L. X. Li and D. Gursoy (2018). "Developments of X-ray fluorescence tomography at 2-ID-E at APS for studies of composite samples." *Microsc. Microanal* **24**(Suppl 2): 520–521.

Aroush, D. R. B., E. Maire, C. Gauthier, S. Youssef, P. Cloetens and H. D. Wagner (2006). "A study of fracture of unidirectional composites using in situ high-resolution synchrotron x-ray microtomography." *Compos Sci Technol* **66**: 1348–1353.

Artioli, G., T. Cerulli, G. Cruciani, M. C. Dalconi, G. Ferrari, M. Parisatto, A. Rack and R. Tucoulou (2010). "X-ray diffraction microtomography (XRD-CT), a novel tool for non-invasive mapping of phase development in cement materials." *Anal Bioanal Chem* **397**: 2131–2136.

Birkbak, M. E., H. Leemreize, S. Frølich, S. R. Stock and H. Birkedal (2015). "Diffraction scattering computed tomography: A window into the structures of complex nanomaterials." *Nanoscale* **7**: 18402–18410.

Birkbak, M. E., I. G. Nielsen, S. Frolich, S. R. Stock, P. Kenesei, J. D. Almer and H. Birkedal (2017). "Concurrent determination of nanocrystal shape and amorphous phases in complex materials by diffraction scattering computed tomography." *J Appl Crystal* **50**: 192–197.

Bourassa, D., S.-C. Gleber, S. Vogt, C. H. Shin and C. J. Fahrni (2016). "MicroXRF tomographic visualization of zinc and iron in the zebrafish embryo at the onset of the hatching period." *Metallomics* **8**(10): 1122–1130.

Bourassa, D., S.-C. Gleber, S. Vogt, H. Yi, F. Will, H. Richter, C. H. Shin and C. J. Fahrni (2014). "3D imaging of transition metals in the zebrafish embryo by X-ray fluorescence microtomography." *Metallomics* **6**: 1648.

Burghardt, A. J., Y. Wang, H. Elalieh, X. Thibault, D. Bikle, F. Peyrin and S. Majumdar (2007). "Evaluation of fetal bone structure and mineralization in IGF-I deficient mice using synchrotron radiation microtomography and Fourier transform infrared spectroscopy." *Bone* **40**: 160–168.

Burnett, T. L., S. A. McDonald, A. Gholinia, R. Geurts, M. Janus, T. Slater, S. J. Haigh, C. Ornek, F. Almuaili, D. L. Engelberg, G. E. Thompson and P. J. Withers (2014). "Correlative tomography." *Sci Rep* **4**: 4711.

Cancedda, R., A. Cedola, A. Giuliani, V. Komlev, S. Lagomarsino, M. Mastrogiacomo, F. Peyrin and F. Rustichelli (2007). "Bulk and interface investigations of scaffolds and tissue-engineered bones by x-ray microtomography and x-ray microdiffraction." *Biomaterials* **28**: 2505–2524.

Cedola, A., G. Campi, D. Pelliccia, I. Bukreeva, M. Fratini, M. Burghammer, L. Rigon, F. Arfelli, C. R. Chang, D. Dreossi, N. Sodini, S. Mohammadi, G. Tromba, R. Cancedda and M. Mastrogiacomo (2014). "Three dimensional visualization of engineered bone and soft tissue by combined x-ray micro-diffraction and phase contrast tomography." *Phys Med Biol* **59**: 189–201.

Cedola, A., M. Mastrogiacomo, M. Burghammer, V. Komlev, P. Giannoni, A. Favia, R. Cancedda, F. Rustichelli and S. Lagomarsino (2006). "Engineered bone from bone marrow stromal cells: A structural study by an advanced x-ray microdiffraction technique." *Phys Med Biol* **51**: N109–N116.

Chukalina, M., A. Simionovici, L. Lemelle, C. Rau, L. Vincze and P. Gillet (2003). "X-ray fluorescence tomography for non-destructive semi-quantitative study of microobjects." *J Phys IV* **104**: 627–630.

Claret, F., S. Grangeon, A. Loschetter, C. Tournassat, W. De Nolf, N. Harker, F. Boulahya, S. Gaboreau, Y. Linard, X. Bourbon, A. Fernandez-Martinez and J. Wright (2018). "Deciphering mineralogical changes and carbonation development during hydration and ageing of a consolidated ternary blended cement paste." *IUCrJ* **5**: 150–157.

Di, Z. W., S. Chen, Y. P. Hong, C. Jacobsen, S. Leyffer and S. M. Wild (2017). "Joint reconstruction of x-ray fluorescence and transmission tomography." *Opt Express* **25**: 13107.

Dorr, A., J. G. Sled and N. Kabani (2007). "Three-dimensional cerebral vasculature of the CBA mouse brain: A magnetic resonance imaging and micro computed tomography study." *Neuroimage* **35**: 1409–1423.

Egn, C. K., S. D. Jacques, M. Di Michiel, B. Cai, M. W. Zandbergen, P. D. Lee, A. M. Beale and R. J. Cernik (2013). "Non-invasive imaging of the crystalline structure within a human tooth." *Acta Biomater* **9**: 8337–8345.

Evans, P., K. Rogers, A. Dicken, S. Godber and D. Prokopiou (2014). "X-ray diffraction tomography employing an annular beam." *Opt Express* **22**: 11930–11944.

Ferreira Sanchez, D., J. Villanova, J. Laurencin, J. S. Micha, A. Montani, P. Gergaud and P. Bleuet (2015). "X-ray micro Laue diffraction tomography analysis of a solid oxide fuel cell." *J Appl Cryst* **48**: 357–364.

Fritz, V., P. Louis-Plence, F. Apparailly, D. Noël, R. Voide, A. Pillon, J. C. Nicolas, R. Müller and C. Jorgensen (2007). "MicroCT combined with bioluminescence imaging: A dynamic approach to detect early tumor–bone interaction in a tumor osteolysis murine model." *Bone* **40**: 1032–1040.

Frølich, S., H. Leemreize, A. Jakus, X. Xiao, R. Shah, H. Birkedal, J. D. Almer and S. R. Stock (2016). "Diffraction tomography and Rietveld refinement of a hydroxyapatite bone phantom." *J Appl Cryst* **49**: 103–109.

Gursoy, D., T. Bicer, J. D. Almer, R. Kettimuthu, F. D. Carlo and S. R. Stock (2015). "Maximum a posteriori estimation of crystallographic phases in X-ray diffraction tomography." *Phil Trans Roy Soc (Lond) A* **373**: 20140392.

Gustafsson, A., N. Mathavan, M. J. Turunen, J. Engqvist, H. Khayyeri, S. A. Hall and H. Isaksson (2018). "Linking multiscale deformation to microstructure in cortical bone using in situ loading, digital image correlation and synchrotron X-ray scattering." *Acta Biomater* **69**: 323–331.

Haase, J. D., A. Guvenilir, J. R. Witt, M. A. Langøy and S. R. Stock (1999). Microtexture, asperities and crack deflection in Al-Li 2090 T8E41. *Mixed-Mode Crack Behavior*. K. Miller and D. L. McDowell. West Conshohocken (PA), ASTM. **ASTM STP 1359**: 160–173.

Haase, J. D., A. Guvenilir, J. R. Witt and S. R. Stock (1998). "X-ray microbeam mapping of microtexture related to fatigue crack asperities in Al-Li 2090." *Acta Mater* **46**: 4791–4799.

Harder, R. and I. K. Robinson (2013). "Coherent X-ray diffraction imaging of morphology and strain in nanomaterials." *JOM* **65**: 1202–1207.

He, F., A. E. Chiou, H. C. Loh, M. Lynch, B. R. Seo, Y. H. Song, M. J. Lee, R. Hoerth, E. L. Bortel, B. M. Willie, G. N. Duda, L. A. Estroff, A. Masic, W. Wagermaier, P. Fratzl and C. Fischbach (2017). "Multiscale characterization of the mineral phase at skeletal sites of breast cancer metastasis." *Proc Natl Acad Sci U S A* **114**: 10542–10547.

Hofmann, F., B. Abbey, W. Liu, R. Xu, B. F. Usher, E. Balaur and Y. Liu (2013). "X-ray micro-beam characterization of lattice rotations and distortions due to an individual dislocation." *Nat Commun* **4**(3774): 2774.

Ignatiev, K., Z. U. Rek and S. R. Stock (2000). "X-ray microbeam diffraction comparison of meso-structures in plates of three aluminum alloys." *Adv X-ray Anal* **44**: 56–61.

Ignatyev, K., K. Huwig, R. Harvey, H. Ishii, J. Bradley, K. Luening, S. Brennan and P. Pianetti (2006). XRF microCT study of space objects at SSRL. *Developments in X-Ray Tomography V*. U. Bonse. Bellingham (WA), SPIE. **SPIE Proc Vol 6318**: 631825-1–631825-7.

Jensen, T. H., M. Bech, O. Bunk, M. Thomsen, A. Menzel, A. Bouchet, G. Le Duc, R. Feidenhans'l and F. Pfeiffer (2011). "Brain tumor imaging using small-angle x-ray scattering tomography." *Phys Med Biol* **56**: 1717–1726.

Jiménez, C., M. Paeplow, P. H. Kamm, T. R. Neu, M. Klaus, G. Wagener, J. Banhart, C. Genzel and F. García-Moreno (2018). "Simultaneous X-ray radioscopy/tomography and energy-dispersive diffraction applied to liquid aluminium alloy foams." *J Synch Rad* **25**: 1790–1796.

Jin, H., K. Ham, J. Y. Chan, L. G. Butler, R. L. Kurtz, S. Thiam, J. W. Robinson, R. A. Agbaria, I. M. Warner and R. E. Tracy (2002). "High resolution three-dimensional visualization and characterization of coronary atherosclerosis in vitro by synchrotron radiation x-ray microtomography and highly localized x-ray diffraction." *Phys Med Biol* **47**: 4345–4356.

Jones, K. W., H. Feng, A. Lanzirotti and D. Mahajan (2005). "Synchrotron x-ray microprobe and computed microtomography for characterization of nanocatalysts." *Nucl Instrum Meth B* **241**: 331–334.

Jupe, A. C., S. R. Stock, P. L. Lee, N. N. Naik, K. E. Kurtis and A. P. Wilkinson (2004). "Phase composition depth profiles using spatially resolved energy dispersive x-ray diffraction." *J Appl Cryst* **37**: 967–976.

Krug, R., J. Carballido-Garmio, A. J. Burghardt, S. Haase, J. W. Sedat, W. C. Moss and S. Majumdar (2007). "Wavelet-based characterization of vertebral bone structure from magnetic resonance images at 3T compared with micro-computed tomoraphic measurements." *Mag Res Imaging* **25**: 392–398.

Laforce, B., B. Masschaele, M. N. Boone, D. Schaubroeck, M. Dierick, B. Vekemans, C. Walgraeve, C. Janssen, V. Cnudde, L. Van Hoorebeke and L. Vincze (2017). "Integrated three-dimensional microanalysis combining X-ray microtomography and X-ray fluorescence methodologies." *Anal Chem* **89**: 10617–10624.

Leemreize, H., J. D. Almer, S. R. Stock and H. Birkedal (2013). "Three-dimensional distribution of polymorphs and magnesium in a calcified underwater attachment system by diffraction tomography." *J Roy Soc Interf* **10**: 20130319.

Lemelle, L., A. Simionovici, J. Susini, P. Oger, M. Chukalina, C. Rau, B. Golosio and P. Gillet (2003). "X-ray imaging techniques and exobiology." *J Phys IV* **104**: 377–380.

Li, T., T. M. M. Heenan, M. F. Rabuni, B. Wang, N. M. Farandos, G. H. Kelsall, D. Matras, C. Tan, X. Lu, S. D. M. Jacques, D. J. L. Brett, P. R. Shearing, M. Di Michiel, A. M. Beale, A. Vamvakeros and K. Li (2019). "Design of next-generation ceramic fuel cells and real-time characterization with synchrotron X-ray diffraction computed tomography." *Nat Commun* **10**: 1497.

Liebi, M., M. Georgiadis, J. Kohlbrecher, M. Holler, J. Raabe, I. Usov, A. Menzel, P. Schneider, O. Bunk and M. Guizar-Sicairos (2018). "Small-angle X-ray scattering tensor tomography: Model of the three-dimensional reciprocal-space map, reconstruction algorithm and angular sampling requirements." *Acta Crystallogr A* **74**: 12–24.

Manzoor, M. A. P., A. K. Agrawal, B. Singh, M. Mujeeburahiman and P. D. Rekha (2019). "Morphological characteristics and microstructure of kidney stones using synchrotron radiation μCT reveal the mechanism of crystal growth and aggregation in mixed stones." *PLOS ONE* **14**: 0214003.

Marien, E., A. Hillen, F. Vanderhoydonc, J. V. Swinnen and G. Vande Velde (2017). "Longitudinal microcomputed tomography-derived biomarkers for lung metastasis detection in a syngeneic mouse model: Added value to bioluminescence imaging." *Lab Invest* **97**: 24–33.

Markey, M. J., R. P. Main and C. R. Marshall (2006). "In vivo cranial suture function and suture morphology in the extant fish Polypterus: Implications for inferring skull function in living and fossil fish." *J Exp Biol* **209**: 2085–2102.

Martins de Souza E Silva, J., J. Utsch, M. A. Kimm, S. Allner, M. F. Epple, K. Achterhold and F. Pfeiffer (2017). "Dual-energy microCT for quantifying the time-course and staining characteristics of ex-vivo animal organs treated with iodine- and gadolinium-based contrast agents." *Sci Rep* **7**: 17387.

Mashiatulla, M., R. D. Ross and D. R. Sumner (2017). "Validation of cortical bone mineral density distribution using micro-computed tomography." *Bone* **99**: 53–61.

McDonald, S. A., C. Holzner, E. M. Lauridsen, P. Reischig, A. P. Merkle and P. J. Withers (2017). "Microstructural evolution during sintering of copper particles studied by laboratory diffraction contrast tomography (LabDCT)." *Sci Rep* **7**: 5251.

McDonald, S. A., M. Preuss, E. Maire, J. Y. Buffiere, P. M. Mummery and P. J. Withers (2002). "Synchrotron x-ray study of micromechanics of Ti/SiCf composites with fibres containing defects introduced by laser drilling." *Mater Sci Technol* **18**: 1497–1503.

McDonald, S. A., M. Preuss, E. Marie, J. Y. Buffiere, P. M. Mummery and P. J. Withers (2003). "X-ray tomographic imaging of Ti/SiC composites." *J Microsc* **209**: 102–112.

McDonald, S. A., P. Reischig, C. Holzner, E. M. Lauridsen, P. J. Withers, A. P. Merkle and M. Feser (2015). "Non-destructive mapping of grain orientations in 3D by laboratory X-ray microscopy." *Sci Rep* **5**: 14665.

Mürer, F. K., S. Sanchez, M. Álvarez-Murga, M. Di Michiel, F. Pfeiffer, M. Bech and D. W. Breiby (2018). "3D Maps of mineral composition and hydroxyapatite orientation in fossil bone samples obtained by X-ray diffraction computed tomography." *Sci Rep* **8**: 10052.

Naghedolfeizi, M., J. S. Chung, R. Morris, G. E. Ice, W. B. Yun, Z. Cai and B. Lai (2003). "X-ray fluorescence microtomography study of trace elements in a SiC nuclear fuel shell." *J Nucl Mater* **312**: 146–155.

Naik, N. (2003). Sulfate attack on Portland cement-based materials: mechanisms of damage and long term performance. Ph.D. thesis, Georgia Institute of Technology.

Naik, N. N., A. C. Jupe, S. R. Stock, A. P. Wilkinson, P. L. Lee and K. E. Kurtis (2006). "Sulfate attack monitored by microCT and EDXRD: Influence of cement type, water-to-cement ratio, and aggregate." *Cement Concr Res* **36**: 144–159.

Newton, M. C., S. J. Leake, R. Harder and I. K. Robinson (2010). "Three-dimensional imaging of strain in a single ZnO nanorod." *Nat Mater* **9**: 120–124.

Nuzzo, S., C. Meneghini, P. Braillon, R. Bouvier, S. Mobilio and F. Peyrin (2003). "Microarchitectural and physical changes during fetal growth in human vertebral bone." *J Bone Miner Res* **18**: 760–768.

Orzechowska, S., A. Wróbel, M. Kozieł, W. Łasocha and E. Rokita (2018). "Physicochemical characterization of mineral deposits in human ligamenta flava." *J Bone Miner Metab* **36**: 314–322.

Owens, J. W., L. G. Butler, C. Dupard-Julien and K. Garner (2001). "Synchrotron x-ray microtomography, x-ray absorption near edge structure, extended x-ray absorption fine structure, and voxel imaging of a cobalt-zeolite-Y complex." *Mater Res Bull* **36**: 1595–1602.

Palmer, B. A., A. Hirsch, V. Brumfeld, E. D. Aflalo, I. Pinkas, A. Sagi, S. Rosenne, D. Oron, L. Leiserowitz, L. Kronik, S. Weiner and L. Addadi (2018). "Optically functional isoxanthopterin crystals in the mirrored eyes of decapod crustaceans." *Proc Natl Acad Sci U S A* **115**: 2299–2304.

Pfeifer, M. A., G. J. Williams, I. A. Vartanyants, R. Harder and I. K. Robinson (2006). "Three-dimensional mapping of a deformation field inside a nanocrystal." *Nature* **442**: 04867.

Pfeiffer, F. (2018). "X-ray ptychography." *Nat Photon* **12**: 9–17.

Poulsen, H. F. (2004). *Three-Dimensional X-Ray Diffraction Microscopy: Mapping Polycrystals and their Dynamics.* Berlin, Springer.

Poulsen, H. F., S. F. Nielsen, E. M. Lauridsen, S. Schmidt, R. M. Suter, U. Lienert, L. Margulies, T. Lorentzen and D. Juul Jensen (2001). "Three-dimensional maps of grain boundaries and the stress state of individual grains in polycrystals and powders." *J Appl Cryst* **34**: 751–756.

Preuss, M., P. J. Withers, E. Maire and J. Y. Buffiere (2002). "SiC single fibre full-fragmentation studied during straining in a Ti-6Al-4V matrix studied by synchrotron x-rays." *Acta Mater* **50**: 3175–3190.

Price, S. W. T., A. Van Loon, K. Keune, A. D. Parsons, C. Murray, A. M. Beale and J. F. W. Mosselmans (2019). "Unravelling the spatial dependency of the complex solid-state chemistry of Pb in a paint micro-sample from Rembrandt's Homer using XRD-CT." *Chem Commun (Camb)* **55**: 1931–1934.

Proudhon, H., N. Vaxelaire, S. Labat, S. Forest and O. Thomas (2010). "Simulation par éléments finis de la diffraction cohérente dans des agrégats polycristallins élastoplastiques." *Comptes Rendus Phys* **11**: 293–303.

Pyzalla, A., B. Camin, T. Buslaps, M. D. Michiel, H. Kaminiski, A. Kottar, A. Pernack and W. Reimers (2005). "Simultaneous tomography and diffraction analysis of creep damage." *Science* **308**: 92–95.

Rau, C., A. Somogyi, A. Bytchkov and A. S. Simionovici (2001). XANES micro-imaging and tomography. *Developments in X-Ray Tomography III.* U. Bonse. Bellingham (WA), SPIE. **SPIE Proc Vol 4503**: 249–255.

Rau, C., A. Somogyi and A. Simionovici (2004). Tomography with chemical speciation. *Developments in X-Ray Tomography IV.* U. Bonse. Bellingham (WA), SPIE. **SPIE Proc Vol 5535**: 29–35.

Renversade, L., R. Quey, W. Ludwig, D. Menasche, S. Maddali, R. M. Suter and A. Borbély (2016). "Comparison between diffraction contrast tomography and high-energy diffraction microscopy on a slightly deformed aluminium alloy." *IUCrJ* **3**: 32–42.

Robinson, I., J. Clark and R. Harder (2016). "Materials science in the time domain using Bragg coherent diffraction imaging." *J Opt (UK)* **18**: 054007.

Robinson, I. K., I. A. Vartanyants, G. J. Williams, M. A. Pfeifer and J.A. Pitney (2001). "Reconstruction of the shapes of gold nanocrystals using coherent x-ray diffraction." *Phys Rev Lett* **87**: 195505.

Rumaiz, A. K., A. J. Kuczewski, J. Mead, E. Vernon, D. Pinelli, E. Dooryhee, S. Ghose, T. Caswell, D. P. Siddons, A. Miceli, J. Baldwin, J. Almer, J. Okasinski, O. Quaranta, R. Woods, T. Krings and S. Stock (2018). "Multi-element germanium detectors for synchrotron applications." *J Instrum* **13**: C04030.

Sangid, M. D., T. A. Book, D. Naragani, J. Rotella, P. Ravi, A. Finch, P. Kenesei, J.-S. Park, H. Sharma, J. Almer and X. Xiao (2018). "Role of heat treatment and build orientation in the microstructure sensitive deformation characteristics of IN718 produced via SLM additive manufacturing." *Addit Manuf* **22**: 479–496.

Schaff, F., M. Bech, P. Zaslansky, C. Jud, M. Liebi, M. Guizar-Sicairos and F. Pfeiffer (2015). "Six-dimensional real and reciprocal space small-angle X-ray scattering tomography." *Nature* **527**: 353–356.

Schroer, C. G., B. Benner, T. F. Günzler, M. Kuhlman, B. Lengeler, W. H. Schröder, A. J. Kuhn, A. Simionovici, A. Snigirev and I. Snigireva (2003a). "High resolution element mapping inside biological samples using fluorescence microtomography." *J Phys IV* **104**: 353.

Schroer, C. G., B. Benner, T. F. Gunzler, M. Kuhlmann, B. Lengeler, W. H. Schroder, A. J. Kuhn, A. S. Simionovici, A. Snigirev and I. Snigireva (2001). High resolution element mapping inside biological samples using fluorescence microtomography. *Developments in X-Ray Tomography III*. U. Bonse. Bellingham (WA), SPIE. **SPIE Proc Vol 4503**: 230–239.

Schroer, C. G., T. F. Gunzler, M. Kuhlmann, O. Kurapova, S. Feste, M. Schweitzer, B. Lengeler, W. H. Schroder, M. Drakopoulos, A. Somogyi, A. S. Simionovici, A. Snigirev and I. Snigireva (2004). Fluorescence microtomography using nanofocusing refractive x-ray lenses. *Developments in X-Ray Tomography IV*. U. Bonse. Bellingham (WA), SPIE. **SPIE Proc Vol 5535**: 162–168.

Schroer, C. G., M. Kuhlmann, T. F. Gunzler, B. Lengeler, M. Richwin, B. Griesebock, D. Lutzenkirchen-Hecht, R. Frahm, E. Ziegler, A. Mashayekhi, D. R. Haeffner, J. D. Grunwaldt and A. Baiker (2003b). "Mapping the chemical states of an element inside a sample using tomographic x-ray absorption spectroscopy." *Appl Phys Lett* **82**: 3360–3362.

Schuren, J. C., P. A. Shade, J. V. Bernier, S. F. Li, B. Blank, J. Lind, P. Kenesei, U. Lienert, R. M. Suter, T. J. Turner, D. M. Dimiduk and J. Almer (2015). "New opportunities for quantitative tracking of polycrystal responses in three dimensions." *Curr Opin Sol State Mater Sci* **19**: 235–244.

Simionovici, A. S., M. Chukalina, B. Vekemans, L. Lemelle, P. Gillet, C. G. Schroer, B. Lengeler, W. H. Schröder and T Jeffries (2001). New results in x-ray computed fluorescence tomography. *Developments in X-Ray Tomography III*. U. Bonse. Bellingham (WA), SPIE. **SPIE Proc Vol 4503**: 222–229.

Simionovici, A. S., B. Golosio, M. V. Chukalina, A. Somogyi and L. Lemelle (2004). Seven years of x-ray fluorescence computed microtomography. *Developments in X-Ray Tomography IV*. U. Bonse. Bellingham (WA), SPIE. **SPIE Proc Vol 5535**: 232–242.

Sinclair, R., M. Preuss, E. Marie, J. Y. Buffiere, P. Bowen and P. J. Withers (2004). "The effect of fibre fractures in the bridging zone of fatigue cracked Ti-6Al-4V/SiC fibre composites." *Acta Mater* **52**: 1423–1438.

Sinclair, R., M. Preuss and P. J. Withers (2005). "Imaging and strain mapping fibre by fibre in the vicinity of a fatigue crack in a Ti/SiC fibre composite." *Mater Sci Technol* **21**: 27–34.

Sottmann, J., M. Di Michiel, H. Fjellvåg, L. Malavasi, S. Margadonna, P. Vajeeston, G. B. M. Vaughan and D. S. Wragg (2017). "Chemical structures of specific sodium ion battery components determined by operando pair distribution function and X-ray diffraction computed tomography." *Angew Chem Int Ed Engl* **56**: 11385–11389.

Stock, S. R. and J. D. Almer (2012). "Diffraction microcomputed tomography of an Al-matrix SiC-monofilament composite." *J Appl Cryst* **47**: 1077–1083.

Stock, S. R., J. Barss, T. Dahl, A. Veis and J. D. Almer (2002). "X-ray absorption microtomography (microCT) and small beam diffraction mapping of sea urchin teeth." *J Struct Biol* **139**: 1–12.

Stock, S. R., F. De Carlo and J. D. Almer (2008). "High energy x-ray scattering tomography applied to bone." *J Struct Biol* **161**: 144–150.

Stock, S. R., K. Ignatiev, P. L. Lee and J. D. Almer (2014). "Calcite orientations and composition ranges within teeth across Echinoidea." *Conn Tiss Res* **55**: 48–52.

Stock, S. R., J. S. Okasinski, R. Woods, J. Baldwin, T. Madden, O. Quaranta, A. Rumaiz, T. Kuczewski, J. Mead, T. Krings, P. Siddons, A. Miceli and J. D. Almer (2017). Tomography with energy dispersive diffraction. *Developments in X-Ray Tomography XI.* B. Mueller. Bellingham (WA), SPIE. **10391**: 103910A.

Stock, S. R., J. Seto, A. C. Deymier, A. Rack and A. Veis (2018). "Growth of second stage mineral in *Lytechinus variegatus.*" *Conn Tiss Res* **59**: 345–355.

Stock, S. R., A. Veis, X. Xiao, J. D. Almer and J. R. Dorvee (2012). "Sea urchin tooth mineralization: Calcite present early in the aboral plumula." *J Struct Biol* **180**: 280–289.

Stock, S. R., A. E. M. Vieira, A. C. B. Delbem, M. L. Cannon, X. Xiao and F. De Carlo (2008). "Synchrotron microComputed tomography of the bovine dentinoenamel junction." *J Struct Biol* **161**: 162–171.

Takahashi, Y., A. Suzuki, S. Furutaku, K. Yamauchi, Y. Kohmura and T. Ishikawa (2013). "Bragg x-ray ptychography of a silicon crystal: Visualization of the dislocation strain field and the production of a vortex beam." *Phys Rev B* **87**: 121201.

Takeda, T., J. Wu, T. T. Lwin, A. Yoneyama, Y. Hirai, K. Hyodo, N. Sunaguchi, T. Yuasa, M. Minami, K. Kose and T. Akatsuka (2006). Progress in biomedical application of phase contrast x-ray imaging and fluorescent x-ray CT. *Developments in X-Ray Tomography V.* U. Bonse. Bellingham (WA), SPIE. **SPIE Proc Vol 6318**: 63180W-1–63180W-2.

Takeda, T., J. Wu, A. Yoneyama, Y. Tsuchiya, T. T. Lwin, Y. Hirai, T. Kuroe, T. Yuasa, K. Hyodo, F. A. Dilmanian and T. Akatsuka (2004). SR biomedical imaging with phase contrast and fluorescent x-ray CT. *Developments in X-Ray Tomography IV.* U. Bonse. Bellingham (WA), SPIE. **SPIE Proc Vol 5535**: 380–391.

Takeda, T., Q. Yu, T. Yashiro, T. Yuasa, Y. Itai and T. Akatsuka (1999). Human thyroid specimen imaging by fluorescent x-ray computed tomography with synchrotron radiation. *Developments in X-Ray Tomography II.* U. Bonse. Bellingham (WA), SPIE. **SPIE Proc Vol 3772**: 258–267.

Tanaka, T., K. Uesugi, A. Takeuchi, Y. Suzuki and T. Iwata (2007). "Analysis of inner structure in high-strength biodegradable fibers by X-ray microtomography using synchrotron radiation." *Polymer* **48**: 6145–6151.

Ulvestad, A., J. N. Clark, O. G. Shpyrko, A. Singer, Y. S. Meng, H. M. Cho, R. Harder, J. Maser and J. W. Kim (2015). "BATTERIES. Topological defect dynamics in operando battery nanoparticles." *Science* **348**(6241): 1344–1347.

Vamvakeros, A., S. D. Jacques, M. Di Michiel, P. Senecal, V. Middelkoop, R. J. Cernik and A. M. Beale (2016). "Interlaced X-ray diffraction computed tomography." *J Appl Cryst* **49**: 485–496.

Vanmeert, F., G. Van der Snickt and K. Janssens (2015). "Plumbonacrite identified by X-ray powder diffraction tomography as a missing link during degradation of red lead in a Van Gogh painting." *Angew Chem Int Ed Engl* **54**: 3607–3610.

Vincze, L., B. Vekemans, I. Szaloki, F. E. Brenker, G. Falkenberg, K. Rickers, K. Aerts, R. V. Grieken and F. Adams (2004). X-ray fluorescence microtomography and polycapillary based confocal imaging using synchrotron radiation. *Developments in X-Ray Tomography IV.* U. Bonse. Bellingham (WA), SPIE. **SPIE Proc Vol 5535**: 220–231.

Vincze, L., B. Vekemans, I. Szaloki, K. Janssens, R. V. Grieken, H. Feng, K. W. Jones and F. Adams (2001). High resolution X-ray fluorescence micro-tomography on single sediment particles. *Developments in X-Ray Tomography III.* U. Bonse. Bellingham (WA), SPIE. **SPIE Proc Vol 4503**: 240–248.

Wilkinson, A. P., A. C. Jupe, K. E. Kurtis, N. N. Naik, S. R. Stock and P. L. Lee (2004). Spatially resolved energy dispersive x-ray diffraction (EDXRD) as a tool for nondestructively providing phase composition depth profiles on cement and other materials. *Applications of X-rays in Mechanical Engineering 2004.* New York, ASME: 49–52.

Withers, P. J., J. Bennett, Y. C. Hung and M. Preuss (2006). "Crack opening displacements during fatigue crack growth in Ti-SiC fibre metal matrix composites by x-ray tomography." *Mater Sci Technol* **22**: 1052–1058.

Wu, J., T. Takeda, T. T. Lwin, N. Sunaguchi, T. Fukami, T. Yuasa, M. Minami and T. Akatsuka (2006). Fusion imaging of fluorescent and phase contrast x-ray computed tomography using synchrotron radiation in medical biology. *Developments in X-Ray Tomography V.* U. Bonse. Bellingham (WA), SPIE. **SPIE Proc Vol 6318**: 631828-1–631828-8.

Yoder, G. R., P. S. Pao, M. A. Imam and L. A. Cooley (1989). Micromechanisms of fatigue fracture in Al-Li Alloy 2090. Aluminum-Lithium Alloys, Proceedings of the Fifth Aluminum-Lithium Conference. J. T.H. Sanders, E. A. Starke, Jr. Birmingham (UK), Materials and Component Engineering Publications Ltd.: 1033–1041.

Zanette, I., B. Enders, M. Dierolf, P. Thibault, R. Gradl, A. Diaz, M. Guizar-Sicairos, A. Menzel, F. Pfeiffer and P. Zaslansky (2015). "Ptychographic X-ray nanotomography quantifies mineral distributions in human dentine." *Sci Rep* **5**: 9210.

Zhu, Z., A. Katsevich, A. J. Kapadia, J. A. Greenberg and S. Pang (2018). "X-ray diffraction tomography with limited projection information." *Sci Rep* **8**: 522.

Zhu, Z., A. Katsevich and S. Pang (2019). "Interior x-ray diffraction tomography with low-resolution exterior information." *Phys Med Biol* **64**: 025009.

Name Index

I

J

K

M

N

O

P

Q

R

T

U

V

W

Subject Index

B

D

E

F

N

O

P

Q

R

T

Y

Z

9781032337388